With the 1st Marine Division in Iraq, 2003

No Greater Friend, No Worse Enemy

by Lieutenant Colonel Michael S. Groen
and Contributors

Occasional Paper

HISTORY DIVISION
MARINE CORPS UNIVERSITY
QUANTICO, VIRGINIA

2006

Other Publications in the Occasional Papers Series

Vietnam Histories Workshop: Plenary Session. Jack Shulimson, editor. 9 May1983. 31 pp.

Vietnam Revisited; Conversation with William D. Broyles, Jr. Colonel John G. Miller, USMC, editor. 11 December 1984. 48 pp.

Bibliography on Khe Sanh USMC Participation. Commander Ray W. Strubbe, CHC, USNR (Ret), compiler. April 1985. 54 pp.

Alligators, Buffaloes, and Bushmasters: The History of the Development of the LVT Through World War II. Major Alfred Dunlop Bailey, USMC (Ret). 1986. 272 pp.

Leadership Lessons and Remembrances from Vietnam. Lieutenant General Herman Nickerson, Jr., USMC (Ret). 1988. 93 pp.

The Problems of U.S. Marine Corps Prisoners of War in Korea. James Angus MacDonald, Jr. 1988. 295 pp.

John Archer Lejeune, 1869-1942, Register of His Personal Papers. Lieutenant Colonel Merrill L. Bartlett, USMC (Ret). 1988. 123 pp.

To Wake Island and Beyond: Reminiscences. Brigadier General Woodrow M. Kessler, USMC (Ret). 1988. 145 pp.

Thomas Holcomb, 1879-1965, Register of His Personal Papers. Gibson B. Smith. 1988. 229 pp.

Curriculum Evolution, Marine Corps Command and Staff College, 1920-1988. Lieutenant Colonel Donald F. Bittner, USMCR. 1988. 112 pp.

Herringbone Cloak-GI Dagger, Marines of the OSS. Major Robert E. Mattingly, USMC. 1989. 315 pp.

The Journals of Marine Second Lieutenant Henry Bulls Watson, 1845-1848. Charles R. Smith, editor. 1990. 420 pp.

When the Russians Blinked: The U.S. Maritime Response to the Cuban Missile Crisis. Major John M. Young, USMCR. 1990. 246 pp.

Marines in the Mexican War. Gabrielle M. Neufeld Santelli. Edited by Charles R. Smith. 1991. 63 pp.

The Development of Amphibious Tactics in the U.S. Navy. General Holland M. Smith, USMC (Ret). 1992. 89 pp.

James Guthrie Harbord, 1866-1947, Register of His Personal Papers. Lieutenant Colonel Merrill L. Bartlett, USMC. 1995. 47 pp.

The Impact of Project 100,000 on the Marine Corps. Captain David A. Dawson, USMC. 1995. 247 pp.

Marine Corps Aircraft: 1913-2000. Major John M. Elliot, USMC (Ret). 2002. 126 pp.

Thomas Holcomb and the Advent of the Marine Corps Defense Battalion, 1936-1941. David J. Ulbrich. 2004. 78 pp.

Marine History Operations in Iraq, Operation Iraq Freedom I, A Catalog of Interviews and Recordings, LtCol Nathan S. Lowrey, USMCR. 2005. 254 pp.

Occasional Papers

The History Division has undertaken the publication for limited distribution of various studies, theses, compilations, bibliographies, monographs, and memoirs, as well as proceedings at selected workshops, seminars, symposia, and similar colloquia, which it considers to be of significant value for audiences interested in Marine Corps history. These "Occasional Papers," which are chosen for their intrinsic worth, must reflect structured research, present a contribution to historical knowledge not readily available in published sources, and reflect original content on the part of the author, compiler, or editor. It is the intent of the Division that these occasional papers be distributed to selected institutions, such as service schools, official Department of Defense historical agencies, and directly concerned Marine Corps organizations, so the information contained therein will be available for study and exploitation.

PCN: 106 0000 0000

Foreword

With the 1st Marine Division in Iraq, 2003, is a unit history written by the participants in the same vein as its predecessors—*The Old Breed*—written at the end of World War II and—*The New Breed*—authored during Korea. It is a narrative describing the actions of Marines in combat during the liberation of Iraq. Portions of the story have been told by embedded journalists—but this full account is told by those who made it happen.

The 1st Marine Division, in concert with the U.S. Army's 3d Infantry Division, captured Baghdad and toppled Saddam Hussein's dictatorship. The division's 28-day "march up" from Kuwait to Baghdad, a distance of 250 road miles, was a remarkable achievement. It represented a validation of the Corps' maneuver warfare strategy, particularly the seamless integration of air into the ground scheme of maneuver and the Marine logistics command's innovative support.

"Blue Diamond," the 1st Division's Operation Iraqi Freedom *nom de guerre*, consisted of some 20,000 Marines and sailors and 8,000 vehicles organized into three regimental combat teams. Designed to be light and self-sufficient, the regiments "conducted the longest sequence of coordinated overland attacks in the history of the Corps," according to Lieutenant General Wallace C. Gregson, then commander of Marine Forces Pacific. The authors of this account were somewhat more colloquial, preferring to state that it "focuses on the collective action of Marines who served as part of the 'Blue Diamond.' It is not a story of each of them, but the story of all of them." Their story is an authentic documentation of the feel, concerns, triumphs and tragedy of the campaign in Iraq.

Richard D. Camp, Jr.
Colonel,
United States Marine Corps (Retired)
Acting Director of Marine Corps History

Preface

In the final analysis, Saddam's regime and its threat could not be defeated except by fighting it. Yet, in the history of war, the two U.S. divisions that carried the brunt of the fighting, the 3d Infantry Division and the 1st Marine Division are not impressive in numbers. History has seen much larger forces. But on the shoulders was placed the burden of liberating the Iraqi people, and planting the seeds of freedom in the Middle East. In the chaos, confusion, and uncertainty of an ever-shifting and always dangerous battlefield, the young men and women who faced this enemy distinguished themselves for their presence of mind, their steadfast commitment to each other, and their willingness to pay the price for our freedom and the freedom of the Iraqi people, a people they barely knew.

The events chronicled here capture the story of the Marines of the "Blue Diamond" as they prepared for war, conquered and army, and liberated a nation. From across America and beyond, they chose the demanding path, to become Marines. These Marines marched in the ranks of this national treasure that we call the 1st Marine Division. And, at a time when timid souls or cynical pundits grew loudest, these men shouldered their weapons and moved without hesitation against the enemy. Our victory was not inevitable. It was the courage, unselfishness, and skill of the young men of Blue Diamond to whom we owe our victory.

Unit histories such as this cannot capture what we will remember of those men we lost. The young smiling faces of these men will carry more inside us than our words can ever convey. For young as they were, they proved themselves to be the very best of warriors. We record history so that others may share our story. We record history so that others may learn from it. We record history to remind us of what happened on these battlefields when we grow old. Most of all, we record history as a monument to our men and their families' sacrifice. We will always remember those we lost. Someday we will smile and laugh when we recall them amongst us, of how we were happy besides them, or even exasperated with them, but how we trusted them always with our dreams and with our very lives.

And as we look today at the men and women of the 1st Marine Division, we can take heart at what we see. The courage to defend our country is still there. That awesome determination to defend our freedoms will never die. In Iraq, those who followed in the footprints of the heroes of Guadalcanal, Inchon, Hue City, and Desert Storm proved themselves worthy to be counted among their number.

No mere narrative can fully capture the efforts, risks, and sacrifices of the men and women of the 1st Marine Division during Operation Iraqi Freedom. No words can capture the tears of family members as they sent their loved-ones off, perhaps for the last time. Nothing we can say will bring back our beloved comrades that made the ultimate sacrifice on the battlefield. Yet, "unsung the noblest deeds die." This is the story of the noble deeds of a special group of Marines who chose to serve their nation in the cause of freedom in a distant place.

Missing from this narrative are many of the individual stories of heroism that were a daily part of this operation. The accounts of squads, platoons, and companies would fill volumes, and cannot be included here. This narrative focuses on the collective action of these Marines who served as part of the "Blue Diamond." It is not a story of each of them, but the story of all of them. Semper Fidelis.

The principle author of this account was Lieutenant Colonel Michael S. Groen. He was assisted in this by the following primary contributors: Brigadier General John Kelly;

Lieutenant Colonels Rob Abbott, Rob Davis, Steve Ferrando, John Maher, Brian McCoy, Sam Mundy, Jim Parrington, Sam Strotman, Mark Toal; Majors Duncan Buchanan, Tom Chandler, Dan Connally, Tim Dzierzek, Brian Gillman, Brian Mangan, Randy Nash, Chris Olszko, Tom Sharp, Ron Spears; Captains George Benson, Joe Plenzler, Brian Rideout, Sherel Ryan, Blair Sokol; and Chief Warrant Officer Mike Holland. This manuscript was prepared for publication at the U.S. Marine Corps History Division by Mr. William S. Hill, Captain Joel P. Braniecki, Mr. Charles R. Smith, and Mr. Charles D. Melson.

James N. Mattis
Major General
United States Marine Corps

Michael S. Groen
Lieutenant Colonel
United States Marine Corps

Table of Contents

"I want you to get yourself spiritually ready for this fight, and to seek forgiveness for what we are about to do to the Iraqi Army."

Chapter 1
Planning and Preparation

The 'White House' at Camp Pendleton, Calfornia; Headquarters for the 1st Marine Division. The Division's logo, the 'Blue Diamond' recalls the Division's first campaign in Guadalcanal and its time spent in Australia during World War II.

Gathering of Commanders

On the quiet afternoon of 2 August 2002, the commanders, sergeants major, and staff principals of the 1st Marine Division gathered in a conference room on-board Camp Pendleton, California. That very morning they had seen the Division's change of command, and this gathering was to be the first opportunity for the new Commanding General (CG) to give his intent. It became immediately obvious he had gathered them for a single purpose; to provide a warning order for the invasion of Iraq. The Division had prepared for combat operations on the Arabian Peninsula for years, and some of these Marines were veterans of Desert Storm. They had planned hundreds of deployments and exercises using an Iraq scenario before, but it was clear that this session was different. During this meeting, the CG laid out a vision of the coming war. By this late date, it appeared that only an unexpected diplomatic cave-in by Saddam could prevent war, and sharing a common vision for the fight ahead was crucial. The leaders gathered here would oversee the preparation, training, deployment, and combat victory of the 1st Marine Division over the Nation's foes in Iraq. The commanders gathered this afternoon were sternly reminded of their responsibility for the destruction of the Iraqi Army should the President call, and were challenged to prepare their Marines to live up to the Corp's high standards in combat. They had no doubt that they would leave a proud legacy for those who came after them.

Iraq 2003 with Governate boundries depicted. The Division's operations were to extend from the ports of Kuwait all the way to Bayji in northern Iraq.

'Blue Diamond' (the collective call-sign for the 1st Marine Division, adopted from the Division logo) had been sharpening its blade in preparation for combat operations in Iraq for some time, and the Marines shared the CG's vision. The vision of this fight was one of speed and lethality, accompanied by the highest moral principles. The Division would wreak havoc on those who opposed it, yet extend a helping hand to those liberated by its actions. The motto, 'No Better Friend, No Worse Enemy' captured the essence of how 1st Marine Division would defeat the Iraqi Regime and liberate its people. Blue Diamond would fight with unprecedented lethality, guided by a strong moral compass and relying on the core values inherent in every Marine. Marines would be ruthless with the enemy, yet retain their humanity. In combat, the Division would fight with an entire faith; each with a faith in himself, and all with a faith in each other. The commanders understood the value of ideals, and would blend the psychology and tactics of war into a single whole. If Napoleon was correct, and the 'moral was to the physical as three is to one,' then the 1st Marine Division would overwhelm their foes with aggressiveness, endurance, and moral courage.

Tactically, the Division would fight lean. This meant eliminating unnecessary logistical drains. All members of the Division would live frugally and with the same quality of life standard as an infantry Lance Corporal. The Division deployed with fuel test kits and funnels to enable them to utilize captured enemy fuel supplies. The Marines would not run vehicles when stationary in order to conserve fuel, and would install special 'gypsy racks' that enabled their vehicles to carry additional fuel and water. Hanging *Flexcells* (fuel bladders) on M1A1 tanks and *FAST* fuel storage devices on Amphibious Assault Vehicles (AAVs) provided commanders with even greater range and decreased reliance on combat trains. These measures and more would extend the Division's logistics culminating point closer and closer to Baghdad.

All Marines of the Division were going tohave the same living conditions as a basic rifleman. Pictured here are Marines manning fighting positions during the operational pause along Highway 1.

To extend the Division's logistical culminating point, vehicles were fitted with external fuel reserves, such as these FLEXCELL fuel bladders on an M1A1.

The Division would fight swiftly towards its ultimate objectives, taking calculated risks along its flanks and rear to strike at the heart of the enemy. Some of these risks would receive criticism to be sure, but the Blue Diamond vision saw beyond the risks of individual tactical engagements to the lives of those saved by a rapid end to the conflict. The commanders pictured a map, with thin blue lines representing maneuver forces, moving in a continuous and relentless attack toward Baghdad, bypassing major urban areas and the enemy's prepared defenses. The commanders pictured support forces, tethered to their supported commands, capable of fighting their way up through bypassed enemy units to deliver supplies. The vision included boundaries that would rapidly snap up as follow-on forces relieved 1st Marine Division of battlespace. This would have the dual impact of keeping Lines of Communication (LOCs) open and sustaining massed combat power at the point of impact. Speed was to be the order of the day; speed of information flow, speed of decision making, and speed of maneuver.

As part of the vision, the Division planned to be the most 'air-centric division in history', crushing the Iraqi indirect fire capability with air power, preserving artillery ammunition for the fights it could not win by air alone. The Division would be well trained and confident in its abilities to operate in a Nuclear, Biological, Chemical (NBC) environment, able to continue the attack under the worst of scenarios. The Division would have integrated logistics capabilities, anticipating demands and maneuvering logistics to the right place and time. Perhaps most importantly, the vision included a Division that had a well-understood commander's intent, from the Commanding General to the youngest Sailors and Marines. This intent would allow the Division to continue the attack in an extended command and control environment under the most arduous of circumstances. The CG's guidance went out to every Marine in the Division, "I want you to get yourself spiritually ready for this fight, and to seek forgiveness for what we are about to do to the Iraqi Army."

Orienting on the Enemy: The Iraqi Battlespace

By the CG's design, operations would be based on sound intelligence. During this early planning stage, the Division G-2, under Colonel Jim Howcroft, provided a detailed understanding of the enemy that enabled the Division to focus its combat power to devastating effect on the vulnerabilities of the Iraqi Army. The Division's intelligence team led the effort to provide a sound basis for planning, allowing the enemy (through the lens of intelligence) to drive the Division's operations. This theme would ring true to the completion of combat operations, as the CG and Commanders refused to become wedded to a single plan, retaining a flexibility that would ensure their success as conditions changed. The Division was unwilling to prioritize the synchronization of operations, preferring a looser construct that used commander's intent to take advantage of opportunities that could be exploited through rapidity of action.

A significant planning consideration was the large size of the Iraqi Army itself. The initial planning efforts called for the 1st Marine Division to attack through the historic Mesopotamian plain of southeastern Iraq, including the strategic city of Al Kut. The Iraqis defended this zone with six divisions of the Iraqi Regular Army and the Baghdad Infantry Division of the Republican Guard. In addition to these conventional forces, the Iraqi Regime could call upon a significant number of Baath party militia, paramilitary troops from the Saddam Fedeyeen (a Regime-loyal group known for its suicide tactics) and Iraqi Special Forces units. Before combat operations were complete, the 1st Marine Division would defeat or bypass elements of nine Iraqi divisions, and would destroy thousands of these paramilitary fighters.

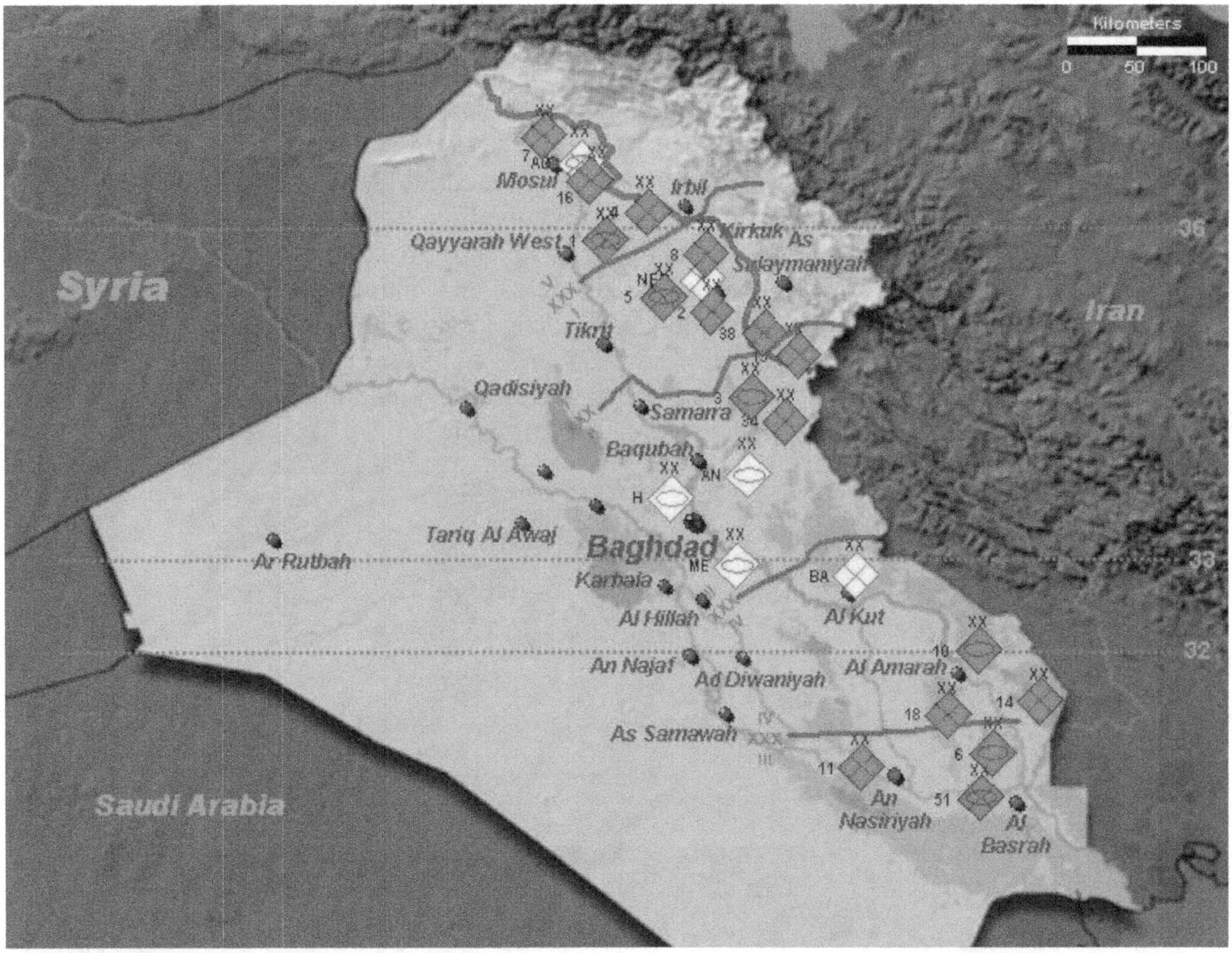

General disposition of Iraqi forces, Fall 2002. The Regular Army (Red) was planned to absorb many of the initial blows of the Coalition assault, while the Republican Guard Divisions (Yellow) were arrayed to protect the Regime in Baghdad and the surrounding urban centers.

Even though only a third the size of the Iraqi Army during the first Gulf War, this force posed a significant conventional combined arms challenge. The Iraqi Army during the first Gulf War fought poorly, but had been defending ground seized by an unjust aggression. This time, the Iraqi Army would be fighting in defense of their homeland. Perhaps even more threatening, the 1st Marine Division would be attacking to remove potential Weapons of Mass Destruction (WMD). It was considered likely that the Regime would use these weapons at some point as Division forces advanced into Iraq. Given the Regime's previous use of these weapons against Iranians and Kurds, the Division was not willing to 'wish away' this capability.

A second major planning factor for the Division's operations was the terrain in southern Iraq. Although the first 50 km of Iraqi territory was the same desert terrain the Division had encountered in the first Gulf War, the marshes of the Fertile Crescent lay just to the north. This terrain had repeatedly stymied British forces attacking north during World War I, and left a British Division stranded in Al Kut (where it was eventually besieged and captured by Turkish forces.) Well-thumbed copies of *The Siege* by Jonathon Braddon (which chronicled the British disaster) were passed around the staff, providing a solid appreciation for the effects of terrain and weather on the British Expeditionary Force. The area between the Tigris and Euphrates Rivers had once been a vast reed marshland, and the home of the Shiite 'Marsh Arabs'. Because of the insurgent threat posed by these people and the terrain, the Iraqi Regime had conducted a massive engineering effort during the 1980s to drain the marshlands. What had been marsh before was now a dense network of drainage canals, muddy fields, and untrafficable terrain. Planning to fight in this terrain would quickly become a challenge to the Division, as the day after the Combined Forces Land Component Commander's (CFLCC) staff assigned I Marine Expeditionary Force (MEF) its battlespace, they also released a mobility study describing almost the entirety of the 1st Marine Division's zone as 'untrafficable.'

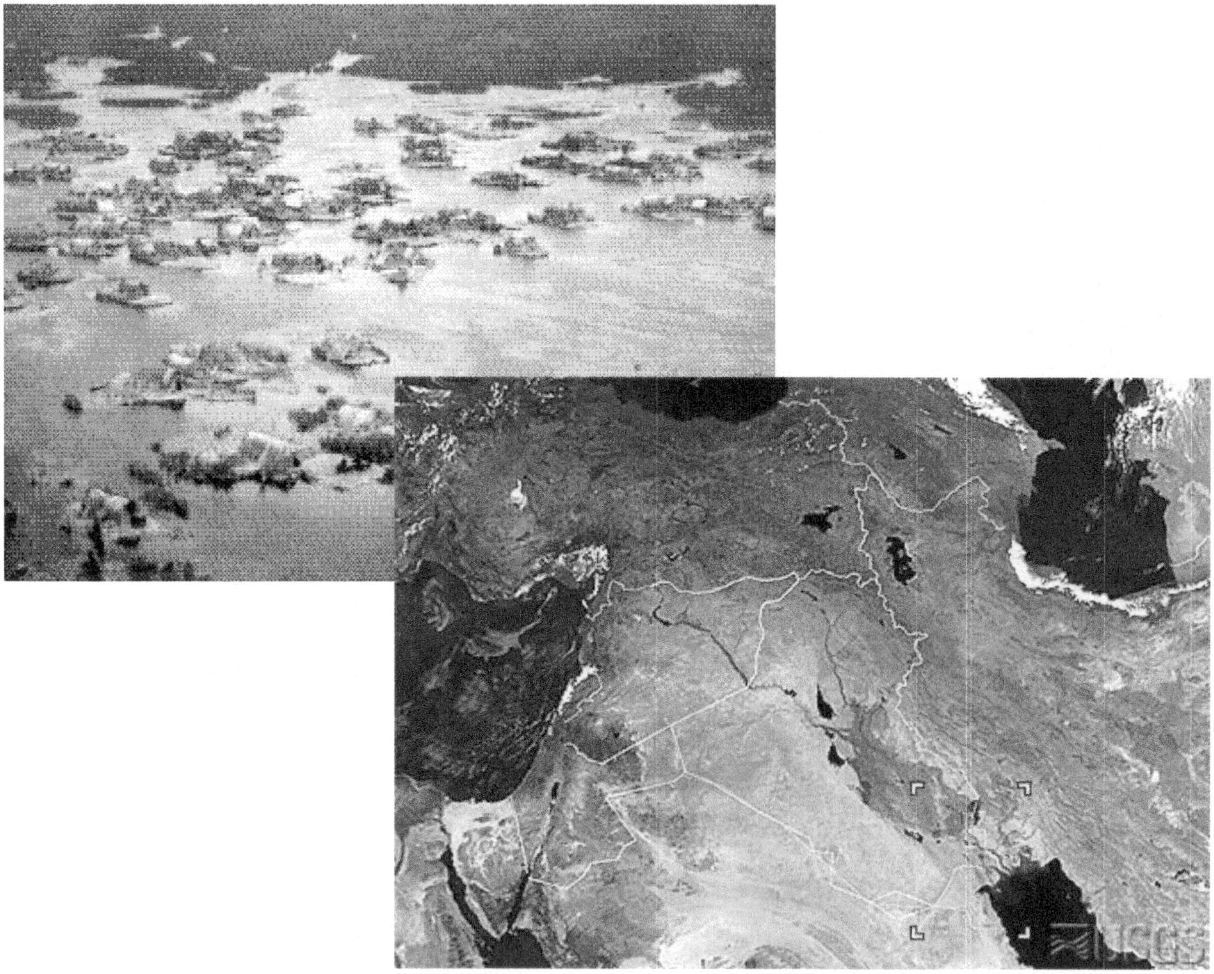

By draining the marshes of Mesopotamia, Saddam's engineers had transformed Marshland to Wasteland throughout Southern Iraq. The Division would be limited to movement along major roads in it's zone.

This difficult terrain posed a significant challenge to off-road maneuver, and served to canalize the Division's attack to the north. The identification of alternative cross-country mobility corridors became a major intelligence challenge. Little was known about trafficability over the vast swaths of former marshland, and even less about the load-bearing capacities of the few roads and bridges through this terrain. In any serious defensive effort, the Iraqis would certainly take maximum advantage of this terrain; destroying bridges, mining culverts, and using inundation warfare to delay the Division advance. During planning, this fight was often characterized as the 'war of the bridges', and nearly every stage of the Division's plan was tied to a water obstacle in some way.

The most pronounced feature in southern Iraq was *Jabal Sanam*, also known as 'Safwan Hill,' or 'Hill 451' for its map designation. The dominating heights of this hill, only 3 km north of the Kuwaiti border, provided near continuous line of sight well into northern Kuwait. It was known to hide several Iraqi observation posts and a signals intelligence gathering facility. This particular terrain feature was also to dominate a great deal of the Division staff's attention later, when faced with the almost impossible task of pre-staging large scale ground forces in Kuwait without giving clear signals of the Division's tactical intent to the enemy.

Jabal Sanam or 'Safwan Hill' was the dominant terrain feature in Southern Iraq, giving the Iraqis a line of sight that extended 30 Kilometers into Kuwait. The Division placed heavy emphasis on the destruction of the Iraqi forces occupying these heights.

To orient the Division to this new topography, the CG directed the G-2 to prepare a southern Iraq orientation map. This map, created by three lance corporals (Robert Lupi, Aaron Lehn, and Andrew Sipe) gave a common baseline to every member of the Division, each of whom would memorize the route names, city names, maneuver corridors and water features of southern Iraq. The Division shared copies of this product out to supporting units, higher headquarters, and even the Commandant of the Marine Corps. The place names depicted on this product would soon have special meaning to the Division, as they would represent the battles fought and won by the blood and sweat of the Marines in combat, and would be the focus of the many efforts of thousands of individuals. The familiarity of the Division Marines with the places on this map would ensure these important places would have strong resonance during the execution of the battle plan.

The Division focused on the vulnerabilities of the large Iraqi force in order to ensure its rapid destruction. There were both physical and psychological weaknesses that could be

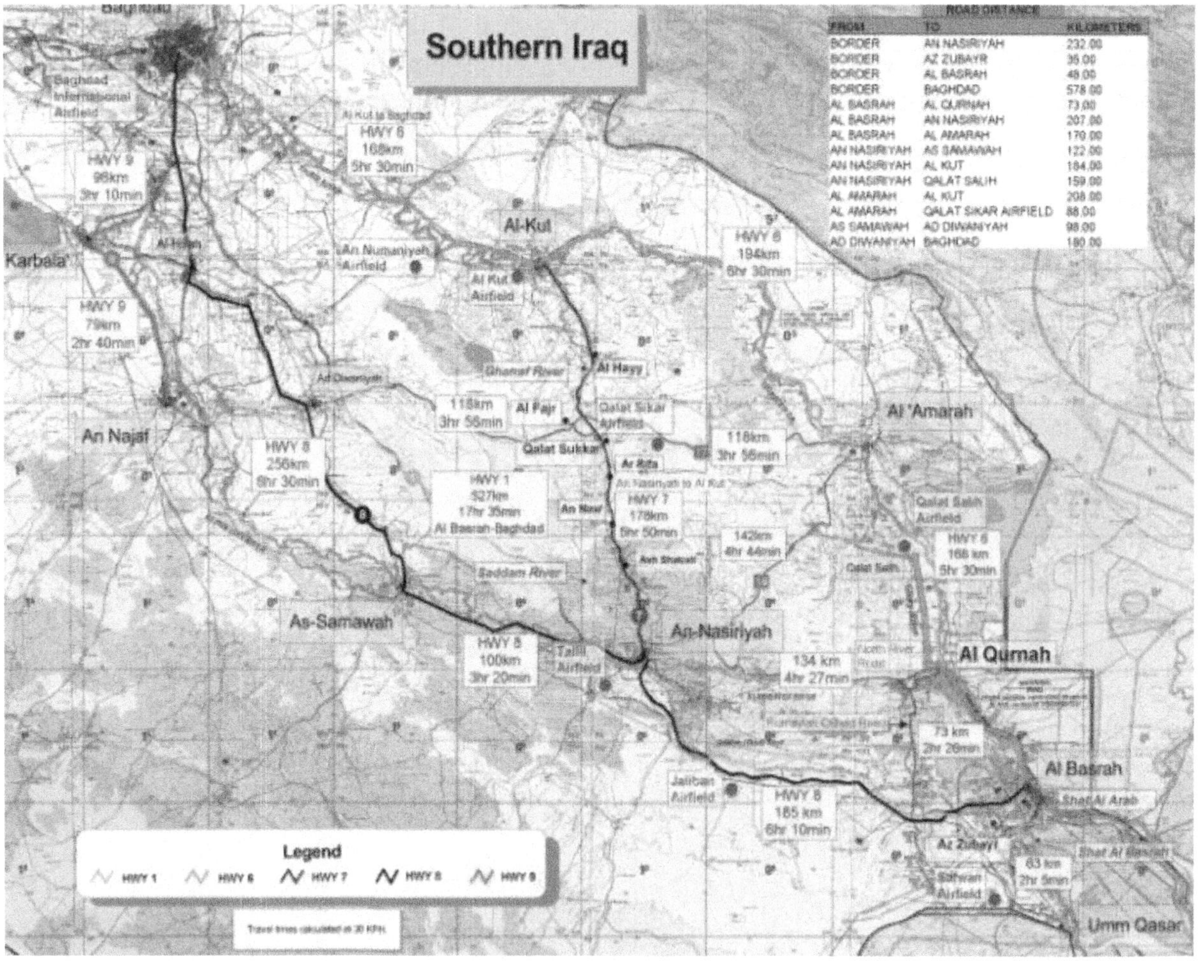

The 1st Marine Division's Southern Iraq orientation map, the 'CG's Special', would prove to be a key familiarization tool for all the Division's Marines.

exploited, and striking them became an important element of the Division plan. It was clear that although large in size, the Iraqi armed forces were poorly trained and equipped. Inadequate logistics and poor equipment readiness rates severely limited the offensive maneuver capability of even the best Iraqi units. The motivation and loyalty of the Regular Army (RA) units were reportedly poor, and their desertion rates high. The Iraqis' ability to command and control depended heavily on orders from a centralized higher headquarters, a system that would quickly break down during a fast-paced maneuver conflict. The Iraqi Regime had a reputation for punishing military commanders who took initiative without consultation. The Regime viewed effective commanders as a threat, and often transferred, punished, or even executed them. This heavy-handed treatment of commanders served to hamstring the maneuver capability of the Iraqi forces almost as much as physical equipment limitations.

Although limited in maneuver, the Iraqis had a robust indirect fires capability, which included a massive number of artillery systems and multiple rocket launchers. If forward deployed, many of these systems could range into northern Kuwait, including the assembly areas the Division would soon occupy. The fires capability of the enemy had its own weaknesses; however, including an inability to acquire targets, rapidly adjust fires, or logistically re-supply if deployed out of garrison. Even with these weaknesses, the Division was careful not to lose respect for the lethality of these Iraqi weapons systems. If the Iraqis used the terrain carefully, they could mount an effective positional defense and draw Division forces into preplanned firesacks and ambushes. The Iraqis could cause significant casualties and mount an effective operational delay by defending in a series of these well-planned tactical positions. Strategically, the Iraqis could then use this delay to continue efforts through diplomacy, employ WMD, or revert to a terrorist campaign. The combination of 'untrafficable' terrain and massed indirect fires might allow even a second-rate Iraqi army to have a strategic impact.

The I MEF Operational Planning Team (OPT), then under Lieutenant General Michael Hagee's guidance, opted to bypass the positional defenses of enemy units in the South (III and IV Corps) and avoid urban areas to the degree possible. The MEF would choose to fight terrain rather than play to the strengths of the Iraqis. The Division viewed this plan as on-target, and a shared appreciation developed among the MEF's major subordinate commands. With this appreciation, the Division concentrated on a scheme of maneuver that forced the enemy to reposition his artillery, and did not allow him adequate time to prepare artillery ambushes and fire sacks.

The Division G-2's well-presented, factual display of the enemy situation revealed the enemy vulnerabilities the Division would exploit, and the enemy strengths the Division would avoid. At the 2002 Division Officer's Ball (held annually to celebrate the Corp's birthday, that year in Primm, Nevada) General Anthony Zinni was the guest of honor. In a discussion of the Iraq situation, he jokingly told the assembled commanders that he would 'disown them' if they didn't handily destroy the enemy's army. These officers understood the Iraqi adversary so well that they only smiled grimly back at the beloved leader. Victory was never in doubt should the President call for the 1st Marine Division. Only the challenge of minimizing Marine casualties and the death of innocent Iraqi citizens gnawed at these men.

Providing Focus

"Everything we do is to be focused on the destruction of the Iraqi Army. Everything. Anything that does not point us to that objective needs to be eliminated."

The CG's guidance laid the groundwork for the preparations of the Division, "Everything we do is to be focused on the destruction of the Iraqi Army. Everything. Anything that does not point us to that objective needs to be eliminated." This focus meant some immediate changes in the Division's peacetime posture. The Division recognized that a focus on fighting the Iraqis would leave some peacetime administrivia by the wayside. Like a turning ship, the Division's momentum shifted from its peacetime direction to a war footing. One of the first (and most applauded) actions was the suspension of extraneous inspections, routine reports, and conferences. Focus meant a renewed emphasis on physical training as well. Every Marine and Sailor attached to the Division was ordered to conduct a minimum of 12 miles of personal physical training weekly, and also to participate in weekly NBC training. The main effort, however, was directed at the minds and spirits of the Marines. The CG's intent was to 'image' the Marines of the Division through every step of the process. The embarkation, planning, deployment, and first five days of combat would all be second nature to the members of the Division, because they would have already 'walked the ground' in their minds. For many of the Marines this would be their first combat experience. The CG wanted these Marines confident, comfortable, and aggressive in the face of the enemy. Destruction of the Iraqi Army began with the construction of a fierce and confident band of warriors in the 1st Marine Division. This construction was based on a strong foundation of three factors, the innate strength of the individuals drawn to serve in the Marine Corps, the Corps' strong traditional and advanced training methods, and the practiced team the previous Division Commander (Major General James Conway) had turned over to the new CG.

Building teams out of these hardened individual warriors was the next order of business. The CG advised his commanders, "Look around you. These are the men you will take to war." He cautioned them to look carefully at their staffs and ensure the team they had assembled was the team they wanted to take to the fight. The CG's intent was to build a fraternity of warriors that would break down any unnecessary formality between commanders and staffs, and between officers and enlisted men of the Division. Rank was respected, but all Marines were accorded the mutual respect due a professional of arms about to enter harm's way. A warrior code took hold, with each member of the Division valued for the contribution of his talents rather than the rank on his collar.

The CG had already inherited and assembled a strong team of officers in the Blue Diamond Staff. Brigadier General (Select) John Kelly, the Assistant Division Commander, was no stranger to division-level operations, coming from his latest assignment as the G-3 of 2d Marine Division. The Division Sergeant Major, Juan Duff, was a rock of stability and a key enlisted advisor. The Staff principals were also all high-caliber individuals in their respective fields, but many of them were newly joined to the Staff. The no-nonsense Chief of Staff, Colonel Ben Saylor, until recently the Commander of 11th Marines, went to work breaking down staff fiefdoms, and developed the fraternity that would lead to success on the battlefield. The G-3, Colonel John Toolan, had been the S-3 of 7th Marines when the CG had been its commander, and had also been a Light Armored Reconnaissance (LAR) Battalion Commander. Colonel Jim Howcroft, the G-2, had years of experience as a military attaché, and little patience for intelligence bureaucracies. His experience in the last Gulf War left him with a passionately held objective of focusing intelligence down to the lowest tactical level. Lieutenant Colonel John Broadmeadow gave the team a strong G-4, with broad-based expertise in logistics as practiced in the type of deep-striking operations the Division planned to execute. His recent Task Force 58 experience in Afghanistan would serve the Division well. Colonel Nick Petronzio, one of the scions of the communications community, and recently the commander of the Division's Headquarters Battalion, led the G-6. Lieutenant Colonel Cathy Powalski arrived fresh from her previous assignment as the 3d Marine Division G-1 and picked up 1st Marine Division operations without missing a beat. The dashing Colonel Jim Lukeman, the G-7 and Division Inspector (and until recently the commander of 2d Battalion, 5th Marines) provided an experienced hand and a trusted agent for the CG for a host of special missions including the rapid fielding of new equipment and a streamlined inspection process focused on embarkation and NBC readiness. Colonel Robert Knapp was the Division's Headquarters Battalion commander, entrusted with the deployment readiness of the Division headquarters and the smooth operation of the Division Command Posts (CPs). He headed a strong team of Military Police (MPs), communicators, truck drivers, logisticians, and even musicians that would be the backbone of successful Division operations.

Not satisfied with a cumbersome field headquarters, the Division Staff immediately set to work on downsizing the Division's CPs. The staff would be small, with no stove-piped experts in any one field. Aggressive 'Marine Air Ground Task Force (MAGTF) officers' manned every post. "Everyone fills sandbags in this outfit," was the expression most heard around the new CP. Speed in decision-making and command and control were to be enabled by release from cumbersome bureaucracies. The intent was to command and control the Division's swift movement with closely integrated air and logistics support as if it were a much smaller unit. The members of the Staff were to think more like a brigade than a Division Staff.

Strong individuals and strong staff teams were critical enablers, and now the Division's organization for combat underwent some changes as well. The CG saw habitual relationships as a conduit for speed on the battlefield, and reorganized the Division into the Regimental Combat Teams (RCTs) with which it would fight. In combat, several 2nd and 4th Marine Division units would join the Blue Diamond team, and were planned into the RCT organization. RCT-1 ('Inchon') was commanded by Colonel Joe Dowdy, who had recently served as the MEF Future Plans Officer. In addition to 1st Marines, RCT-1 included 2d LAR Battalion, major elements of 4th AA Battalion, and 2d Battalion, 23d Marines, a superb 4th Marine Division battalion. Colonel Joe Dunford commanded RCT-5 ('Grizzly') consisting of 5th Marines, 1st LAR Battalion, and 2d Tank Battalion. RCT-7 ('Ripper') was commanded by Colonel Steve Hummer, and included 1st Tank Battalion and 3d LAR Battalion. The 'Cannon Cockers' of 11th Marines, commanded by Colonel Mike Marletto, were integrated into the RCT's in habitual direct and general support relationships, as were elements of the Division's 3d AA Battalion (reinforced by 2d and 4th AA Battalions), 1st Combat Engineer Battalion, 2d Combat Engineer Battalion, and 1st Reconnaissance Battalion. Thanks to the foresight and aggressive initiative of the 1st Force

Service Support Group (FSSG) commander, Brigadier General Edward Usher, the FSSG provided its own task-organized representation, with assigned logistics units providing direct support down to the regimental level.

Two weeks after the initial orientation, the CG gathered his four regimental commanders, principal staff, and separate battalion commanders. In a quiet setting at the recently built MEF Operations Center, the CG continued to build the command team for the Division. This 'sweat lodge' meeting had the same solemn contemplation, mutual respect, and shared commitment as a meeting of tribal chieftains joining their tribes for battle. There were clear lines of command, but the assembled chieftains were there to gain consensus on the Division's vision, each committing himself to the fight ahead. The fraternity of shared risk and common vision grew. The CG provided his analysis of the upcoming fight, and imaged the commanders through the preparations he expected them to undertake. The assembled commanders were entrusted with the preparation of the Division for war with the final commission, "Gentlemen, we are now in the province of war. Consider every week your last week of peace, and apply your time accordingly."

"Gentlemen, we are now in the province of war. Consider every week your last week of peace, and apply your time accordingly."

Operational Planning

First Marine Division had always played a key role in plans for the defense of the Arabian Peninsula. Now, however, the Division changed its focus from the defense of the Arabian Peninsula to the eradication of the Iraqi Regime. August marked the arrival of a new G-

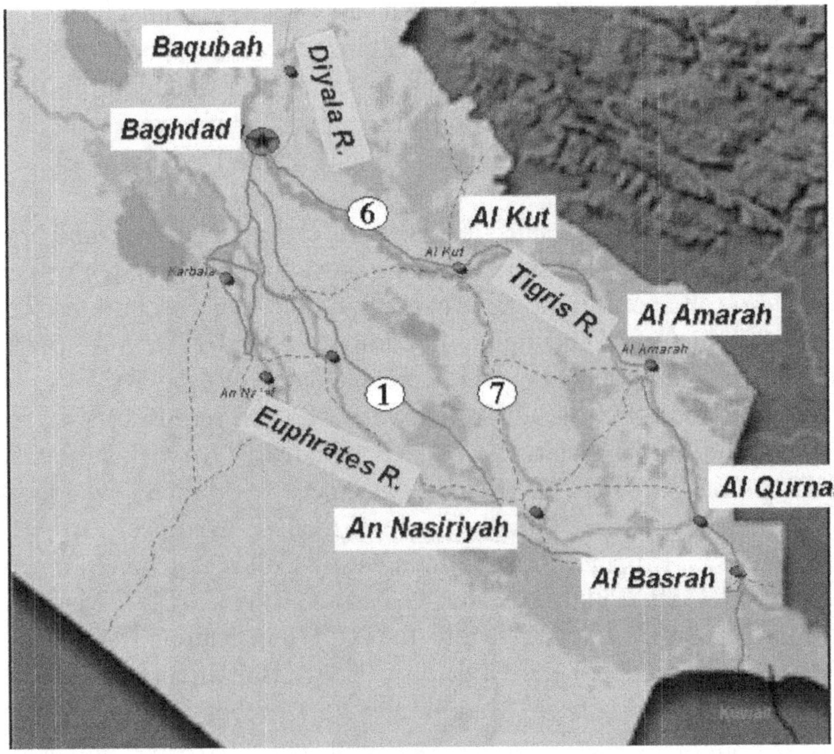

Some of the major Routes, Cities, and Waterways that marked the Division battlespace. These places would become household words by the time the Division was through with them.

3 Plans Officer, Lieutenant Colonel Paul Kennedy (replacing Lieutenant Colonel John Mayer who assumed command of 1st Battalion, 4th Marines.) He joined an already experienced planning team including Lieutenant Colonel Mike Groen from intelligence, Lieutenant Colonel (USA) Gary Smythe from Division fires. Communications, engineer, air, and logistics planners rounded out the team. The 1st Marine Division, as the major Ground Combat Element (GCE) of I Marine Expeditionary Force (I MEF), would fight as part of the ground scheme of maneuver developed by the CFLCC. The CFLCC was a high-level headquarters that coordinated the land actions of Marine, Coalition, and US Army ground forces in southern Iraq. The CFLCC commander was Lieutenant General David McKiernan, and the headquarters was primarily staffed with Army personnel, although there were a number of Marine representatives and liaison officers. As the CFLCC plan matured, 1st Marine Division underwent a number of mission changes, often with short notice. Before crossing the Line of Departure, the Division was to plan and re-plan for operations against An Nasiriyah, Al Basrah, the South Rumaylah oilfields, Highway 7 to Al Kut, the Tigris River, Highway 6, and along the unfinished Highway 1 corridor. Planning would continue with the Diyala River crossing, Baghdad, Baqubah, Tikrit, and Bayji added before combat operations ceased.

Seemingly every combination of unit and objective was explored, and each of these

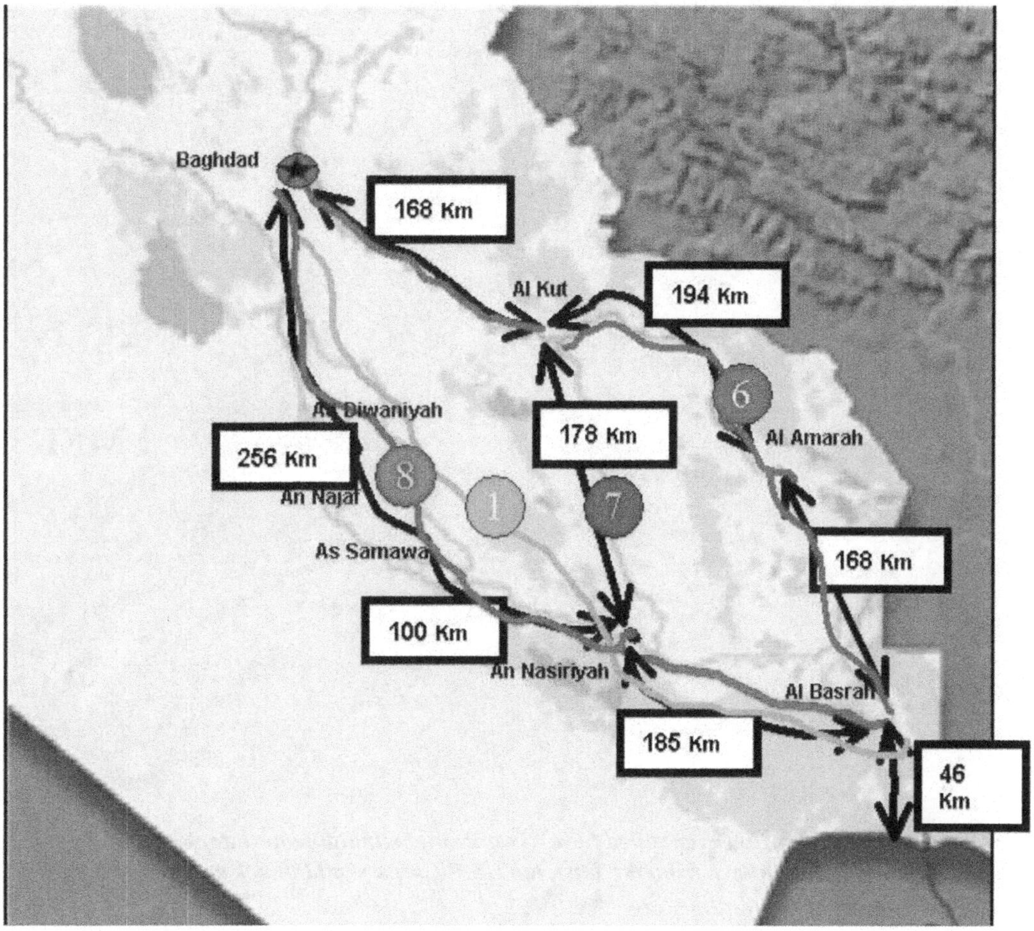

Road distances between major cities in southern Iraq generally split the battlespace into three 100 mile segments between the Kuwaiti border and Baghdad. Movement was restricted to major highways due to canals and irrigation ditches.

sequentially proposed scenarios required detailed planning. The staff paid special attention to the possibility of operations in the former marsh areas between Al Basrah and Al Kut, and noted the significance of rivers and bridges to every proposed course of action. Although none of the plans developed during this cycle were executed as written, the staff gained an invaluable understanding of the factors that would determine success or failure in the Division's coming fight. Major General Conway, the new I MEF Commander (who relinquished command of the Division in early August) quoted General 'Ike' Eisenhower on several occasions when referring to this rapid-fire planning environment, "The plan was useless, but the planning was invaluable." These words would prove prophetic, as the Division's eventual scheme of maneuver bore only slight resemblance to any of the original plans. The Division was able to maintain tempo and flexibility solely by a high degree of situational awareness and a full understanding of the commander's intent.

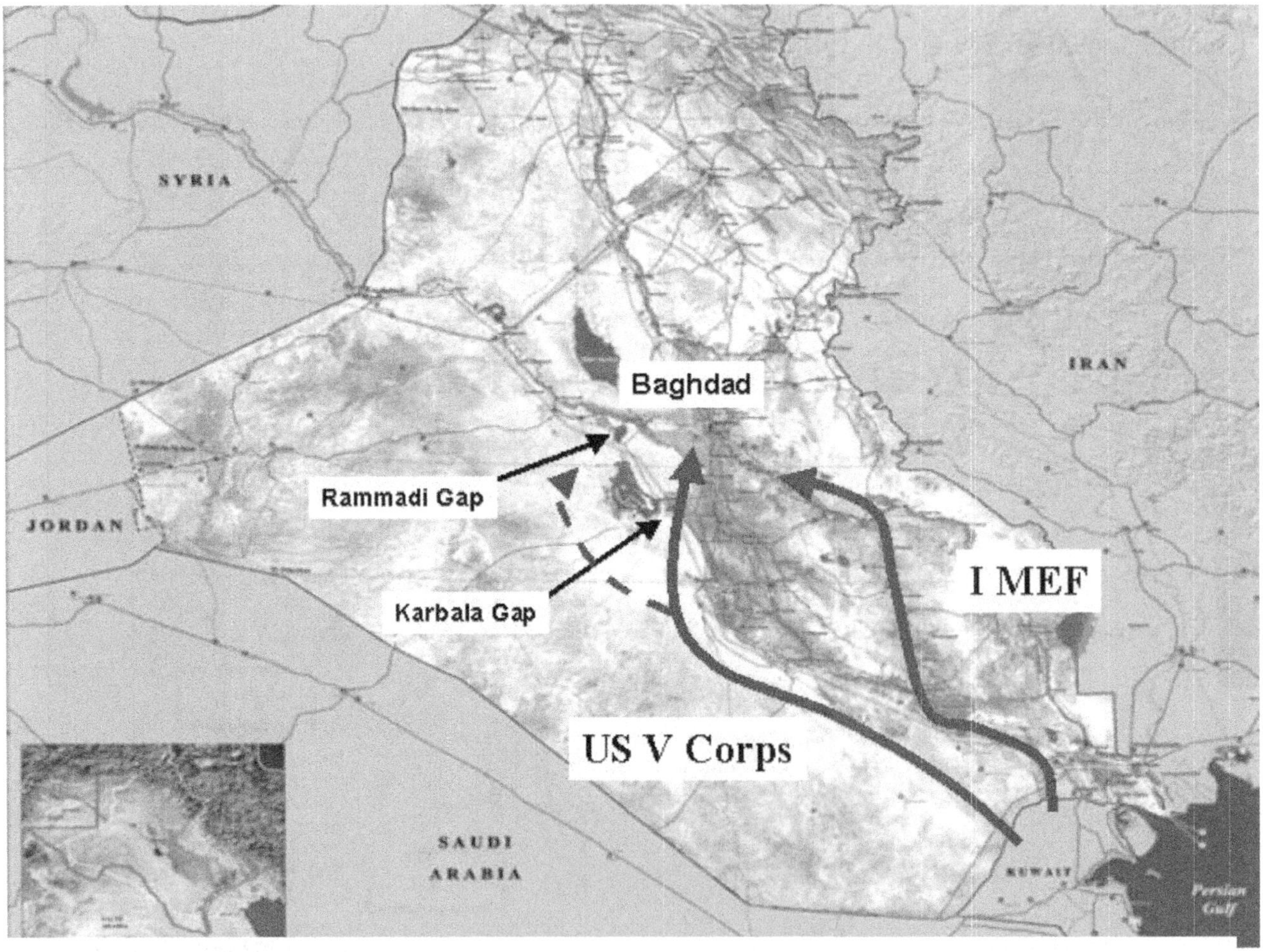

The basic CFLCC Scheme of Maneuver called for a 'Two Prong' simultaneous attack on the Regime. With I MEF conducting a supporting attack from the East, the US V Corps would attack straight into the heart of Baghdad from the South.

Like many armies with rigid centralized control, the Iraqi Army would likely collapse if cut off from their source of strength, in this case, Baghdad. The limited mobility of the Iraqi forces made them extremely vulnerable to a campaign of rapid maneuver, and the preplanned Iraqi defensive positions would only be effective if attacked from the expected direction and with a slow tempo. The Division planners worked aggressively to identify potential actions, timings, and routes that would allow the Division to seize its initial objectives and sustain its momentum to Baghdad, without playing to the enemy's strength. The G-2 assessed that the Iraqi divisions from III and IV Corps arrayed along the Iran/Iraq border did not pose a significant maneuver threat and could be safely bypassed. Once cut off from Baghdad, these units would be marginalized, and would likely surrender. With I MEF as a planned supporting effort to a V Corps attack to the west, 1st Marine Division could be most effective by defeating the Baghdad Republican Guard Infantry Division in Al Kut, the enemy's operational Center of Gravity in southern Iraq. This was forecast to be a fight of some significance, as it would mark the first engagement between Coalition forces and the Republican Guard. 1st Marine Division intended

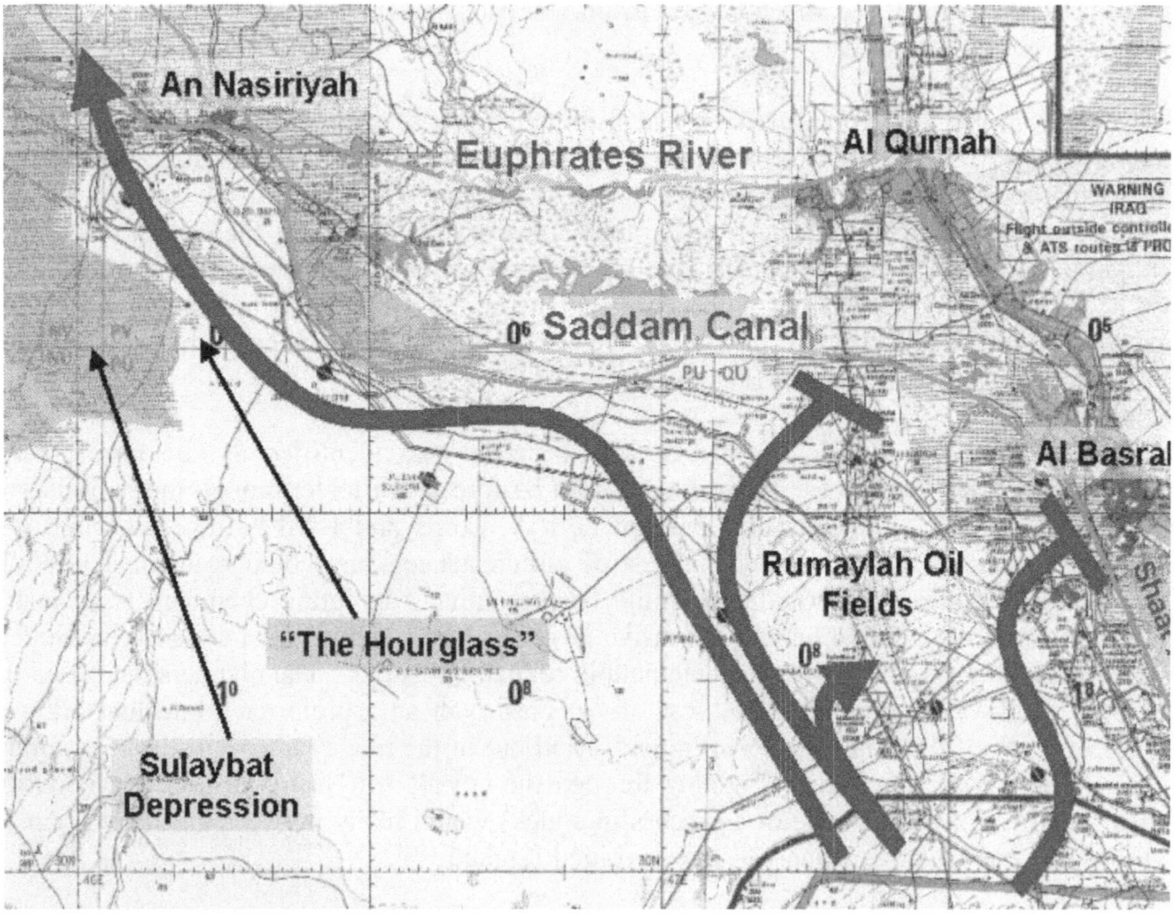

The basic premise of the 1st Marine Division's planned maneuver included bypassing major urban centers, and rapidly gaining a bridgehead over the Euphrates River in order to speed the attack on Baghdad, not getting 'bogged down' in heavy urban fighting.

to destroy the Baghdad Division in detail, as an example to the rest of the Republican Guard, and those remaining loyal to the Iraqi Regime. The Division had met some of these units in combat during the first Gulf War, and the CG reminded the planners that those Iraqis who had faced 1st Marine Division before "were not eager for a rematch."

Despite the sometimes-chaotic planning environment (due to uncertain deployment schedules and operational constructs that changed with the diplomatic situation) the seeds of the Division's basic operational concept were planted in this initial intelligence assessment. The Optional Planning Team refined the Division's concept of destroying Iraqi forces where necessary, but bypassing them where possible. In keeping with General Hagee's guidance, the Division would fight terrain rather than prepared defensive positions as long as it offered a faster route to Baghdad. The Division planned to bypass large elements of the Iraqi III and IV Corps, and quickly get to the north side of the Tigris River. Once the Division was across this strategic river, the Baghdad RG Infantry Division would be isolated, and ripe for destruction. Cut off from Baghdad, the remaining defenders in the south could be reduced piecemeal, with fewer casualties, and at the Division's chosen pace. In a coordinated action with the US Army's 3d Infantry Division, the 1st Marine Division would threaten to enter Baghdad itself in order to destroy the Regime's true seat of power.

By late Summer 2002, the Division had formulated a base plan, with the bypass of major elements of the III and IV RA Corps as a fundamental premise. It was the Division's assessment that by succeeding in building and maintaining momentum to Baghdad, the Division would likely defeat large elements of the Iraqi armed forces without engaging them with direct fire. The plan was to destroy the 51st Mechanized Division in the Az Zubayr area, destroy elements of the 11th Infantry Division in the An Nasiriyah chokepoint (formed by a number of canals, bridges, and untrafficable terrain) then block the remaining forces to the east while the Division continued the attack to Al Kut and Baghdad.

The 'hourglass' chokepoint formed at An Nasiriyah was identified as key terrain. All attacking forces from south of the Euphrates would be forced to trickle through the hourglass to get to the north side of the River. Passing both V Corps and I MEF forces through this chokepoint would be difficult under the best of circumstances, and could result in battlefield friction-induced delays. The Coalition would be presenting a tempting chemical target to the enemy after breaching one of his key defensive positions (the Euphrates), and careful battlefield circulation planning would be a critical element in reducing this risk. The planners identified the unfinished Highway 1 bridge to the west of An Nasiriyah as a preferred route that allowed movement north of the Euphrates River quickly, and out of the range fans of artillery remaining in the An Nasiriyah urban area. Fighting through the city of An Nasiriyah (and the enduring tasks of pacification in the wake of a successful attack) would likely pose a significant drain on the Division's combat power, and was to be avoided if practicable.

In anticipation of the second stage of the Division attack, the planning team refined their assessment of the attack routes from An Nasiriyah to Al Kut. The intelligence assessments of trafficability off the major highways in southern Iraq were quite bleak, presenting challenges for the attack to destroy the Baghdad Division in Al Kut. The Division's battlespace did not include the Highway 1 corridor at this time, but the Division recognized the necessity of finding a way to

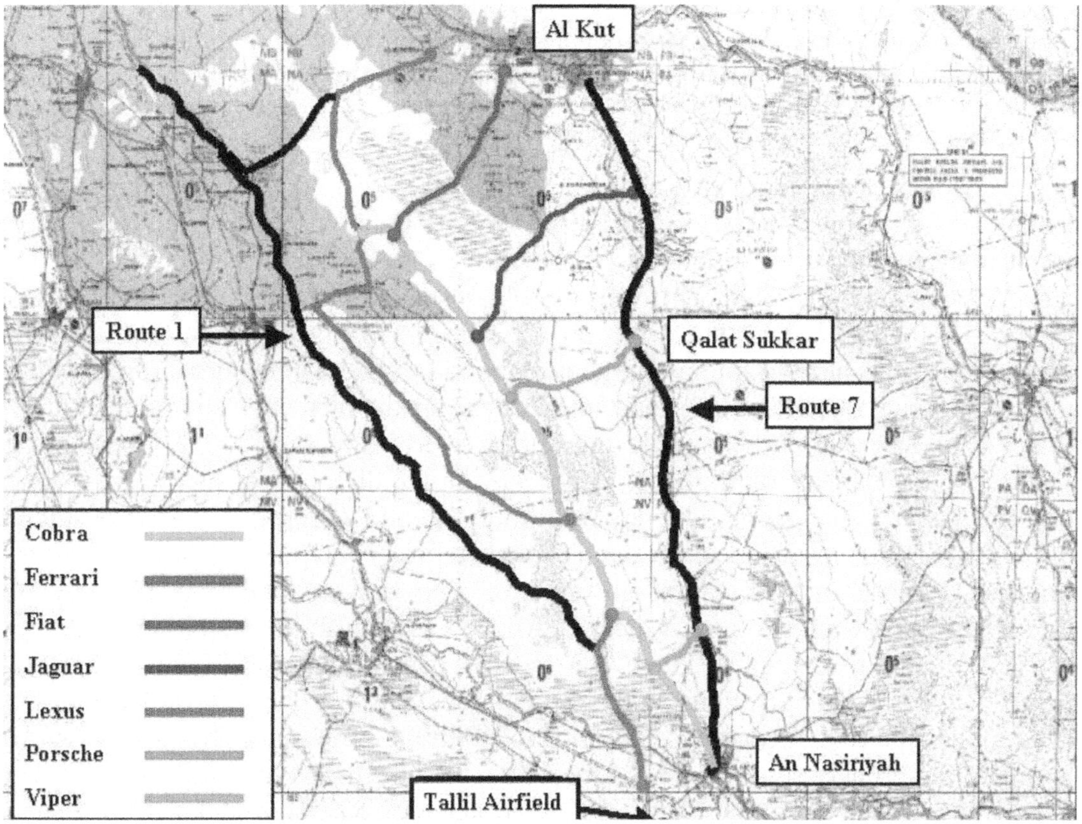

Numerous alternate routes (of dubious quality) were developed to facilitate a means to maneuver the Division if the primary routes were not viable.

envelop the Al Kut defenders by using fix and bypass tactics. If the Division was constrained to an attack up Highway 7, it would be forced to attack as a Division in column and play directly into the Iraqi strengths along an expected avenue of approach. This was an unacceptable scheme of maneuver. The Division worked aggressively to identify alternatives, and the poor maneuverability along Highway 7 became a significant catalyst for some of the creative solutions the Marines crafted to maintain the tempo of the attack. From creative engineering solutions to KC-130 highway landing strips, the solutions originally developed for the Highway 7 problems were later implemented across the battlespace as a way to gain speed in the attack.

Repeatedly, terrain experts told the Division that any maneuver off Highway 7 was not possible, thus a frontal attack or localized envelopment of Al Kut appeared to be the only feasible schemes of maneuver. Because of the responsive high-quality support the Division experienced from the Marine Corps Intelligence Activity (MCIA) in the past, the Division turned to them again for an analysis of alternatives for routes to the north. The product was a 'route finding' study vice a 'mobility analysis', and took as a precondition that some route through the challenging terrain existed. An MCIA team led by Mr. Art Kohler and Captain Sean Braziel conducted the route finding analysis and presented the results to the Division. Together, the Division and MCIA were able to identify three potential bypass routes using combinations of canal roads, farm roads, and limited segments of cross-country movement. Although not suitable for large convoys or Combat Service Support (CSS) elements, these bypass routes offered some limited maneuver options. The Division still assessed that it could defeat the Baghdad RG

Infantry Division by cutting them off from the capitol, leading to their rapid collapse. The MCIA identified routes were critical to this concept, and the Division planning effort continued using these limited bypass opportunities as the planned basic maneuver routes for the Al Kut fight.

There was the usual competition of tactical concepts among the Major Subordinate Commands (MSCs) of the MEF. One of these planning issues faced by the Division was the requirement to secure a Forward Operating Base (FOB) for 3d Marine Aircraft Wing (MAW) aircraft and 1st Force Service Support Group (FSSG) sustainment forces. The MEF chose Qalat Sikar as the site for this FOB. The Division was concerned that it lay 30 km to the east of Highway 7. Using this airfield for sustainment required a round trip excursion from Highway 7 of over 60 km, along a single access road into and out of the airfield. The momentum of this sustainment would be due east, opposite the northwesterly direction of the Division's attack. This airfield, well to the east of Highway 7, was also in the direction of the Iraqi 10th Armored Division in Al Amarah, the operational reserve for the III and IV Corps. From its position to the north of the major river obstacles, this potentially pivotal Iraqi force could move to reinforce the defenses at Al Qurnah, An Nasiriyah, Highway 7, or even Al Kut. The MAGTF Staff Training Program (MSTP) Red Cell had moved the 10th Armor Division in this fashion during the April 2002 MEFEX. More troublesome, the 10th Armor Division could readily move to interdict an FOB located at Qalat Sikar. Given its material limitations, the ability of the 10th Armor Division to mount a long distance maneuver was questionable, but the proposed FOB site lay nearby, and right on one of the primary routes the 10th Division could take to move to the west. Even with a limited maneuver threat, the vulnerability and importance of a FOB at Qalat Sikar demanded a combat force sufficient to deter or defeat the 10th Division should it sortie west. To address this issue, the Division's scheme of maneuver incorporated the 1st Reconnaissance Battalion in an unconventional 'economy of force' role. Instead of being used in the standard 'team operations' role in front of the Division's advance, Recon Battalion was prepared to fight in larger units, forming a 'picket line' to the east of Qalat Sikar, observing each of the possible routes the 10th Armored Division could take in a potential attack into the Division's east flank. Recon Battalion's elements were augmented with additional Forward Air Controllers (FACs) to ensure that any sortie attempt of the 10th Armored Division would meet a swift and deadly response by 3d MAW.

Preparing for the multiple wet and dry gap crossings that the Division's scheme of maneuver entailed was a major element of the plan, and one that cut across multiple lines of responsibility. The Division planning team hosted a 'River Crossing OPT' that brought together elements of intelligence, maneuver, engineering and logistics from across the MEF. 1st Combat Engineer Battalion (commanded by Lieutenant Colonel Paul Cyr), 8th Engineer Support Battalion (commanded by Lieutenant Colonel Rick Nelson), and the MEF Engineer Group (commanded by Admiral Charles Kubic) all participated. The Division planners drove this OPT as a way to 'jump start' engineer planning by providing them with a nominal ground scheme of maneuver as a base plan. A detailed assessment of potential gap crossings revealed over 220 bridge or culvert crossings on Highway 7 alone. An additional 55 or more were on each of the MCIA bypass routes. From an enemy perspective, each of these served as a potential ambush and delay point, the very sort of asymmetric tactic that a second-rate military might use to level the playing field. The use of captured Iraqi bridging equipment was even considered as a way to reduce the

Division's dependency on bridging assets. The OPT served as an extremely useful review of the potential engineering requirements, and was a valuable tool to assess likely enemy actions to the attack north from An Nasiriyah.

Force Deployment Planning

An experienced deployment team was assembled under the deputy G-3, Lieutenant Colonel Clarke Lethin, and the Plans Chief, Gunnery Sergeant Timothy Brimhall. This team looked at all possible ways to get forces into theater to support the new operational construct. Gunnery Sergeant Brimhall implemented a plan to reassign Military Occupational Specialty (MOS) 0511s (MAGTF planners) from the G-3 Plans section down to the regimental level to decentralize the enormous task of building the Division's Time Phased Force Deployment Data (TPFDD). The Marine Corps' assignment of 0511s to the Division and subsequently by the Division down to the regimental staff allowed the Division to quickly produce solid deployment plans for forwarding to the MEF.

Early planning began for the employment of two of the Division's RCTs: one using gear from the Maritime Prepositioned Ship Squadron (MPSRON), and the other using amphibious shipping. This plan called for an offload of MPSRON-2 by RCT-7 (with MPSRON-1 by an RCT

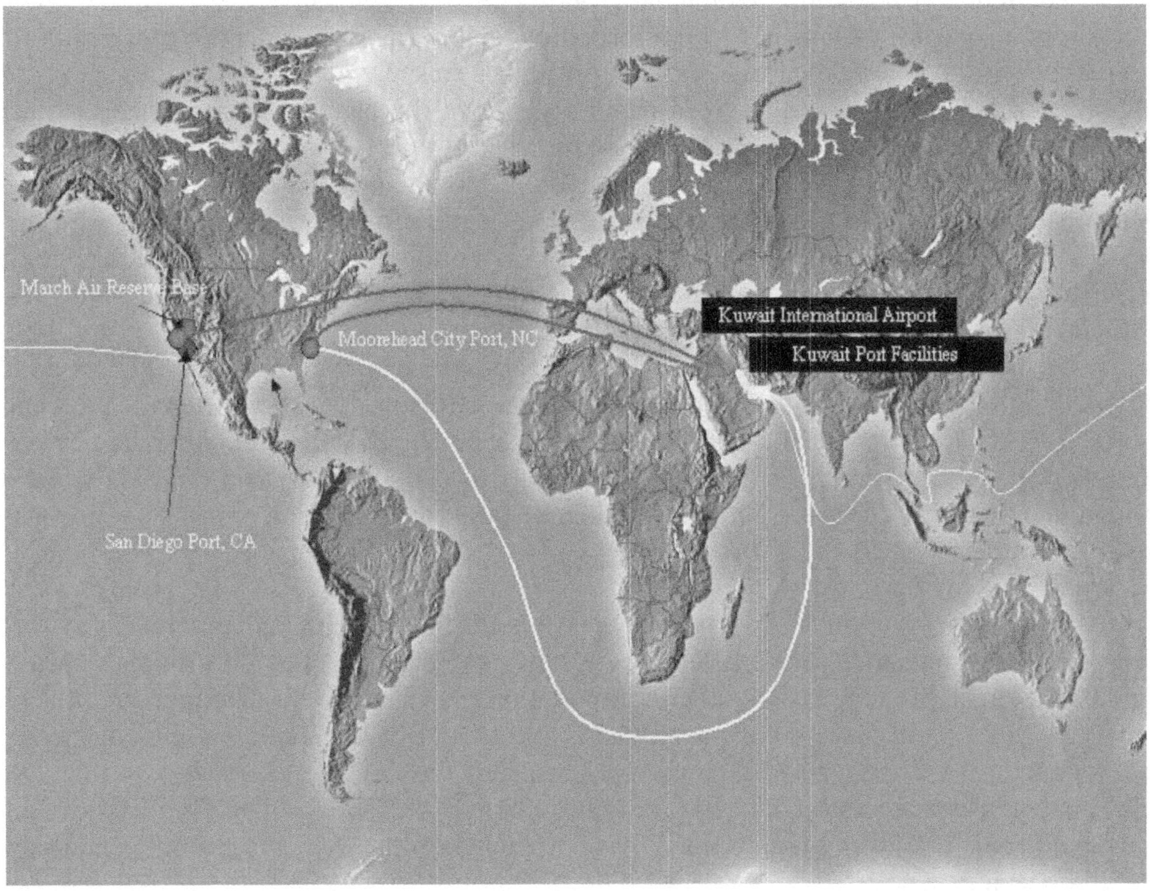

Units were to deploy to Kuwait from both East and West Coast ports, bases and air stations

from the 2[d] Marine Division) followed by the arrival of the amphibious RCT (or MEB) some weeks later. Getting a desired two additional RCTs into theater required additional lift, and a variety of methods were explored to avoid placing the bulk of this requirement on strategic air. Teams of logisticians explored options as diverse as extending the deployment window, using contracted cruise ships, and massing amphibious shipping to support the deployment of additional brigades. With the switch from a reactive 'defensive' force to an 'offensive' force, the nature and composition requirements changed.

Continued updates to the force deployment plan occurred. Eventually, the first two RCTs of the force would both come from 1st Marine Division (instead of one each from 1st and 2[d] Marine Divisions) and take advantage of the two Maritime Prepositioned Ship Squadrons (MPSRONs) for the bulk of their equipment, utilizing strategic and commercial air to transport personnel and other specialized equipment not contained on the MPS shipping. The force flow planning was updated to reflect MPSRON-1 gear supporting RCT-5 and 2[d] Tank Battalion. The Division needed to make the total force as mobile as possible, and requested all available AAV companies, both active and reserve, from throughout the Marine Corps. Planners also applied all available truck companies to the TPFDD, and updated the plan as refinements continued. The last RCT from the Division to arrive in theater would travel by amphibious shipping. By the original timeline, the first RCT would arrive and be ready for combat in fifteen days from the order, with the rest of the Division completely closed within 30 days.

The continuously changing mission, size, and deployment windows for 1st Marine Division forces were a significant challenge to the force-flow and operational planning processes. Since most of the information related to operational planning was 'Originator Controlled' (ORCON), only selected individuals could be 'read-in' to portions of the plan. As a result, eager Marines from across the Division built deployment plans based on only broad generalized guidance of what to take and what to leave behind. No mention of mission specific tasking was possible, and what was understood often changed rapidly. Strong leadership, social energy, and patience on the behalf of the entire Division team allowed for the building of a solid TPFDD despite these challenges. Planners at I MEF worked throughout this period to clarify basic decisions, but the MSCs had to remain highly flexible. It was clearly time to get the staffs of the Division's RCTs read in to the plan. In September 2002, the CG assembled a briefing team, and brought all battalion and regimental commanders assigned to the Division into the information exchange. The briefing sequence included an 'Intelligence Roadshow' brief, an operational overview, and a discussion of the CG's intent. This briefing had a tremendous impact on building situational awareness among the Division's commanders and staffs, during a time when this information was still very closely held

As the plan evolved, there were no less than seven TPFDD planning conferences at the US Transportation Command, at Scott Air Force Base (AFB) near Belleville, Illinois. Here, the assembled experts put together a global transportation plan that would take men and machines around the globe in massive numbers. Close by Scott AFB, the people of Belleville were the manufacturers of the most prolific transportation means the Division would use in the coming fight, the new desert combat boot. In just one example of the strong mutual pride between the American people and their Marines, the Division planners were welcomed with open arms to a tour of the factory.

Ships from all three USMC Maritime Prepositioned Ship Squadrons (MPSRONS) supported 1st Marine Division during Operation Iraqi Freedom by delivering the vital equipment and vehicles that were pre-staged throughout the globe.

Fire Support Planning and Preparation

Regardless of which route the Division took, the CG's guidance made the role of fire support in the coming fight clear. The enemy's tactical center of gravity was his indirect fire weapon systems; artillery, rocket launchers, and surface-to-surface missile systems. The mission of Fires was their destruction. The BM-21 mobile rocket launchers, elusive, easily camouflaged, and capable of firing in seconds a salvo that could saturate a grid square, would be relentlessly hunted down. The batteries of G-5 and GHN-45 howitzers, capable of outranging the Division's own M198s, and other medium range artillery would be located, identified, and attacked. The ideal prosecution of these targets was by 3d MAW aircraft before the Division closed within their range, but many would survive the air strikes, hidden in palm groves or withdrawn into urban centers for concealment and protection. For those, retribution by reactive counterfire would be swift and certain. The Iraqi Regime's FROG-7, Astros, and Ababil missile launchers topped the list of priority targets due to their ability to range the Division's Tactical Assembly Areas (TAAs), breach sites, and bridges from extreme standoff ranges. All of these systems were also capable of delivering WMD.

Hostile artillery is the natural enemy of artillerymen. The Division Fire Support Coordinator, Colonel Mike Marletto, and the fire supporters of the 11th Marines had been preparing for this fight for years. While the Division's organic artillery would remain the workhorse of the counterbattery fight, the fires planners knew they would have to leverage the capabilities of Army Multiple Launch Rocket System (MLRS) and Marine Air. Speed, again, would be the measure of effectiveness and the Division would stack the full capabilities of combined arms in its favor. The goal was to prepare the way for the RCTs to advance unimpeded, so that the momentum and combat power of the Division could be sustained all the way to Baghdad. This would be accomplished by an aggressive Division deep fight, destroying artillery and maneuver formations before the RCTs ever closed with them.

The Division G-3 Fires and Air sections, together with 11th Marines, had invested considerable time and effort in the development of techniques for controlling the Division's general support fires with airborne controllers, but the work remained in its infancy. The Division had also begun experimenting with a 'quickfire' capability to bring air fires against counterbattery targets. The wholehearted support of Major General James Amos, Commanding General of 3d MAW, now energized the dialogue between Wing and Division planners. Captain Rylan C. Bennigson of the Division G-3 Air Section, Captain Joe Deigan, the 11th Marines Air Officer, and Chief Warrant Officer Dave Thomas, the 11th Marines Radar Officer, were chief among the authors of the Division's new quickfire procedure. The quickfire procedure developed used the 11th Marines Target Processing Center (TPC) to provide radar data to the Direct Air Support Center (DASC), Division Air, and to the Division Fire Support Coordination Center (FSCC). In order to ensure redundant lines of communication and to expedite target processing, an Air Support Liaison Team (ASLT) was assigned to the 11th Marines CP. Captain Mark Halverson, 11th Marines S-6, and Major Roger Stanfield of the Division Communications Company engineered the critical communications link between the counterfire headquarters and the DASC. Automated fire support technology and training with the new systems progressed to the point where a hostile weapon, detected by counterbattery radar and selected for attack by the TPC, could be engaged by air within minutes. If within range, surface artillery could respond immediately to provide suppressive fires while aircraft from the stack were pushed to attack the target.

The full support and cooperation of 3d MAW ensured that assets required to validate the concept were available. In November, 11th Marines, the Division FSCC, and Marine Air Control Squadron Three (MACS-3) cooperated in the first live-fire rehearsal of the quickfire procedure at Camp Pendleton's Zulu and Whiskey Impact Areas. This exercise validated the concept, with artillery response times and the aviation targeting cycle gradually going faster as the procedure was practiced.

Third Marine Aircraft Wing provides unprecedented support to the Division as evidenced by the marking scheme of this F/A-18 D From VMFA-225.

Intelligence Planning

Intelligence planning was driven by a recognition that the enemy situation drove the Division's operations, and the G-2's synthesis shaped how that enemy situation was perceived. The Division would receive intelligence information from a tremendous variety of sources, and it would be a key challenge for the G-2 operations section (headed by Major Tim Oliver, Master Sergeant Jay Joder, and Gunnery Sergeant James Cecil) to convert this information into intelligence. The speed at which the Division hoped to move meant that there was little time for a 'paralysis of analysis'. The CG's intent was clear, and getting useful synthesized intelligence quickly into the minds of his commanders was the measure of effectiveness. The passion for dissemination on the part of Colonel Howcroft helped ingrain this mindset across the Division's intelligence effort.

In September 2002, the Division initiated an Iraqi 'Understudy' Program. This program was designed to match an officer of the Division staff with each of the Iraqi division and corps commanders the Division expected to face on the battlefield. The officer would research the birthplace, tribal affiliation, education, and professional track record of each of these Iraqi commanders with the hope that the Division could better anticipate their actions on the battlefield. The G-2 prepared a detailed assessment of the capabilities of each of the Iraqi units to complement the personal information about the commander, and the understudies met regularly to discuss the tactics, capabilities, and likely reactions of their respective Iraqi generals. This program was an innovative way to get into the minds of the commanders faced by the Division.

The intelligence community had a good handle on the military capabilities of the equipment the enemy possessed, but insight into the moral aspect of the fight was in short supply. Would the enemy commanders turn against Saddam if the Division attacked? Would RG and RA units cooperate on the battlefield? Who was likely to fight, and how aggressively? In marked contrast to the amount of information published in open sources about American commanders, the tightly controlled Iraqi society produced little information on anyone but Saddam Hussein. There was exasperatingly little data available on Iraqi military commanders. As a result, the effectiveness of the Understudy Program in understanding individual personalities was a difficult thing to measure. For its efforts, however, the Division gained a cadre of officers well schooled in the tactics, techniques, and capabilities of the Iraqi Army. The Understudy Program allowed for concentrated research efforts and a tactical focus on the enemy that might otherwise have been diluted. In execution, many of the Iraqi units that the Division had so carefully analyzed seemingly 'evaporated' in the face of 1st Marine Division's combat power, and the enemy commanders seemingly had little control over the actions of their units.

During this period, the G-2, with the assistance of Colonel Dave Larson at MCIA, arranged for a former Iraqi General officer (now living in the US) to visit and speak to the understudies, the Division Staff, and subordinate staffs. This officer painted a picture of an Iraqi military in a serious state of equipment disrepair. The requirement to use a large portion of the armed forces to maintain internal order left the Iraqi regime without an effective combat reserve. The Regime had saturated the army (RA and RG) with informants and political officers, creating an environment of distrust that permeated the armed forces. The palpable disgust with which the professional military and educated segment of the population held the Saddam Hussein Regime could serve as a source of popular support if the Division attacked, but was only one of many social dynamics at work. Much to the chagrin of the assembled officers, he advised an attack using only the western desert corridors to Baghdad, and dismissed out of hand the idea of an attack through the former marsh areas (he did not know, nor was he told, that this was the battlespace allocated to 1st Marine Division.) He made repeated references to the instability of Saddam's decision-making process, in which carefully developed plans were disrupted by irrational and impulsive decisions at inopportune times.

The oil infrastructure of southern Iraq was a complex maze of pipes, pump houses, and hazardous Gas Oil Separation Plants (GOSPs). All of these facilities would need to be secured by Marine forces to ensure Iraq's oil production capability would remain intact to provide the 'new Iraq' an economic base.

This information confirmed much about what the Division had previously assessed about the Iraqi armed forces. The Division tried to use the General's description of the decision environment to help understand what the trigger for chemical weapons employment might be. His predictions for the brittle nature of the enemy's defenses (i.e., rapid collapse if confronted directly), cooperation by portions of the citizens, and the larger political environment would later prove accurate. At the time, the Division accepted his comments as one of many possible scenarios, possibly meant to influence as well as inform, and continued to plan for a wide range of possibilities, including more dangerous assessments from other sources.

By October 2002, the G-2 planning team had developed a more mature picture of the Iraqi military situation, one that was decidedly more complex. One of the issues of concern was the importance of selected elements of the Iraqi southern oil infrastructure. As the Division's understanding of this complex system matured, oil specialists began to describe the Rumaylah oilfields as a series of interactive nodes. Some of these nodes were more important than others and were subsequently identified as special targets for Division operations. The Division's own troop-to-task analysis revealed that seizing all of these nodes would quickly drain the combat power of the entire Division, and the G-2 worked with I MEF and the CFLCC headquarters to narrow the list to the smallest set of 'critical' nodes possible.

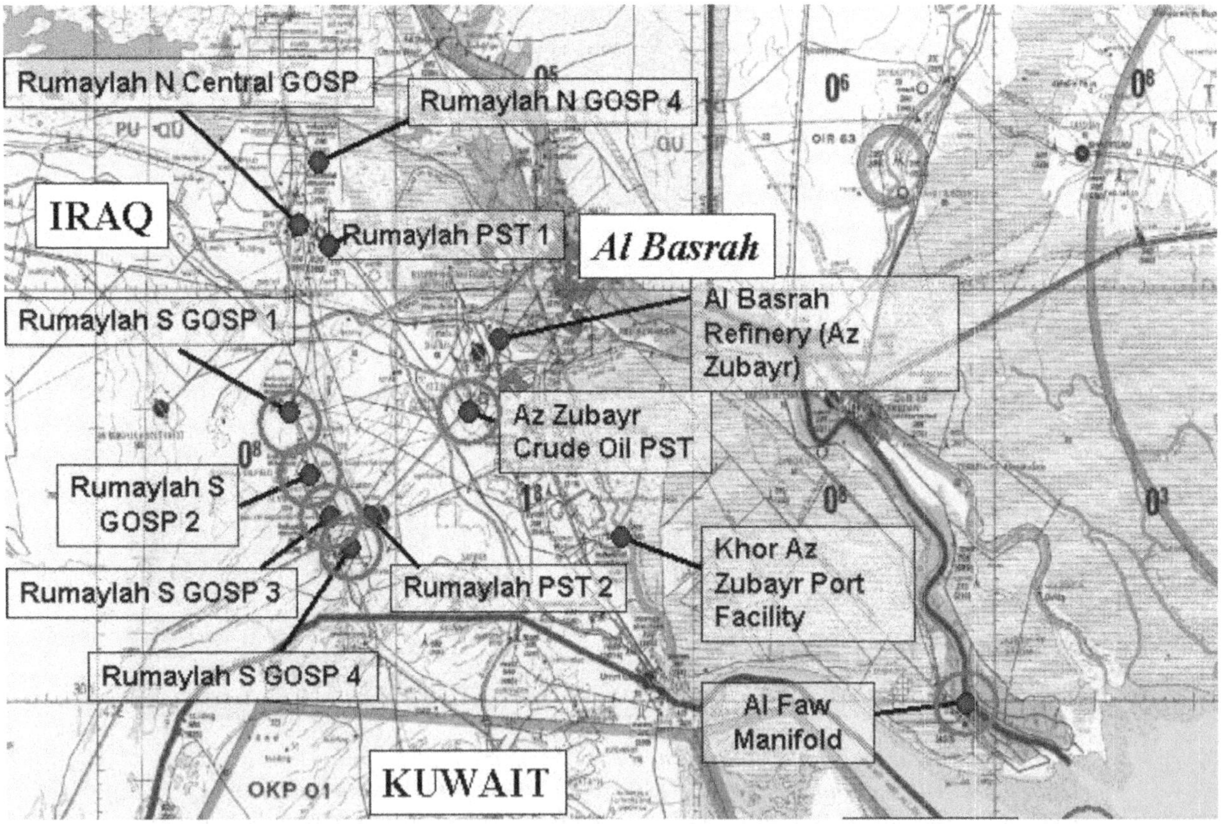

The South Rumaylah oilfield infrastructure, accounting for nearly two-thirds of Iraqi oil production, is composed of several critical nodes. Oil was pumped from wells in South Rumaylah to major pumping stations and refineries in Az Zubayr and Al Basrah.

Inundation warfare was another issue of concern in the Iraqi battlespace. The Iraqis had a well-developed doctrine for the use of water as a weapon, and had done this on a massive scale during the Iran-Iraq war, where they used water diversion to deny the Iranians avenues of approach. It seemed clear that if the Iraqis were to seriously defend along the limited avenues of approach available to the US, that inundation warfare would be a natural element of their defense. The An Nasiriyah area seemed especially vulnerable to this tactic, as well as the Highway 7 corridor between An Nasiriyah and Al Kut. MCIA's geospatial specialists conducted a detailed assessment to balance some of the alarmist reporting that was being circulated. MCIA sent out a briefing team headed by Mr. Jim Hill, who laid out the true nature of the possibilities. His assessment indicated that 'tactical' inundation was likely (the use of local water control infrastructure to deny areas of terrain), but the likelihood of 'catastrophic' inundation was remote (i.e. the large scale destruction of dams and barrages to release massive amounts of water into the system.) Modeling worst-case scenarios on the Iraqi 1954 flood, the Division identified alternate maneuver options if the Iraqis chose to use water as a weapon.

The manpackable Dragoneye UAV system would prove invaluable for decentralized intelligence collection at the Battalion level.

In late October and early November 2002, the G-2 was also able to work with Marine Corps Systems Command (MARCORSYSCOM) to accomplish the rapid acquisition of laptop computers loaded with the Falconview software for 3D terrain visualization. By MARCORSYSCOM's proaction, these laptops were fielded down to the battalion level across the Division, giving every maneuver commander the capability to 'fly-through' his objectives prior to the attack. The G-2 also worked with the Marine Corps Warfighting Laboratory to coordinate the deployment of prototypes of the Dragoneye UAV system in advance of a formal system fielding. These systems proved very useful in the close fight, as they were beholden only to the requirements of the owning maneuver commander, and were not co-opted or tasked by higher headquarters.

Nuclear, Biological, and Chemical (NBC) Warfare

The Division contemplated the coming fight against an enemy who was expected to employ WMD. This threat demanded that Division personnel develop patterns that made donning Mission Oriented Protective Posture (MOPP) gear second nature on the battlefield. During September, the Division began 'Gas Mask Wednesday' to instill this concept in all its members. Gas masks were worn all day on the body and donned for a 30-minute period.

Chief Warrant Officer Stacy Jeambert, the Division NBC Officer supervised an aggressive training effort to demystify operations in the NBC environment. Sergeant Taryne Williams, the Division's NBC operations chief, conducted a series of briefings that stressed proper wear, maintenance, and storage of MOPP gear. Familiarization training with the detection and decontamination assets each unit maintained was conducted. Instilling confidence in the Division's protective and detection equipment contributed greatly to the individual Marine's fighting effectiveness. The confidence with which the Division entered a probably chemically contaminated battlefield was a direct result of these efforts. Saddam could potentially employ a multitude of chemical or biological weapons against the Marines, but they would be prepared. From running obstacle courses in MOPP IV, to conducting a reenlistment ceremony similarly attired, the Division made itself ready.

As part of a thorough demystification training program, Marines spent a great deal of time in their MOPP gear. Meals were especially difficult.

Training for War, Exercises in CONUS

Almost two years before Operation Iraqi Freedom, the events of September 11, 2001 gave special impetus to the Division's preparations. On that very day, the Division was conducting a preparatory planning session for a series of combined arms training exercises in Twenty-nine Palms, based on an Iraq scenario. The exercise planners present were eager to get into action against those responsible for these heinous acts, and considered aloud whether the Division should change the focus of the exercise. Colonel Steve Hummer, the Commander of 7th Marines, calmly recommended that 1st Marine Division should change nothing in their preparations, "When this all shakes out, our role in this will be to attack and destroy those responsible for this, using a well trained combined arms team. The best thing the Division can do for our Nation right now is to keep training. Our turn will come."

> ***"When this all shakes out, our role in this will be to attack and destroy those responsible for this, using a well trained combined arms team. The best thing the Division can do for our Nation right now is to keep training. Our turn will come."***

He was right. The 'Desert Scimitar' series of exercises formed the foundation for successful execution of Division Command and Control in the combat operations to come. In anticipation of the requirement to conduct command and control over long distances, conduct deliberate river crossings, and spend extended periods in the field, the Division carefully crafted the Desert Scimitar exercise to contain all of these challenges. The plan called for an exercise in two parts; a defense against two attacking Republican Guard Divisions, followed by a rapid exploitation attack over a long distance culminating in a major river crossing operation. The distances covered in the California desert during the exercises were hauntingly similar to what would be expected of the Division a year later, and environmental restrictions encountered in the US mirrored real terrain restrictions in Iraq. The exercise scenario included objectives such as seizing key oilfields, isolating large cities, traversing impassable terrain, and conducting major river crossings.

During these exercises, the Division headquarters practiced multiple passages of control between Main and Forward CPs and gained valuable experience in providing command and control over large distances. The staff exercised satellite communications, long distance radio communications, and developed a battle rhythm for sustained 24-hour combat operations. The Division G-4 stood up Unit Movement Control Centers to control battlefield circulation and movement on main supply routes. The exercise culminated in a simultaneous two-site, expeditionary crossing of the Colorado River. All of these tasks would be critical in the fight the Division would conduct in Iraq less than a year later. The Division was able to arrange for training between Twenty-nine Palms and the Yuma Proving Grounds by negotiating land use with a variety of private and governmental agencies. The patriotism and cooperation of the citizens in the training area was inspirational, and made a difference in their Marines' ability to fight just one year later.

Aggressive Training Schedules

The regiments and battalions of the Division struggled with doing the additional preparatory work in getting ready for this particular conflict, while trying to maintain elements of normal training plans, and juggling the requirements for larger events such as MEU workups, Combined Arms Exercises (CAX), and the Unit Deployment Program (UDP). The combined training and operational tempos were brutal, and drove plans for 2 day weekends or summer vacations off the table. Even so, the Marines of the Division did not bemoan their fate. Marines, recognizing that their skills would soon be tested against a potentially tough opponent, were eager to hone their skills. In true warrior fashion, the Marines of RCT-1, RCT-5, and RCT-7 each shouldered their unique additional burdens without complaint. Their actions well before the conflict ever started were nothing short of heroic.

Seventh Marines was an example of the tremendous amount of work all members of the Division were accomplishing in preparation for this conflict. The Regiment had originally been scheduled to participate in CAX 5 & 6 in March to May 2002. The demand for Regimental participation in the Desert Scimitar series of exercises bumped this plan to CAX 1 & 2 in October to November 2002. The Regiment made the adjustment to their training plan and conducted the Desert Scimitar series, a September Division Command Post Exercise (CPX), a double CAX, and the Steel Knight exercise in December. They did all of this while simultaneously supporting Division-level Operations planning, NBC Defense training, and a vigorous syllabus of embarkation planning and assessment.

Individual battalions played heavily in the preparatory actions. One example was 3^{d} Battalion, 7th Marines, who trained in Camp Pendleton in May in the Basic Urban Skills Training - Assistant Instructor (BUST-AI) program to build a cadre of Military Operations in Urban

Terrain instructors, then used those skills in Millennium Dragon 02 and an Urban CAX in July and August 2002. In September, the Battalion participated in the Division's CPX, then was evaluated in an NBC Defense Marine Corps Combat Readiness Evaluation (MCCRE). With no rest, the Battalion served as the Ground Combat Element (GCE) for the October/November CAX.

Another fine example was 1st Battalion, 7th Marines. This battalion dealt with the unpleasant manpower implications of returning from a UDP rotation. Lieutenant Colonel Christopher Conlin planned to lose half of his well-trained and experienced Marines, and replace them with new personnel, within a matter of months. The Battalion established an impressive series of junior leader's schools, as well as participating in the BUST-AI training and the Steel Knight exercise. Layered on top of these training events was an extended cycle of planning and re-planning for combat operations in Iraq. Because the mode of employment for the battalion was uncertain (a limitation in the equipment carried in the MPSRON), the battalion had to plan and train for employment as a heliborne force, a mechanized force, and as truck-mobile infantry. Battalion training regimes for each of these methods of employment were developed and implemented.

First Battalion, 4th Marines, faced similar shortfalls in manpower. Having returned from deployment in June 2002, they assumed duties as 1st Marines' Training and Support Battalion, providing over 200 Marines to support the Fleet Assistance Program (FAP), the Camp Augmentation Program (CAP), and the Camp Guard. Coupled with the natural attrition of Marines reaching their end of obligated service, the battalion's on-hand troop strength was a fraction of what it once was. Lieutenant Colonel John Mayer, who served as the Division's Plans Officer, assumed command in August 2002 and charged his commanders and staff to "get your outfit in order". He directed the training priority as command post exercises, small unit leadership, NBC defense readiness, physical fitness, and combat marksmanship. The battalion's forward and main command groups and logistics train conducted weekly Command Post Exercises (CPXs) to refine their techniques and procedures. Select members of the staff also participated in the Mountain Warfare Training Center's Staff Planning Course in December 2002.

1st Reconnaissance Battalion outfits for HMMWV-mounted patrolling instead of its traditional dismounted team employment.

First Reconnaissance Battalion Trains for a New Mission

With the arrival of the new CG, came a new concept of employment for the Division's Reconnaissance Battalion. The Battalion had focused its training prior to this time on team operations, operating as the Division's eyes and ears near the limits of indirect fire range, and observing key Named Areas of Interest (NAIs). Over the year prior, the Division had worked to integrate the reconnaissance teams into the Division deep fight, inserting small teams to observe enemy artillery and troop concentrations so long-range fires could be brought to bear decisively against these targets.

The Iraqi battlespace was not conducive to this scheme of reconnaissance, and a new concept was developed. The CG's vision was to use reconnaissance as an independent maneuver element, providing reconnaissance-in-force and direct action capabilities on the flanks of the Division. One particular mission for the Battalion would be to set up a 'picket line' to the east of the Division, protecting the Division's flank as it moved up Highway 7. There were some significant changes in store for the Reconnaissance Battalion as they adjusted to the shift in employment strategy. Lieutenant Colonel Steve Ferrando shifted his battalion into high gear in preparation for the new missions. The Battalion began to re-equip, including armored vehicles (HMMWVs), heavy guns, and vehicle weapons mounts.

Recon battalion revamped its training plan to include heavy weapons. New communications systems were required to support battalion operations. VHF had not been a priority for the battalion S-6, but now they needed to support a number of new battalion nets. VIASAT cards for long-haul data communications were also integrated. The battalion's command and control mechanisms underwent some changes as well. According to Lieutenant Colonel Ferrando, "I just pointed them in the right direction, and those guys always made it happen."

Planning for Rotary Wing Aviation Support – Teaming with 3^d MAW

At no point during the tumultuous build-up for deployment did 3^d MAW fail to be an active part of the solution to the Division's challenges. The responsive support provided by 3^d MAW was instrumental in everything from Unmanned Aerial Vehicle (UAV) employment to casualty evacuation planning. Marine Air Groups 16, 29 and 39, provided most of the Division's rotary wing support. MAG-16 and MAG-29 were planned to provide support from the decks of amphibious ships, while MAG-39 would support from Ali Al Salem Airfield in Kuwait. Mission analysis determined that each Division CP and each RCT would require a minimum of one UH-1 for Command and Control support, preferably two. In addition, each RCT would require ready CASEVAC capability in the form of two CH-46Es. These assets would be in direct support of the RCTs and would be co-located with each Regimental CP. With guidance from the CG and through the OPT process, it was determined that a habitual relationship between RCTs and squadrons in direct support was essential for mission success. Squadrons vied for which RCT they would support. MAG-39 squadrons were assigned with HMLA-267 supporting RCT-1, along with HMM-364. HMLA-169 and HMM-268 supported RCT-5. HMLA-169 and HMM-364 supported RCT-7. Division general support was provided by HMLA-269 from MAG-29.

Creative and Proactive Logistics Planning

During the planning phase, the CG's guidance, "Speed equals success," drove the logistics preparations for combat. The Division's logistics team, headed by Lieutenant Colonel John Broadmeadow, was up to the challenge of supporting an aggressive scheme of maneuver over great distances with extended and vulnerable lines of communication. Unconstrained by doctrinaire approaches, the logistics planners set out to develop a concept of operations facilitated by smaller staffs, less lift, and occupying the smallest possible footprint that could still support large scale operations. The Division CG made it clear that he would accept risk to ensure the maintenance of a rapid operational tempo, and units would operate at times with less than three days supply of food, fuel, or ammunition. Marines following in support would, at times, have to fight their way up to re-supply the Division's combat trains, handling enemy raids and ambushes in the rear area themselves. The desired end state was a light, lean, lethal organization for combat that could fight and win on a 'logistics-light' diet.

The Division looked internally and made the first of many changes by altering the size and composition of the G-4 staff. Composed of 120 Marines and Sailors, (including the Medical section), the G-4's Logistics Operations Center (LOC) occupied several tents-full of space and several trucks-full of lift. Division logisticians assessed staff functions by how they impacted current operations commanded and controlled at the Division Main CP. If a function supported a Division's maneuver element, G-4 mobilized it and included it in the Main CP, but left other functions in the Division Support Area. Watch standers would have specific technical or commodity expertise but also have the capacity to function as MAGTF officers – working in logistics functional areas adjacent to their own specialty. This approach reduced the Main LOC staff to 31 multifunctional Marines and Sailors, responsible for logistics operations, movement control and medical operations. Using reach-back capabilities, the complementing functions of supply (including ammunition), maintenance management, food service, mobility and embarkation could set up at key nodes to affect liaison with supporting agencies and push sustainment forward. As a result of this reorganization, the LOC was small enough to be physically incorporated into the tentage of the Main CP, and was able to provide integrated decision-making information to the CG while maintaining full situational awareness of current and future requirements. Anticipating future requirements, and finding ways to meet them before they became a problem, would prove critical to the Division's swift attack to Baghdad.

Similarly creative initiatives originated across the Division. The commander of 3d Assault Amphibian Battalion, Lieutenant Colonel Rob Abbott, anticipated that repair parts for Amphibious Assault Vehicles (AAVs) that had not received system upgrades (those embarked aboard MPF ships) would not have mature Class IX assets available in the Marine Corps' Supply System. As the remaining fleet of older AAVs went through these upgrades, serviceable parts removed from upgraded AAVs were retained and provided to the Battalion. These parts were subsequently categorized and prepared for embark in support of the Division's AAVs on MPF shipping. The result was a high state of AAV readiness throughout the fight.

Lance Corporal David Evans and Corporal William W. Robertson from Weapons Company, 2d Battalion, 1st Marine Regiment, 1st Marine Division, fire the Javelin Anti-Tank missile.

In increasing the efficiency of the logistics support to the Division, the staff was careful to ensure that they did not sacrifice effectiveness. The confidence of commanders and their units in the logistics system was paramount. In one instance, several key leaders in the Marine Corps expressed concern over the lack of training allowances for the Javelin weapon system. The Ordnance Readiness section sought and received approval to conduct a live fire of the weapons system as a benchmark for the weapon's advertised performance. The round missed. They obtained a second round, fired it and recorded another miss. This raised several questions regarding the issue of 'break glass in case of war' type weapons. Working in concert with MARCORSYSCOM, Division logisticians and operators developed the solution set to improve the weapons system's performance and bolster crew and commander confidence. The effectiveness of those solution sets would be seen when the Division deployed to theater and began rehearsals for combat.

A New Concept For Combat Service Support

Throughout the early planning and preparation phase, the G-4 worked with the 1st FSSG and its Brigade Service Support Group One (BSSG-1), which later became Combat Service Support Group Eleven (CSSG-11) for wartime operations in support of the Division. Together, they developed agile, capable, Combat Service Support (CSS) units. The challenge was to provide a logistical element with shared situational awareness with the supported commands, which could proactively calculate logistical needs and have them out the door before a unit even registered a request. BSSG-1's staff of 32 Marines, commanded by Colonel J.J. Pomfret, would form the nucleus of this new CSS organization. They understood that logistics would be the critical link to keep the 1st Marine Division moving without an operational pause. Logistics had to be mobile enough to keep up with the rapidly advancing force, and flexible enough meet changing missions. It was important that CSS units execute as a maneuver element of and be fully integrated into the 1st Marine Division.

1st FSSG's answer to the challenge was a light, fast-moving, highly capable Combat Service Support Company (CSSC) in direct support of each RCT. CSSCs would answer to the RCT staff for sustainment, and maneuver in concert with the RCT's combat trains. They planned to carry food, water, fuel, and ammunition packages tailored to the RCT's mission. Supporting the CSSCs was Combat Service Support Battalion Ten (CSSB-10), which kept an additional three to five days of supply on hand, and was maneuverable enough to keep up with the rapid pace of battle. Tying the two together was the headquarters of CSSG-11. They were the critical link coordinating logistical requirements from the Division, and ensuring the replenishment of the CSSCs and CSSB-10 by the rest of the FSSG.

For the first time in recent history, the 1st Marine Division would fight with its own integrated CSS, and the Division embraced the new CSS elements as part of its own task organization. Colonel Pomfret was considered a maneuver commander of the Division, and his staff was thoroughly integrated. This 'family-building' approach was the model for all elements working with Blue Diamond. From embedded reporters to bridge companies, everyone who was going into the fight with 1st Marine Division was made part of an extended and tightly integrated family.

Planning For Casualties

In order to plan for the evacuation, movement and treatment of casualties, the Division had to estimate the number of casualties expected. The Division G-1 Operations Officer, Captain Sherel Ryan used a variety of tools to estimate the total number of casualties based on intelligence estimates of the fight ahead. The initial estimation of casualties was below 5% of the total expected strength of the Division. This assessment was used by the Division Surgeon's office as well as the FSSG to begin their preliminary planning of medical supplies, field hospitals, and re-supply.

The planned rapid advance of the Division's forces also required special consideration with respect to casualty evacuation (CASEVAC). The distances that air evacuation platforms would travel and communicate presented challenges to timely evacuation. The Division began planning with the Navy's medical support, the Wing's lift support and the FSSG's administrative and logistical support branches in order to build a swift and efficient system for the extract of potential casualties from as far forward on the battlefield as the tactical situation allowed. Two dedicated air evacuation platforms were provided in direct support of each RCT. Additionally, the U.S. Army would have UH-60 CASEVAC birds that the Division could task if needed. Speed of evacuation to a fully capable medical facility was the measure of effectiveness. To refine techniques and increase the speed of this process, the Division conducted two CASEVAC drills at the Marine Corps Air Ground Combat Center (MCAGCC) at Twenty-nine Palms, California. The distances and conditions provided by this training base would prove a near perfect match for those the Division would operate under in Iraq only months later. The CG stressed that, as part of our commitment to the Marines of the Division, it was imperative we build trust and confidence in the medical system. The best of American warriors would receive the best of American combat medical support.

Personnel Augmentation Planning

Planning for personnel augmentation to fill both existing shortfalls and additional requirements based upon the war plan presented unique challenges. Initially, the G-1, Lieutenant Colonel Cathy Powalski, identified personnel shortfalls based on the 1st Marine Division's current Table of Organization (T/O). Critical manpower shortfalls to T/O were evident early on as battalions just returning from deployments, most notably 1st Battalion, 7th Marines (1/7), were at 65% of T/O due to heavy attrition (which normally follows post deployment under the Corps' cyclical readiness paradigm.) This policy caused many units to fall to 'non-mission capable' manning levels for readiness reporting status. There were shortfalls in NCO ranks across the board, especially in low density MOSs. Some of the Division units prepared to go in harm's way well short of full strength. Trusting in the spirits of their Marines to compensate for their fewer numbers, commanders focused on preparing the Marines they had, and put their faith in Lieutenant Colonel Powalski to find as many Marines as possible to fill the ranks. The G-1 worked non-stop, using many creative innovations to find ways to increase the Division's endstrength.

The Secure Mobile Anti-jam Reliable Tactical Terminal (SMART-T) provided improved data communications to the Regiments of 1st Marine Division.

Manpower Management Enlisted Affairs (MMEA) did assist in limiting the impact caused by personnel shortfalls. They redistributed School of Infantry (SOI) graduates programmed for 2d Marine Division to reconstitute 1st Marine Division's critically short Battalions. They allowed extensions in 90-day increments for those who did not want to reenlist for four years, but wanted to fight alongside their fellow Marines. Where feasible, they also modified approved retirement dates for those who wanted to stay for the fight.

One of the most critical personnel shortfalls was in trained Forward Air Controllers (FACs). The G-3 air shop began efforts to staff FAC billets to T/O, but by the end of January 2002, deficiencies still existed. Manpower Management Officer Assignments (MMOA) branch would not meet these requirements, as no deployment orders were yet issued. 3d MAW once again stepped forward as the Division's best friend in the fight. 3d MAW began to look at ways to fill every gap to set the team up for success. The Wing also began to set aside training days to ensure these FACs, once on board, would be brought up to their potential capabilities. Ultimately, the Division identified the need for an additional 18 FACs and Air Officers. This number would be sufficient to staff the Division and subordinate commands for combat operations. The Division would continue the fight for FACs until the end of February 2003.

Communications Planning

The G-6 conducted detailed communications planning and preparation to support the Division, taking maximum advantage of the Desert Scimitar exercises earlier in the year. This exercise series closely resembled the communication distances that the Division would experience in Iraq. Colonel Nick Petronzio directed the planning efforts of Major Tom Davis, Major Pete Copeland, Chief Warrant Officer Bruce Gere, and many other Marines who would take these lessons learned and apply them to future communications support. For example, the Desert Scimitar exercise demonstrated the importance of radio retransmission teams for extending Line-Of-Sight communications. It also identified the importance of High Frequency (HF) radio and Ultra-HF (UHF) Tactical Satellite communications. This training was further enhanced through the CG's mandated weekly HF and UHF Tactical Satellite radio training for all Division units.

One of the Division's major communication enhancements was the arrival of the Secure Mobile Anti-Jam Reliable Tactical Terminal (SMART-T). This equipment was fielded to support the Division and RCTs. It was a secure data transmission system that allowed email, tactical phone, and data connectivity between major command and control nodes. With SMART-T, the Marines of the Division possessed an organic satellite data transmission system with the operational agility and speed of support that could keep up with the Division's high operational tempo. This new technology significantly enhanced the Division's flexibility to provide phone and data support without concern for terrain limitations or extended retransmission strategies.

The Division coordinated with the MEF G-6 for combat operations support planning as well as for training exercises in theater. The planning requirements centered on communications support for a Division Forward CP in Kuwait. This would allow the Division to have a Command and Control element in theater with communications connectivity firmly established in advance of any major Division deployment. During these joint planning sessions the G-6 and Major Chris Nelson laid the groundwork for what became the Division's wartime communications architecture. One communicator victory was the successful argument for two different space segments, one for each AN/TSC-93, allowing simultaneous data traffic for both the Forward and Main Command Posts. This allowed for uninterrupted, sustained data communications across the Division, even when one or the other Division CP was displacing.

The G-6 recognized that the standard Table of Organization for communicators in the Division was not adequate to support the additional challenges the Division would face. Multiple Command Post echelons, widely spaced Life Support Areas (LSAs) a major Division Support Area (DSA), hordes of liaison officers that arrived without communications equipment, and other challenges stretched Headquarters Battalion's Communications Company to the limit in both manpower and equipment. One creative solution was the activation and integration of the entire Communications Company from the 4th Marine Division. The addition of these Marines made the difference for mission accomplishment in the face of many of these non-standard challenges.

The G-6 resolved many of the communication problems encountered during earlier CPXs, and incorporated the solutions into subsequent training exercises. Of particular note, the 9th Communications Battalion attachments to the Division satellite and terrestrial microwave teams began to rehearse their habitual relationships. SMART-T and Single Channel Radio training internal to the Division began to pay off as communications links were established and maintained with a higher degree of reliability. Through practice, the Division also conducted field Video Tele-Conferencing (VTC) with a high degree of reliability.

Public Affairs Planning

The CG recognized the importance of the Public Affairs (PA) function, and continually hammered home the fact that, by every action, the Division would write its *own* headlines in the press. The Division also recognized the truism, "left unsung, the noblest deeds will die." The Division's recently arrived Public Affairs Officer, Captain Joe Plenzler, was faced with the daunting task of creating a PA section from scratch.

From the start, the PA section established a new focus for working with the media. The Marines dismissed words such as 'handle,' 'escort,' and 'manage' when talking about media operations. The CG exhorted commanders to win hearts and minds and adopt the media in the same fashion as a new private, rapidly integrating embedded journalists into the Blue Diamond family and give them complete access. He encouraged his Marines and Sailors to use interviews as an opportunity to show their courage to the world. The Division viewed the media, in every facet, as an entirely winnable constituency due to the strength of the example the Marines would make in the upcoming fight. The PA section paid special attention to the local Southern California media that would be critical in keeping the friends and family members back home informed of the Division's activities and honored without question any habitual relationship between a unit and reporter when determining the assignment of embedded media.

Preparing for embedded media was a significant undertaking accomplished during the planning phase. The PAO formed Public Affairs Liaison Teams (PALTs) specifically to address the Office of the Secretary of Defense's mandate to 'embed' international and national journalists within tactical units. The PA section split into 4 PALTs, allocated to each Regimental headquarters, while the PAO maintained a presence at the Main Command Post and Division Support Area. Each PALT consisted of 2 NCOs whose primary mission was to assist the regimental commands in facilitating the embedded media access and reporting. The PAO set a media embed cap of 80 journalists, approximately 4 per battalion, but adjusted this upward later with the support of the battalion commanders. Many major news organizations contacted the Division for embedded media opportunities, and many experienced war correspondents that had worked with Marines in Afghanistan were brought on-board, as well as novice reporters experiencing combat for the first time. The Central Broadcasting Service (CBS) television requested to bring along their own commercial HMMWV, hard-wired with satellite transmission equipment, allowing them to transmit live from anywhere in the world.

Deployment Readiness

In mid August, the CG tasked the staff to develop a Pre-Deployment brief to capture his mission, intent, and expectations. Recognizing the tremendous power of a Marine Division with a clear understanding of the commander's intent, the CG initiated a program to visit every unit and review his intent with them personally. Every staff function helped assemble a comprehensive briefing to 'image' every Marine and Sailor from their home duty station to Kuwait, and through their first five days of combat. The Staff touched on themes of aggressiveness, initiative, and the meaning of 'No Better Friend, No Worse Enemy.' The Staff hoped to reinforce to every rifleman in the Division how to destroy the enemy while maintaining his humanity. The objective was to allow them all not only to survive, but also to *thrive* in combat.

Junior enlisted Marines and a few company grade officers, the primary target audience, were selected to 'murder board' the initial drafts. The PAO supervised the briefing's preparation with a focus on images and photos, and short pitches by several different briefers. The brief was 90 minutes in duration and mandatory for all Marines and Sailors prior to deploying in harm's way. By all accounts, the briefings were very successful. According to one Lance Corporal, "This was the only briefing the Marines ever sat through in an auditorium that they actually enjoyed." From October 10, 2002 to January 13, 2003, the briefing team gave over 30 briefs, reaching nearly every member of the First Marine Division. The team also flew to Camp Lejeune, N.C. in the first week of December to brief the 8,000 Marines and Sailors of II MEF, as a way to welcome them into the Division team, or provide information to those deploying under other commands.

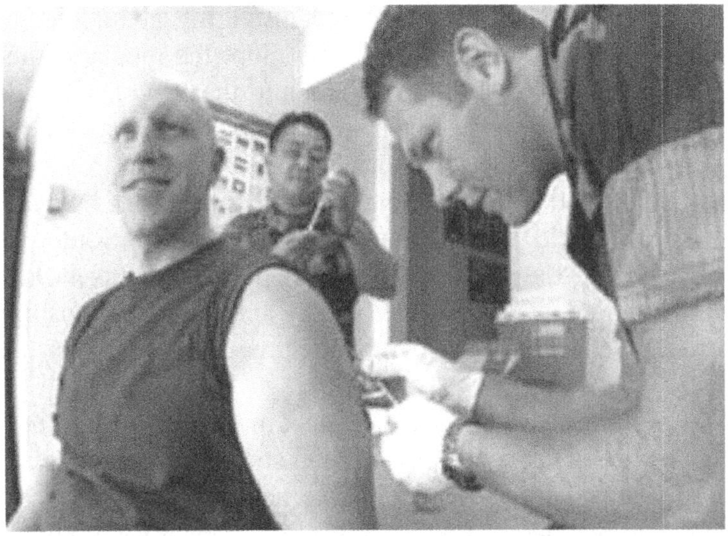

Marines received a series of inoculations, including Anthrax and Smallpox shots prior to deployment.

While a select few participated in operational planning, every Marine and Sailor in the Division was actively engaged in preparations for force deployment. With the prospect of war on the horizon, unit deployment readiness and embarkation training received renewed emphasis. The Marines and Sailors of the Division were made ready through a combination of physical and mental preparations.

One of the physical preparations was manifested in the CG's policy of 'Guardian Angels'. The Marines of the Division went nowhere or did anything without at least one ready Marine with a loaded weapon providing overwatch to their activity. The practice began even before deploying to Iraq, and continued until well after the last Marines had returned to CONUS. The presence of one visible Guardian Angel, always in full gear and with weapon at the ready, provided a visible reminder to any would-be attacker that these were 'varsity' warriors. This was practiced from the departure from Camp Pendleton and Twenty-nine Palms to the streets of Baghdad and beyond. At every checkpoint, the Marines doing the searching or checking identification never had to worry about their backs being covered. The Guardian Angel concept proved a source of confidence for the Marines and a source of intimidation for any threat that might wish to test the preparedness of 1st Marine Division.

Mentally, the Marines and Sailors of the Division were made tough. With attitude as a weapon, the Marines would be prepared to face down the Iraqi army in a battle of wills. The Marines were reminded of A. A. Vandegrift's (a 1st Marine Division commander during World War II) commentary, "Positions are seldom lost because they are destroyed, but almost invariably because the leader has decided in his own mind that the position cannot be held." The Marines were ready to dominate the Iraqis in a contest of wills. Anticipating a potentially long and mentally exhausting campaign, the Division also planned to support the psychiatric health of the Marines. Regimental Recuperative Centers (RCCs) were planned-for as a way to limit psychological casualties and alleviate mental and physical exhaustion. Located just off the front lines, these RCCs would give local commanders the ability to keep Marines physically and mentally ready for sustained combat operations. It was planned that no Marines would be evacuated further back than these RCCs for non-physical conditions.

Tough but humane was the desired attitude for the Division Marines. This was best articulated in the Division motto: "No Better Friend, No Worse Enemy." The depth of warrior ethos in the Marines was neatly summed up in that one statement. Mature, compassionate, and lethal; the Marines of 1st Marine Division went into Iraq with the full confidence that they would so dominate their foes that they had no fear of showing compassion. Later, in the streets of cities across Iraq, the Marines would demonstrate time and again that a confident Marine walking among the people was more endearing and effective in winning the hearts and minds of the civilian population than soldiers riding inside armored vehicles or ensconced behind barbed wire.

Deployment of the Forward Command Post and Planning Team

In October of 2002, the Division formed its Forward CP to deploy to Kuwait. There had yet been no public decision to employ military force against Iraq, but by moving the Forward Headquarters and key planning staff into theater early, the Division would have a solid footing for participation in theater planning efforts, identification of assembly areas, and reconnaissance of Aerial Ports Of Debarkation (APODs), Surface Ports Of Debarkation (SPODs), and other facilities required. The efforts of the Forward CP in the upcoming deployment of the entire Division were critical to Blue Diamond's later success, and marked a transition in the Division's preparations for combat. The initial phase of planning and preparation ended, as the Division began to take action to flow forces into theater and closely integrate planning efforts with the MEF and CFLCC.

Chapter 2
Force Flow and Planning Refinements

The Commanding General and Marines of the Division Forward Command Element spend a motivated Christmas 2002 in Camp Commando, Kuwait.

The arrival of the Division Forward CP and the Division planning cell marked the beginning of the Force Flow phase of the operation. The US Administration, exhausting diplomatic options before exercising military force, remained heavily engaged with the United Nations. Lacking approval and a clear deployment order for operations against the Iraqi Regime in late November, there was, understandably, a great deal of uncertainty surrounding the force composition, timing, and objectives of potential combat operations. Having the Division planners and a small CP forward allowed for the beginning of integrated planning among CFLCC, I MEF, US 3d Infantry Division, 1st UK Armoured Division, and 1st Marine Division. Although the progress toward finalizing operational plans in the early days of this stage appeared slow, the networking and shared situational awareness among the stakeholders in the coming battle was invaluable. Ultimately, this added measurably to an in-depth understanding of the MEF Commander's intent. Perhaps more importantly, the presence of Division Marines in Kuwait gave 1st Marine Division a 'leg up' on preparations for the Reception, Staging, Onward movement, and Integration (RSOI) process that would bring over 20,000 Marines, Sailors, and Soldiers of the Division into theater.

There was a clear requirement to have the GCE present at these early stages of the planning process, because it was clear that the majority of the planning would be done in theater by the MEF staff. Although no deployment order had been given, there had been an approved Request For Forces (RFF) that required Marine forces in Kuwait to participate in a series of exercises. Upon the Division's careful review, the auspices of RFF-164 implied Division representation for the Internal Look and Lucky Warrior training exercises in Kuwait. The Division chose to deploy a small Forward CP and planning staff in support of the exercises, riding on the coattails of the deploying MEF staff. The first main-body echelon of the 1st Marine Division Forward CP arrived at Kuwait City International Airport on 18 November 2002. They joined a smaller advanced quartering party who had been in country since early November. Lieutenant Colonel John Broadmeadow led the 28 Marines and a Navy Corpsman to their home for the next three months. A 70-man security force from 1st Battalion, 4th Marines, commanded by Second Lieutenant Andrew R. Jones, also deployed

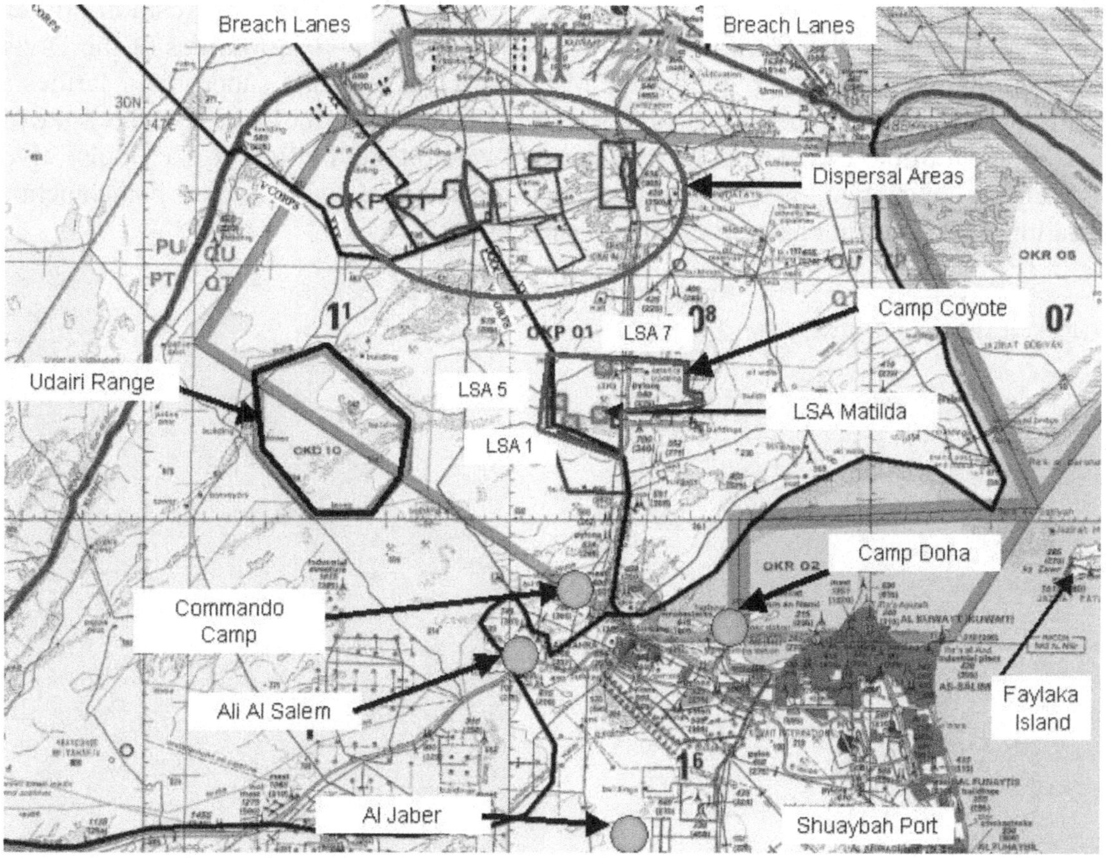

Locations of the major camps, Life Support Areas, and training areas where Marines from I MEF and its subordinate elements would prepare for the attack into Iraq.

to Camp Commando to provide force protection for the I MEF Command Group. These Marines provided force protection for the Camp until the arrival of the battalion's main body on 24 February 2003.

Commando Camp, located about 35 km northwest of Kuwait City at the foot of the Mutlah Ridge would be home base for much of the planning for Operation Iraqi Freedom. Ironically, it was located within a few kilometers of the first Gulf War's 'Highway of Death' that had marked the end of the Iraqi occupation of Kuwait. The lonely terrain evoked vivid images of the last time the Division had fought the Iraqis, and there was a strong sense of a job unfinished that provided a daily focus to the warfighters now assembled. The MEF Headquarters Group was located at this camp, the garrison compound of the Commando Brigade of the Kuwaiti Land Forces (KLF). The camp still bore the scars of the Iraqi occupation, left unrepaired by the Kuwaitis as a reminder of deeds past. In a gracious gesture of support that set the tone for future cooperation, the Kuwaitis made room for the US Marines.

The Marines of the Division Forward CP were assigned to one remote corner of the compound. Early efforts by Gunnery Sergeant Jason Topp of the advanced party had literally 'moved mountains' of trash and sand in order to make a spot for the Division CP and billeting areas. First Lieutenant Richard Fisher, the Forward's Camp Commandant, had the Marines turn-to on making this spot livable, and soon the camp was complete with its own Guardian Angel post, sandbagged bunkers, and living tents. A homemade gym, made with a combination of bricks, fenceposts, and cast-off Air Force equipment was quickly made operational. Assuming this position would be occupied for several months, the Marines made themselves at home. The Marines of the Division Forward CP were only a small group among the MEF, MEG, FSSG, and supporting Marines also bivouacked in the camp, and a special camaraderie developed among the Marines who knew they were the lead element of the Blue Diamond Marines that would take the fight to the Iraqis. Even a special 21st century adaptation of the Bushido code was adopted, with severe 'punishment' for whining or failing to display an aggressive warrior spirit.

1st Marine Division Forward CP occupied a small corner of the larger I MEF base at Camp Commando, Kuwait.

The decision to deploy the Division Forward CP for CENTCOM's annual rehearsal of the defense of Kuwait against an Iraqi invasion would pay huge dividends. While the remainder of the Division trained at home stations and prepared for force deployment, the Forward CP's participation in Lucky Warrior and Internal Look provided a foothold in MEF and CFLCC planning meetings where the Division's operational planners would strongly advocate the ground combat perspective. Advance parties from all functional areas including logisticians, communicators, and engineers seized opportunities to reconnoiter aerial and seaports of debarkation, routes, and tactical assembly areas.

The presence of the Division Forward CP in Kuwait also provided a 'reach forward' capability that brought the Division one step closer to destroying the enemy should the President call. The members of the staff in Kuwait sent a constant stream of information from planning meetings and site surveys, while responding to numerous specific requests for information from Division units back in California. During one especially arduous task of TPFDD tracking, the Forward's Corporal Justin Duplain was asked to contact the Division Operations Center (DOC) back at Camp Pendleton to answer some time-sensitive questions. He proceeded to establish a secure Internet chat dialogue with Corporal Chad Rhyner in the DOC at Camp Pendleton. Within minutes, questions and answers were being exchanged in real-time by the two echelons of the Division staff. The solution, a testament to the technology the Division would bring to the fight, is more remarkable because it was simply second nature to a new generation of Marines. Any who harbored doubts about the 'New Breed' of Marines were quickly educated. The Marines of this generation were made from the same stuff as those who had gone before, trained killers who could set up an intercontinental data exchange without skipping a beat.

Intelligence Planning - The Changing Enemy Picture

By December, the enemy had made only limited preparations for the coming attack. Major units remained in the vicinity of their garrisons and no overt defensive obstacles or fortifications were identified. The G-2 assessed from a variety of sources that the Iraqi Regime did not think the threat of a ground attack was imminent and had refrained from early deployment of forces as a serious logistics drain that they could not sustain. Because there were no major changes in the enemy's disposition or material status, intelligence analysis focused on refinement of the Division's assessment of border obstacles, sensitive sites requiring exploitation, keen attention to terrain along the few avenues of approach, and preparation of detailed support packages for each critical oil node identified in the Division's zone. Lieutenant Colonel Groen and the rest of the forward deployed intelligence team were able to go directly to the higher headquarters sources for much of this information, saving valuable research time.

The intelligence community's assessment of the South Rumaylah oil fields also began to mature to a point that specific critical maneuver objectives could be identified. A strategic goal was to capture the Iraqi oil infrastructure intact in order to use these revenues for the rebuilding of Iraq. Restoring the hope represented by these oil revenues to the newly liberated people of Iraq would be among the Division's proudest legacies. The eventual strategic goal was the production of over two million barrels of oil per day, most of which would be produced by the South Rumaylah oilfields. Seven critical nodes of the southern oilfield system were identified, all of which were with in 50 km of the Kuwaiti border, and all of which were in the proposed Division zone. The G-2 tried to accommodate the sudden requirement for Division Marines to become keen students of the oil services industry. The Division learned to identify Gas-Oil Separation Plants (GOSPs), Intermediate Pumping Stations (IPSAs), storage facilities, and offshore terminals; and understand all of their functions.

Securing four critical GOSPs and a pumping station complex (later known as the 'Crown Jewel') in the southern oil infrastructure became key elements of the Division plan. Detecting and understanding the threat to these critical nodes from Iraqi military activity or sabotage became a key Priority Intelligence Requirement (PIR). Integration of imagery intelligence, human intelligence, signals intelligence, and open source intelligence was used to ascertain the extent of preparations the

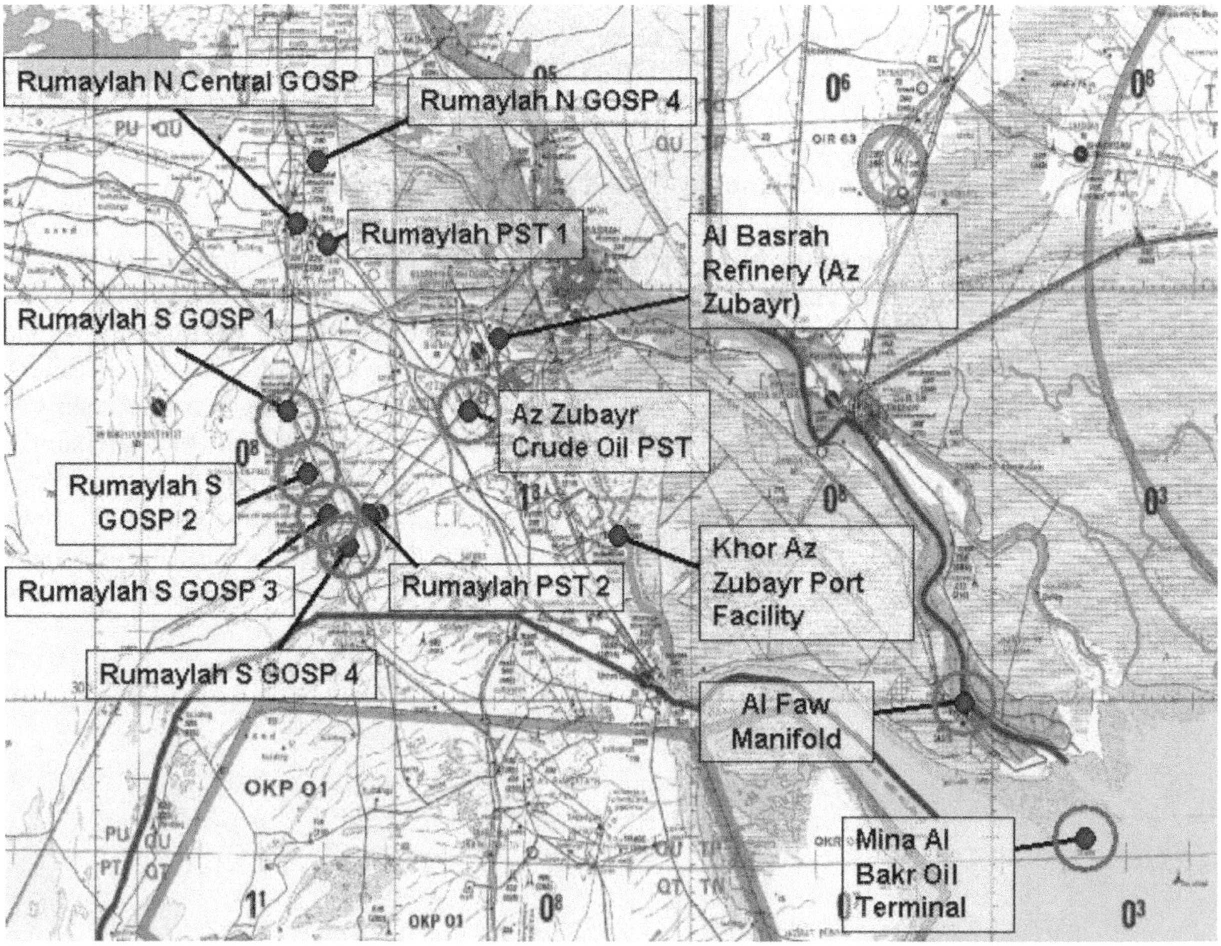

Seven critical oil infrastructure nodes are identified (red). These nodes would require years to rebuild if damaged, or could be used to cause massive environmental damage if left in the hands of the Iraqis. Marines were tasked to secure these nodes during the initial stages of the war.

Iraqis had made. The careful timing and coordination of the seizure of all of these oil nodes intact required close cooperation among all elements of the MEF. Although a supporting effort, the Division's early tasks were to have strategic implications. General Franks, the CG of the US Central Command (USCENTCOM) described the seizure of the South Rumaylah oil infrastructure intact as "equal in strategic priority to the seizing of Baghdad."

The border obstacles on the Kuwaiti side of the border were another significant obstacle to attacking forces, and were mostly inside the UN-demilitarized zone. The obstacles consisted of a rear berm and anti-tank ditch, an electrified fence, and a forward berm and ditch along the actual border. There were no indications of similar continuous obstacles or minefields on the Iraqi side of the border. Although derelict minefields and areas of unexploded ordnance from the fist Gulf War were identified, these had been largely cleared in the now agricultural areas that predominated in the eastern half of the Division zone. Later, when the Division attacked across the border, little evidence was found of any border obstacle preparation other than a few areas of hastily scattered mines.

Marines inspect the Kuwaiti border defenses near Umm Qasr. These included berms and anti-tank ditches (above) and electrified fence (below)

Another aspect of the border defenses identified by the G-2 was a series of Iraqi border guards, intelligence collection, and visual observer posts. These were strung across the border in order to provide reporting on the activity of US forces, monitor US/Kuwaiti communications, and detect Coalition aircraft transiting Iraqi airspace in support of Operation Southern Watch (OSW). From atop the dominating heights of *Jabal Sanam*, Iraqi visual and signal observers could detect and report the Division's activity in northern Kuwait. Alerted to the presence of the Iraqi observers and intelligence collectors on the hill, the Division was able to incorporate the destruction of this key observation post into the early attack planning. To preserve tactical surprise, the neutralization of several of the border guard posts that lay astride the Division's planned avenues of attack would also be required. G-day (the start of the ground attack) would not be a good day to be an Iraqi border guard.

During December, MCIA gave the Division a detailed assessment of tactical inundation possibilities in zone. By tracking the levels of precipitation in northern and central Iraq and monitoring the water levels in the various dams and reservoirs, MCIA was able to conduct a reasoned approach to the issue of inundation warfare. The reservoirs that fed the Euphrates water system were only partially full, and had very small output capacities (comparable to draining a bathtub with a straw.) More troublesome, however, was the risk of tactical inundation from the Tigris River, in the heart of the Division battlespace.

As most of the precipitation in Iraq fell in the northern part of the country, the northern reservoirs held a significant capacity of water that could be released into the Tigris River system. Releasing too much water directly into the Tigris would flood Baghdad itself, but the MCIA analysts had found a potential way the Iraqis could create flood conditions without resorting to this drastic measure. If the Iraqis were to raise the water level of the Tigris to the maximum capacity it would bear through Baghdad, then inject another large amount of water via the Diyala River to the east, significant flooding along Highway 7 or around Al Kut might occur. There was natural precedent for this action in a major flood that struck the region in 1954. In this flood event, the banks of the Tigris were breached near Al Kut in order to save Baghdad, and the entire Gharraf River basin was flooded. If the Iraqis had studied their own history as carefully as MCIA had, they may have already integrated this course of action into the defense of Al Kut or Baghdad itself. Once again, the advantages of speed in the Division gaining a foothold on the far side of the Tigris were made clear. The Marines had Baghdad in their sights, but how much mud would be on their boots when they arrived remained to be seen.

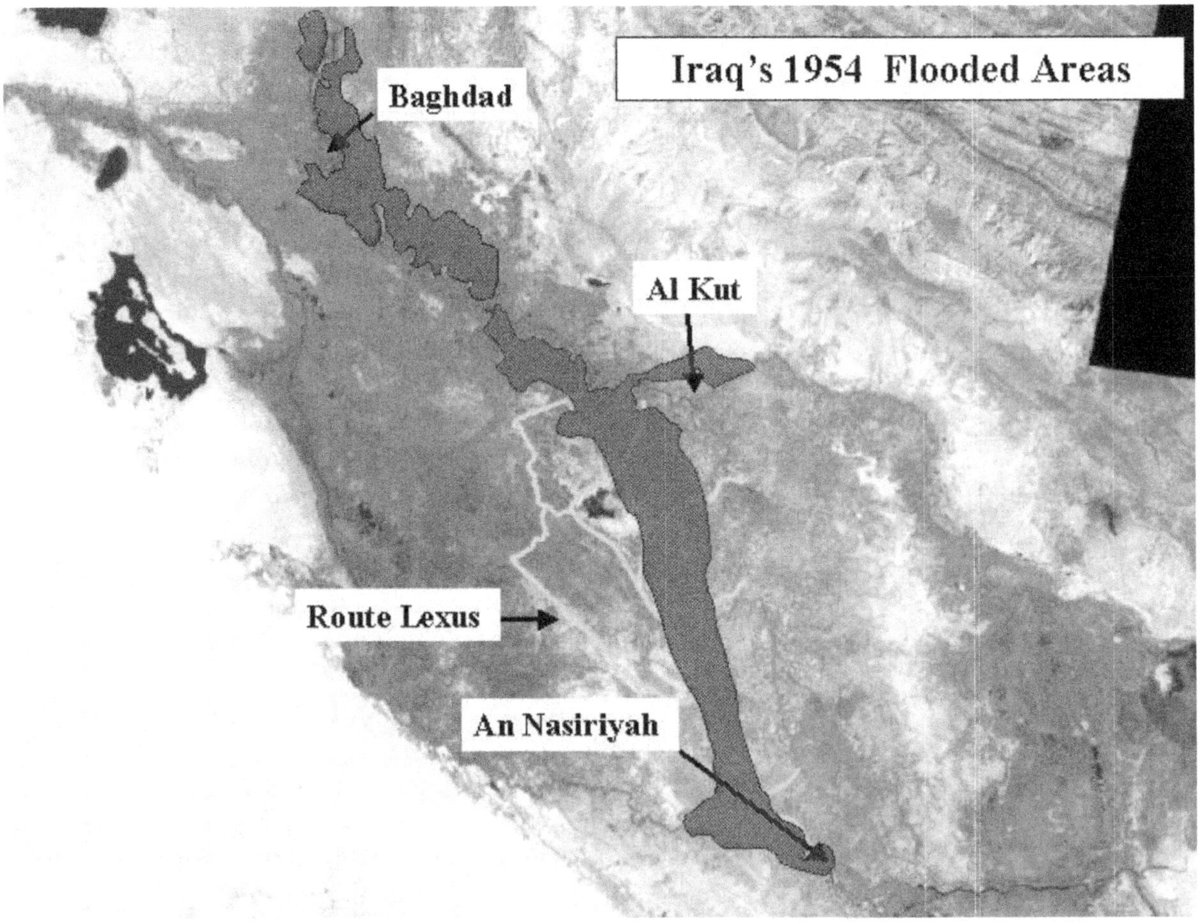

The 1954 Flood along the Tigris and Gharraf Rivers inundated much of the Division's battlespace, including its major avenue of approach along Highway 7. The potential for deliberate flooding of the rivers in 1st Marine Division's battlespace was an ongoing concern.

What the intelligence estimates did not reveal was the magnitude of the Regime's weapons holdings. Huge caches were hidden in every area of the country, but it was only after the Division closed on these facilities that the full magnitude of the distribution of tons of weapons and ammunition throughout the country came to light. Perhaps most disturbing was the large caches discovered in hospitals, mosques, and schools. Almost every one of these facilities revealed a weapons cache upon inspection. The Iraqi Regime had prepared themselves to fight a long-duration guerilla war, but had seriously miscalculated the amount of support the populace would provide. The US intelligence community and its technical sensors were well equipped to find enemy order of battle (tanks, artillery, defensive positions), but failed to detect these hidden weapons caches. Additionally, the Division still had more questions than answers about the enemy commanders the Division would fight. The military background of Iraqi commanders and their willingness to fight for Saddam's Regime remained an open question.

Operational Planning – The Baghdad Option

In early December, the Division Planning cell conducted a preliminary planning effort for an attack on Eastern Baghdad. The Division had already planned through its attack to Al Kut, and now the prospect of a follow-on mission to support the US Army in Baghdad was raised. CFLCC proposed participation of the Division in the fight for Baghdad as one of the potential follow-on missions to the attack on Al Kut. Several iterations of this plan were discussed, including cross-attachment of Marine Corps infantry units to the Army's V Corps (surprisingly, a US Army Mechanized Infantry Division has fewer infantry dismounts than any single regiment of 1st Marine Division.) The cross-attachment proposals were not given serious consideration by the Marine leadership, but the implications for stabilization operations would be felt later, as the Army needed to relieve foot patrolling Marine rifle squads with Bradley fighting vehicles and M1A2 tanks. Eventually, the Baghdad planning stabilized with the Tigris River as a proposed boundary between the MEF and V Corps, and the Division prepared a plan for combat operations in eastern Baghdad.

From the East, Baghdad has a series of natural and manmade obstacles that serve as natural defensive lines. The terrain outside the Baghdad urban area is very difficult to traverse due to the high density of agricultural fields, bisecting canals, and rivers. This limited mobility terrain surrounds the city, and makes maneuver to attack positions near the city very difficult, as the terrain transitions to a complex urban area. Once through this limiting terrain, an attacking force must cross the Diyala River, which runs north to south along the eastern edge of the city. This high-banked river would be difficult to cross, especially if defended with massed indirect fires. Across the Diyala, an attacker reaches the outskirts of the Baghdad urban area. To the north, on the outer ring of the eastern Baghdad urban area, are the remains of an ancient flood-diversion canal that serves as a 'moat' of sorts. The ancient canal and its crossings were a second natural obstacle, and were expected to be incorporated into any city defense. Once inside the 'moat', the dense urban area served to severely limit maneuver corridors to the center of the city. The final obstacle belt was another drainage project known as the Army Canal. The Army Canal bisected East Baghdad from Northwest to Southeast, and formed a natural final defensive line.

The G-2 identified the Rasheed Military Complex, positioned just across the Diyala River near its confluence with the Tigris River, as key terrain. This military/industrial area on the southeastern side of Baghdad would allow an attacking force to get into the heart of the city by moving along the natural barriers instead of across them. The Rasheed area contained a large military base with an airfield and several WMD sensitive sites. It was expected, naturally, that this area would be strongpointed by the enemy in an urban defense, as the value of this 'hinge' could not go unnoticed by the Iraqi commanders. Other key terrain identified included the downtown riverfront area that allowed direct fire onto the Presidential Palace area in western Baghdad, and the large Shia ghetto known as 'Saddam City'. There were also over 20 WMD-related sensitive sites in East Baghdad alone that would likely require securing.

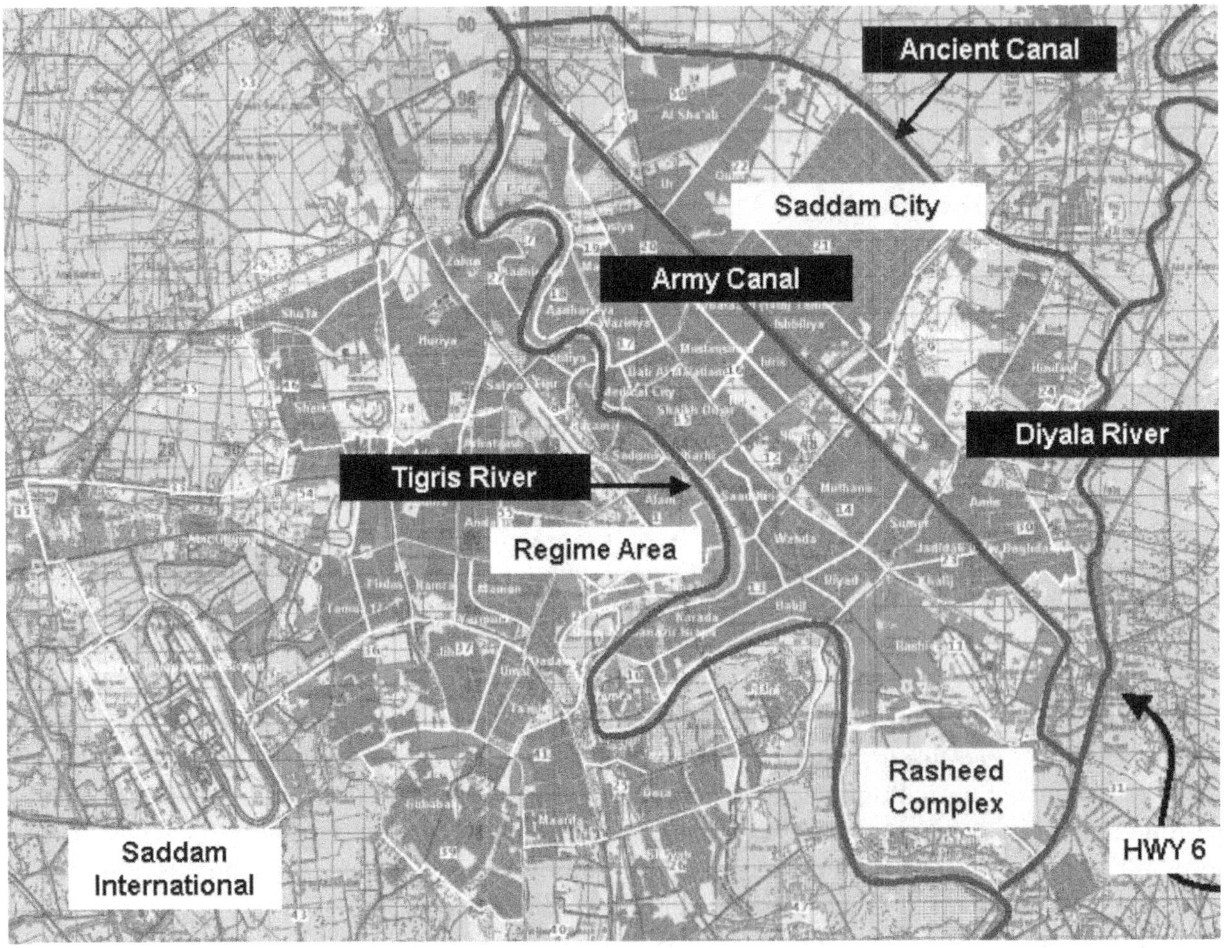

Eastern Baghdad's canals and waterways form a 'moat', presenting a series of obstacles for any attacking force.

CFLCC established a well-reasoned numeric zoning system for the Baghdad battlespace, primarily to control fires. The Division G-2 adopted this same convention to make preliminary recommendations for attack routes into Baghdad. The recommended routes allowed quick access to key terrain without fighting through urban neighborhoods. These attack routes would also naturally position forces at key intersections and neighborhoods that would allow control of the city in the stabilization phase. The Division's contingency Baghdad plan called for a fixing force to attack up the Highway 6 avenue of approach into the Rasheed area, while an enveloping force entered the city from the north and east. To the northeast, Saddam City was occupied by (presumably) sympathetic Shia Muslims, but its urban canyons would prove too dense for any attacking force. Instead, the Division chose two lesser-developed areas on either side of Saddam City that provided relatively speedy access to the Division's objectives downtown. Using speed and natural Marine aggressiveness, the Division would stab into the heart of Baghdad and disrupt the cohesion of the disparate defending elements

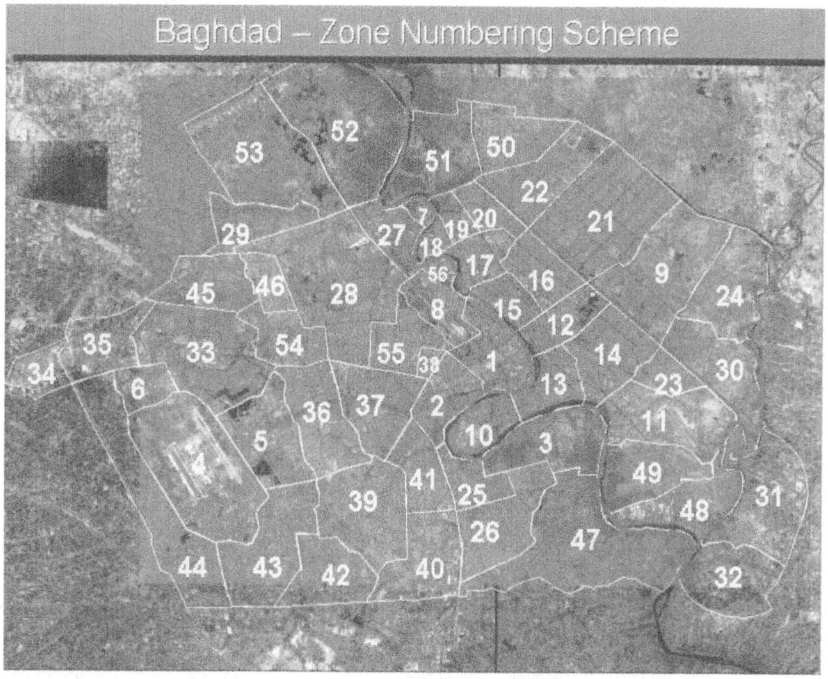

Baghdad is divided into numbered sectors. This numbering scheme was to facilitate battlespace coordination and shaping fires.

For Fires planning, the key Baghdad issue was the CFLCC commander's need to maintain unity of command in a shrinking battlespace. There were tremendous collateral damage, airspace control, and fratricide risks as disparate attacking units closed on Baghdad from multiple directions. How to provide unity of command for these units was a significant challenge. The MAGTF concept seemed a natural fit, and the Division and MEF argued for retention of the MEF's combat power as a cohesive unit (instead of parceling out Marine forces to Army command and control.) I MEF could work within the confines of CFLCC's fire control measures, with its forces intact, supporting V Corps which would retain unity of control of both air and ground fires within the Baghdad Restricted Operating Zone (ROZ). In the final plan, both I MEF and V Corps would have their own zones in Baghdad. Close liaison and permissive fire support control measures would facilitate the rapid engagement of enemy forces across the MEF-Corps boundary by both air and indirect fires. The DASC, however, would ultimately relinquish control of the airspace to the V Corps Air Support Operations Center (ASOC) as the constricted air space required a single coordinating headquarters.

The G-2 laid out the expected layered defense plan for the city, and identified the Al Nida Republican Guard Armor Division as the primary defender of the approaches to Baghdad from the East. The Al Nida Division was expected to defend the outer cordon of Baghdad, on the east side of the Diyala River. The planners crafted a script for the Al Nida's demise through shaping fires and a series of local envelopments. West of the Diyala River, the urban defense consisted of the most loyal of the Regime's forces, the Saddam Fedeyeen, internal security troops, and the Special Republican Guard (SRG). The planners began to study these forces in more detail. The G-2 pointed out the ill-suited nature of many of the forces that made up the defense, and

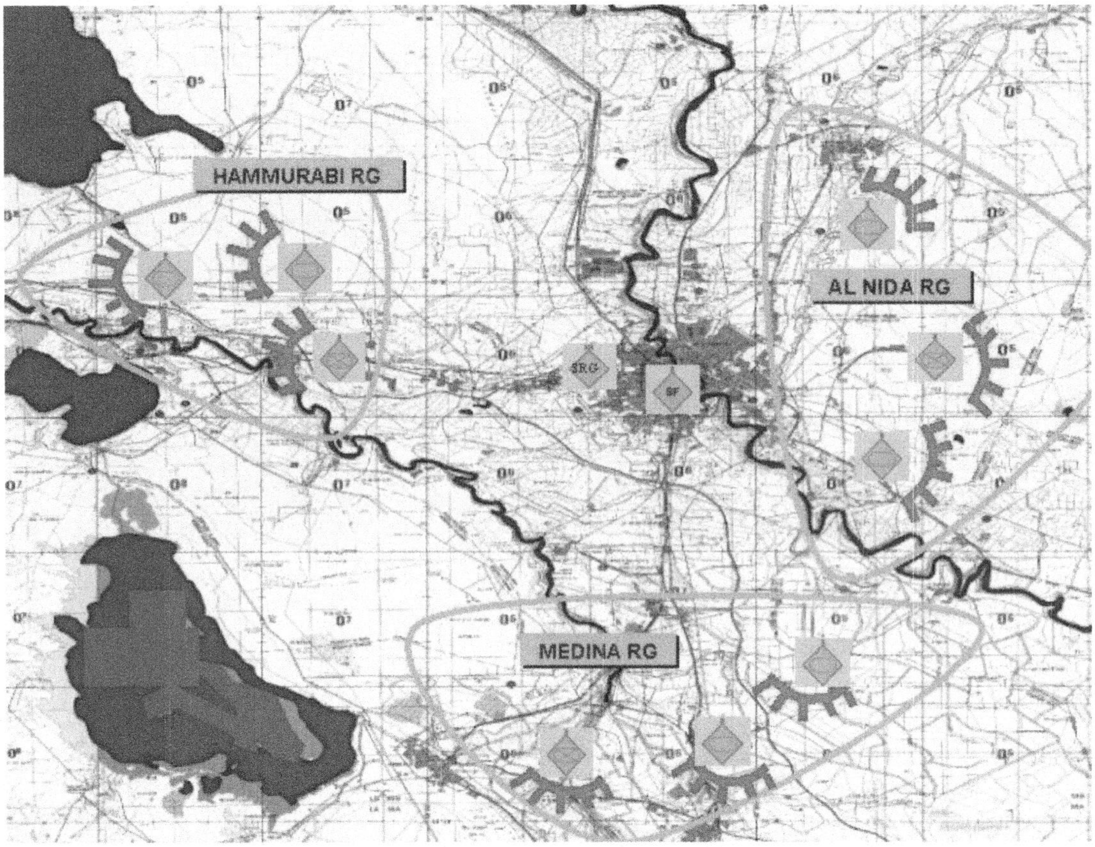

Iraqi Republican Guard divisions guard the primary avenues of approach leading into Baghdad. This planned outer ring of defense had integrated armor, infantry and artillery.

predicted there would be isolated strongpoints located throughout the city, with little coordinated action among them. If they chose to fight, these (at most) company-sized units would have to be isolated and reduced by the Division in a potentially difficult urban battle.

As quickly as the Division plan for Baghdad was completed, the battlespace shifted once again, leaving the Division without a specified mission in Baghdad. By the US Army's new plan, the Marines would play no part in the Baghdad fight. The planning efforts undertaken in these early days, however, proved invaluable when the Division later arrived at the gates of eastern Baghdad prepared to pursue and destroy a crumbling enemy defense. By then, the Division had far outrun supporting intelligence. The time the Division spent planning this operation in December, however, gave them the situational awareness required for the rapid continuance of the attack. This was a clear case where deliberation in planning resulted in speed on the battlefield, precisely as Lieutenant General Conway had anticipated.

Operational Planning – Refinement of the Base Plan

Seizure of Iraq's South Rumaylah oil infrastructure intact would have a tremendous impact on the ability of the Coalition to stabilize and rebuild Iraq after victory on the battlefield. To the Coalition forces, the seizure of this oil infrastructure rivaled the seizure of Baghdad in importance. The Iraqis recognized this fact, and had made public statements that they would take action to ensure the oil infrastructure did not fall into US hands. The Iraqis had created an ecological and financial disaster twelve years before by destroying the Kuwaiti oilfields, and they threatened to do the same to their own if they were to be lost to the Coalition anyway. The Division was faced with a dilemma. Clearly, the Iraqis on Safwan Hill and their agents in Kuwait would be aware of any operational move made by the Division. Because of the guaranteed 'operational' compromise, it would be necessary to generate 'tactical' surprise if there was to be any chance of securing the oil infrastructure intact. The MEF planners began to review the planned timing and sequencing for seizure of the Mina Al Bakr Offshore Terminal (MABOT) and manifold assemblies on the Al Faw peninsula, as well as the seizure of the GOSPs and Pump Stations in the South Rumaylah fields. Any attempt to seize one would certainly signal the Iraqis to begin the planned destruction of the other, thus simultaneous seizure of all the critical nodes was required to preserve the entire system.

The Division did not know what the Iraqi's trigger for destruction of the oilfield infrastructure would be. If the trigger was ground maneuver, a deception plan would have to be crafted. If the trigger was the beginning of a Coalition air campaign, there was a risk posed by any delay between the onset of air operations and the ground attack. In the earliest stages of operational planning, there had been a four day gap (imposed by force flow constraints) between when forces would be sufficiently available to conduct air operations and when limited ground operations could begin. As planning progressed, and the force flow constraints were overcome by events, the residual four-day planning gap was never closed. The commencement of air operations (A-day) and the planned ground attack (G-day) remained four days apart. If this timing was executed as planned, the Iraqis would have at least a four day warning that a Coalition attack had begun, easily giving them time to destroy the critical infrastructure. Division planners argued vigorously for a simultaneous air and ground attack. The objectives of the air campaign were strategic, and not mutually exclusive with simultaneous ground action in the South Rumaylah. Since shaping of military targets in the south was virtually nonexistent in the four day air campaign, there was no increased risk to ground forces conducting a simultaneous attack. The Division accepted the risk that a rapid seizure of the oilfields would mean less shaping fires on the enemy defenders, but balanced that with the risk to the strategic objective of gaining the oilfields intact. The arguments made by the Division and I MEF planners were initially unsuccessful, leading one planner to note, "We can't care about this infrastructure more than higher headquarters does; if they are aware of the risks, the destruction is a cost they must accept." Even after significant lobbying at the highest levels, the situation remained unresolved until the last moment, causing hasty planning sessions as the Division's forces were crossing the Line of Departure months later. The Division prepared to go into the attack as early as A-1 day, and as late as A+4 days. In consonance with I MEF, the Division did lobby for and receive permission to conduct pre G-day security zone operations, and maneuver operations in Kuwait that would condition the enemy to movement in the vicinity of the borders. These actions would enable the early insertion of reconnaissance assets or a Light Armored Reconnaissance screen in advance of the lead attack elements. The intent was to blind the enemy and provide some degree of tactical surprise regarding the Division's timing, main effort, and specific objectives.

'Capitulation' of enemy units was another topic of much discussion at this time. Higher headquarters had made a distinction between an organized unit action to cooperate with the Coalition and the conventional surrender of enemy individuals or defeated units. One of the central themes of the Information Operations (IO) campaign was to allow the Iraqi commanders the opportunity to signal their cooperation with Coalition forces (capitulate) in order to avoid their destruction. Although this concept gained some level of support at high levels, the distinction was somewhat lost on the Marines of the Division, who necessarily had to focus on the wholesale destruction of the enemy should he fight. The care taken by the IO planners in their attempts to ensure that enemy units were advised of the timing and some of the planning for their destruction also struck many in the Division as an unnecessary risk. There was even talk of offering the Iraqi 51st Mechanized Division in the Rumaylah oil fields a specific window of opportunity to capitulate, which would have given the Iraqis clear advanced notice to begin oil infrastructure destruction. Marveled one member of the Division, "We can't win this war by being nice. If they need killing we are going to kill them. If they don't, then they will be treated appropriately." During the actual attack, these enemy units that did not desire to fight simply melted away, and there was no capitulation of significance among forces in the Division zone.

FRONT

FRONT

BACK

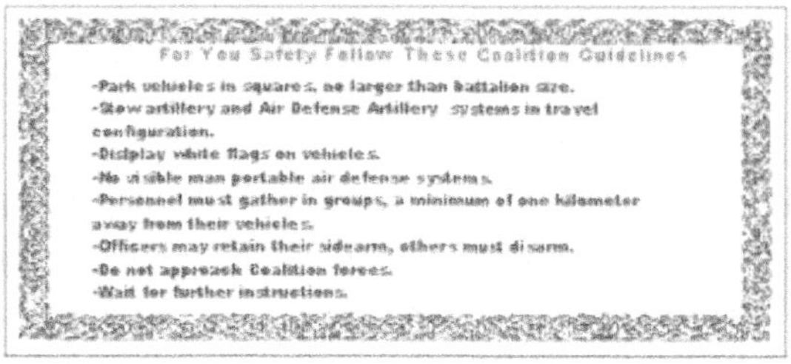

BACK

Detailed 'Capitulation' instructions were distributed to Iraqi units by leaflet. Instead of following this complex procedure, many Iraqi soldiers simply ditched their uniforms and equipment and just walked away.

Another refinement of the base plan was a re-tasking of the Combat Engineer Battalion (which would be made up of elements of both 1st and 2d CEBs.) There were multiple engineering tasks that would have to be accomplished in support of the opening attacks. Breaching the obstacles on the Kuwaiti side of the border was only the first. Obstacles on the far side of the border remained a question mark. Although the Iraqis had not been observed making overt defensive preparations near the border, the Division was reluctant to declare the border area free of mines or obstacles. In any case, the attack corridors the Division planned to use crossed the defensive areas of the Iraqi Republican Guard during the first Gulf War. In effect, the battlespace given to the Marines required maneuvering through a twelve year old impact area. Even with no recent mining of these areas, the potential for residual impacts from abandoned minefields and unexploded ordnance was high. Because of the uncertainty surrounding the disposition of minefields on the Iraqi side of the border, the CEB would not be involved in the initial breach beyond supervision of the preparation, position, and number of breach lanes for the Division. Instead, the Division would hold them ready for minefield and obstacle breaching north of the border, where speed, cohesion, and efficiency would be critical. The Division arranged for the Kuwaitis to breach their own obstacle belts, to be accomplished in the last days preceding H-hour. Additional support would be provided by US Navy Seabees from the MEF's Engineer Group (MEG), and USMC Engineer Support Battalions.

Planning for a heliborne assault capability also began early in the planning process. Recognizing the terrain impacts on ground maneuver, the Division expected that helicopter assaults of key bridges, interdiction of routes, and turning movements of enemy defensive positions could be necessary during the coming campaign. The Division established a close relationship with the planners of both MAG-39 and MAG-16. For example, planners from 1st Battalion, 4th Marines working with embarked staff from MAG-16 while enroute to Kuwait aboard the USS BOXER (LHD-6), conducted a series of planning and helicopter load exercises to determine the maximum number of assault Marines that could be loaded into a CH-53E helicopter. They also refined the procedures for conducting CASEVAC in a NBC environment. Collectively, the Division planners developed a battalion-sized lift package that could be applied nearly anywhere, provided the established criteria for this type of mission were met. Several potential objectives were considered for this mission package. The Qalat Sikar Airfield, potential crossing sites of the Tigris River, An Numaniyah Airfield, Al Qurnah, oil infrastructure nodes, even various airfields near Baghdad (in case of a rapid collapse scenario) were all considered as possible objectives for this method.

Another potential early objective for a heliborne assault was the Rumaylah oil field highway bridge over the Hammar marsh. The bridge had proven difficult to destroy during a recent MEFEX exercise, and had become known as the 'Alamo Bridge'. Outside of Al Basrah, this bridge was the only link between the Rumaylah oilfields in the south and the preponderance of combat power from the Iraqi III and IV Corps. Without it, the enemy could only attack via Al Basrah. If the enemy wished to reinforce the Rumaylah defenses, support would likely come from the 6th Armored Division over the Alamo bridge the northern Rumaylah oil fields. The need to block any reinforcing efforts of the 6th Armored Division might be accomplished by helicopter assault. The Division conducted an OPT to investigate this option, which resulted in a detailed list of decision points. The Division and 3d MAW collectively compiled a comprehensive list of go/no-go criteria for a helicopter insert, and applied the criteria to this situation. Among the criteria were the time required to conduct a force linkup, the presence or absence of Surface to Air Missiles (SAMs), Anti-Aircraft Artillery (AAA), and Man-Portable Air Defense (MANPAD) threats. The presence of armor or mechanized units in the objective area, along with the effects of Iraqi artillery upon the landing force were other significant conditions to consider. Although the risks involved in the Alamo Bridge operation might prove to be too high, the detailed planning for the helicopter assault and the teaming with 3d MAW planners was in place to speed execution on the battlefield should conditions permit.

Developments in Fire Support Planning

The planned concept of fires was straightforward. Fires would begin by shaping the enemy artillery before the RCTs crossed the international border. A battalion of BM-21s and a battalion of GHN-45s, both chemical-capable, topped the hit list drawn up by the Division Targeting Board. The 51st Division's own supporting artillery would be next in line for destruction, including two battalions of medium-range Type-59-1 and M-46 towed howitzers, and three battalions of D-30 or M114 towed howitzers garrisoned around Az Zubayr. As part of the shaping fires, a massive airstrike would eliminate the threat of enemy observation from *Jabal Sanam*. After successful completion of these early fire support tasks, the focus of shaping would shift to the artillery of the 6th Armored Division and additional III Corps artillery located north of Al Basrah, in Al Qurnah and Ad Dayr.

At H minus 1 hour, as the Engineers' Operational Combat Detachments (OCDs) put blade to the northern berm, 11th Marines planned to engage the surviving artillery in Az Zubayr with a massive 30-minute preparation. Near-simultaneously, Iraqi border guard posts within visual range of the breach sites would be attacked by fixed wing aircraft and Cobra gunships and destroyed. Finally, if the Iraqi 6th Armored Division showed indications of preparing to move south, the Alamo bridge would be cratered, mined, or destroyed, and the 6th Armored relentlessly interdicted.

Operational Planning - The Opening Gambit

With some predictability beginning to take shape in the force flow planning, and with the relative timing issues at least partially resolved, the Division began the final stages of operational planning refinement. By mid-December, the Division planning team had a well-developed scheme of maneuver and timing for the attack that was ready to be integrated with the other elements of the MAGTF. From a comment attributed to the 3d MAW Commander, the plan came to be called the 'Opening Gambit'. The Opening Gambit covered the first 96 hours of the fight, ending with the Division poised to continue the attack north of the Euphrates River. The concept of operations for the Opening Gambit began with the elimination of Iraqi reconnaissance assets on *Jabal Sanam* (Hill 451), and other selected border guard posts in a counter-reconnaissance fight designed to blind the enemy. This would begin with a series of air strikes designed to turn 'Hill 451' into 'Hill 441.' The strikes would be followed by the insertion of a Force Reconnaissance platoon to ensure no enemy observers remained alive in a position to report the Division's activity. Light Armored Reconnaissance (LAR) units would ensure the other Border Guard posts were neutralized. A series of preparatory fires on enemy artillery would then be conducted, with the intent of destroying all remaining indirect fire assets the enemy could bring to bear against the Division's maneuver elements. The Division predicted that a simultaneous A and G-day would eventually be authorized, and planned a thorough preparatory fire syllabus to compensate for an anticipated lack of air shaping fires.

Colonel Steve Hummer's RCT-7, as the main effort, planned to attack through the Safwan corridor with the objective of securing the Az Zubayr pump station complex (the 'Crown Jewel')

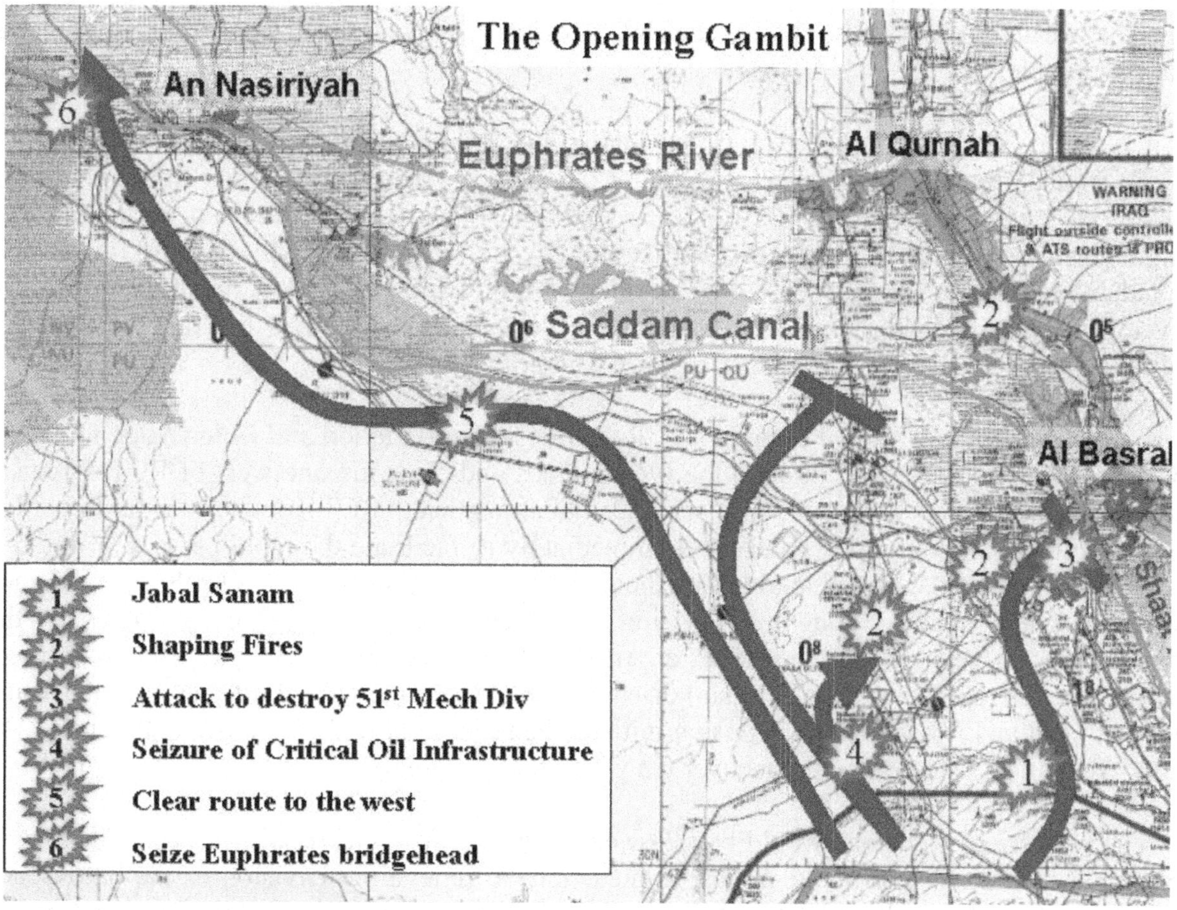

The 'Opening Gambit' operational plan was a team effort with MEF, 1st FSSG, 3d MAW, and Division planners all contributing to the effort.

intact. In order to control this key oil node, the RCT would attack to destroy elements of the 51st Mechanized Division that remained south and west of the Shaat al Basrah waterway. Simultaneous with the RCT-7 attack, Lieutenant Colonel Stacie Clardy's 3d LAR Battalion planned to maneuver north of Safwan hill to clear Position Areas for forward displacing artillery, then move east toward the Shaat al Basrah to pose a threat to the enemy's 32d Brigade from the southeast. If the 32d Brigade were to put up a strong defense, this supporting attack would clear approaches that would enable Cobra gunships to fly against the 51st Division positions at dawn without flying into the rising sun.

On the West of the Division's zone, Colonel Joe Dunford and his RCT-5 would launch Lieutenant Colonel Duffy White's 1st LAR Battalion ('Highlanders') as a screening force, then attack to seize the Southern Rumaylah oil infrastructure by moving along the western side of the GOSPs and peeling off company-sized forces to secure critical nodes. If the heliborne assault criteria previously developed were met, RCT-5 would also launch a battalion-sized heliborne assault to a blocking position on the south side of the Alamo Bridge (alternatively, this force would attack to their blocking position on the ground.) This force would establish a blocking position oriented north. The remainder of RCT-5 would quickly link up with the inserted force to effectively block any reinforcements to or escape of the 51st Mechanized Division in the Rumaylah area. If the blocking action were delayed, RCT-7 would need to refuse their northern flank before they turned east against the 41st Armored Brigade and the Az Zubayr Pumping Station Complex.

Failure to block the bridge would risk losing the advantage of momentum. The Division engineers and fire support planners prepared a contingency plan to crater the south side of the bridge, effectively stopping all traffic across it if the blocking force did not arrive in time. Although it was on the Restricted Target List, the Division appealed to MEF Force Fires to plan to take the bridge out if the situation developed unfavorably. 3d MAW targeting experts developed a plan to take it out. The MEF received CFLCC's approval to drop the bridge if required.

Lieutenant Colonel Steve Ferrando's 1st Reconnaissance Battalion and Colonel Joe Dowdy's RCT-1 would follow RCT-5 through the breach lanes and clear in zone west of RCT-5 toward the Jazair ammunition storage facility and the Jalibah airfield. RCT-1's attack would position them to quickly exploit success and move immediately to facilitate the rapid passage of the rest of the Division across the Euphrates in the vicinity of An Nasiriyah. The Reconnaissance Battalion would operate along the Division's northern flank, ensuring no surprises as the Division turned west toward the bridges at An Nasiriyah. Due to the assumed chemical threat and the need to rapidly (within 24 hours) disengage the 1st Marine Division from Az Zubayr and the South Rumaylah once 1st UK Division initiated its relief in place, it was essential that RCT-1 and 1st Reconnaissance Battalion uncover the western zone rapidly.

The rapidity of the attack would be followed by an equally swift relief in place. The 1st UK Armoured Division would move almost immediately to relieve the Division of the battlespace south and west of the Shaat al Basrah waterway, enabling the Division to quickly continue its attack toward the Euphrates. As RCT-5 and RCT-7 were relieved by the British 16th Air Assault and 7th Armoured Brigades, respectively, they would also clear in zone to the west, ensuring the battlespace was safe for the massive influx of logistics and support echelons that would soon follow. Harkening back to the days of Genghis Khan, the Division's measure of effectiveness for clearing this battlespace would be the ability of, "a naked virgin with 30 pieces of silver tied to her saddle to ride her horse down the street with not a hair on her head being molested."

The CG repeatedly reminded the Division planners that this was to be the most air-centric Division in the history of warfare. Consequently, the Division staff had developed a close relationship with the 3d MAW staff. These two MSCs of I MEF had worked extremely closely together through months of planning, and saw the fight from the same perspective. The close relationship between the Wing and Division had been evident from the first planning sessions in southern California, and the first socialization of the Opening Gambit plan was naturally with the Commander and Staff of the 3d MAW. On 28 December, the Wing and Division commanders, and their combined staffs, assembled at the 3d MAW headquarters at Al Jaber Air Base, Kuwait, and conducted a detailed review of the timing and objectives of the Opening Gambit. Members of the I MEF G-3 staff, including Colonel Larry Brown and Colonel Fred Milburn, attended to provide 'rudder guidance' from the MEF Commander's intent. This meeting proved to be a pivotal session in the transition of the Opening Gambit from the planning stage to execution. The respective staffs of the 3d MAW and the Division walked away from this session with a clear vision and execution guidance, able to immediately begin the detailed operational planning that made the Opening Gambit a success. This teamwork between Marines of each MSC began at the highest levels, and bred confidence and trust among the staffs. Eventually, the confidence shared between the Grunts and the Wingers was to prove a significant force multiplier in combat operations. Division planners were able to immediately begin the detailed operational planning which resulted in a 48-page execution matrix that spelled out in detail the movement of forces across the line of departure and the specific details on how the ground and air forces were to support operations that would make the Opening Gambit a success. The execution checklist described the sequence of events that had to take place, but made only indirect reference to event duration. Commanders would execute their tasks on the timelines allowed by the events on the ground.

The Division was invited to conduct a similar integrated planning session with the MEF Commander. The Division and Wing were confident of their ability to accomplish the MEF taskings. Now it was up to the MEF Commander to decide if the Opening Gambit met his guidance and intent. On 31 December 2002 the briefing was held at Camp Commando. With Colonel Toolan orchestrating the brief, Lieutenant Colonels Groen and Smythe presented the Opening Gambit in detail to the assembled Commanders of I MEF, 1st Marine Division, 3d MAW, 1st FSSG, MEG, and other elements of the MAGTF. Again, the plan received favorable reviews. With an approved attack plan in hand, the Division staff spent a quiet New Year's Eve in Camp Commando, preparing themselves for the transition from a paper plan to steel on target.

Getting the Force to the Fight

In early December, Division representatives attended the MARCENT War Council in Bahrain. During this session, the fluid nature of planning efforts was noted, as certain basic decisions had not yet been made at higher levels. The Operation did not, as yet, have defined tactical objectives, force flow approval, authorization for reserve activation, or other supporting documentation, making the march to a detailed plan more dependent on assumptions than desired. It was clear to the Division that even if the detailed plans had not been approved at the highest levels, combat operations were going to take place in Iraq at some point in the near future. The Division lobbied for more aggressive force posturing in theater, and the beginning of the flow of Marine forces to join the significant Army forces that had already been deployed to Kuwait. This proved to be challenging for many, as some outside the operational chain were not universally convinced that combat operations would take place. These holdouts were hedging their support of the Division's preparation efforts with missions elsewhere in the world and in rhythm with peacetime measures of assignment and procurement. To the Division's good fortune, MARCORSYSCOM, under Brigadier General Bill Catto, did not fall into this category. They continued to be the Division's strongest advocate for bringing proven capabilities on-line rapidly, and continued to move aggressively to equip the Division with the latest equipment that could be brought to theater and integrated prior to crossing the line of departure.

Previously, force flow planning had been based on a specific Operational Plan's (OPLAN) Time Phased Force Deployment Data (TPFDD). The careful force flow sequencing for this OPLAN had been updated over the course of the summer of 2002, providing a master script, written for a planned force flow sequence under a specific deployment scenario that included a defense of Kuwait. The Division deployment planners, under Lieutenant Colonel Clarke Lethin, had built the flexibility into this TPFDD that would allow for its adaptation to a new force flow sequence. By packaging the TPFDD by force packages consisting of RCTs, supporting arms, supporting logistics, and a slice of Division C2, the data was made flexible enough to be used in a variety of force flow scenarios. This foresight now proved invaluable as force flow planning took a new turn in December. Instead of deploying the planned TPFDD into theater, forces began to be approved incrementally under the Theater Combatant Commander's RFFs. The RFFs were centered on a desire for specific combat units, and often did not include the required supporting packages from the Division, Wing, and FSSG. Also, an ad hoc combination of modes of transportation would be required. The decision was made to get the MPSRONs in early, fly in the Marines, and use Fast Sealift Ships (FSS) to carry what would have been the Fly In Echelon equipment of the RCTs. What looked straightforward on paper would leave a trail of sleepless nights for logisticians and combat commanders from California to North Carolina to Kuwait City. The unfortunate consequence of a changed deployment strategy was the partial disruption of the carefully linked logistics, communications, fires, and maneuver packages that existed in the OPLAN TPFDD force flow. Under the RFF process, force flow was driven more by available lift, and much of the careful precision and integration that went into planning the TPFDD was lost. Marine embarkation planners now talked in terms of the 'Mother of all Deployment Orders' (MODEP), and the 'Son of MODEP' (SOMODEP), and scrambled to re-document their units force flow requirements for the new modes of transport. The impact of missing key enablers lost in the RFF process would be felt right through the combat phase, with drivers arriving after their vehicles were offloaded, and some CSS units arriving much later than required. Additionally, commanders and staffs focused inward on detailed embark plans at a time when the CG wanted them focused on the enemy and their tactical missions.

On 11 January, the President released the long anticipated Deployment Order, RFF-177. The Division's main body began deployment to Kuwait. With the flow of Marine forces approved, attention was turned to the task of moving more than 20,000 Division Marines and their equipment to theater. Preparing for their arrival was another significant challenge and the Division Forward CP immediately turned to the task.

Continuing Mission – Security in the Pacific

Three of the Division's battalions were not included in the deployment roster to Kuwait, as they were required to fill the Division's responsibilities under the Marine Expeditionary Unit (MEU) and Unit Deployment Programs (UDP). A fourth battalion from the Division was deployed to Kuwait as part of the 15th MEU. Supporting artillery, Light Armored

Reconnaissance (LAR), and Assault Amphibian support accompanied these battalions. Under the UDP, 2d Battalion, 4th Marines (2/4), Commanded by Lieutenant Colonel Andrew Schlaepfer; and 2d Battalion, 7th Marines (2/7), commanded by Lieutenant Colonel James Cameron, were deployed to Okinawa. As the planning continued from August 2002 through the end of the year, the status of these two battalions and their rotation was the subject of keen interest at MEF and Division levels. As with any potential major deployment, the disruption to unit rotation cycles under the UDP is significant. In this case, as the months rolled by without a deployment order to Kuwait, serious consideration had to be given to either leaving these units in place or replacing them with units from CONUS and sending them straight into combat. By Christmas Eve, the requirement to rapidly deploy the rest of the Division to Kuwait drove the decision to freeze UDP units in place. By this time, the advance parties from 2/4 and 2/7 had already returned to CONUS to coordinate the return of these units home. With the decision to leave these units in place in Okinawa, the advance parties returned to the Western Pacific, rejoining their comrades who braced for an extended rotation on the island, fulfilling the Division's responsibilities in the Pacific theater for the duration of Operation Iraqi Freedom. For these Marines, a six-month rotation to Okinawa stretched to a deployment of almost a year. Although disappointed at missing the combat action, these Marines recognized the role they played in stabilizing another important portion of the globe. With the war on terrorism in full swing, the efforts of these Marines were vital to the National effort and were never forgotten by their comrades in the Division.

In addition to these two battalions, the Division would also maintain its support to the Marine Expeditionary Unit (MEU) program. The 2d Battalion, 1st Marines (2/1) would join the fight in Iraq, but as a Battalion Landing Team (BLT) under the command and control of the 15th MEU. The 1st Battalion, 1st Marines (1/1) would remain in Camp Pendleton, completing the workup cycle for the 13th MEU rotation. The 3d Battalion, 1st Marines would return from its own 11th MEU assignment, and be turned around in 30 days to redeploy with RCT-1. The Division's responsibilities across large segments of the globe precluded all elements of the Division from participation in ground combat in Iraq. As part of the MEU and UDP programs, each of these battalions would remain a vital part of the Nation's war on terrorism, and accomplished their assigned missions with pride and professionalism. Between the UDP and the MEU programs, a full four infantry battalions were unavailable for combat when the Division was called.

In order to provide sufficient infantry for the Division's impending mission in Iraq, 2d Battalion, 23d Marines was reactivated. This battalion had already served on active duty for nine months as a post 9-11 domestic US Quick Reaction Force (QRF). Added to RCT-1's troop list, it brought that regiment to full strength. Motivated and well-trained, the 2/23 Marines and Sailors reported back to active duty only weeks after returning to civilian life without a murmur of discontent. These Marines truly lived up to the highest standards, and their later combat contributions would be remarkable.

Creation of the Division Administrative Center

In early January 2003, Gunnery Sergeant Marco Rico of the Division G-1 did an administrative assessment of the personnel receiving process in Kuwait. He realized that Marines in theater would need more than just accountability, they would need Reporting Unit

Code (RUC)-level administrative assistance for pay, promotions, unit diary assistance, legal assistance, emergency leave, and other administrative support far beyond the capabilities of the small staff currently located forward. Before the rest of the Division arrived in Kuwait, an administrative structure needed to be in place. On 30 January, Lieutenant Colonel Powalski and a small administrative team departed Camp Pendleton to set up a Division Administration Center (DAC) at Commando Camp in Kuwait. The DAC, comprised of administrators from each RCT, would handle the bulk of administrative matters in concert with 'reachback' administrative support to assist every Marine within the 1st Marine Division. Unit administrators from the RCTs would be co-located, but not consolidated. Using this unique arrangement, unit integrity and administrative processes consistent with the guidance of their immediate Commanding Officers would stay intact, but all of the administration Marines would benefit from their mutual experience and shared workspace in case of a requirement to surge in support of a crisis situation. The various unit representatives would work together on several issues to meet a common goal, 100% personnel accountability and administrative support for all Marines attached to 1st Marine Division, to include attachments from the Army's 1st Field Artillery, 2d Marine Division units, 4th Marine Division units, and individual augments. Equipment shortages, especially for computer and communications assets, proved critical shortfalls during the first few weeks of the DAC's operation, as this equipment arrived later in the force flow. The DAC found a stable work area using space gladly donated by the MEF G-2. With the assistance of their fellow Marines, and by the exceptional efforts of the Division G-6, the G-1 ensured that the DAC had communication capabilities consistent with their needs. 'Reach Back' and 'Push Forward' administration became a concerted effort daily between the G-1 Marines in Iraq, Kuwait and California. The CG's vision of warfighting units having only a small administrative staff as they moved forward (with the majority of administrative work completed in the Division Support Area) was achieved.

"Sappers Up" – Combat Engineers Organize For Combat

Operation Iraqi Freedom would see something different when it came to Combat Engineers. Forward thinking engineers started with a fresh sheet of paper as they figured out the best way to organize engineer support for the Division. The Marine Corps would combine both 1st and 2d Combat Engineer Battalions (CEBs) into a single 'Engineer Battalion'. Contingency planning between 1st and 2d Combat Engineer Battalion (the two active duty battalions) and 4th Combat Engineer Battalion (the reserve combat engineer battalion) had begun to take shape in October 2002. The Engineer battalion commanders jointly recommended merging battalion staffs, and providing reinforced line companies directly to the RCTs to give the fighting units an immediate and robust combat engineer capability. After presenting this course of action to the Commanding General of 1st Marine Division, Lieutenant Colonel Paul Cyr, the commanding officer of 1st CEB, was selected to be the CG's principal engineering advisor as the Division Engineer, while Lieutenant Colonel Mike Micucci, the commanding officer of 2d CEB, would actually command the combined combat engineer battalion.

All agreed that this organization would provide unity of command, best support the RCTs, place the majority of the Division's engineer capability forward, and at the same time retain a sizeable engineer capability still in general support of the division. The engineer support companies held in general support of the division could be used to reinforce one or more RCTs with additional breaching capabilities, be used in Military Operations in Urban Terrain (MOUT), assist in cutting expedient combat roads and trails, or provide survivability or other combat support. In addition, engineer sections would be provided to 1st and 3d LAR Battalions and 1st Reconnaissance Battalion to assist in bridge classification, performing obstacle and minefield reconnaissance, and other combat engineer support. Engineer support company detachments reinforcing the RCTs would include at least six M9 Armored Combat Earthmovers (ACE), two or more bulldozers, three dump trucks, armored HMMWVs, a maintenance contact vehicle, and the operators and mechanics required to provide ample support.

By mid-February 2002, all the engineers were deployed to Kuwait, and their equipment soon followed. 1st and 2d CEB completed their merger, and a new personnel structure emerged that would run the combined command. The combat engineer letter companies from both CEBs were attached to the RCTs. C Company, 2d CEB went to RCT-1; B Company, 1st CEB went to RCT-5; and both C and D Company, 1st CEB went to RCT-7. The Engineer Battalion would retain one combined Headquarters and Service Company and two separate engineer support companies, one from each of 1st and 2d CEB. Throughout February, CEB created a staff that would be divided between CEB forward command post, main command post, and combat train. Having a forward and main command post would allow the battalion to echelon forward and support varied and potentially independent missions. By 10 March, all sealift ships supporting the engineers had arrived, and the combined battalion had sufficient equipment to be combat-ready.

Building a Home for the Division in Kuwait – Camp Matilda

When the Forward CP staff arrived in Kuwait, the expanse of desert that would eventually become home to the Division's more than 20,000 Marines and Sailors (as well as the growing number of Soldiers that had joined the Division's ranks in a variety of combat specialties) was bare ground and blowing sands. The forward Division Engineer, Major Dan Longwell, had begun the terrain management planning in late summer of 2002, but it had not progressed past circles drawn on a map by the time the Division Forward CP arrived in country. Raising the camps would require a significant outlay of funds, and making this commitment was a large step with strategic overtones. Over the course of three months, the plan for Tactical Assembly Area (TAA) Coyote would mature into four Life Support Areas (LSAs) that provided food, shelter, shower, and sanitation facilities to units as they arrived in country. The Division's LSAs were collectively referred to as 'Camp Matilda'.

The transition was not without challenge. Competing demands, conflicting priorities, terrain management issues, and bad weather all conspired to slow LSA (Logistical Support Area) construction. The demands of building camps of sufficient size, and the lack of assets readily available in Kuwait to accomplish the construction made this a challenge. In the end, a handful of Marines were successful in applying the social energy and sweat equity needed to give the Division a home. In days, reminiscent of the 1st Marine Division's training in Australia during World War II, the Division began to establish itself at Camp ('Waltzing') Matilda in Kuwait for final combat staging and rehearsal.

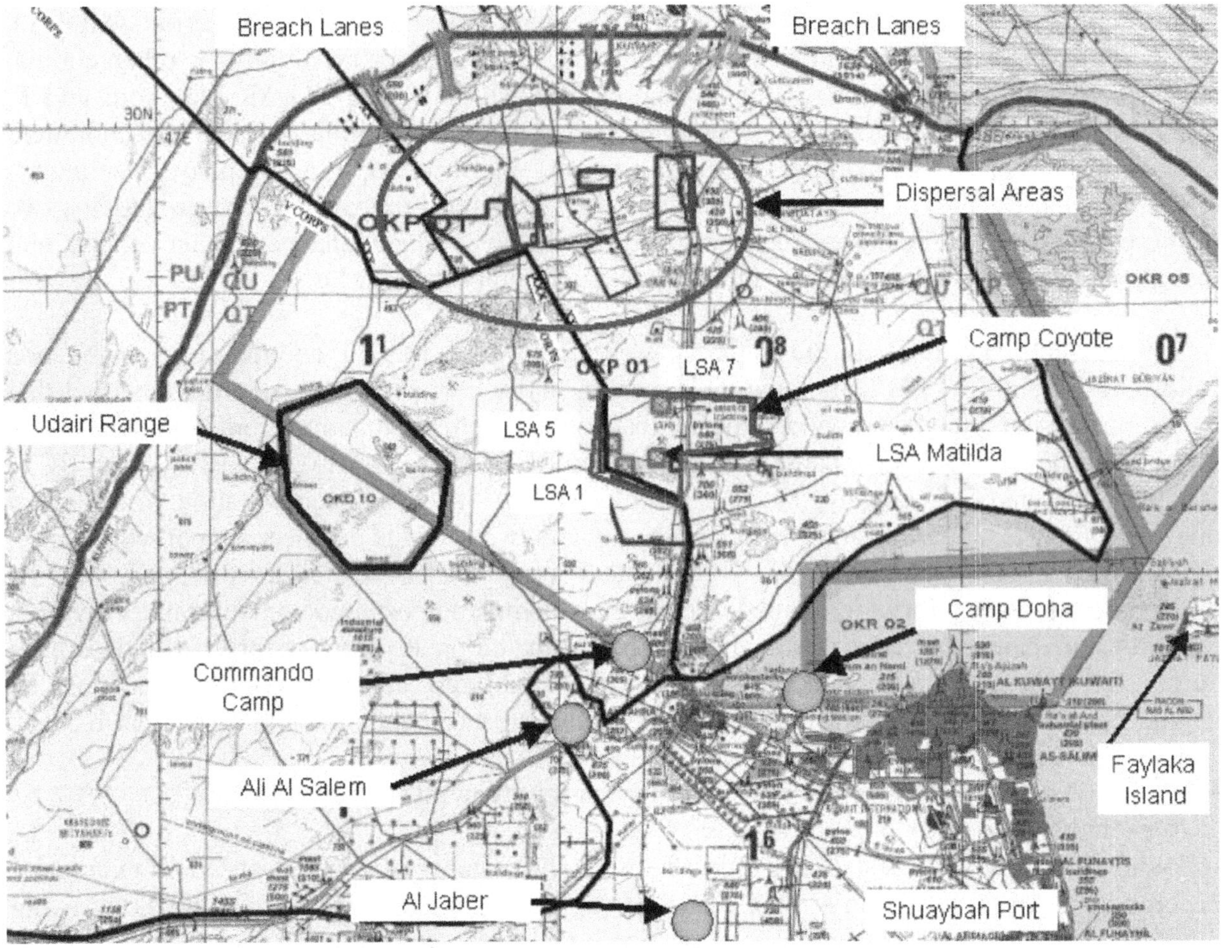

Camp Coyote in central Kuwait was the home of the Division prior to crossing the Line of Departure. It was near the Udairi training range complex, and the Division's planned Dispersal Areas

The area easily met the Division's needs for training ranges, living areas, and storage space. Driven by the likelihood that units would arrive in Kuwait and could immediately step off into the attack, the Commanding General's intent was to live as austerely as possible during the short time period that units would be there. In November, the plan consisted of circles drawn on a map within TAA-Coyote and unit names assigned to the circles. A reconnaissance on the ground of the circles drawn on the map resulted in a modified plan. Coyote was not a vacant stretch of land as suggested by CFLCC, but contained garrison and training areas of the Kuwaiti Land Forces (KLF), as well as a large number of Bedouins.

In December, CFLCC released a directive establishing a common living standard for all forces in Kuwait. The impact to Army forces was minimal, since they billeted in established camps that easily met the new quality of life standard. For the Division, however, this was a fundamental shift. The austere TAA that would support units for a short while before crossing the line of departure were upgraded into a more permanent LSA requiring significantly greater engineering efforts and financial expenditure than previously planned. The staff commenced planning to respond to the change. Colonel Toolan and Lieutenant Colonel Broadmeadow joined Major Longwell in assessing and prioritizing the Division's requirements.

Four carefully considered large cantonment areas replaced the circles hastily drawn in the initial plan. The Commanding General's intent remained tactically focused, and no LSA would have the elaborate berming or wire obstacles associated with a defensive mindset. LSA development anticipated the flow of Division forces into theater, so LSA-Ripper, supporting RCT-7, would be the first LSA to be completed. LSA-Matilda would be built next for the Division Headquarters Battalion, 11th Marines, and separate battalions. This was followed by the development of LSA Grizzly for RCT-5, and (eventually) LSA Inchon for RCT-1. In keeping with the CFLCC's intent, the Division planned for tents with decking for billeting; electricity for work and living spaces; showers, portable chemical toilets and dumpsters for hygiene maintenance; kitchen areas for preparation of hot meals; and gravel laid on main routes throughout the camps to keep down the dust. Similar initiatives began within the MEF's other MSCs, including the recently joined 1st UK Division and the 15th MEU (SOC). The re-allotment of real estate within TAA-Coyote was conducted at strongly argued meetings. In the end, the 1st UK Division and 15th MEU (SOC) set up near the camp of the 6th Kuwaiti Brigade, east of Highway 80. The FSSG set up south of the Kuwaiti brigade, east of Highway 80, while the Division maintained the areas west of Highway 80.

The Kuwaiti Ministry of Defense (MOD), and Ministry of the Interior (MOI) were fully supportive of the Division and endeavored to help wherever possible. But as it is in most large bureaucracies, decisions at the top are not rapidly or completely disseminated down to the execution level. Cutting the berm that runs from the border with Iraq and down Highway 80 to make entrances into the LSAs was such an instance. On two occasions, contractors cut the berm into the LSAs only to be stopped by local MOD officials who not only halted their work, but also made them restore the berm to its original condition. Lacking a translator, Major Longwell could only rely on a document written by CFLCC in English on one side and Arabic on the reverse explaining what he was trying to do. Similar situations arose when MOD officials asked for and then confiscated third country nationals' passports (most of the contract workers were not Kuwaitis.) Major Longwell spent several nights at MOD Headquarters at Ali As Salem retrieving workers' documents so they could return to work. Once the information and language obstacles had been breached, MOD and MOI personnel were obliging and enabled work to continue

Life Support Area Matilda was one of four Life Support Areas the Division occupied while in central Kuwait.

From the signing of the first contract to the first Marine arriving at LSA-7, contractors had only two weeks to show results. There was plenty of competition for scant contractor resources from among the FSSG, TF Tarawa, and other camps. While one of the sub-contractors began the task of road improvement, numerous others began construction of decking, erection of tents, installation of electrical generators and showers, and limited berming around the LSAs. In an assembly line fashion, as one project was completed at one LSA, that subcontractor moved to the next LSA in priority and began the process again. The net effect was simultaneous development of the first three LSAs by multiple sub-contractors under the direction of the Division Engineer. The MPF offload added further pressure to the situation since the Arrival and Assembly Operations Echelon (AAOE) area had to be set up in the desert near the LSAs in order to receive the units' equipment and prepare it for combat. Available funding was severely strained, and work-arounds were required. LSA-1 was developed to house RCT-1 but relied on habitability suites for amenities and LSA-5's showers and chow hall. Soon, MSE representatives were flown in to supervise the completion of their own LSAs. 4-5 Marine teams began arriving around 1 February and provided the link between the contractors and the using unit. For their part, contractors completed portions of LSAs just as Marines scheduled to occupy them arrived. Final improvements to the route leading to the AAOE area wrapped up as the first truck from the offload rolled up to the gate.

In each LSA, commanders had 60 large tents for billeting Marines, 22 medium tents for billeting SNCOs and Officers, and 22 smaller tents for other billeting. 15 shower units, 200 portable chemical toilets, 5 generators, 50 dumpsters, and 1 food preparation area completed each facility. Marines would have to transit to the Udairi Range Complex to fire large caliber weapons but found room within the TAA to fire small caliber weapons.

Embarkation, Deployment, and Off-load Operations

The ability to get to the fight is a critical element of combat power, and the ability to get a Marine Division across the globe and into a fight quickly is one of the Nation's most remarkable capabilities. Critical to all was the adaptation and flexibility displayed by the MEF and its components in the reception, staging, onward movement and integration of all RFF forces. Without the benefit of a mature theater logistical command and without sufficient transportation assets, the components (including the Division with its Forward CP) refocused efforts to bring these forces in. The Forward CP established convoy security teams, airfield reception teams, surge drivers for all MPS assets and much more in order to receive and stage personnel and equipment. The Division staff focused on sequencing the right forces with the right gear into theater, while the units of the Division focused on packaging, weighing, and marking their

MPF stern offload of an AAV at ports in Kuwait. The preponderance of Division equipment was transported by ship.

Air Force C-5 loading at March Air Reserve Base, California. Many Marines and some equipment were transported overseas to Kuwait by strategic air.

equipment. Conducting strategic mobility exercises in the fall of 2002 was a means to bring the two efforts together to assess how ready units were to respond to a deployment order. The Assistant Division Commander (ADC), Colonel John Kelly, and the Deputy G-3, Lieutenant Colonel Clarke Lethin, became intimately involved in the embarkation preparation of the Division, lending focused urgency to the process. Through unit briefs they injected discipline into the embarkation process and ensured unit plans reflected the nature of the fight the Division was about to undertake – light, lean, lethal – down to the pallet-load. Units laid equipment out on parade decks in stick order, weighed loads, and measured vehicles. Because of the flexible nature of RFF deployment, the units had to do double duty to ensure their equipment would fit on either ships or aircraft (as the means of deployment was not certain.) When DEPORD 177A was released in January, deployment planning and preparation was validated as the Division departed home stations for Kuwait. Time spent planning and inspecting Unit Line Number (ULN) serials during the previous months paid off in speed and flexibility of deployment when the Division deployed by RFF instead of by the timeline of the TPFDD.

Deploying the force by a combination of military and civilian modes of transport proved challenging. In January, ships from all three MPSRONs sortied to the CENTCOM AOR to provide the first suites of Marine equipment. U.S. Navy Amphibians, Fast Sealift Ships (FSS), and contracted civilian carriers followed to provide additional capability. The majority of the Division's Marines and sailors flew on commercial aircraft of the Civil Reserve Air Fleet (CRAF) while high-value organic equipment flew on military transports. RCT-1 steamed out of San Diego harbor aboard amphibious ships as part of Amphibious Task Force - West (ATF-West).

In early January, groups of Marines headed to the Indian Ocean, the Mediterranean Sea and Kuwait to start the MPSRON offload process. Offload Preparation Party (OPP) Marines linked up with MPSRON ships in Diego Garcia and Sigonella, Italy to check equipment. They ensured gear was ready to drive down the ramp when the ships docked in Kuwait. Survey, Liaison, and Reconnaissance Party (SLRP) Marines flew into Kuwait to ensure Ash Shuaybah Port and the Kuwait Naval Base (the two ports of debarkation) were ready to receive ships, and Kuwait City International Airport was ready to receive military aircraft. On 17 January, the advance party of CSSG-11 arrived in Kuwait to support the large MPF offload. In total, eleven ships with over 5,000 pieces of rolling stock and 7,500 of containers were offloaded. Augmented by convoys from CSSB-10, CSSC-115, and CSSC-117, the eleven-ship offload was completed in 40 days, two weeks ahead of schedule.

Division logisticians worked hard to expedite the deployment. The Deputy G-4, Lieutenant Colonel Jeff Horigan, designed and stood up the Division Operations Center (DOC) at Camp Pendleton. The DOC became the clearinghouse for force flow information and the link to ongoing operations in theater. At the unit level, Major Carl Small and the Division Embarkation team did yeoman's work assisting deploying units in reordering and reconfiguring their information. From data entry to assistance at the Sea and Aerial Ports of Embarkation (S/APOE), Major Small's team maximized available lift to allow the Division to close into theater as quickly as possible. At Kuwait City International Airport, Gunnery Sergeant Manuel Johnson became the single point of contact to receive Division units and marry them up with transportation for movement to their LSAs. Gunnery Sergeant Johnson was one of a number of unsung heroes who scrambled to move mountains of equipment and thousands of vehicles off the ships and into the hands of the warriors who needed them.

The 7th Marine Regiment was the first Division unit to deploy. As the primary Maritime Prepositioned Force (MPF) GCE, they planned to link-up with the gear being offloaded from the MPSRONs. 7th Marines had conducted numerous exercises and inspections to be able to quickly leave Twenty-nine Palms, fly to an operating area, and marry up with assets from MPF shipping. In the weeks following 9-11, 7th Marines had conducted a two-ship MPF offload during exercise Bright Star in Egypt. During the second week of January, the advanced party of 7th Marines arrived and prepared to offload equipment from the ships of MPSRONs 2 and 3, which had closed on Kuwait. The first two ships – Motor Vessels (*MV*) *Hauge* and *Lopez* – arrived on 15 January before the Survey, Liaison, and Reconnaissance Party (SLRP) had arrived. With typical can-do spirit, Colonel Toolan assigned Major Andrew Petrucci (the 7th Marines planning liaison officer to the Division Forward in Kuwait) to the task. With a pickup team of only 15 Marines, Major Petrucci organized the offload and staging of RCT-7's equipment with a makeshift Arrival and Assembly Operations Element. There was nowhere near the full complement of drivers needed to offload the vehicles, either from the ships or from the flatbed trailers on which they arrived at their desert staging areas. Marines from the Division Forward CP responded to the call for drivers and were quickly bussed down to the port to assist. 'Cooks, bakers, and candlestick makers' were all drafted into service to drive Amtracks, MTVRs, and HMMWVs all over Kuwait. On 20 January, the draftee drivers were returned to Camp Commando when the remainder of the shore party arrived. By 24 January, 7th Marines added equipment from the *MV Anderson* and US Naval Ship (*USNS*) *Stockham*. The first main body battalions arrived by air on 24 January. The Marines of 3d Battalion, 11th Marines, 1st Tank Battalion, 3d LAR Battalion, 1st Combat Engineer Battalion, and 3d AA Battalion that combined with 7th Marines to form RCT-7 soon followed. By the time the *MV Button* and *MV Bonnyman* completed offload on 27 January, the bulk of the combat power of RCT-7 was in Theater. On 1 February, Colonel Hummer told the ADC that RCT-7 had closed the force. It took only two weeks to go from garrison locations to a combat ready RCT in LSA-7.

The deployment that started as a single RCT in Kuwait, quickly grew to add a second RCT, and eventually included the entire Division. Fifth Marine Regiment was the next to go. Without an MPF-focus in daily operations, 5th Marines relied on intensive educational seminars and strong leadership to successfully get up to speed on MPF operations. The Marines of Grizzly had taken a crash-course in MPF operations from their comrades in 7th Marines. This training included a visit to the Blount Island Command, where first hand information on the contents and offload procedures for specific ships was gained. Chief Warrant Officer Brian Spooner led the 75 Marines from the regimental OPP to Kuwait to accomplish the download of MPSRON-1. Their first elements departed Camp San Mateo in late January and joined the Division in Kuwait. *MV Pless* and *MV Kocak* brought in the first loads of what would be over 17,000 items of equipment offloaded for RCT-5's use in combat. Working closely with the shore party, the OPP downloaded the *MV Obregon* and *USNS Martin* ahead of schedule, finishing up with the *MV Bobo*. They received other units from Camp Pendleton, Twenty-nine Palms, and Camp Lejeune to eventually form RCT-5. By 15 February, RCT-5 was staged with elements of 2d Tank Battalion, 1st LAR Battalion, 1st CEB, and 3d AA Battalion. The RCT had gone from garrison to assembly areas in Kuwait in weeks, and, with their brethren in RCT-1 and RCT-7, would go on to write Marine Corps history in Baghdad a few weeks later.

Marines from 1/4 conduct live fire training aboard the USS Boxer. The long transit gave RCT-1 time to build habitual relationships between its organic units and supporting attachments.

RCT-1 prepared to load seven amphibious ships that would take them half way around the world as RCT-5 and RCT-7 prepared to fly in and link up with MPF. With the release of a naval message on 6 January, Amphibious Task Force (ATF) West was created. In addition to the equipment of RCT-1, 1st LAR Battalion loaded all but one company of its LAVs and equipment on the amphibious ships (1st LAR received one company's worth of LAVs from MPSRON shipping, but was required to ship the rest of the battalion from Camp Pendleton.) Marines staged equipment at Naval Station San Diego and the National City commercial port. Onload of the *USS Bonhomme Richard, USS Boxer, USS Cleveland,* and *USS Comstock* began on 10 January. With the assistance of the ATF West Embark Officer, Chief Warrant Officer Mark Ivy, the *Cleveland* and *Comstock* were loaded in two days. Division Marines, ship's crew, and Navy staff began loading the next ship before the last was finished. The *Pearl Harbor*, *Anchorage*, and *Dubuque* loaded simultaneously, all finishing at the same time as the *Boxer* and *Bonhomme Richard.* ATF West pulled away from the pier on 17 January as families waved and news crews filmed. During transit, the RCT-1 Marines continued their combat conditioning by conducting innovative training that optimized the available space onboard ship. The RCT conducted daily communications exercises among the units onboard the various ships of the ATF to improve the regiment's radio and data connectivity reliability. Several commanders' conferences were conducted to refine ashore training schedules, combat procedures, and image the leadership through the initial actions of the Opening Gambit. Reversing the training cycle, 1st Battalion, 4th Marines (1/4) conducted the majority of their live fire, weapons handling, NBC, and physical conditioning training at night. The Marines and Sailors from Weapons Company, 1/4 instructed the Marines from CSSC-111 on weapons handling, individual protective measures, and other essential combat skills. Their transit would take over a month and include major mechanical problems with the *Bonhomme Richard* that delayed its arrival. Commodore Rome and his Sailors lived up to the great reputation they had with their Marine brothers. These hard-working, mission-oriented Sailors got their ships to Kuwait despite many challenges. On 25 February, the first ships pulled pierside at the Kuwait Naval Base and began to offload. In four days the offload was complete, and by 1 March RCT-1 stood combat ready at LSA-1.

Eleventh Marine Regiment and Headquarters Battalion faced unique challenges throughout the deployment. As the unit responsible for providing artillery support to the maneuver elements, 11th Marines had elements traveling by both air and amphibious shipping, and would marry up with MPF equipment once there. Deployed by both air and sea, the regiment was equipped with a combination of MPF and organic equipment. In Kuwait, the regiment gained (from 10th Marines) a Counter Battery Radar (CBR) Detachment (two Q46A Firefinder radars, a Target Processing Center (TPC), and 23 Marines) and two cannon batteries (India 3/10 and Romeo 5/10). Additionally, it attached the 1st Field Artillery Detachment (1st FAD) from the U.S. Army XVIII Airborne Corps in Fort Bragg, North Carolina (two Q-37 radars, a TPC, and 24 soldiers.) By 13 March, Colonel Marletto informed the CG that 11th Marines had their full complement of howitzers and counter-battery radars ready for the fight.

For Headquarters Battalion, the challenge lay in maintaining sufficient staffs in CONUS and Kuwait to command and control units in both locations. As February drew to a close, Headquarters Battalion re-organized to provide truck detachments to each of the RCTs. Following this redistribution, they were fully staffed for combat operations. The *Soderman* completed download on 13 March and provided the remaining assets the Division needed in the fight to Baghdad. It was the last of four FSS and seven contracted commercial ships that enabled the force to close.

In Kuwait, the RSO&I tracking responsibilities fell to Corporal Justin Duplain, who kept the Forward staff informed about the progress of the TPFDD and arriving forces. His duties included identifying force flow discrepancies, and communicating this to the DOC at Camp Pendleton. Corporal Duplain's additional duties included briefing the CG on the daily progress of RSO&I during staff meetings, as well as serving as the Forward CP's Plans Chief. Considering the rank of individuals on parallel staffs with similar responsibilities, his weighty responsibilities were testament to the trust and confidence that the CG put in the young Marine NCOs who turned plans and ideas into action on the battlefield. Once, when the good Corporal had missed a few staff meetings because of demanding duties, the CG asked where he had been. Without skipping a beat, Corporal Duplain answered, "Well, I've been too busy to brief you General!" He was, and the CG knew it.

Fires Planning

As the Commander of 11th Marines, Colonel Marletto was also the Division Fires Support Coordinator. Going into this operation, Colonel Marletto's guidance to his staff and subordinate commanders was simple: "We must mass to kill the enemy at every opportunity, no pinpricks." His intent was equally succinct, and left no doubt with anyone, regardless of rank, as to the 11th Marines purpose, which was to "Protect the Marines and sailors of the 1st Marine Division from the effects of enemy indirect fire systems." These straightforward words resonated throughout the 11th Marines and were backed up by the regiment's actions during the attacks from northern Kuwait to Al Kut, Baghdad, and Tikrit.

The members of the G-3 Fires cell that deployed with the Forward CP possessed a breadth and depth of fire support expertise that represented the Division's commitment to the philosophy that fire support is enabled by building partnerships. Lieutenant Colonel (USA) Gary Smythe,

the Assistant Division Fire Support Coordinator led the team. For an Army Officer, the Marines of the Division thought Lieutenant Colonel Smythe was all right. He was the single Division officer with the 1st Marine Division patch on his shoulder, and he carried the serious Fire Support Coordination responsibilities on his shoulders as well. The Division Target Information Officer (TIO) was Chief Warrant Officer Quint Avenetti, a targeting expert. He was the first 0803 Target Acquisition Officer to be assigned to the 11th Marines Fire Support Coordination Section since the Marine Corps restructured the 0803 Military Occupational Specialty career path. The decision to place greater experience and technical expertise in the TIO billet was deliberate, and recognized the importance of the relationship between intelligence and fires. The team was soon joined by Master Sergeant Brad Lee, the Division Observer Liaison Chief, whose years of fire support experience and knowledge of automated systems would be critical. The Division's Air Officers rounded out the Fires team, with Lieutenant Colonel "Iron" Shank, Major "Grams" Gramuglia, and Captain "Benny" Bennigson. Both "Grams" and "Benny," who had each been with the Division for over two years, had earned the trust and respect of their comrades over the months of preparation that had already gone into this fight. These Marines had demonstrated a thorough knowledge and tactically focused mindset that made them superior in their billets. Iron's arrival in late 2002 brought another solid member to the team.

The Marines of the Forward Fires team seized the opportunity afforded by the CFLCC exercises to test the Division's fire support communications nets and automated fire support coordination systems. They continued this training during a Division Forward CPX and displacement exercise held in the third week of December. While AFATDS and its companion PC-based software, Effects Management Tool (EMT) were a focus of training, the Marines also received instruction by the MEF Information Management Officer (IMO) and contractors in the use of Command and Control Personal Computer (C2PC) and Automated Deep Operations Coordination System (ADOCS). Although not a system of record, ADOCS was used extensively within CENTCOM to automate the deliberate and reactive targeting processes. This would be the first war in which fire support coordination was largely digital, and many lessons would be derived from the experience.

In addition to planning and coordination, the Fires cell was equipped to conduct artillery reconnaissance and survey. At the request of Lieutenant Colonel Phil Lark, the I MEF Force Fires Plans Officer, Lieutenant Colonel Smythe and his team conducted an artillery reconnaissance of Kuwait's Bubiyan Island on 27 December, to determine whether artillery could be feasibly positioned there in support of an assault on the Al Faw Peninsula (as artillery on Bubiyan could range across the narrow Khor al Abdulla waterway separating Kuwait from Iraq.) The Royal Marines planned to conduct the assault early in the campaign, and Brigadier Gregory, Chief of Royal Artillery, and Commanding General of the 1st UK Armoured Division's Offensive Support Group (OSG), accompanied the team. They surveyed Bubiyan Island from the shore to 35 km inland and confirmed the suitability of the single bridge to the island for heavy vehicles and towed artillery. The team determined that the only suitable firing positions from which to support an assault were the Kuwaiti coastal defense artillery positions. Located at approximately 2 km intervals along the marshy northern shore of the island, each of the built up positions could accommodate a battery of howitzers. This venture marked the beginning of collaboration between the Brit gunners and the Cannon Cockers of 11th Marines, which would be to the advantage of both during the war.

Continuing Collaboration with UK 1st Division

UK headquarters and planning elements arrive at Camp Matilda.

The 1st Marine Division and the UK's 1st Armoured Division planners met a few days later to work out a detailed plan for the Relief in Place (RIP). The swift RIP of the Al Basrah area would be critical to sustaining the momentum of the attack. By this time, the planners expected A-day sometime during the first week of March, and the British planners were counting on the arrival of the 7th Armoured Brigade in time for LD. It was agreed that the relief would be conducted from east to west, beginning with RCT-7 and the 7th Armoured Brigade in the Az Zubayr area. This was to be followed by RCT-5 and 16th Air Assault Brigade around the GOSPs. To facilitate close coordination, the UK 1st Division's tactical command post would co-locate with the 1st Marine Division Forward CP for the assault. In addition, a Royal Air Force Liaison Officer from the RAF Regiment supporting the 1st UK Division would move forward with MWSG-37 to assume control of the FARP site at Safwan Airfield.

The plan for the relief of artillery would remain fluid almost up to G-day. At the time of the RIP coordination meeting, the British planners believed that only two batteries of the 3d Royal Horse Artillery (3 RHA) would be ready by G-day, and both would initially be required to support 3d Commando's assaults on the Al Faw Peninsula and Umm Qasr. Depending how the fight went in that quarter, there was a chance—rather slim—that the 7th Armoured Brigade might not have the support of its affiliated artillery regiment when the RIP in RCT-7's zone occurred. In that case, the Americans would leave a general support artillery battalion and counterbattery radar behind to continue to engage enemy artillery in the Al Basra area under the tactical command of the British division. This was offered in good faith, since the Marines could not hand over the counterbattery fight if the 1st UK Division lacked a capability to match whatever indirect fire threat remained in Az Zubayr and Al Basrah. Teamwork between the two divisions was so strong that neither would consider any activity that did not work to the benefit of the other. The two divisions shared remarkably similar perspectives on almost every issue, and mutual admiration and trust were the order of the day.

On the other hand, the British fire supporters had indicated from the beginning that they were eager to attach the light gun batteries of 7th Para Royal Horse Artillery (7 RHA) to the US for the Opening Gambit. They explained that 16th Air Assault Brigade would not be committed until the RIP, and artillery should not be kept in reserve. In a first meeting between the two artillery commanders, Brigadier Gregory confirmed this intention to Colonel Marletto, even though there was some question at the time whether London would permit cross-attachment of units below the brigade level. From the standpoint of facilitating the RIP, however, attaching 7 RHA to the 2d Battalion, 11th Marines (2/11) made sense, especially if it came with the British weapons locating radar. If the plan was approved, 16th Air Assault Battalion's affiliated artillery regiment would already be in place when it came up to relieve RCT-5, and the British radar—having been integrated into the counterbattery fight—would have built up its target lists and situational awareness.

The US planners coveted 3 RHA's self-propelled AS-90s with their long-range ammunition. In the end, the timeline for G-day would slip enough to prevent all four of 3 RHA's batteries from arriving in time for the fight. Thus, the 3d Battalion, 11th Marines (3/11) would cross the Line of Departure (LD) with only two batteries of the AS-90s attached, while 2/11 was reinforced by two batteries of light guns. As 3 RHA's remaining batteries were freed from the fight on Al Faw and Umm Qasr, they would move up to join the British command element with 3/11. This would get the 1st UK Division's artillery and radar into position—and into the fight—ahead of the British maneuver brigades, allowing the RIP to proceed seamlessly from the standpoint of fire support. Meanwhile, the American artillery would be free to lean toward the Division's upcoming fight in the west. Sierra Battery 5/11, with the 15th MEU, would support 3d Commando's assault on the Al Faw Peninsula from Bubiyan Island. Collaboration between Marine artillerymen and Brit gunners would be a hallmark of the Coalition under I MEF.

Intelligence Planning

The G-2 continued to spend a great deal of time and effort negotiating the composition, command relationships, and force flow requirements of supporting intelligence attachments. What the Division was attempting to create (through the attachment of supporting intelligence capabilities) was a full complement of systems, all directly responsive to Division influence. This included secure communications, topographic systems, and a host of '-INT's (Intelligence specialties such as Signals, Imagery, and Human intelligence, SIGINT, IMINT, and HUMINT, respectively.) Through these systems, the Division would be self-sufficient enough to satisfy its own intelligence needs through organic collections while remaining plugged-in to the higher intelligence architecture. This proved to be a tremendous success, as the Division relied on its own intelligence gathering, analysis, and dissemination during most of the fast-paced operations that were to follow.

The Trojan Spirit II special communications system provided critical access to Intelligence products to each RCT and the Division CP.

The Division's five deployable Trojan Spirit II (TS II) satellite data communications systems provided critical secure connectivity to each RCT, enabling instant access to all supporting intelligence information from anywhere on the battlefield. Against a host of obstacles, I MEF's

Intelligence Systems Officer, Major Mike Scheiern, had fought successfully for the assignment of TS IIs down to the RCT level. The planned fast-paced maneuver of the RCTs made a fast moving, flexible secure communications system like the TS II an important requirement. The Division's allowance, however, was for a single TSII, and there were not many in the Marine Corps. With key support from 1st Radio Battalion, I MEF crafted a TS II distribution plan that borrowed assets from 1st Radio Battalion, 3d MAW, MARCENT, and 2d Radio Battalion. Pushing digital data to the RCTs at up to Top Secret (TS) and special classification levels was required, and the Trojan Spirits would be critical as both a data and voice capability during the fight. For SIGINT, the Division had the collection capabilities of nine attached SIGINT Support Teams, three Mobile Electronic Warfare Support System (MEWSS) vehicles (LAV-mounted SIGINT collectors) and four Operational Control Elements that provided connectivity and analysis for Theater and National SIGINT collectors. For imagery, the Division could reach back through the SIPRNET to archived imagery located in Imagery Product Libraries at CENTCOM, CFACC, CFLCC, and the 1st Intelligence Battalion. With its two organic imagery interpreters, the Division also had a minimal capability to do U2 wet-film exploitation using a deployable field light table. Additionally, the Division had direct feed of both Predator and Pioneer video downlinks through the Global Broadcast System (GBS) at Division Headquarters, and three Remote Receive Terminals (RRTs) from the VMU squadrons. Communications with the UAV pilots and mission planners was typically conducted via SIPRNET chatroom, supplemented by formal post-mission reports from the UAV squadron. The Division would have seven attached Human Intelligence Exploitation Teams (HETs) providing interrogator and translator support, as well as a small control element at the Division Headquarters. Because of the limited utility of the Marine Corps' obsolete unattended ground sensor equipment, the Division had no sensor support for this operation. The Division's own organic MOS 0261 Topographic Marines and equipment provided critical topographic support.

Arranging for this wide array of systems from a variety of supporting units proved exceedingly difficult, as each supporting attachment had to be individually sought-out, negotiated, or coordinated. Each supporting unit seemed to have unique command and control relationships, making their integration a challenge. Major Scheiern, working with the Division's Major Tim Oliver, enabled the Division to get the assets it needed to be self-sufficient. His assistance in pushing these assets even further down to the RCTs also had an exceptional impact. It became clear in subsequent operations that having intelligence support organic or under tactical control of the Division was critical, as intelligence support by a centralized higher headquarters became very tenuous once combat operations began. The Division demonstrated the requirement for an integrated organic UAV capability, access to theater and national imagery collections, SIGINT reporting, HUMINT reporting, local 3-D battlespace visualization capability, and access to the analysis products of higher headquarters through a high-speed circuit such as the Trojan Spirit.

Sensitive Site Exploitation (SSE) was to prove another significant preparation effort. The national intelligence community had identified a multi-tiered listing of potential hide-sites for chemical, biological, nuclear, or missile activity. These listings included sites that ranged from active research centers to sites of possible historical storage of prohibited items as far back as the Iran-Iraq war. A separate listing of potential war-crime evidence, mass graves, or leadership locations was also developed. CFLCC had established an exploitation methodology that

Division riders join the crew of this P3-AIP aircraft to conduct battlefield surveillance and reconnaissance during the build-up phase of the operation.

included Mobile Exploitation Teams of experts that could properly investigate and exploit a sensitive site. The Division G-2 worked to identify and prepare objective folders on the sites in zone that were most likely to contain WMD-related materials.

The Division enjoyed the support of an additional intelligence collection capability beginning in December 2002. The P-3 aircraft was being used in a surveillance role of maritime targets adjacent to and inside the Division zone. The P-3 Anti-surface warfare Improvement Program P-3 (AIP) had fitted the aircraft with sensors that were tremendously helpful in giving commanders situational awareness of their battlespace. Many of the Division staff had recent experience with the P-3 (AIP) in Afghanistan as part of Task Force 58. Exploiting the lessons learned from that recent experience, a program of Division riders on the P-3 aircraft was begun. Coordination with the squadron to utilize Al Jaber or Ali al Salem (AAS) in Kuwait allowed for them to pick up Division riders in country and observe targets of interest to the Division while flying routes that supported the P-3's primarily naval mission. Thanks to Major General Amos' support, ramp space and other 3d MAW support to the P-3s was forthcoming at AAS. The Division liaisons worked directly with the Commander of Combined Task Force 57 elements operating out of Bahrain. CTF-57 and the VP squadrons in Bahrain were to prove essential partners in the success of this program. The teamwork extended to such an extent that the CTF-57 arranged for the basing of a large P-3 detachment at Ali al Salem Air Base in Kuwait to more readily support the Division. This forward basing gave the aircraft tremendous loiter time over the target areas. The Division rider team, under Colonel Jim Lukeman, served as 'Guardian Angels' for Division forces moving to contact, providing critical indications and warnings to maneuver commanders in real time. The value of this aircraft as a collection platform in support of Marine operations was clearly demonstrated as ground operations pushed beyond the littoral area. The P-3 was able to push ever deeper into Iraq to provide day or night coverage for the GCE. To some extent, the Division suffered from the success of the P-3 as a surveillance platform and its Division rider program. As word of the capabilities of this platform began to get out, competition for access increased. The 1st UK Division, NSW elements, Force Reconnaissance, and even the Army's 3d Infantry Division all vied for the attention of a limited number of aircraft and crew. In another example of superb cooperation among the units making up the Coalition, the asset was shared by all, and soon teams of riders were aloft every time the P-3 was available. Throughout, the initiative and commitment of the Sailors of CTF-57 under Commander Robert Lally were commendable and in the finest traditions of the Navy-Marine team. Vice Admiral Keating, the

Coalition Forces Maritime Component Commander (CFMCC) provided his total support, enabling this critical combat multiplier to make its influence felt. The P-3's technical capabilities, coupled with Colonel Lukeman's intimate knowledge of the Division's scheme of maneuver, was a confidence-building combination that increased the speed of tactical decision-making and maneuver against the enemy.

With the Division's expected tasking to attack up the Highway 7 corridor, the Division G-2 requested high-volume production of detailed urban IPB products for the small towns along Route 7. These requirements were negotiated with the MEF's Intelligence Battalion. Noting the lack of a comprehensive map-based obstacle overlay product for the opening attacks, the Division G-2 also initiated action to develop a comprehensive map product that included pipelines, power lines, oil wells, and agricultural areas. In a strong display of support, CFLCC quickly adopted this product development effort as their own, and developed a complete 'Attack Corridor Study.' The 'Attack Corridor Study' was eventually to prove one of the most useful and highly desired intelligence products for the entire operation.

A 3^d MAW F/A-18D with an ATARS imagery pod installed. This aircraft provided imagery in support of the MEF's targeting effort.

Third MAW had also recently received the imagery pods of the Advanced Tactical Air Reconnaissance System (ATARS). This imagery system was fitted into the F/A-18D to provide a tactical photoreconnaissance capability, one that had been lacking in the Marine Corps since the retirement of the RF-4B. The G-2 began to integrate ATARS imagery collection into the Division's intelligence architecture. Because of the very recent fielding of this capability to 3d MAW, the promise and capabilities of this platform were not fully realized due to a number of platform, data connectivity, and exploitation constraints. Because of the multi-role nature of the F/A-18D, every ATARS photoreconnaissance mission came at the expense of a deep strike, FAC (A), or CAS sortie. Achieving intelligence collection coverage of GCE maneuver and Intelligence targets required cost/benefit analysis within the MEF. Many ATARS sorties were also required to support 3^d MAW targeting assessment. Noone appreciated 3^d MAW's targeting efforts more than the grunts in the Division, but this left very few sorties to collect intelligence to meet Division requirements. A number of the remaining sorties were passed up to the CFACC for tasking, forcing the Division to compete within the larger Coalition environment to receive ATARS support from assets organic to the MEF. Few targets of primary interest to the GCE were exploited and disseminated, but the future potential for this platform was clearly demonstrated.

On 13 January, the Division attached an eight-man Joint Surveillance and Target Acquisition Radar System (JSTARS) Common Ground Station (CGS) team from the Army's 513th Military Intelligence Brigade. These soldiers, led by Staff Sergeant Adrian Flores, were to have a significant impact on the Division's combat operations, and were a welcome addition to the Division team. The CGS maintained data connectivity with an airborne sensor platform, and provided real-time Moving Target Indicators (MTIs) and Synthetic Aperture Radar imagery directly to the Division G-2. This data would be used at critical junctures of planning and combat operations to determine trafficability of routes, detect condition of key bridges, analyze terrain to find mobility corridors, and detect large unit movements. From the outset, Staff Sergeant Flores and his team proved they possessed the same fighting spirit and can-do attitude of the Marines they were supporting. These soldiers would serve with distinction through the entire campaign, and would be commended by the Division CG after the fighting was over.

Visit By the Red Cell

On 21 January, I MEF provided the Division staff the opportunity to sit down with an MCIA-sponsored 'Red Cell', led by Colonel Dave Larson. The Red Cell team combined Iraq experts from across the Intelligence Community and brought them to Kuwait to review the Division's plan from the enemy's perspective. The Division Forward Staff had the opportunity to discuss the capabilities of the enemy with this team, and gained significant understanding of the nature of the Regime and the capabilities they would likely employ on the battlefield. The Red Cell's review of the Division Scheme of Maneuver contained no significant surprises. The Red Cell noted the likely Iraqi use of humanitarian disaster, tactical inundation, and an urban-centric defense by conventional and paramilitary forces. The picture they painted was of a highly capable and increasingly challenging enemy as the Division advanced closer to Baghdad. According to the Red Cell, this enemy would allegedly use integrated fires, flooding, and chemical weapons to add to the challenges already posed by the difficult terrain. This reinforced the Division's expectation for a tough conventional fight, especially closer to Baghdad.

The Red Cell and the Division staff agreed that an asymmetric defense would be the Regime's best strategy, and their most likely recourse. Paramilitary forces, non-traditional defenses in the urban areas, a 'scorched earth' policy beginning with the Rumaylah oilfields and the use of chemical weapons might each be attempted. The group noted that isolated attacks could probably not be stopped, but the Regime would have a difficult time sustaining a coordinated asymmetric defense. Sustaining such an effort would require a well-trained force and sophisticated orchestration of effort, capabilities that the Iraqis did not have. The Division's 'No better friend, no worse enemy' policy would also seek to remove the motivation for resistance once segments of the population came under Division control.

Iraqi units were infiltrated and controlled by Regime overseers, and would not likely capitulate or surrender before US forces were closing on their positions on the ground. For the Iraqi soldiers, it was only when the balance of fear shifted from what the SSO would do to them to what the firepower of the Americans might do to them that surrender became an option. This same concept of a 'tipping point' with the civilian population was also discussed. The US had not intervened when the Shias rose against Saddam in 1991, and thousands were murdered by the Regime. Even though the Iraqi populace was vehemently anti-Regime, they were not likely to begin to aid the Coalition attack until it was clear that the US was committed to absolute Regime removal on the ground, and had the upper hand.

One final topic of discussion was the personality of Saddam Hussein. Saddam had kept himself in power by keeping his options open, then making brash decisions when he felt cornered or thought he saw an opportunity. The early use of chemicals would remove his flexibility. Saddam would likely recognize his inability to mass chemical fires for an 'area denial' effect, making them most effective as a weapon of terror. On the world stage given to Saddam by the international media, the *threatened* escalation to the use of chemical weapons was probably as effective as the weapons themselves. On the battlefield, these weapons would have limited effectiveness against well-trained US Marines. The Division G-2's line of reasoning led to the assessment that it was unlikely that Saddam would use chemical weapons in the opening days of the fight. This was contrary to some other assessments that predicted the use of chemical weapons 'early and often'. Regardless of the assessment, prudence would dictate that the 1st Marine Division be on guard against this eventuality throughout the campaign.

The unstable personality of Iraq's dictator Saddam Hussein made it difficult to predict his actions.

Teaming with the 1st UK Armoured Division

One of the rewarding opportunities of the Operation was the close linkage between the 1st UK Division and the 1st Marine Division. The 1st UK Division, led by Major General Robin Brimms, had a shared understanding with the Division about what tasks existed, and the general order to complete them. Major General Brimms preferred not to push fragments of his combat power into the fight piecemeal under the Marine Division control. Rather, 1st UK Division planned to keep them ready to relieve the Division as rapidly and coherently as possible. This would enable the Marine Division to quickly disperse to the west, allowing not only preserved momentum and speed, but also a reduced chemical target. The Marine Division CG agreed fully with this concept. The Brits sent over one of their finest officers, Major John Brookings as a liaison to the 1st Marine Division CP. The quick-witted Major Brookings, with true Marine-like irreverence and humor, soon made himself a welcome and valued member of the Division team. At the Division's regimental command posts Major Andrew Flay (UK 7th Armour Brigade) joined RCT-7, Major Cameron Day (Royal Artillery) joined 11th Marines, and Major Frasier (16th UK Air Assault Brigade) joined RCT-5.

The Brits planned to have 3 Commando (with the US 15th MEU TACON) operating on the Al Faw peninsula, the 7th Armoured Brigade relieving RCT-7 in the Zubayr area, and the 16th Air Assault Brigade relieving RCT-5 in the Rumaylah area. Because UK forces were deploying to Kuwait on their own timeline, during planning there was little certainty about what the 1st UK Division would consist of when it actually came time for a RIP to be conducted. In fact, forces would still be flowing into Theater when the Division crossed the Line of Departure. A possible problem might arise if most of the 'Desert Rats' of the UK 7th Armoured Brigade were not available to relieve the Division in time. In this case, the Marine Division would have to destroy all vestiges of the 51st Mechanized Division and close off all the bridges over the Shatt al Basrah waterway. These were tasks that the Division would have otherwise handed off to the 7th Armoured Brigade once the South Rumaylah oil field infrastructure was secure. As it turned out, the Desert Rats would be sufficiently closed to commence the RIP on schedule within 24 hours of the Marines crossing LD, even attacking in concert near Umm Qasr, and handily finishing off those elements of the 51st Mechanized Division that the Division left undestroyed.

Coordination with the Kuwaiti Land Forces

The Division also began coordination with the Kuwaitis soon after the arrival of the Forward CP. Colonel Toolan made contact with the Kuwaiti Ministry of Defense (KMOD), Kuwait Land Forces (KLF) and the Office of Military Cooperation-Kuwait (OMC-K) to discuss cross obstacle breaching, maneuver training areas, terrain walks of the border area, and potential pre-assault preparation tasks. These early discussions set the tone for follow on planning and set the Division up for success during future combat operations. Major Clark Watson served as the Division's liaison to the KLF 35th Brigade. In between dinners of fresh lamb and cold drinks, he was to play an important role (Major Watson took a good deal of ribbing from his Division comrades, subsisting on Meals Ready to Eat (MREs).) His efforts were instrumental in the working relationship that was established with the KLF, and led to a mutual understanding of the Division's desires and movement plans while in Kuwait.

In order for the Division to occupy Attack Positions near the border, it was necessary to conduct a Forward Passage of Lines (FPOL) with the two Kuwaiti Brigades that had responsibility for border security in the Division's zone. The Kuwaiti Land Forces, especially the 6th and 35th Brigades were proud units, who wanted to assist the Americans and carry out their missions as full partners. The Division understood the strong national pride of the Kuwaiti people, and worked eagerly with the KLF to find ways to achieve the mutual goal of a rapid and safe FPOL.

The breaching of the Kuwaiti border obstacles became a central topic of discussion between the KLF and the Division. The Division requested that the Kuwaiti allies create twelve 300 - 500 meter lanes through the southern and northern obstacle belts. After coordination between CFLCC and the KMOD, the Kuwaitis hired a contractor to do the job. A Kuwaiti contractor diminished all border obstacles along the Kuwait-Iraq border in 1st Marine Division's zone, including the Southern berm and tank ditch, an electrified wire obstacle, and a northern berm and ditch. The Kuwaiti contractor opened all lanes in the southern berm obstacle by creating a hole in the berm and filling/compacting the ditch. The contractor made lanes in the wire obstacle at the breach sites by pre-cutting and preparing the fence. This method left the fence in place, with the appearance of being intact, yet the Marines would be able to easily open the fence by hand when the time came. The Kuwaiti contractor opened the northern berm and ditch obstacles in the same manner as the southern set.

The biggest impediment to early opening of the berm lanes by the Kuwaitis was the late closure of our forces and the danger of alarming the Iraqis who could observe the berm cut locations from their Observation Posts on *Jabal Sanam*. If they reacted to the berm cuts by commencing destruction of the oilfields before the Division had sufficient forces to attack, the key oil infrastructure could be lost.

Major Hassan – A Combat Multiplier

Major Hassan - Senior Kuwaiti Liaison Officer to the UN Border observer mission and to Coalition forces. He was to play a key role in building the Division's situational awareness prior to the start of combat operations.

Few individuals had a greater impact on the success of the Division's preparations than Major Hassan of the Kuwaiti Ministry of the Interior. This imposing figure had worked the murderous Kuwait/Iraq border area for over a decade, and was one of the most knowledgeable advisors the Division was to gain. Major Hassan was introduced to the Division Commander as a Kuwaiti Border Guard officer, who could discuss the activities of the Iraqis, Kuwaitis, and the UN along the border. It soon became clear that Major Hassan had tremendous insight into the situation along the border area. With a charming personality and an unveiled passion to assist the Marines in taking out the Saddam Regime, he soon was considered a friend by many members of the Division. Over the course of months, Major Hassan was to provide access to sensitive border areas to enable the Division's attack route planning and leader's reconnaissance. He provided key intelligence of Iraqi patterns of activity in the border area, and described the plight of the poor Iraqi people oppressed in southern Iraq. With Major Hassan as a guide, many of the Division's commanders and key staff members got a close look at the border obstacles, and routes that would take them the initial steps in the march to remove Saddam. Major Hassan proved a true friend of the 1st Marine Division, and an important agent for the liberation of the Iraqi people. In August 2003, he received a commendation from the Division CG for his invaluable advice and coordination. More importantly, he earned the fondness and respect of the Marines who served alongside him.

Force Protection and the Atmosphere in Kuwait

Even in the friendly country of Kuwait, the Marines maintained a very strong force protection posture. The Division Marines were tough, and already tough to kill. If terrorist elements were to infiltrate into Kuwait and attack the Division prior to crossing the Line of Departure, they would have to get past a series of alert Marine sentries and Guardian Angels with weapons ready.

The vast majority of Kuwaitis were supportive of the US presence in their country and understood the mission and the reasons for it. The entire Middle East region, however, retained political, religious, and anti-western discontent for a vocal and violent few, largely focused on the Israel-Palestine issue. US forces in the region were a naturally high-profile target for these groups, and the Division continued to present a hard-target for terrorists.

There were a number of incidents that influenced the Division's careful approach to force protection. Even before the arrival of the Division Forward CP in Kuwait, one Division Marine, Corporal Antonio J. Sledd from Battalion Landing Team 3/1 (3^{d} Battalion, 1st Marines) was killed, and another wounded, by a terrorist attack. This attack occurred during an 11th Marine Expeditionary Unit (MEU) training exercise on Fayhlaka Island, Kuwait. The incident brought to light the critical importance of Force Protection while deployed in Theater, and reminded Division Marines of the importance of having Guardian Angels even while operating in an allied country. Although difficult to see at the time of his death, Corporal Sledd's sacrifice would have a significant impact on setting the conditions for keeping his fellow Marines alive during Operation Iraqi Freedom. Another contemporary incident in theater involved soldiers from the 101st Airborne Division arriving in Kandahar, Afghanistan on strategic lift aircraft. Because the soldiers were not permitted to carry ammunition on the military aircraft, they were left defenseless when they came under fire upon arrival. The Division raised these concerns to higher headquarters, and insisted that Marines deploying into this elevated force protection risk area must be equipped to defend themselves. With CFLCC's assistance, travel regulations were modified to allow the Marines of the Division to travel with their T/O weapons and self-protection ammunition, including on civilian contracted carriers.

Incidents continued after the Marines arrived in Kuwait. Another shooting incident involving two US Army soldiers and a deranged Kuwaiti policeman occurred in November, and two US civilian contractors were killed near the entrance to Camp Doha in January. These incidents justified the Division's alert policy of multiple vehicle convoys with armed Marines for all travel. Convoy drills, immediate action drills, and counter ambush techniques were rehearsed by all Division Marines.

In addition to the number of isolated shooting incidents, there were other suspected terrorists incidents. The Iraqi Intelligence Services (IIS) were known to attempt operations in Kuwait. In February it was discovered that the IIS had aggressively solicited food service workers of contract food providers as a possible way to introduce toxic or lethal diseases to large numbers of Marines. These incidents would give pause to advocates of cutting food service Marines, given the potential mischief to food supplies posed by contract workers with close access. Bomb threats were routine, and all vehicles entering Camp Commando and the TAA were carefully searched.

Logistics Planning and Rehearsal

As units received weapons and vehicles from MPF shipping and their own equipment via strategic airlift, modifications and upgrades that began in garrison continued in the field. The Division Ordnance Readiness team, led by Major Greg Frich, was instrumental in working with USMC and theater agencies to source assets that enhanced both the Division's lethality and survivability. After establishing close liaison with the CFLCC at Camp Doha, Chief Warrant Officer Rod Fiene procured nearly 4,000 Combat Identification and Thermal Identification Panels (CIPs and TIPs, respectively) for the Division. Marines from 1st and 2d Tank Battalions attached CIPs to their M1A1s to identify the tank and reduce the chances of fratricide. Remaining Division units attached TIPs to wheeled vehicles for the same purpose. Similar initiatives were pursued in the design, development, and installation of the Blue Force Tracker and Mobile Data Automated Communications Terminal (MDACT) systems. Working in collaboration with Major Thad Trapp of MARCORSYSCOM and the USMC weapons station in Crane, Indiana, Major Frich was responsible for coordinating the modification of the Division's Shoulder-launched Multi-purpose Assault Weapon (SMAW) systems. In total, technicians ruggedized the sighting systems of over 720 SMAWs to allow them to better withstand combat and to accommodate firing of the Novel Explosive (NE) round (a thermobaric blast warhead with tremendous explosive power.) On this and numerous other issues, the staff of MARCORSYSCOM worked tirelessly in support of the Division and its Marines, ceasing only when units moved to Dispersal Areas (DA), and Attack Positions (AP). From the earliest days of planning through the Division's movement into its final positions for the attack, MARCORSYSCOM's unstinting support was felt daily. The CG and staff of MARCORSYSCOM earned the Division's gratitude many times over.

With equipment on hand and modifications ongoing, providing the units with resources for training and combat was the next order of business. The meticulous and comprehensive planning efforts of Lieutenant Colonel Kiel Gentry, the 11th Marine Regiment S-4 Officer, and Captain Tim Collins, the Division Ammunition Officer would prove crucial to the artillery regiment and RCTs' success during combat. These Marines developed a plan that sequentially staged, drew and fired ammunition by lot number in order to provide the greatest accuracy for the Regiment's seventy-five M198 howitzers. Working with all of the Division's units and 1st FSSG ammunition technicians, the G-4 ammunition section designed and pre-staged numerous ammunition re-supply packages, each tailored to a particular mission.

The Division G-4 sourced Humanitarian Daily Rations (HDRs) and distributed two cases to each vehicle in the Division. At their discretion, commanders would distribute the HDRs to alleviate food shortages or as a demonstration of goodwill to Iraqis caught up in the war and in need. The Marines intended to give the Iraqi people an introduction to a different kind of army, one that bore no resemblance to Saddam's oppressors.

In keeping with the Commanding General's intent to be the most air-centric Division in history, and the CG, 3d MAW's intent to rapidly open Forward Arming and Refueling Points (FARPs) in the Division's wake, the G-4 initiated planning with Marine Wing Support Group -

37 (MWSG-37), commanded by Colonel Mike Anderson. The quick thinking planners at MWSG designed Rotary Wing FARP-opening packages designed to extend time on station of the Hueys and Cobras. The Division's G-4 Plans Officer, Lieutenant Colonel Jeff Moyer, and

A logistics convoy moves across the open desert of northern Kuwait prior to the Division's movement to Dispersal Areas. Planning for logistics support over limited routes was a significant challenge to the Division's planners.

the MWSG-37 Operations Officer, Lieutenant Colonel Fred Blish, worked to integrate these FARP packages into the tactical columns of the RCTs, allowing for quick set up and extending the reach of the Divisions air arm. FARPs would be placed as far forward as possible, primarily to provide fuel and ammunition to rotary wing aircraft flying in support of maneuver elements. As the Division continued to push further north toward Baghdad, extending the tenuous lines of communication to Kuwait, FARPs would take on the secondary mission of providing a node into which critical classes of supply could be flown.

CSSG Planning and Rehearsal

BSSG-1 officially became CSSG-11 on 10 December 2002, marking the initiation of an integrated CSS element within the Division. In less then one month CSSG-11 grew from a cadre planning staff of 32 Marines to over 1800. Planning and training continued to focus on the requirement for speed over the long distances necessary to sustain the 1st Marine Division. Detailed attention was given to the logistical footprint each CSSC would require to support their assigned RCT's. The CSSC's and CSSB-10's organization were designed with the correct equipment to support sustainment. Also, specific attention was given to calculating exact requirements and developing capabilities sets for each class of supply. CSSG-11's staff coordinated with each Battalion to develop re-supply blocks which included medical supplies, repair parts, battery blocks, ammunition, and emergency repair parts. This eliminated the requirement for a line-by-line supply request that consumed valuable time. The ammunition technician's emergency re-supply packages were pre-built and pre-staged to fly in at a moments notice.

Working with the Division G-4, CSSG-11 developed a Logistic Tasking Order (LTO) to streamline the logistics request and tasking effort. The LTO was designed to take rapid requests, consolidate requirements, and then produce an order that would lead to mission accomplishment. To reduce the number of requests that had to be submitted by the battalions, the LTO took recurring requests such as food, chow, and water then automatically inserted the requests into the system. The LTO would be distributed to each of the CSSCs, the CSSB, and to Division G-4 to allow all units to know in advance where and when their supplies would arrive.

On 27 January, the CSSG-11 main body landed, hitting the deck running. They provided immediate CSS to RCT-7 and the Division headquarters within the first five days of arrival. An Ammunition Supply Point (ASP) was constructed immediately to receive, hold and distribute ammunition to Division units. The ASP ran around the clock up until a few days prior to crossing the border, making sure that there was no shortage of steel when the targets presented themselves.

Internally, CSSG-11 began intensive unit training for its over 1,800 Marines from 8 different USMC units, the USMC Reserves, the US Navy, and the US Army. The intensive training program culminated with an abbreviated Marine Corps Combat Readiness Evaluation (MCCRE) for each unit. The MCCRE scenario was designed to rehearse the first 96 hours of the war and focused on establishing and disestablishing RRPs, convoy operations, night movements, actions on enemy contact, force protection, long range communications, and integration with supported units.

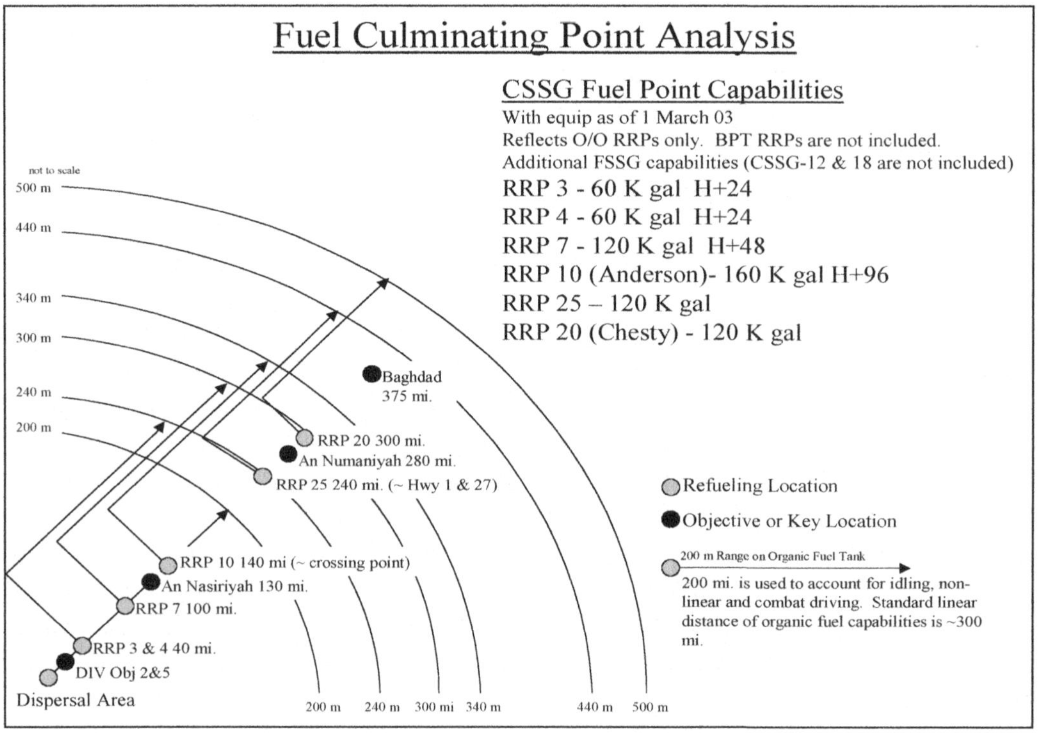

In Medical Services, the Forward Resuscitative Surgery System (FRSS), pioneered by Navy Captain H.R. Bohman and Lieutenant Commander T.R. Bilski, had come of age after initial conception in 1997. After limited use in Bright Star and in Afghanistan with TF-58, the full capabilities of the mobile surgical system were about to get the first major test in battle. Prior to departing CONUS, two FRSS teams spent a month at Los Angeles County Hospital (Navy Training Trauma Center) sharpening skills that would save lives in the months to come. After arriving in country with two full teams, the staff of the FRSS spent the month of February training medical augments and reservists to form an additional four teams. Over 200 hours of intensive training were devoted to lectures and hands-on training on new equipment. In addition, six Care Teams were trained to accompany critical medevac patients using ground and air transportation. Designed to provide initial life saving surgical care, the FRSS would become the most significant medical innovation to be validated during the war.

The role of CSSG-11, hastily formed and cutting new doctrinal ground, would be critical during the fighting. The habitual relationship between the CSSG elements and the Division it supported was to be a combat multiplier that connected the Marine Logistics Command effort in Kuwait to the Division's design to push deep and fast into Iraq. Its success was a testimonial to the vision of Brigadier General Ed Usher and Colonel J.J. Pomfret; and the commitment of their Marines and Sailors to support the 1st Marine Division.

Command and Control

The first major effort for the G-6 was to negotiate the allocation of the Marine Expeditionary Brigade Command Element equipment on board the ships of MPSRON 1 and 2. The Division needed the unallocated gear (there were no MEB command elements) to provide redundant, long line communications over extended distances and multiple camps. The Deputy G-6, Lieutenant Colonel Brian Barton, attended the conference and was able to have 26 additional communications vehicles assigned to the Division, overcoming many of the most critical communications deficiencies. This war proved what the Division's communicators had known for years, that a Marine Division's peacetime equipment allowance would not support Division-level combat operations. The requirement to support multiple CPs, a Division Support Area, LSAs, CONUS-based forces, liaison officers, and other add-on communications requirements aggravated an already tenuous communications equipage and manning situation.

As communication requirements were refined, it was clear that the G-6 and the Headquarters Battalion's Communications Company manning levels were not adequate either. It was necessary to support the CG's Jump CP, Division Forward CP, Division Main CP, and the Division Support Area (DSA), while also having personnel support at Camp Pendleton to support reach-back capabilities. With the superb assistance of the G-1, the G-6 began the process of identifying reserve augmentation requirements, to include personnel and equipment support from the 4th Marine Division's Communications Company. Concurrently the process of requesting augmentation for some key specialty personnel was begun. Despite the G-1's constant efforts, it was not until 15 February that Major Tom Sobey was temporarily assigned to the Division as the data officer. Unfortunately, a data chief and technical controllers were never assigned. Had it not been for the last minute augmentation by the 4th Marine Division Communication Company's personnel and equipment, the Division communicators would not have been able to support many of the Division's communications requirements in combat. As always, the Reserve Marines came through like champions once the system finally authorized their mobilization and deployment.

The Division received a number of new communications technologies just prior to deploying for combat. These systems, though appreciated, arrived at the Division with little training, logistics, or employment concepts. Some arrived only days before the Division crossed the Line of Departure, leaving barely enough time to find room on already packed vehicles, and scant time for integration.

The MDACT and the Army's Blue Force Tracker were two new technologies that provided Position Location Information (PLI). This was an operational assist in order to provide friendly 'blue' tracking on C2PC, which provided commanders a Common Operational Picture (COP). These two systems automatically provide 10-digit position information to the COP from vehicles with the equipment installed. The MDACT used the line of sight EPLRS radio for the communications path while the Blue Force Tracker was a satellite-based system that supported PLI without terrain or distance concerns. The Blue Force Tracker became the system of choice throughout the Division, not only for the PLI provided, but equally because it provided a capability to send pre-formatted and free text messages. Both systems were successful in finally allowing for real time automated input to the COP vice having the Division staff 'fat finger' the positions of friendly units based upon out-of-date unit position reports. The Division received and distributed 319 MDACT and 104 Blue Force Trackers. Installation of systems into the Division's vehicles and the training of Marines on both systems continued until the end of February.

The Global Broadcast System (GBS) was a receive-only satellite system that was fielded to the Division only one month prior to deploying from Camp Pendleton. This new capability allowed for receipt of CONUS-based cable network news services (CNN, and Fox News) as they broadcast, which meant that the Division could receive real time news services at its deployed location. This proved to be a good source of intelligence, and kept the Division's situational awareness high. The GBS was augmented with additional decryption equipment that allowed receipt of Predator UAV real time imagery. Two Division Marines were trained in the basic setup and operations of the GBS ground station equipment, but the Division required the augmentation of Corporal Nathan Dutton from 9th Communications Battalion to ensure that both the GBS and decryption equipment remained operational. As with all new systems that were pushed to the Division, the training of Marines to install, operate, and maintain it was key to ensuring the reliability of the equipment. Gaining the required expertise on these systems often fell to the motivation and intellect of the

Mobile Digital Automated Computer Terminal. This equipment was installed in vehicles throughout the Division to provide immediate position reporting of friendly forces as well as rapid transmission of free-text messages.

individual Marine, as training opportunities were limited. Without the trained personnel borrowed from 9th Communications Battalion, GBS (and other new systems) would not have been such a success.

Use of commercial videophones and video teleconferencing suites became more the norm than the exception for both the Division Main and Forward CPs. Despite receiving the new equipment just prior to deploying, and having limited training opportunities, it functioned very well. The Voice Over Internet Protocol (VoIP) Phone was used far more than the video teleconferencing suite and provided reliable voice conferencing service for up to nine locations simultaneously. Since the VoIP phone transmission path was via dedicated SIPRNET bandwidth, it allowed for reliable connectivity over hundreds of miles in support of the geographically separated CPs, although at the expense of the rest of the users on the network. This new technology set a new standard for supporting commander and staff interaction.

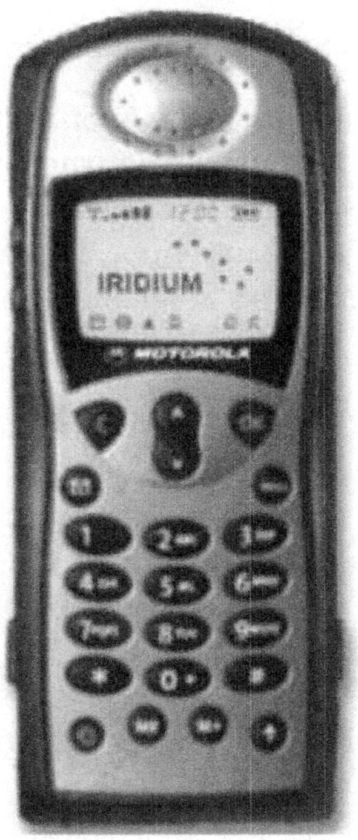

Commercial satellite telephones, like this Iridium phone, were a valuable addition to organic systems for tactical communications redundancy.

For over a year prior to deploying, the Division G-6 had begun the process of procuring and activating Iridium phones for use throughout the Division. Initially the focus was to provide each Commander with an Iridium phone. With the flexibility and reliability of these phones it became apparent that many more were desired to provide an alternative means of voice communication throughout the Division. Starting with seventeen, the Division's appetite for Iridium phones grew until there were nearly 100 phones activated and supporting the Division. At a total cost of more than three hundred and fifty thousand dollars for these phones (to include the secure calling capability) this represented a significant investment in communications by commanders throughout the Division. The investment was well worth it when considering the successful use of these phones to prop-up the Division's aged tactical radio communications capabilities.

The CG directed the Division Information Management Officer, Lieutenant Colonel Norm Cooling, to eliminate all unnecessary reports required from subordinate commands and to streamline the remaining reporting requirements. To aid in this process, a web-based strategy was employed to increase the availability of information to all of the Division's units. In conjunction with the IMO, the G-6 Information Systems section led by Major Chris Nelson, 1st Lieutenant Spence Hanneman, and Staff Sergeant Bradley Hall developed a plan for the activation of the first ever deployed 1st Marine Division secure World Wide Web site. The efforts put forth to procure the necessary hardware, train personnel, and coordinate the communications connectivity at the DSA was instrumental to the Division's success in activating the site by 24 February.

As Web services came on line at the DSA it became apparent to Major Sobey that an alternative was available to solve E-mail forwarding problems experienced by key personnel (CG, ADC, Chief of Staff, etc.) as they moved between Division CPs. Using the skills of Lieutenant Colonel Glen Edisen, Staff Sergeant Hall, and with the support of Major Roger Stanfield, email servers were allocated to the DSA. These servers, for secure and non-secure email, used the same communications links as the Web Server but allowed consistent and reliable e-mail service for the entire Division Battle Space while simultaneously providing a single stable receipt point for e-mail forwarding from Camp Pendleton. E-mail was critically important to the modern battlefield for commanders, key staff officers, and Division CPs. This initiative was highly successful in meeting this operational imperative. Headquarters Battalion Communications Company tackled a series of like issues in preparation for combat operations.

The number of Liaison Officers, full staff participation, and the additional requirements for radio, telephone, and data support threatened to overwhelm the communication support available. The computer support network needed to be changed from a simple 'flat network' to a more complicated series of Virtual Local Area Networks (VLANs), which took time for the Marines to learn. With practice and the implementation of new procedures, the Communication Company Marines were able to streamline setup and operation of the Division Main CP to six hours rather than the twelve to twenty four hours it had previously taken.

Personnel Issues - Assigning Troops to Tasks

Once deployment decisions were made, the Division faced the task of getting its Fleet Assistance Program (FAP) and Force Protection Marines back from the Marine Corps Bases (to which they were temporarily assigned.) On 6 January, at the Division's request, the MEF published a plan for returning Marines that were currently serving in these billets. The Division implemented a plan to ensure these Marines were included in the deployment preparations of their combat units right along with non-FAP personnel. Base Commanders at both Camp Pendleton and Twenty-nine Palms gave urgent, whole-hearted support to this time-sensitive issue. This was consistent with their continued strong support of deployment requirements, Key Volunteer Networks (KVN), casualty notification, and a host of other requirements. The home base and station commanders scrambled to fill the void left by deploying combat units, providing unwavering support throughout the campaign.

On 8 January, Headquarters Marine Corps released Marine Corps Bulletin (MCBUL) 1900 establishing a 'Stop Loss/Stop Move' personnel policy. This stabilized personnel numbers and mitigated many of the problems with planning and tracking for Permanent Changes of Station (PCS), End of Active Service (EAS), and retirements. Although required to stabilize deployment planning, the policy was not without impact on individual Marines. Many Marines had to cancel new job opportunities, freeze career changes, cancel retirement ceremonies, or extend on the Unit Deployment Program (UDP). The fidelity, valor, and patriotism of the Marines asked to give the Corps 'just a little more' were noteworthy. The Division CG did not hear a single request mast from the many hundreds of effected Marines who unselfishly put the needs of the Corps and country ahead of their personal wishes. Always leaning into the next fight, Lieutenant Colonel Powalski and her team began efforts that anticipated early release of these Marines once their talents were no longer required.

Public Affairs Preparations

At this time, the 1st Marine Division Public Affairs Office (PAO) was informed that a large number of media personnel that had worked with the Marine Corps in Afghanistan had asked to accompany the Marine Corps should they deploy to Iraq. The PAO, led by Captain Joe Plenzler, started building a positive relationship with them and putting together a 'by name' list of requested media to embed in 1st Marine Division in the event of combat operations.

Marines show their courage to the world through embedded media personnel.

A retired Marine, Colonel Dick Camp, reminded the Division that during the Korean War, a Saturday Evening Post journalist had accompanied a squad of Marines from Camp Pendleton through 18 months of combat. The reporting provided the American public an up-front and personal account of the sacrifices junior Marines made for their country. Now, ABC's *Nightline* funded a similar effort that resulted in a weekly broadcast of Fox Company, 2d Battalion, 5th Marines (F/2/5) on the nation's airwaves as they readied for deployment. The ratings for the show significantly increased as a result of the weekly broadcasts. As the unit deployed, *Nightline* increased the number of broadcast episodes to two per week. The reports were so popular that ABC began re-broadcasting the stories on *Good Morning America* and other news shows. Other news outlets followed smaller Division units, including squad-sized elements from 7th Marines. The collaboration between journalist Mike Cerre, the PAO, and F 2/5 brought the fine young Marines and Sailors of 1st Marine Division right into the living rooms of America.

The Division PAO embarked on a media blitzkrieg hosting more than 400 reporters in 40 days - blanketing the media with 1st Marine Division stories. The PAO put together media opportunities, often hitch-hiking with the media due to the scarcity of available vehicles. In one instance, a media convoy from Camp Commando visited 1st Tank Battalion at Udairi Range during a full-blown Arabian sandstorm. With visibility at less than 50 feet and battling 40 - 60 mph winds, the tanks couldn't see their targets on the range. The story of the Marines operating in harsh weather conditions carried the headlines and taught the reporters about the tenacity of our young Marines and a good lesson of what to expect should they decide to embed with the Division for combat operations

Concurrently, the Division PAO began exploring the feasibility of many TV media's request to bring along hard-wired, satellite transmission capable, 4-wheel-drive, diesel, desert colored vehicles and night vision qualified drivers to support their broadcasts while embedded with the Division. The PAO determined that allowing the media to use their own vehicles would benefit the Division by ameliorating some of the logistical burden on the receiving units (a TV crew generally requires 15 large boxes of lift and space to haul their gear) and allow the media to broadcast "live on the fly" in combat. The Division PAO requested that I MEF appeal the USCENTCOM prohibition on media vehicles in embedded units. Most TV media opted to assign vehicles to their embedded correspondents in the hope that the verdict would be positive and fielded the vehicles on short term embeds with Division units. On all early tests, they performed soundly.

On 10 March, the Division PA section met with the 80 reporters assigned to embed within tactical units at the Hilton Hotel in Kuwait City and began the media integration process. Media members were issued NBC suits, masks and nerve agent antidote kits. They received anthrax and smallpox vaccinations and briefings on ground rules, organization, and missions. The Division PA section worked hard to answer hundreds of questions, and ensure the media were prepared to embed. During one briefing a reporter skeptically asked, "Really, how close are you going to allow me to get to the front lines?" Captain Plenzler retorted, "I can put you in the back of an AAV with 18 pissed-off grunts, drive you within 300 meters of the enemy objective and send you in the assault while the Marines you are with storm the enemy's trench lines and drive their bayonets into the hearts of the enemy." The room went silent. You could have heard a pin drop. "Is he serious?" whispered one skittish reporter. The Division would later deliver, in scores.

> ***"I can put you in the back of an AAV with 18 pissed-off grunts, drive you within 300 meters of the enemy objective and send you in the assault while the Marines you are with storm the enemy's trench lines and drive their bayonets into the hearts of the enemy."***

On 11 March, the Division assembled their 80 reporters and transported them to LSA-Matilda to integrate them into the Blue Diamond family. Each battalion assigned a Marine to be the 'media buddy' for each correspondent. The PAO made it clear that, "In the 1st Marine Division the media do not have 'handlers' or 'escorts'. If you want an 'escort' take $500 and go to Palm Springs. Here in this Division we are going to adopt you media members and show you the ropes. You are going to be part of the family."

The following morning, the media accompanied by their 'media buddies' and PA Liaison Teams in LAVs, 7-ton trucks, and HMMWVs bid their farewell to LSA-Matilda and started their odyssey. The regiments and battalions each conducted their own welcome aboard and assigned the media to their tactical units.

The Division PA section continued to prepare for combat operations and moved with the Division Main to its tactical assembly area. Early on March 19, the Division received the directive from CENTCOM Public Affairs that no media vehicles were to be allowed on the battlefield. The Division PAO contacted the regiments to pass the bad news. In the chaos that ensued, units attempted to re-load TV media equipment onto jam-packed tactical vehicles. Some journalists opted to stay put, and follow the units in their own vehicles despite the restriction. There was an eye-opening lesson for future combat operations, as this decision made the difference between headlining on the evening news and receiving sporadic coverage at best. For example, the US Army's embedded media, whose media vehicles found their way to the fight regardless of policy, dominated the airwaves due to their ability to broadcast on the fly. Marine embeds who relinquished their vehicles faced significant challenges in getting their product out. They eventually adapted and overcame most communications difficulties with the hard work and assistance of the Marines.

Camp Commando, Kuwait, grows in size as more forces are deployed to theater.

Chapter 3
Adjustments to the Plan and Combat Rehearsals

The Commandant of the Marine Corps (and former Commanding General of I MEF), General Michael Hagee, speaks to Marines as they prepare for war

Now that the Marines were in-theater, the requirement to remain alert was even more important. The terrorist murder of two Army civilian contractors just outside the Camp Doha base, and repeated terrorist threats, heightened the security posture. The Iraqis had certainly attempted to infiltrate agents into the areas of Kuwait where US forces were staging, and were likely monitoring the actions of all US forces in the country. While waiting for the Division's secure communications equipment to arrive, the Division was forced to use couriers for secure communications between sites. Units still conducted non-secure business over cellular telephones operating on the civilian network, but even unclassified sensitive information was at risk of compromise. As the Division neared the final weeks before combat operations were to begin, the CG re-emphasized the force protection guidance he had provided to his Marines in August. Due caution was exercised in all activity, including armed Marines traveling in multi-vehicle convoys, Guardian Angel use, and secure Entry Control Points (ECPs) to all LSA areas. The Division carefully monitored all terrorist threat indicators. The Marines considered themselves 'hard targets', and were prepared if the enemy chose to initiate hostilities early through a terrorist-style attack. By their discipline, the Marines sent a message of intimidation to the Iraqis on the other side of the border. Despite the heightened security posture, however, the priority of effort was clearly on combat preparation. The Division focused on being 'No worse enemy' as it prepared to send a shockwave through the Iraqi army.

As expected, the operational plans continued to change even well after the Division arrived in Kuwait. The Division took these changes in-stride, and was careful not to be too wedded to a base plan. Planning now took a back seat to training and rehearsals as the Division planned to win this fight through 'brilliance in the basics'. The tactical, physical, and moral readiness of the individual Marine was going to be the harbinger of success, not reliance on a scripted plan. Mental agility, habitual relationships, and well-honed tactics, would earn the Division the speed necessary to reduce risk. Division leaders worked to reinforce the aggressive spirit and mental toughness of their Marines during the last weeks of preparation. Whatever the Iraqis were planning, the Marines of the Division would be prepared to crush any resistance in their path.

The Iraqis Welcome the New Year

The presence of the Marines in Kuwait exacerbated an already grim manpower situation for the Iraqi chain of command. Iraqi desertion rates accelerated as enemy forces heard the distant booms of US artillery firing at Udairi range during the quiet desert nights. As desertion rates continued to rise, the Regime began to take more drastic measures to enforce discipline. Reports of arrests and executions of Iraqi deserters increased. Despite the enemy's poor morale, however, the Iraqis were beginning to make some visible preparations for the coming Coalition attack. The Iraqis clearly expected an extended air campaign prior to actions on the ground, and focused their preparations on increasing their ability to survive such an attack. Over the month of January, extensive trenching and canal extensions were dug in Al Kut, An Nasiriyah, and Baghdad. Like raindrops on a pond, the surface of Iraq was soon dotted with revetments, both new and old. Survivability revetments appeared just outside the garrisons of every Republican Guard (RG) unit, and Regular Army (RA) units began to make survivability moves outside their own garrison positions. The RG Divisions around Al Kut and Baghdad gave every indication that they intended to survive a US bombing campaign by dispersal, frequent movement, and by digging in. The Marines of the Division were very impressed with the prowess demonstrated by the Iraqi engineers. The Iraqi soldiers might not have a reputation as great combat infantrymen, but their engineers were a marvel. A marked increase in the use of decoy artillery pieces in many of the new revetments was also noted in the Rumaylah and Al Kut areas. The Iraqis had been on the receiving end of US air power before, and it was clear that they had learned some lessons. There was a large body of information available from open sources regarding the recent successes the Serbians had achieved in eluding US air power, and it looked like the Iraqis had done their homework.

Kuwaiti oil wells burn in 1991 after the Iraqi Regime set them ablaze. The possibility of the Iraqis executing a similar 'Scorched Earth' campaign in Iraq was a significant threat.

Curiously, little evidence existed of integrated, prepared defensive positions. All of the revetments observed were haphazardly arranged, not seemingly oriented to defend from one particular direction or another. Apart from the revetment digging, activity levels among Iraqi units seemed to be no greater than seasonal norms. If the Iraqis were preparing for an air attack, why were they not also preparing ground defenses? Some intelligence assessments began to make sweeping extrapolations from the limited data available. The existence of additional revetments in a given area, for example, was assessed as an indication that the unit doing the digging was about to be massively reinforced. Alarmist reporting of this type began to increase, and reports of Iraqi suicide units or terrorist forces operating in Kuwait were frequent. Reports surfaced on a variety of heinous tactics to be used by the Iraqis including electrified water obstacles, poisoned water supplies, widespread inundation warfare, chemical-laced fire trenches, and suicide commando units. In actuality, none of these tactics were ever used effectively against Blue Diamond (either because of Iraqi inability or the unexpected rapidity of the US ground attack.)

The status of the South Rumaylah oilfield infrastructure became a daily obsession for the watching Division. As visibility of these strategic objectives continued to increase, conflicting reports of whether or not they were wired for destruction were received almost daily. The Division received reports from several seemingly reliable sources that indicated that the oilfield infrastructure was not rigged for destruction. Other sources, just as reliable, referenced the arrival of Regime forces, explosives smuggled across the Syrian border, or conventional defenses as imminent risks to the infrastructure. Every intelligence report was carefully scrutinized for indications of Iraqi mischief in the oilfields.

At this time, there were also no indications that any Iraqi units had caught-on to the complex capitulation instructions that were formulated for them by staffs at US higher headquarters. Although individual and mass surrenders were anticipated, there were no reports that any organized Iraqi units planned to lay down their arms or coordinate actions with the Coalition. Multiple source reporting did indicate a continued rise in desertion rates among Iraqi units, and gave indications that some Iraqi forces were planning to surrender at earliest opportunity. In many units, officers began to punish soldiers for wearing civilian clothes under their uniforms. In other units, the officers demanded bribes to allow soldiers to remain in a leave status, and took to wearing civilian clothes themselves. Regular Army units reportedly had to be locked out of their barracks in order to get them to conduct night training. Even the Republican Guard was not immune to morale problems, as the rank and file of many RG units were fearful that they would receive 'special attention' from the US air campaign. Yet, despite the morale problems, the Regime reportedly maintained tight control over the military through intimidation and fear. Stories abounded of Regime execution squads murdering uncooperative or insufficiently motivated soldiers. Fear of Regime reprisals appeared to be the most significant motivator for keeping military forces in-line, and the Division assessed that many Iraqis would fight only out of their fear of their own Regime.

During February, the Iraqi high command was attempting to execute the elements of some defensive plan, but it was difficult to ascertain exactly what that plan was. Small-scale deployments of battalions and battery-sized artillery units were noted in the vicinity of Al Basrah and Al Faw. In early January, the RA 14th and 18th Infantry Divisions showed indications they were deploying from their Al-Amarah garrisons to the Al Faw peninsula and the Al Qurnah regions, drawing the attention of US leaders at all levels. The deployment of the IV Corps infantry to strategic defensive ground was logical, especially given their limited mobility. If the Iraqis did not make strategic moves during this permissive environment (prior to the commencement of US air strikes) they would probably lose their chance. Given the penchant of Saddam Hussein for using the international press as his own intelligence tool, the Division was not surprised to note that the Iraqi artillery and infantry units began to appear in the Al Faw peninsula just a few days after the international press reported a planned British amphibious assault in this location.

It was not just the Iraqi RA that was moving. The Division noted throughout January that the Regime was reinforcing the southern zone just across the border from the Marines. This reinforcement included elements of the RG, the Mujahadin E Kalq (an Iranian dissident group beholden to the Iraqi Regime), and Fedeyeen troops. These forces had much higher loyalty to the Regime, and were likely sent to bolster the stamina of the RA in zone. The internal security situation in the south appeared to be deteriorating, with anti-Regime opposition seething among the Shia inhabitants. The Regime needed to do something about the security environment in the South, especially in the Al Basrah and An Nasiriyah areas. The Regime forces were reportedly spending 75% of their effort controlling internal dissent, and only 25% preparing for a ground invasion. Baath party officials were relocating to schools and mosques, and the military

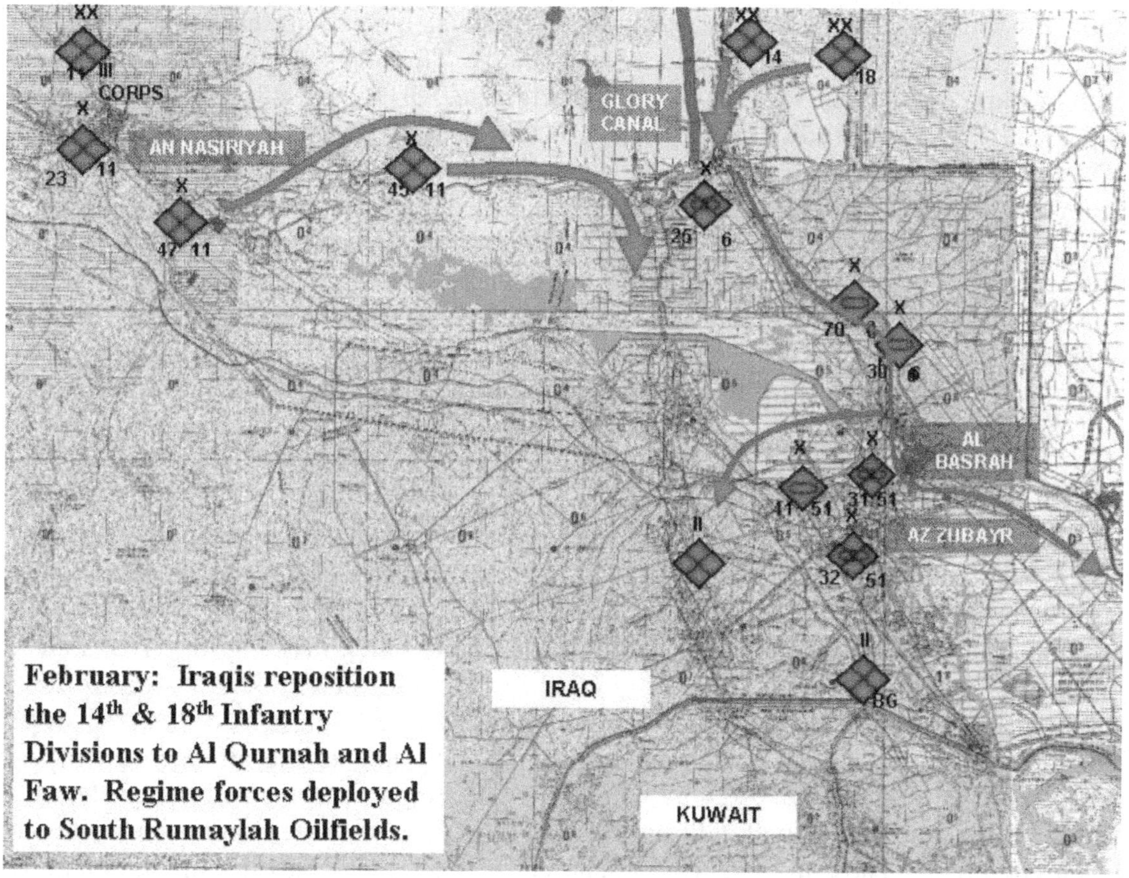

In late February, the Iraqis began to reposition Regular Army forces in the South, but there was still no discernable coordinated defensive plan.

was enduring more frequent officer purges. This context helped explain much of the Iraqi's apparent inaction. Even with the presence of US forces in Kuwait, the Regime likely thought that a ground invasion was not inevitable. Perhaps the Regime knew it could not prevail on the ground, so was taking every precaution to ensure that the mere threat of an invasion did not precipitate a popular uprising. If so, many military forces in southern Iraq were tied down in the vicinity of their garrisons because that is where their internal security mission was most pressing. As the Regime shored-up its internal security situation, it began to threaten tribal leaders, military commanders, and the southern populace with retaliation for any failure to provide a vigorous defense against the Americans. The Republican Guard also continued to give every indication that it would give battle closer to the Regime centers in Baghdad and Al Kut.

The BM-21 Multiple Rocket Launcher System was a highly mobile Iraqi missile system that could rapidly mass fires.

On 2 February, Tariq Aziz, the Iraqi Foreign Minister, promised a massive 'zero hour' missile attack if the US invaded Iraq. Regime statements in the public press also included references to missile counterattacks against US forces in Kuwait. Later in the month, indicators of the redeployment of Surface-to-Surface Missiles to southern Iraq appeared, lending credibility to this threat. It was possible the Iraqis were preparing to offset their poor conventional defensive capability by using preemptive chemical attacks against US forces. The combination of an increased chemical threat, and the possible repositioning of potential delivery systems within range of the Division's assembly areas was received with some attention. Later, the Division was to detect even more specific threats, and would drive the targeting cycle against these missile systems. The Marines of the Division remained alert, but now steeled themselves for the possibility that they might have to fight their way up to the border under the long-range artillery of the enemy.

Developments in Fire Support Planning

The Division became familiar with the 'killbox' system, a geographical reference established by the Combined Forces Air Component Commander (CFACC). The Killbox Interdiction (KI) reference grid system was defined by drawing parallel lines along the 00-minute and 30-minute lines of latitude and longitude, dividing the AO into a patchwork of squares approximately 55 km on a side. Each killbox was further subdivided into nine smaller sections approximately 18 km on a side, referred to as 'keypads' (from the practice of designating them according to the arrangement of the numbers on a telephone keypad.) A number of fire support planners advanced the idea that killboxes made the Battlefield Coordination Line (BCL) obsolete on the modern non-linear battlefield. These officers argued that the BCL was not necessary in this situation, since the MEF could open and close killboxes wherever it chose to focus or deny the fires of the MAW, including Close Air Support (CAS). This concept was generally referred to as the 'KIck-CAS' CONOPS, taking its name from CENTCOM's concept of operations for the employment of KI and CAS. To facilitate the permissive engagement of high-payoff targets with air short of the BCL, the Division's planners turned to killboxes to meet the requirement.

With killboxes as a centerpiece of its fire support coordination plan, the Division became very comfortable with them. For example, many FACs who did not receive the Special Operating Instructions (SPINS) of the Air Tasking Order (ATO) used killbox corners as geographic references for the control of CAS, instead of more conventional measures like holding areas and contact points. In fact, the general familiarity with this system eventually rose to the point that regimental air officers contacted the DASC directly to open up killboxes in their zones short of the BCL, posing a new battlespace management challenge to the Division FSCC. Fires planners came up with a prioritized target list (eventually numbering about seventy targets), which was planned to be pasted into the final hours of the last pre-hostility Operation Southern Watch ATO. This would allow for a planned sequence of shaping fires, even if a short decision timeline did not allow for wholesale ATO rewriting. MAW planners later indicated that they would need about eight hours to service these targets. In the case of a simultaneous A-day and G-day, the MEF would be prepared to make up for the deficit of shaping by using CAS sorties to prep the battlefield. Accordingly, the MAW planned to 'spike' to generate a high volume of CAS sorties during the first ATO of the Operation. As with maneuver, timing was critical because of the need to achieve tactical surprise and simultaneity of

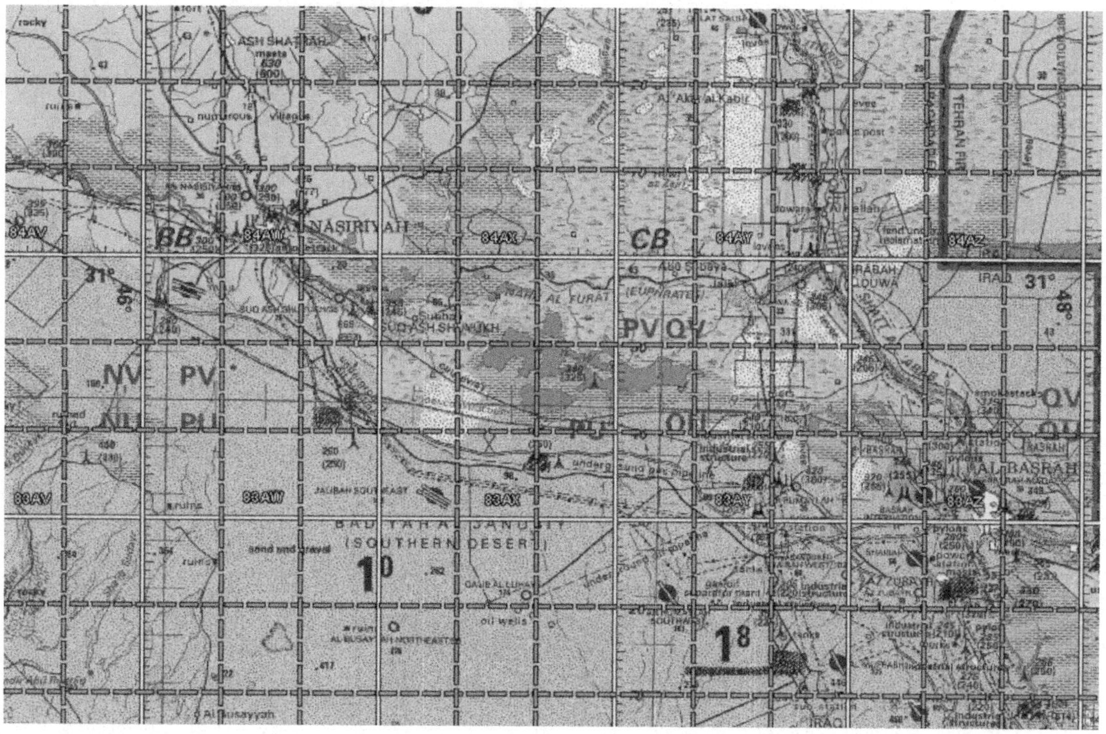

The 'Killbox' Grid System divided the battlespace into a two-letter grid system, with subdivisions similar to a numeric keypad. The grid was used to standardize target reference information.

action across the CFLCC objectives east and west of 1st Marine Division's zone.

The entire targeting process was significantly hampered by the thousands of sites placed on the Restricted Target List (RTL) at the theater level, based on general characterization alone. This meant that many Iraqi combat units and tactical objectives were protected by a somewhat arbitrary restriction on shaping fires. Some of the most important Division shaping objectives had been placed on the RTL by this nation-wide generalized approach, with no coordination with tactical units or apparent thought to specified objectives. The Targeting Guidance Working Group, and the detailed planning that followed, allowed the early clearance of Division shaping targets on the RTL, and prepared the Division well to adapt swiftly to a rapidly changing situation as the final hours counted down to G-day.

Operational Planning Update – The Highway 1 Corridor

On 9 February, the Division was to benefit from a fortuitous change in battlefield geometry. The unfinished Highway 1 corridor had previously been in the V Corps zone, just to the west of the 1st Marine Division zone. When complete, Highway 1 was going to be Iraq's Baghdad to Basrah expressway. During the previous summer, the Division had explored the possibility of using this battlespace to enable the envelopment of Al

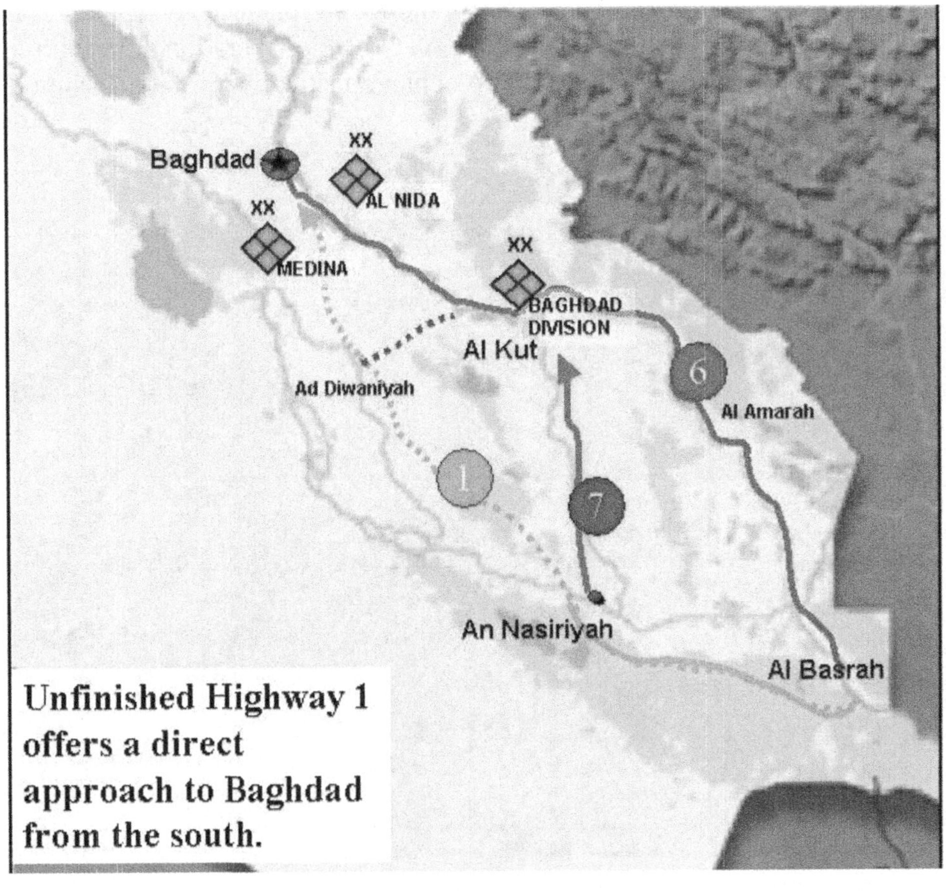

Highway 1 offered a better route to threaten Baghdad from the south in support of the main effort. The boundary shift that gave this unfinished highway to the Division was a critical factor in the successful attack.

Kut, but had been rebuffed by CFLCC. The Army plan had changed, and the Army V Corps planned to shift much of its combat power to the western attack corridors in their zone, leaving this major avenue of approach unused on their eastern flank. Out of concern for the main effort's open flank, CFLCC offered the Highway 1 corridor and the surrounding battlespace to the MEF. The MEF planners solicited the opinion of the Division, as it was not known if the incomplete highway was suitable as a maneuver corridor. The Division planners were dumbstruck at their good fortune. Where the Division scheme of maneuver had been previously dependent on a tenuous combination of goat-paths, canal roads, and cross-country marches, the Division was being offered a well-developed highway bed, large sections of which were paved. Using JSTARS Moving Target Indicators, recent U-2 imagery, HUMINT reporting and map studies, the G-2's Sergeant David Auwen, Sergeant Christopher Cloutier, and Lance Corporal Marcellus Moore did an exhaustive study to independently verify the viability of the Highway 1 corridor. They found it to be well-suited for a maneuver force. Iraqi civilian traffic was known to use the route, and it seemed to offer solid trafficability for military vehicles. Better still, the route could be improved to carry large-scale logistics traffic with some effort. With the recommendation of the G-2, the Division requested and received the Highway 1 corridor battlespace.

Based on this intelligence and new battlespace, the Division reconvened the OPT and briefed the CG on 11 February. There was an immediate recognition among the Division's tactical planners that this avenue of approach could significantly improve the Division's speed to Baghdad. Opening the Highway 1 corridor would also allow the Division to disperse in its maneuver to the Tigris over multiple avenues of approach; presenting less of a target for WMD, and reducing reliance on a single route (Highway 7). The CG made it clear that his ideal was for each RCT commander to have his own attack route. In addition to offering additional maneuver flexibility, a portion of Highway 1 north of Ad Diwaniyah had even been turned into an airstrip by the Iraqis, complete with a control tower and a collapsible railing in the median. This airstrip, the Sheikh Hantush highway airstrip (usually shortened to 'Hantush' or 'Hantush airstrip') was in a perfect position to serve as a KC-130 serviced FARP. If the airstrip could be seized and defended, it would provide a critical refueling point for forces before they crossed the Tigris. The result would be the Division getting across the Tigris with tanks nearly full, ready to push on to Baghdad.

There was another advantage to Highway 1, made clear when the Division Intelligence planner mapped out the detailed laydown of the outer defenses of Baghdad. This analysis revealed an interesting gap in the enemy's assessed indirect fire support plan. The Iraqis had pulled the Al Nida Division tight up against the Diyala River, blocking the routes to Baghdad from the east. At the same time, the Baghdad RG Division had gone to ground in Al Kut, establishing a positional defense primarily oriented to the east and south. In between, there was a 50-kilometer gap between the indirect fire range fans of the two defensive positions. At this stage in the planning, the Iraqis had not yet attempted to close that gap. The 'seam' between these defensive belts would make the Baghdad RG Division even more vulnerable to Blue Diamond's planned enveloping attack, and now the Highway 1 corridor offered an ideal way to exploit it.

Highway 27, which branched off Highway 1 near the Hantush airfield, offered a narrow, paved route that ran all the way to the Tigris River Bridge in An Numaniyah.

Operational Planning Update - Running the Seam

The Division planners immediately went to work to exploit the opportunity of running the seam between the enemy's defensive belts. The Highway 1 corridor stabbed straight at the underside of Baghdad, perhaps opening up an attack route for the Division to arrive in Baghdad even faster than previously thought possible. The Iraqis did not have any positional defenses blocking this route. In the enemy's eyes, the unfinished Highway 1 roadbed probably did not appear to be a credible attack corridor for American forces (similar to many US assessments.) The narrow Highway 27, slicing to the northeast between the defensive belts of the enemy, probably seemed even less credible, as it went in a direction away from Baghdad. For the Division, the opportunity to attack an unprepared enemy from an unexpected direction seemed too good to pass up. Feinting an attack up Highway 1 would at least fix the Medina Division in support of US V Corps (the CFLCC main effort). The plan offered better support to the main effort, and an opportunity to achieve tactical surprise. By using Highway 1, the Division could satisfy the CG's intent of offering alternative corridors for the RCT commanders, and could use the 'fix and bypass' tactic against the Baghdad Republican Guard Division in Al Kut.

The Blue Diamond CG and G-3 review the operational plan in light of the changing enemy situation on the ground.

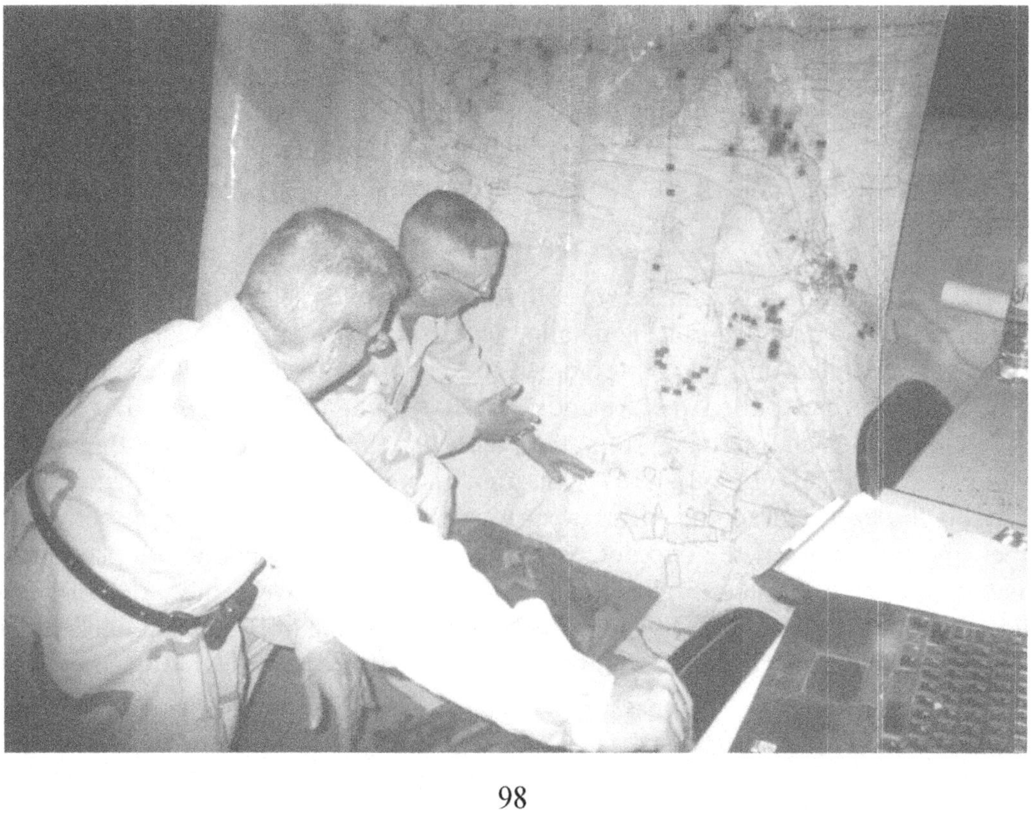

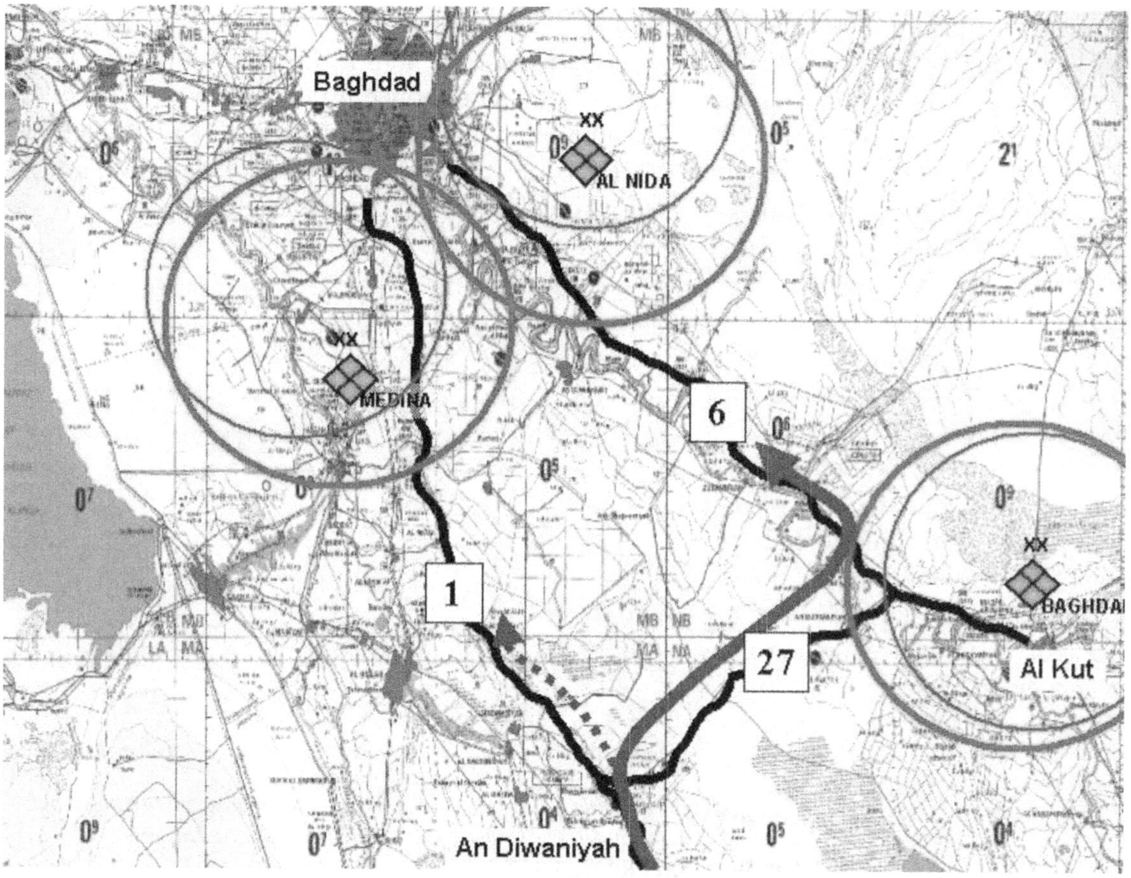

The Enemy's defensive belts had left a seam in indirect fire weapons systems coverage. Blue Diamond recognized and exploited this seam to get across the Tigris quickly with minimal casualties.

Blue Diamond modified the attack plan to take advantage of the opportunity. RCT-1, attacking up the Highway 7 corridor, would telegraph the Coalition's intent to move up the expected avenue of approach from An Nasiriyah to Al Kut. This would reinforce the Baghdad Division's predisposition to defend Al Kut to the south, effectively fixing this Division in place (The Baghdad Division was also prepared to defend against an expected Coalition advance up Highway 6 from the east.) By attacking this division by fires, it would drive the Iraqis to return fire with their own indirect fire systems, opening them up to detection by 11th Marines Counter-Battery Radar, and subsequent destruction by 3d MAW. While this occurred, RCT-5 would continue the attack up the unfinished Highway 1 to the limits of the Division's battlespace in order to pose a threat to Baghdad from the south. This would aid the main effort by forcing the Iraqis to commit forces to defend against the Marines' advance, making these troops unavailable to defend further west where the main American attack would come. The Iraqis were not expecting a major advance up the unfinished highway, and had made no discernible preparations for its defense. The attack from this unexpectedly large force would have the effect of fixing the Medina RG Armored Division should it try to mass against the US V Corps. LAR would reinforce this deception by continuing to press north along Highway 1 beyond the point where the rest of the Division would turn to move against the Baghdad RG Infantry Division and cross the Tigris River. RCT-5 would make the turn to the northeast on

Highway 27, and secure crossing sites over the Tigris. RCT-7 would then pass through RCT-5 and also 'run the seam' between the indirect fire fans of the Iraqi defensive belts. Resupply would be conducted near the Hantush airfield once it was opened for use by 3[d] MAW. Once across the River, RCT-7 would be in position to turn the Al Kut defenses to the west, and destroy the Baghdad RG Division. The Division would rapidly mass its combat power along Highway 6 and be in a position to drive-on to Baghdad.

While still in Kuwait, Division communications Marines work with a variety of newly fielded systems, including the SMART-T.

Command and Control (C2) of the Division presented the planners with a different set of challenges. With two RCTs moving rapidly up Highway 1, and a third RCT moving along the Highway 7 axis, there was concern in some quarters that the Division would not be able to maintain C2 of all three RCTs over extended operating distances as far as 100 km. The G-6 conducted extensive terrain studies and developed a detailed plan for retransmission sites for all Division priority 1 radio nets (Intelligence, Fires, and Maneuver.) This analysis ensured that the reach of the Division's line-of-sight communication capabilities was maximized. The Division also supplemented these nets by utilizing HF Radio, UHF Tactical Satellite radios, SMART-Ts, Blue Force Trackers, and Iridium satellite telephones to maintain communications with the RCT CPs moving up the two routes. After doing the 'impossible' for so long that it had become routine, the Division communicators set themselves to the task of laying out the C2 architecture that would allow the Division to keep on its pace to Baghdad. Approximately one month later, these communicators would have to deal with the wrath of Mother Nature as well, as they successfully implemented the plan on the battlefield.

Operational Planning Update – The Eastern An Nasiriyah Bridges

Using both the Highway 1 and Highway 7 corridors north of the Euphrates was possible, but the chokepoint at the single Highway 1 bridge site west of An Nasiriyah would seriously impede the Division's tempo. The forces bunched on the approaches to this 'hourglass' chokepoint would also be a tempting target for the enemy's employment of chemical weapons. On the north side of the Euphrates, natural factors conspired against any cross-corridor movement. There were no connecting routes between Highway 1 and Highway 7 until the Division had maneuvered all the way to Ad Diwaniyah, 150 km to the north. There, Highway 17 connected the two major avenues of approach, including a critical bridge over the Saddam Canal, which flowed between the two highways. Additionally, there would be additional friction and congestion generated if the Division conducted a major Forward Passage Of Lines (FPOL) near the An Nasiriyah area, a cost to speed that the Division was not willing to pay. To rely exclusively on the Highway 1 bridge for the entire Division's maneuver would be a gamble.

An exhaustive search had found no suitable routes over the Euphrates to the east of An Nasiriyah. The Division planners revisited the earlier work that had been done on the bridges in An Nasiriyah itself, and found that the easternmost bridges in the city offered a direct avenue of approach across the Euphrates River and Saddam Canal, and directly onto the Highway 7 corridor. Geographically, this was an obvious choice and General Conway agreed. If the Division could secure these bridges intact, and the urban environment permitted, this was a preferred route for the attack force using Highway 7. As the Division understood the intent of the US Army's 3d Infantry Division at the time, the Army would clear the An Nasiriyah urban area south of the Euphrates in their defeat of the Iraqi 11th Infantry Division and their seizure of the western Highway 1 bridge. If this occurred, the enemy in An Nasiriyah would likely be subdued enough to allow Division combat forces to pass over the eastern bridges in the city. The Division G-6 worked with the MEF G-6 to ensure that the frequencies, nets, and call signs of the US 3d

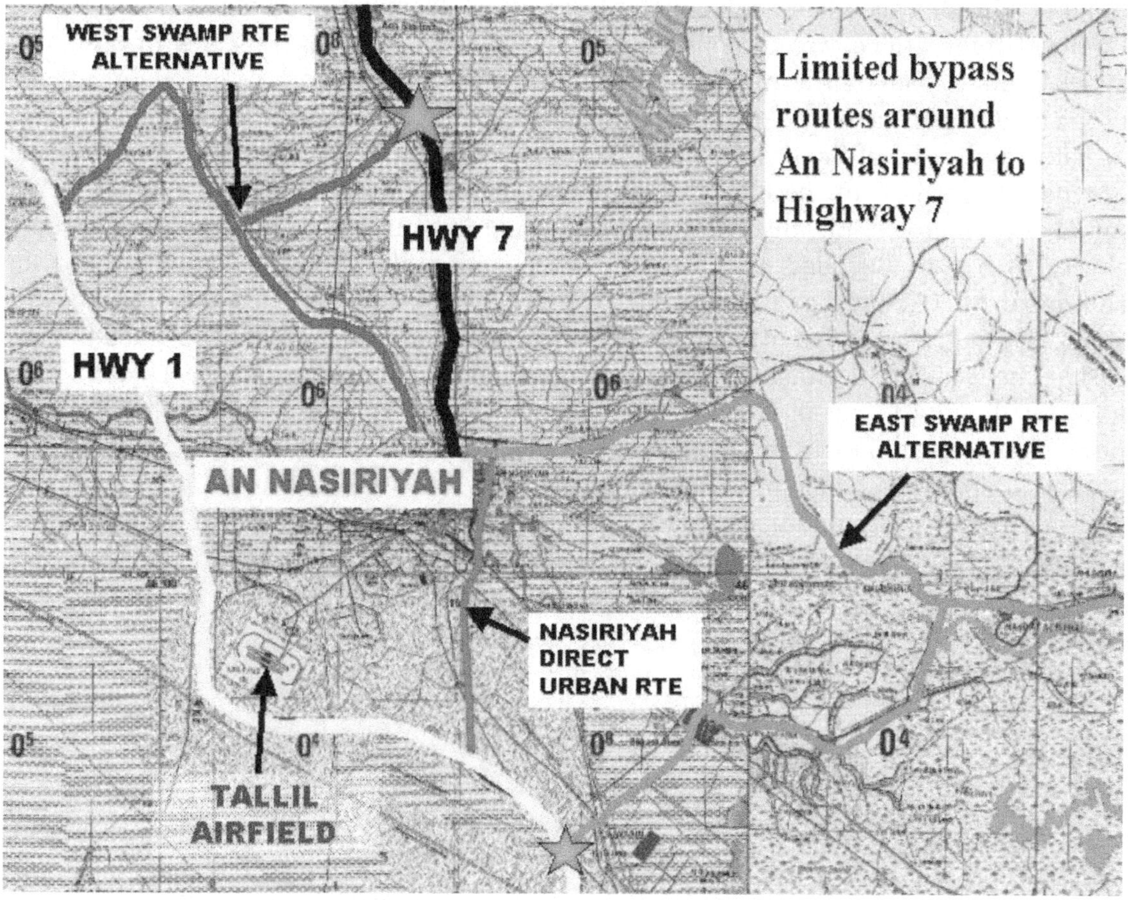

The swampy ground along the Euphrates on either side of An Nasiriyah left little room for maneuver. The most direct route would be along the eastern edge of the urban area, but this would entail greater risk.

Infantry Division were pre-planned in order to conduct detailed coordination on the ground. A decision point and decision criteria were inserted into the Division plan to allow for rapid exploitation of the eastern bridges if circumstances allowed. If the securing of the city was delayed, the decision could be made based on the level of risk. Task Force Tarawa, a MEB-sized force with a full headquarters staff, under General Richard Natonski, was given the mission of securing the crossings in An Nasiriyah and stabilizing the urban area. The Division would carefully coordinate its passage through An Nasiriyah with this force.

Operational Planning Update – Envelopment of Baghdad and the Outer Cordon

In late February, the Division was asked to consider a different role in the Baghdad fight. In this scenario, the Division was to participate in the establishment of an outer cordon around Baghdad instead of a direct assault to secure the city. By the proposed plan, forces from an outer cordon would contain Regime forces within, and conduct lightening strikes into the urban area. The fight for Baghdad would still belong to V Corps, but the Division would isolate the city from the east and north. This would prevent interference with US actions in Baghdad by Iraqi elements outside of the city, and deny the enemy the ability to withdraw from Baghdad to make another last-ditch stand at Tikrit. This scheme ran counter to the Blue Diamond operational vision for the Baghdad fight, which was to continue the attack in an unrelenting series of sustained attacks, rather than a series of raids. Raiding from and returning to a cordon outside the urban area might encourage the enemy by allowing him to remain in control of the field of battle at the end of the action, forcing US forces to repeatedly re-attack over the same ground. Regardless of the urban strategy eventually pursued, plans were made for an initial cordon around the city. Key to establishing this outer cordon was the Division's ability to bypass Baghdad to the east. Weeks prior, the Division had requested an additional MCIA study of routes from Al Kut to Baghdad. Termed the 'Hook Two' study, the routes presented offered some limited opportunities to envelop Baghdad to the east. The planners decided to recommend that one RCT cross over the Diyala River north of the city to put in blocking positions on both Highway 2 and Highway 5, the principle escape routes from Baghdad. Another RCT, attacking along the Highway 6 avenue of approach, would provide uninterrupted pressure on the Baghdad defenses from the south, setting the conditions for the third RCT to attack into the urban area using the friendly corridors along Saddam City. The Division continued to develop this preliminary plan, but waited for higher headquarters to define what role Blue Diamond would have in the Baghdad fight.

Division Rehearsals – Imaging the Plan

Even before Division forces began to flow into Theater, the Division had embarked on an aggressive plan to rehearse for the coming fight. The CG's intent was to 'image' the Division through the fight in a number of ways. Training events like the pre-deployment briefings had accomplished personal imaging. Intelligence Roadshow briefings had provided operational imaging. An important next step of the process, however, was to image the Division through some of the complex tactical maneuvers that

The Blue Diamond principal staff members review the details of the plan while still in Camp Commando, Kuwait.

would be required. Some creative mechanisms were developed. From a detailed terrain model at Twenty-nine Palms to plastic Lego blocks in a Camp Pendleton parking lot; from computer simulation to colored unit T-shirts in a Kuwaiti desert sandtable; the Marines of the Division put their collective focus into preparing for the fight. These rehearsals and combat preparations continued even as forces flowed into theater. In some cases, units had to participate by VTC. With force flow and deployment concerns nearing resolution, the Division refocused on shattering the Iraqi Army's cohesion. With the same momentum they were determined to achieve against the Iraqis, the Division entered their final rehearsal period.

Imaging the Marines through the first five days of combat was an important component of victory, and one of the CG's personal priorities.

Division Rehearsals – September CPX

One of the last live-fire training events the Division would have collectively, was a Command Post Exercise (CPX) conducted in Twenty-nine Palms in September 2002. This was a difficult exercise to conduct, as the Division was seemingly overwhelmed by the preparatory requirements associated with deploying the Division to war. In fact, since Seventh Marine Regiment was in the final preparation phase for their upcoming Combined Arms Exercise (CAX) 1 and 2, they were represented by First Tank Battalion in the CPX. This four-day exercise was crucial, however, in finalizing procedures and building confidence in the Blue Diamond team.

The September 2002 Command Post Exercise was an opportunity for the many new members of the Division and Regimental staffs to exercise Division Command and Control in a live fire environment. Many of the procedures refined here in Twenty-nine Palms would later be used to crush the enemy in Iraq.

The objectives of the CPX were in line with the planning efforts to date. The G-2 and G-7 produced a fictitious enemy scenario that exactly mirrored the Iraqi threat and some of the challenges the Division's attack would pose. Previous Division exercises had concentrated on integration of fires, maneuver, and intelligence. This CPX added command and control rehearsals to these other tactical tasks. One of the more significant of these objectives was the reconfiguration of the Division CP. Division-level C2 is a difficult thing under the best of circumstances. Now the Division would have to prepare to conduct a rapid attack over multiple corridors at distances exceeding 600 kilometers from their initial LD. Innovative thinking was required to lighten the lift requirement of the Division Forward and Division Main CPs, and to modify the Division Jump CP to meet the expectations of the new CG.

Marines from Headquarters Battalion rehearse personal and vehicle NBC decontamination procedures as part of the September CPX.

Colonel Ben Saylor, the Division Chief of Staff, and Master Gunnery Sergeant Bennett Woods, the Division Operations Chief, took the lead in redefining the CPs. Over time, and with the added presence of dozens of liaison officers from a variety of units, these had grown into large beasts, with hundreds of vehicles. The CP, with all its supporting and associated attachments was too cumbersome and too difficult to move quickly and efficiently. Worse than its physical size, however, was a loss of focus on the tactical action at hand. Colonel Saylor believed the CP could be reconfigured in order to restore a 'killing focus' to the actions of the Division Staff. Through hours of drills and rehearsals, he worked mightily to refocus the Division CP on supporting combat operations. He coordinated with the key staff members to establish a Main CP configuration whose centerpiece was a U-shaped watch floor, oriented on a series of shared situational displays. The term "Killing U" was coined to describe the setup inside the Combat Operations Center, a term that reflected the primacy of shaping the enemy as a mission of the Division staff. The Chief continually stressed that if the Division Staff was doing its job, the RCTs would have a much less costly close fight. Knowing the MEF was focused even further out in the shaping fight, the Division fight knitted the deep and close actions within the MEF to create a single battle.

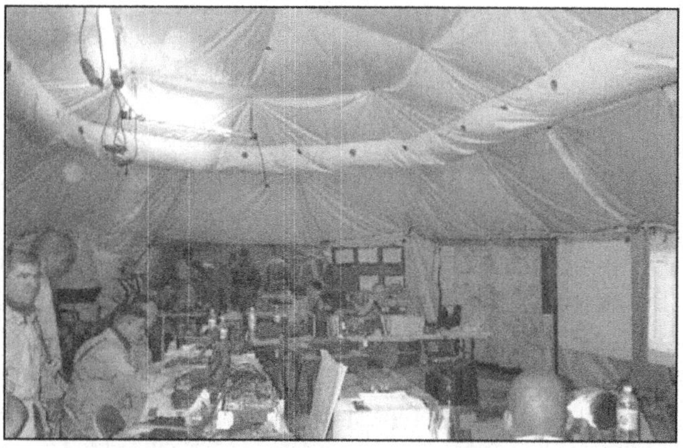

The 'Killing U' of the Division COC was formulated and rehearsed during peacetime training events. This repeated training resulted in a smoothly functioning COC in combat only months later.

The Division CP also began a concerted focus on getting lighter. Every vehicle and function was analyzed to see if it could be shifted to the Division Support Area, or eliminated altogether. The G-6 did yeoman's work defining a communications architecture that would wholly support the Division's movement over vast distances, with widely separated maneuver elements. The passage of control between the Division Main and Forward CPs was practiced repeatedly, as this would be critical in maintaining the momentum of the Division's attack to Baghdad. As a first test of the SMART-T in field operations, the CPX was a success. Reliable data communications were established to regimental level using these new systems, building confidence and demonstrating a reliability and ease of use that would be crucial to combat operations. The exercise reinforced the need to acquire more off-the-shelf communications equipment to supplement the aging 'green gear' the Division currently possessed. The use of secure Iridium phones, for example, was developed and exercised as a backup communications method, and the requirement to purchase many more Iridium satellite telephones was identified.

In the final analysis, the Division and regimental staffs were overwhelmingly positive in their assessment of the time spent during this training. Perhaps most important was the confidence building that began to occur. Coming right after the summer personnel rotation cycle, all of the staffs had new players serving in critical billets. The opportunity for these new members of the Division to gather around the sand table and walk through the operation under the tutelage of the CG was important. The Division used the opportunity to image the assembled staffs through the fight once again, ensuring a solid understanding of what remained to be done. This exercise had also tested of the metal of the Division and regimental staffs, and the CG was pleased with the results. This newly focused team was up to the challenges of enemy and terrain it would face only months later.

Division Rehearsals – Division MOUT Seminar

The Iraqis had made a considerable amount of noise in the press regarding their intent to conduct an urban-centric defense. While it was difficult to separate Iraqi rhetoric from reality, urban combat would certainly be one of the realities of the coming fight. The threatened Iraqi plan was to draw attacking forces into the cities, where they could lessen the technological and fires superiority advantage of the Coalition. The possibility of this course of action had strategic implications as well. Collateral damage, civilian casualties, and the loss of Coalition lives could create negative feelings at home in the US and abroad. The Iraqis had, no doubt, studied the results of the American fight in Somalia, as well as the Russian experience in Chechnya. These recent experiences stood as a reminder that individual Marines, well-trained and well equipped for an urban fight, would make a critical difference to the Division's success on the Iraqi urban battlefield. Accordingly, the Division hosted a Military Operations on Urban Terrain (MOUT) Seminar in December 2002 to consider the implications of the urban fight for the Division and supporting units.

The I MEF plan to bypass urban areas, developed by Lieutenant General Hagee and embraced by Lieutenant General Conway, would help to avoid some of the worst implications of the proposed Iraqi urban-centric defense. Still, fighting in the urban terrain of Baghdad and selected other cities would inevitably occur at some point in the war. Lieutenant Colonel Sean Magee, an enthusiastic reservist recently activated for this fight, worked diligently to set up the Division's MOUT Seminar on short notice. A fine example of Marine Reservists eager to get into the fight, McGee had volunteered for mobilization, and put together a superb seminar. This was an important preparatory event for the Division, and the CG flew back to California from Kuwait to attend. Much of the ambiguity about how the Division would plan to fight in Baghdad was resolved over the three days of the seminar, with nearly all regimental, battalion, and supporting commanders present. The event included the input of subject matter experts from various Marine Corps agencies including the Marine Corps Warfighting Lab (MCWL), Marine Aviation Weapons and Tactics Squadron (MAWTS-1), and MARCORSYSCOM. Attendees included staff members from the Division, 3rd MAW, and I MEF. RAND corporation advisors also attended. These experts gave the commanders and planners of both the Division and MEF insight into the specific tactics, techniques, and procedures that would allow them to win on the urban battlefield. The assembled Marines discussed the implications associated with MOUT, built on established Marine Corps doctrine, and developed mitigation strategies suited specifically for the urban areas in the Division's AO. Guest speakers, Generals Martin Steele and Norman Smith discussed their combat experiences in Hue City to help focus the commanders and staffs. The contributions of these veteran Marines and their combat insights were enormously helpful in preparing a new generation of Marines for this type of operation.

Immediately following this seminar, the Division's actions to acquire some of the necessary communications equipment for this fight picked up steam. The aging 'green gear' communications equipment the Division possessed was not suitable for modern urban combat, and the G-6 coordinated with the Warfighting Lab to develop communication alternatives. A major initiative pursued was to provide a more robust communications capability between the armored and infantry units, and one proposed solution was to procure British-made Personal Role Radios (PRRs). These personal communications devices would be used throughout the Division to augment and interface with the Division's tactical radios. Working in conjunction with Mr. Steve Fisher and the Marines of the Marine Corps Warfighting Lab, these radios were procured to augment the Division's communication assets. They would prove very useful for infantry squad support, operations in built up areas, security force operations, and convoy support. Generals Hailston and Conway wholeheartedly supported and resourced this critical buy. The G-6 received and distributed 3,443 radios throughout the Division. These PRRs received the universal acclaim of the Marines who used them.

The MOUT symposium dialogue could only be validated by practice in execution. Earlier training had occurred at the urban complex located at the old Victorville Air Base in Southern California. Two of the Division's maneuver battalions had put their Marines through the rigors of urban combat at this facility. This training, dubbed the Urban Combined Arms Exercise (UCAX), was highly effective, and much praised by the battalions who had the opportunity to attend. Major Mike Samarov, the Operations Officer from 3d Battalion, 7th Marines (in the vernacular of the day) described it, "UCAX is Money! Every scenario we trained to in Victorville, we ran into during combat operations." This theme was repeated by members of the Division who had participated or observed the training. Participants described the skepticism they originally had when faced with some of the scenarios presented in the UCAX. Militias competing for power on the streets, breakdown of civilian authority, unruly crowds at food distribution centers, car bombs, and snipers hiding in crowds were among many of the 'questionable' scenarios that were found later to directly mirror situations the Marines were faced with on the ground in Baghdad. The Division set out to provide this opportunity to all its maneuver battalions, and established a unit rotation. After sending through one additional battalion (2d Battalion, 5th Marines), the deployment order came and units began to head to Kuwait. Although the training evolution was cut short, those who participated in the UCAX generally felt that this was the best training they could have received for this fight.

Division Rehearsals - The Lego Drills

While the Division Forward CP made its way to Commando Camp in November, planning efforts continued in CONUS to ready the rest of the Division for deployment. The remaining G-3 staff members led by the Deputy G-3, Lieutenant Colonel Clarke Lethin, began to formulate the plan for the first tactical movement exercise that would affectionately become known as the 'Lego Drill'.

The tactical ground movement required during the coming fight presented a particular challenge to the Division. The restrictive terrain would limit travel to existing roads for much of the battlespace. Self-induced 'friction' caused just by moving the Division's over 5000 vehicles and more than 20,000 personnel over the restricted avenues of approach would be severe. Enemy actions could only make the situation worse. Since reliefs-in-place or convoy reconfiguration on the fly would induce even more friction, they would be nearly impossible. Careful attention had to be paid to organization for combat, convoy configuration, and battlefield circulation planning. If something was likely to be required at the point of attack, its place in the convoys would have to be established early on so it would be able to be employed forward rapidly. Engineering assets, NBC decontamination equipment, communications gear, artillery, and logistics support (to name only a few) all jockeyed for position in the column. The potential for traffic jams to seriously impact the Division's tempo, and to put lives at risk at bottled-up chokepoints, made rehearsals necessary. Doing these on a scale that would allow the Division to visualize the challenges was problematic.

The CG presented the staff with an idea to help the Division image its way through this challenge. Inspired by a local southern California attraction, the CG decided using toy Lego blocks to represent every vehicle in the Division would be a fine way to visualize the maneuver challenges. Using these plastic blocks on an appropriately scaled terrain model would give an appreciation to all the commanders of the challenges posed by moving massive numbers of vehicles down the limited roads available. Legos were available in a variety of sizes, and were color coded as well. Specific colors and sizes would be assigned to a unit's vehicles, then the blocks would be attached to corresponding plates that would represent each of the units of the Division. A scaled terrain model was built to replicate the major terrain features in southern Iraq to include the enemy order of battle, avenues of approach, objectives, and existing control measures. For the drill, each battalion walked through its scheme of maneuver, moving their Lego pieces in the proper sequence, limited by traffic flow and movement rates dictated by conditions. This allowed the staff, commanders, and operations officers to image through the sequence of movement for the Division and its supporting units, gain understanding of the battlespace, and provide a forum to validate the current task organization. It also identified potential friction points, and helped establish priorities of movement according to the desired sequence of each RCT. The drill also helped forecast when the Division would be ready to assume battlespace from the RCTs, and turn it over to a relieving force.

Heading the 'toy' department were Lieutenant Colonel Rob Whitters and Lieutenant Colonel Steve Santa Ana, who bought over 6,000 Lego blocks to depict most of the Division's vehicle pool, and those of supporting attachments. The G-3 and G-4 began to supervise the task of assembling the entire Divisional task organization, one vehicle at a time. Putting an entire Division together in this manner presented the opportunity for each RCT to gain in-depth understanding of the size and composition of their planned attachments, supporting units, and other forces on the battlefield. The RCTs gained a collective understanding of just how massive a full Marine Division was, and how drawn out the convoys would be over the limited road network through the muddy terrain of southern Iraq. This drill foreshadowed the complexity of the challenges posed by the Division's plan to run up an unexpected avenue of approach over terrain that even the enemy questioned for trafficability. The Lego Drill also cemented the tactical belief that forward passages of lines must be avoided at almost any cost in order to maintain tempo.

The first Lego Drill took place on 5 December 2002 in front of the Division CP. The palm-shaded parade deck in front of the 'White House' offered the perfect setting for the 1st Marine Division to rehearse yet another action against the Nation's foes, even if the security issues posed by the outdoor setting were difficult to deconflict. Accomplishing meaningful training while also practicing good Operational Security (OPSEC) were difficult missions to reconcile throughout the preparation and rehearsal phases. The G-2 constructed a scaled terrain model of southern Iraq using athletic field chalk, and commanders at all levels attended. The drill was successful in depicting the challenges the Division faced, and graphically demonstrated some of the particular choke points that would slow the Division's rapid advance. Questions on battlespace management were discussed and answered. For each traffic jam of plastic blocks, the training audience was forced to ask itself, "Who owns the battlespace? Where exactly are the boundaries? Who will act as the referee when fog and friction came into play?" The team-generated answers to these questions were to greatly facilitate the Division's movement when it executed this same plan with real M1A1s, AAVs, and MTVRs (7-ton trucks.)

A second Lego Drill was held on 10 January at Camp Pendleton's LZ Horno. The audience, much larger this time, included key personnel from the Division, its supporting units, and higher headquarters. By this time, the plan had gone through several refinements in Kuwait. The CG again flew back from the Forward CP in Kuwait to attend this important training event, and took Lieutenant Colonel Paul 'Uriah' Kennedy, the lead planner, with him. Together, they talked the audience through the most recent changes to the operational plan. The battlespace, task organization, and sequencing of the Opening Gambit had all been recently defined by the planning staffs in Kuwait, and were now briefed to the Division and regimental staffs at large. RCT commanders back-briefed the CG on their understanding of the scheme of maneuver. This second drill allowed for a common understanding of all the refinements that had taken place up to that point, and allowed members of supporting and adjacent staffs to visualize their own role in the coming fight.

These drills shaped the actions that would take place on the ground in Iraq in a few short months. For example, as the collective members of the MAGTF team talked through the sequence and the integration of assault support, Lieutenant Colonel Bradley Lowe, the Marine Aircraft Group-39 (MAG-39) Operations Officer, brought up the fact that the AH-1's were oriented to the east in their support of RCT-7's attack on the 51st Mechanized Division (should it hold out longer than anticipated and was still fighting at dawn.) This presented a particular challenge, as the pilots would be attacking into the morning sun. The Division's plan was changed accordingly, with the addition of a 3d LAR supporting attack from the southeast to open a corridor for the Cobras, allowing them to orient to the northwest, mitigating the blinding effects of the sun.

Although of great value, the scale of the Lego Drill was very small, making it difficult for spectators to identify the players. For detailed attack planning, the scale was appropriate. For visualizing the operation as a whole, however, it was confusing for the spectators once the players began to move their tiny pieces on the vast terrain model. In order to keep track of the different players there was a need to personalize each 'team' on the field. Captain Warren Cook, the CG's aide, came up with an idea to use color and number coded jerseys to represent the elements of each RCT. Each RCT 'team' would take to the field with its own color, and subordinate battalions would wear the number that corresponded to their battalion identifier (e.g. 3[d] Battalion, 5th Marines would wear RCT-5's color with the number 35.) A similar scheme was established for independent battalions and supporting elements. (Captain Cook would have a number of brilliant suggestions throughout planning and combat operations that would have a significant influence on the Division's success. Much more than an aide, he was a trusted advisor for the CG and members of the Staff.) The Division Staff hurriedly purchased colored T-shirts, and Lieutenant Colonel Lethin hand-carried them all the way to the Forward CP in Kuwait. The first 'Jersey Drill' would take place in LSA-Matilda during February.

NBC Rehearsals – This is for the Birds

Back in Kuwait, as part of the NBC rehearsal plan, every Marine and Sailor in the Division attended a mandatory Mission Oriented Protective Posture (MOPP) gear exchange and refresher NBC classes. These classes included instructions on the use of NBC detection equipment, types of agents, and instructions for establishing a decontamination site. This training took on a special seriousness with the Marines of the Division given the current intelligence assessments that assumed a chemical attack. Members of the Division paid close attention to this training, which 'demystified' the whole NBC environment. For perhaps the first time in Marine Corps history, Marines eagerly stood in line for a mandatory training event. The Marines of the Division fully expected to have to operate in a chemical environment, and were determined not to let it slow them down.

One risk the Division prepared for was the effect of petroleum-based vapors on the detection systems used by the Marine Corps. Although these systems provided accurate and timely agent confirmation, they also often gave false positive readings when in contact with certain vapors and liquids associated with petroleum products. Contact with petroleum products was a defining feature of the Division's battlespace, and this limitation was significant. Any false reading required forces to increase their MOPP level, pulling on masks and gloves that degraded the performance of the Marines. The Division needed a redundant, reliable chemical detection system that would mitigate false cueing, and give greater confidence to Marines standing down from a heightened alert posture.

The task of finding a solution for this deficiency fell on Chief Warrant Officer Stacy Jeambert, the Division NBC officer. A creative NBC professional, he decided that the assistance of an additional species was required for this weighty responsibility, and advocated the use of a 'Sentinel Species'. Like a canary in a coalmine, a living, breathing creature could be used to provide early indications of NBC agents in the air, even in a petroleum-contaminated environment. The Division expanded into the poultry business and purchased 200 chickens to aid in its NBC detection capability. This

Sentinel Species would likely show symptoms of chemical agents well before detection kits identified the threat.

The integration of these new 'marines' into the Division was difficult. Several of the birds disappeared immediately, followed by suspicious cooking fires in hidden corners of the camp. Others made their escape from the hands of inexperienced chicken wranglers and 'deserted' into the desert. Most, however, died before they got to see any combat action due to the extreme climatic conditions. Apparently, the birds sold to the Marines by an enterprising young Kuwaiti were civilized laying hens and not 'yard birds'. When exposed to the daily living conditions of the Marines, these civilized animals quickly broke down and died.

NBC Pigeons proved to be a 'Sentimental Species' that could survive the tough living conditions of the Marines.

The Division's next biological experiment was with a more robust and hardy species, the homing pigeon. These birds were fielded to the RCTs and separate battalions to reinforce their organic detection capabilities. The survival rate of these birds was much higher because they were used to living in the desert heat (and they had much less meat.) The Marines even took to naming the birds. 'Pidgeodo', under the watchful care of Corporal Joshua Hardy from Headquarters Battalion, even made it all the way to Saddam's palace in Tikrit before he was relieved of his duties and mustered out of the service. The Sentinel Species concept was validated, and instilled additional confidence in the Marines' ability to operate in a contaminated environment. The skepticism and humor with which this employment was met also provided valuable comic relief at a time of heightened tension. A sense of humor was a critical aspect of courage as the Marines prepared to attack into the unknown.

'Pideodo', an NBC Pigeon cared for by Corporal Joshua Hardy, survived all the way to Tikrit before being mustered out of the service.

In the event that units did come under chemical attack, they would conduct self-decontamination to the extent possible. If supporting assistance was required, the task would fall to the Operational Decontamination Unit, task organized under Lieutenant Colonel Rob Abbott's 3^d^ AA Battalion. In addition to the Division's organic NBC decontamination assets, 3^d^ AA Battalion had received the 101st Chemical Company (-) from the US Army.

This company specialized in equipment decontamination, and was well organized and equipped. Immediately upon being briefed on the scheme of maneuver and concept of operations, the 101st Chemical realized that they would also be expected to do decontamination of personnel as well. After mission analysis and detailed planning, they were told to be prepared to decon chemical casualties, evacuate chemical casualties from AAVs, and bury contaminated remains. Each new mission got a resounding "*Hooah*" and the company reviewed the procedures in the manuals or made it up as they went along. At one point the Soldiers of the 101st asked, "*While we are doing decon, who is providing our security?*" The response went, "*Well, in the Marine Corps we have a saying that every Marine is a rifleman...welcome aboard.*" The hard-chargers from the 101st gave another "*Hooah*" and embarked on a two-week crash course on infantry training, immediate action to convoy ambushes, and basic tactics. The Company Commander, Captain Kerr, and First Sergeant Edick were committed to mission accomplishment, and what the soldiers lacked in formal infantry training, they made up for in enthusiasm. A strong bond developed between the two services and by the end of the campaign, the Marines of the AA battalion were referring to the 101st Chemical as the '*Devil Doggies*'.

Another addition to the Marine's task organization was six M93A3 Nuclear, Biological, and Chemical Reconnaissance Systems, or 'Fox Vehicles'. Six decon sections from the Joint NBC Regiment of the 1st UK Division were also made available. These units would provide a robust and accurate detection capability. Collectively, the NBC team would be prepared to accurately detect any chemical or biological agents employed by the enemy and conduct operational decontamination as required.

Rehearsal of Concept – Jersey Drill Number One

Marines observe the first 'Jersey Drill' near LSA Matilda in central Kuwait. The Jersey Drill helped key participants in the Opening Gambit visualize their role in the attack.

The first comprehensive rehearsal for the Opening Gambit was held on 7 February in Camp Matilda, Kuwait. Positioned on a gently sloping hillside nestled between sandstone outcroppings, the natural arena was made into a miniaturized version of southern Iraq and Northern Kuwait, at a scale large enough for human players to walk through the steps of the attack. This was the first of the 'Jersey drill' series that was born from the Lego Drills conducted earlier in CONUS.

In stunned disbelief, the hard-charging terrain model detail, directed by Master Sergeant Mark Choates and Staff Sergeant Ken Costine, watched as the boundaries of the sand table were laid out, 80 meters on a side. The team quickly designated the project the 'mother of all terrain models' and commenced digging. The hard charging team spent several days at LSA-Matilda constructing a terrain model depicting major terrain features, objectives, enemy units, and control measures for the first 96 hours of the operation, down to the Kuwaiti berms and the electric fence. Representatives from all Division units attended, as well as key leaders from 3^{d} MAW, the MEG, 1st Force Reconnaissance Company, the UK 1st Armoured Division, Task Force Tarawa and the MEF.

Individual commanders represented their units by donning their respective color and number coded jerseys, and then talked through their initial organization, task, and purpose. The RCT commanders each talked through his scheme of maneuver while the subordinate battalion representatives simultaneously walked the terrain model. Through a series of exercise turns, the entire Opening Gambit was rehearsed. The exercise successfully oriented participants and attendees to the Division's scheme of maneuver and reinforced the Commander's intent.

The Jersey drill also helped the situational awareness of senior commanders. Major General Stalder, I MEF Deputy CG, Major General Amos from 3^{d} MAW, and Rear Admiral Kubic from the MEG each emphasized the critical importance of these rehearsal events. General Amos noted that he would have more assets than any other Wing in history to support I MEF ground forces, and was glad for the opportunity to gain situational awareness for his aircrews. Major General Stalder noted the importance of the drill in helping to integrate I MEF planning and efforts.

Training and Calibration in Kuwait

The Division continued a robust plan of combat training, weapons calibration, and weapons zeroing at the extensive ranges in western Kuwait. The Udairi range complex constituted a significant portion of the land area of western Kuwait. It took an integrated, detailed effort from division, regiment, and battalion staffs to coordinate training on these ranges. Deconflicting training with Army units and the Kuwaitis proved difficult at times, but with a common goal of combat readiness, all parties worked through the issues of airspace coordination, surface danger zones, and safety limits. In many cases, the RCTs coordinated to set up their own small arms ranges just north of the LSAs. This allowed the Marines the opportunity to fully integrate and test the new weapons systems they had recently received.

Marine snipers calibrate their weapons on hasty firing ranges set up near the LSAs in central Kuwait.

One of the significant issues for the Division was calibration of artillery and tank systems. 11th Marines conducted a full-court press to ensure accurate ammunition powder/projectile and muzzle velocity data was obtained, a calibration process that ensured the combinations of ammunition and firing conditions produced the most accurate artillery fires possible. Calibration required firing rounds of each expected type, with various propellants, and carefully measuring muzzle velocities using radar. This can be a tedious process, but is a critical requirement for first-round accuracy in war. Later, in combat, the Marines of the Division learned it was well worth the time and effort to achieve the last critical points of accuracy when required to silence hidden enemy mortars, urban ambushes, and other targets. These were often prosecuted at 'danger close' ranges, and with a high potential for friendly or civilian casualties.

For tankers, this calibration was equally important in order to ascertain the firing characteristics of the various combinations of ammunition and conditions. For both artillery and tank ammunition, however, the Division faced significant challenges in getting war reserve ammunition released for the calibration process. Because of concerns with expending war reserve stocks prior to combat and environmental concerns with depleted uranium ammunition expenditure, getting tank ammunition released for calibration proved difficult.

A similar issue arose with the Tube-launched Optically-tracked Wire-guided (TOW) antitank missiles. The Marines had conducted all their training using the TOW-IIA, with the TOW-IIB reserved for war stocks. When the Marines opened the crate to fire TOW-IIB's for the first time, there were a number of misses. This caused significant concern that the new missiles were defective in either software or hardware. Repeated firings could not determine the cause. Finally, CWO Charles Colleton, the RCT-5 Weapons Officer, discovered the cause of the misses. Because the Marines had only fired the TOW-IIA in training, they had not adjusted to the different aim point required when firing the TOW-IIB. Because the TOW-IIB uses a significantly different flight profile and attack mechanism, using the sighting method the Marines were used to with the TOW-IIA caused the missile to overfly the target. Although easily corrected once discovered, this incident was a warning of the risk of reserving new weapons for combat only and not fully integrating them into the Marine's experience. 'In case of war, break glass' weapons systems can have dangerous unintended consequences.

Each of the battalions trained their TOW Marines to overcome this potential gunnery error. Chief Warrant Officer Thomas Parks, the Weapons Officer for 1/4, trained two Marines, Sergeant David Smith and Sergeant David Kaufer, to serve as 'TOW Snipers'. These Marines received focused training on TOW-IIB gunnery and were allocated all of the battalion's TOW-IIB missiles, while the remaining TOW gunners were allocated TOW-IIA missiles. Iraqi troops and foreign paramilitaries would soon know firsthand the tactical prowess of these two Marines.

Marines test-fire a Javelin missile in Kuwait. This would be one of few opportunities many gunners would have to gain hands-on experience with this lethal new weapon.

The Combat Engineers knew that combined arms rehearsals were the key to any successful deliberate breaching operation. While the breach of the Kuwait border obstacles was expected to be lightly opposed, the breach of the Iraqi border obstacle system was more uncertain. To ensure the quality and thoroughness of the Division's preparations, CEB constructed a full-scale mock up of the Kuwaiti border obstacles near LSA-Ripper. Each Obstacle Clearing Detachment (OCD) performed combined-arms rehearsals through the breach lanes with their supported battalions. The rehearsals allowed the engineers and infantry to practice the complexities of a deliberate breach and work out timing and control issues prior to combat. During this period, the new Towed Assault Bridges (TABs) arrived in theater, and CEB and MARCORSYSCOM trained RCT-5 and RCT-7 on the employment of this new type of bridge over wet and dry gaps.

Marine Engineers created a full-scale mockup of the border obstacles to rehearse breaching and bridging techniques.

The training also had a material cost on men and machines. The Division carefully balanced the desire to train with the necessity of keeping all of the AAVs, tanks, and other equipment ready to reorient to the north. Maintaining the AAVs was a special challenge, as the Division was in the process of transitioning from the old AAV7A1 '*straight leg*' track, to the upgraded model, the 'RAM/RS'. Old 'straight leg' parts were no longer stocked in the supply system in anticipation of the completion of the upgrades.

Many of the vehicles contained on MPS ships, however, were the old models. For the year prior, 3[d] Assault Amphibian Battalion had systematically built up very large Pre Expended Bins (PEBs) of parts. These stocks of high usage, low cost items also included large stocks of oil, fuel, and air filters. The Amphibian Vehicle Test Branch (AVTB) also gave the AA battalion all of its 'straight leg' parts to add to their stocks a few months prior to their deployment. The battalion also stockpiled parts scrounged from AAVs going to the depot for upgrade. The battalion embarked and mobile loaded all these parts and expendables and brought them into theater. The Division immediately began low mileage mechanized infantry training in the local Kuwaiti desert, designed to limit wear and tear. Despite the low mileage, this hard training expectedly took a toll on the MPS amtracs that had sat in storage on ships at sea. The AAV Company and battalion maintenance crews worked often around the clock to keep all these vehicles in top running condition. The supply system for repair parts in Kuwait was not functioning anywhere near capacity and the only parts to be had were the ones the battalion had embarked themselves, or parts selectively interchanged off of dead AAVs. A month later, 3[d] AA Battalion's parts and supplies were down to only 10% remaining. Because of a lack of resupply in theater, a good share of the items they had painstakingly laid in for combat operations had been required just to ensure the Division was ready to cross the LD at a moment's notice. The repair parts challenge, identified early on, would continue to bedevil Division and CSS Marines in the months ahead.

The Desensitization Plan

The Division's actions in Kuwait would likely be observed and reported by a well-developed Iraqi intelligence network, including observers on *Jabal Sanam.* Nothing the Division did in Northern Kuwait north of the LSAs would go unobserved by the enemy. By the same token, this network was susceptible to deception, or saturation of the network with repeated reports of US forces repositioning in Kuwait. The G-2 laid it out quite simply, "Since we cannot hide, we must deceive." The Division crafted a 'desensitization' plan to accustom the Iraqis to large-scale ground movements in Kuwait. This desensitization plan became a significant theme of the Division's planned rehearsals prior to crossing the Line of Departure.

Since late February, the Division had been conducting rehearsals across Northern Kuwait. Combat elements of the Division in transit between the LSAs and Udairi range areas began to take detours through more northern areas. Small units of Marines became a regular presence near the Kuwaiti berms (At one point, this presence became too aggressive when the United Nations (UN) observers complained vigorously to the UN Headquarters, and alerted the international press. After this, Marines pursued their reconnaissance in a less overt manner.) The Division also took the opportunity to habituate the Iraqis to the presence of the Pioneer UAV, as missions began to fly over southern Iraq.

The Division continued to lean forward, and took advantage of each day of diplomatic maneuvering as an opportunity to train for military action. This training included several CP displacements for the Division Forward and Main CPs as well as all of the RCTs and separate Battalions. During combat operations, the Division planned to employ the Main and Forward CPs in a leapfrog manner, each displacing as far forward as possible in an attempt to extend communications ranges as the Division moved into and through Iraq. As each CP displaced in succession, it would travel just behind the lead element of the column to ensure Division C2 was continuous. The idea of 'desensitizing' the Iraqi border forces became an operational reality, as the Division conducted a CP displacement exercise on 23 February.

Although moving the CPs and maneuver units forward to the border area then moving back to the LSAs was designed to have the affect of desensitizing enemy observers both in Kuwait and on *Jabal Sanam,* these movements to and from the training areas also provided excellent opportunities for all CPs to refine their movement plans, convoy procedures, CP setups, communications support plans, communications equipment installations and operation. During this drill, the Division moved the Forward CP to a position in northern Kuwait. Once the Forward had assumed control, the Division Main CP displaced to a point even further to the north, at the outer limits of the range fans of the Iraqi's long-range tube artillery (and well within the line-of-sight from *Jabal Sanam.*) After operating for several days from this location, the Division Main retrograded to Camp Matilda. The Division Forward followed, returning to Camp Matilda on 1 March. As they departed these northern Kuwait areas, the Marines registered the terrain in their minds. All knew that the next time they stood on this ground, it would be to direct the Division across the LD into the face of the enemy.

Even as war tensions heightened, Marine Corps business continued as usual. Here, a promotion ceremony is held at the Division's Forward CP, very near the Iraqi border.

The Division Forward CP exercises in northern Kuwait as part of the 'Desensitization' plan. This isolated stretch of desert was referred to by the Marines as 'Camp Nowhere.'

As part of the desensitization plan, the Division developed an Emissions Control (EMCON) plan that ensured there would be no unexpectedly abrupt change in the level of the Division's communications during any portion of the rehearsal phase. Lieutenant Colonel Brian Barton, the Deputy G-6, drafted a series of EMCON standards. These EMCON Standards were designed to control the Division radio emissions to support the Division scheme of maneuver while also providing the ability to mask movements by limiting the Iraqi Signals Intelligence gathering ability. Prior to the commencement of hostilities, the Division also implemented the I MEF G-6 Information Assurance plan called 'River City'. This plan imposed three different security levels on non-secure data transmission within the MEF. The rotating restriction on the non-secure data was designed to keep a level of uncertainty on the network as a primary defense against hackers from outside of the Division and I MEF.

Final Preparation Efforts - Intelligence

As a result of the successful teaming effort with 3d MAW at the Jersey drills, another successful teaming effort between the Wing and Division was born. Major General Amos had been a tremendous advocate for the Division to date, and the employment of VMU was no exception. Through the strong support of Gen Amos, the VMU employment scheme was finalized, with both VMU's flying in direct support of the Division as they seized their initial objectives. As a result of the shared vision, a special relationship was established between the Division and both of the Marine Corps Pioneer Unmanned Aerial Vehicle (UAV) squadrons (VMU-1 and VMU-2.) These squadrons would be operating in direct support of the Division for much of the operation, leapfrogging forward to fly off improved airstrips, paved roads, and even dirt strips. The cooperation and aggressive spirit of the VMU's were to play an important role in supporting the Division fight. The Division G-2 convened a special UAV planning conference, where the VMU staffs and the Division G-2 were able to devise a support scheme that included the attachment of Remote Receive Teams (RRTs) directly to two of the Division's maneuver RCTs and the Division CP. This placed an additional burden on the G-6 to establish an unplanned communications link to the VMU Squadrons in order to tie them into the Division's voice and data systems, as they would typically be too far forward to receive support from 3^d MAW during combat operations. For the Opening Gambit, this allowed for a direct support relationship between each attacking RCT and a supporting VMU squadron. An airfield was identified near Camp Matilda in central Kuwait that would allow the Pioneers to cover the Division's initial objectives. An airfield was identified in Camp Matilda in central Kuwait that would allow the Pioneers to cover the Division's initial objectives. In March, permission was granted for UAV operations over Iraq, and the VMUs began providing the Division with their first real-time look at the battlespace. Having both VMUs fly in support of the Opening Gambit was essential. The Division was willing to accept the corresponding gap in Pioneer coverage afterward, as the VMU squadrons displaced from the southern airfield to positions that could support the attack further to the north.

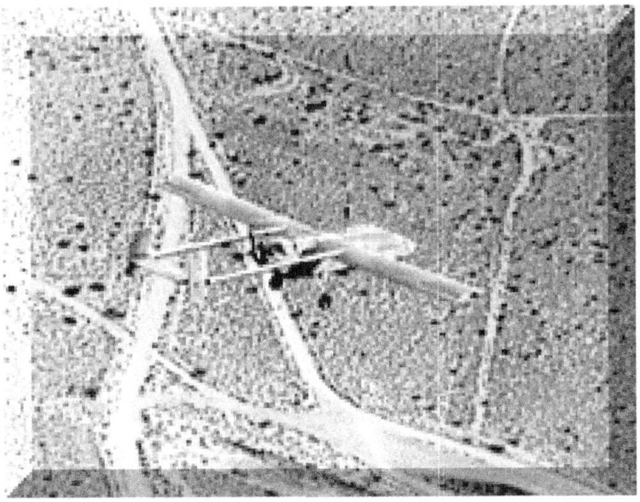

The Pioneer UAV system was used extensively in direct support of the Division. The VMU squadrons gave the Division its first look at the battlespace in real time.

Final Preparation Efforts - Force Flow

As February turned to March, units anxiously awaited the arrival of Follow On Echelon (FOE) shipping to round out their combat requirements. The Division had received most of its combat equipment from MPS ships, amphibious shipping, or inter-theater airlift, but critical components remained on FOE shipping. The FOE was carried on a mix of US Naval Service (USNS) and contracted ships, and included much-needed 1st and 4th Marine Division communications equipment, the balance of weapon systems, and a large portion of unit logistics trains. RCT 1, in particular, looked forward to the arrival of the AAVs that would mechanize two of their battalions. Faced with the possibility that the FOE shipping would not arrive before G-day, Division planners redistributed medium lift trucks from RCT-5 and RCT-7 to motorize all of RCT-1's battalions. This belt tightening had a ripple effect, as the RCTs took most of the trucks from their logistics trains. The burden of carrying their now-stranded supplies fell to CSSG 11 who also awaited assets from the FOE. Colonel Pomfret's gung-ho team took this challenge in stride, just as they had the previous ones. Much to the relief of operators and logisticians across the Division, the arrival of the *USNS Soderman* on 12 March and the *MV Greenlake* on 17 March eased this burden by bringing over 140 AAVs and hundreds of trucks. The continuing diplomatic dialogue had provided time that the Marines took full advantage of. The much needed assets that these and other FOE ships brought enabled the Division to move itself in a single lift and supported the equipment requirements of multiple CPs operating independently of each other. This capability would be essential in the fight on the road to Baghdad, unleashing significant Division fighting power.

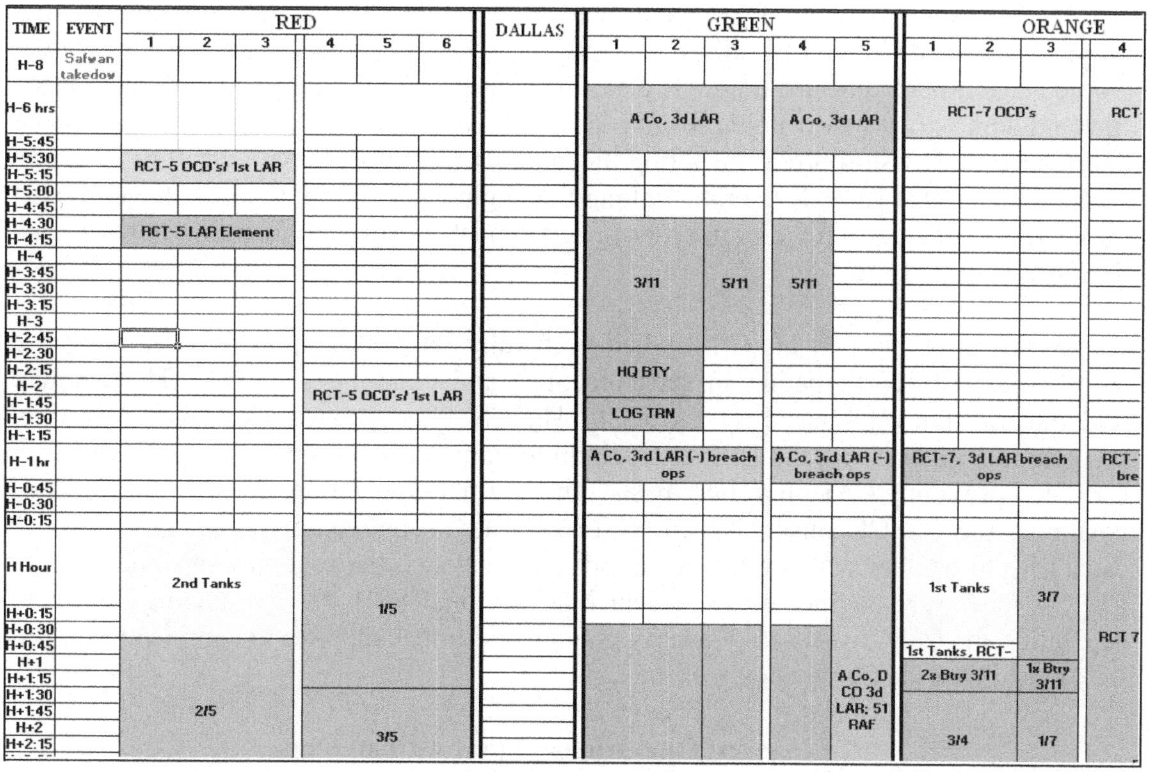

The G-4's Breach Lane Movement Matrix gave traffic management and logistics Marines a reliable indicator of the sequence of units crossing the breaches.

Final Preparation Efforts - Logistics

The G-4 joined in the staff effort to refine the first hours of the Opening Gambit as depicted on the Execution Checklist. Led by Major Chris Snyder, the G-4 produced the breach lane movement matrix for flow of units through the breach into Iraq. This single page document graphically depicted when, and by which route, every US and UK unit would cross the LD. It would prove especially useful for tactical movement control Marines who could know at a glance how to deconflict movements at their breach lane.

It became clear that fuel would be the Division's greatest concern in the first 48 hours, so the G-4 staff reviewed and updated its fuel culminating point analysis. Taking each unit's equipment density list, they projected fuel usage in order to more precisely place Rapid Refueling Points (RRP). With more than 5000 Division vehicles (plus those non-Division vehicles from units in direct support of the Division) consuming 220,000 gallons of fuel daily, precision was crucial. Set up in an assembly-line fashion, the RRPs were important as pre-staged fuel distribution points, positioned at critical points in the fight. RRPs would serve as a transfer points for fuel, water, food, and ammunition from the CSSG to the supporting CSSCs and their units. This rigorous analysis revealed the need to focus on RRP-7 near Jalibah airfield in order to sustain the Division's momentum during the Opening Gambit. CSSG-11 displayed the first of many instances of operational flexibility and readjusted their concept of support. The addition of Highway 1 to the Division's scheme of maneuver not only provided new maneuver opportunities, but also provided CSSG-11 options for locating and establishing RRPs that Iraqi forces would have a hard time locating. It also obviated the requirement to send half-loaded fuel and ammo trucks over unimproved roads that would have severely limited the speed of advance by Division forces. Shifting the anticipated Division main effort to Highway 1 also dictated that RCT-1, operating along the Highway 7 corridor, would have to exist on a radical 'logistics light' diet in order to accomplish its mission of fixing the Baghdad RG Division in Al Kut.

By this time, most of the CSSCs had been fully integrated into the RCT they were supporting, fully participating in staff meetings and planning sessions. The habitual relationships thus established were crucial. The CSSCs were also doing their own final preparations for crossing the LD. In addition to completing combat skills training, they had been providing CSS on a daily basis. In the final days before leaving the dispersal positions, the CSSC's ensured each RCT was fully supplied, giving them a 48-hour capability to operate without any re-supply. To enhance the logistics capability of the CSSCs, they were augmented by a detachment from CSSB-10. By adding this extra capability, the CSSCs were able to carry enough additional supplies to completely sustain the Division for the first 96 hours of the fight.

Current Operations - Final Adjustments

By late February, the Division was combat-capable, and prepared to respond to the President's direction, or to react if the Iraqis decided to initiate actions themselves. RCT-7 and RCT-5 were both combat capable, each having at least three mechanized/tank battalion maneuver elements (plus supporting capabilities.) RCT-1 was still steaming to Kuwait on schedule and 11th Marine Regiment had two battalions with five batteries each and its organic counter-fire capability.

When the offload of the FSS was completed on 13 March, the Division had the capability to mechanize two additional infantry battalions, and it gained the requisite mobility for both the Division's Main CP and 11th Marine Regiment's Main CP. The additional 'Cannon Cockers' allowed 11th Marines to reorganize into four battalions. Additionally, the US Army's 1st Field Artillery Detachment with two 'Firefinder' Q-37

At Camp Matilda, the CG, ADC, and Regimental Commanders make final preparations.

radars joined the 11th Marines team. By the first of March, RCT-1, 1st CEB, and 1st Reconnaissance Battalion had also become combat ready with the offload and integration of their remaining gear from the FSS West ships and the arrival of ATF West. The arrival of additional FOE shipping would continue to enhance the Division's combat capabilities. Piece by piece, the Division's combat power came together at the 11th hour. The Marines could always use more time to integrate and train with fully equipped units, but were now ready to fight when the call came.

When the offload of the FSS was completed on 13 March, the Division had the capability to mechanize two additional infantry battalions, and it gained the requisite mobility for both the Division's Main CP and 11th Marine Regiment's Main CP. The additional 'Cannon Cockers' allowed 11th Marines to reorganize into four battalions. Additionally, the US Army's 1st Field Artillery Detachment with two 'Firefinder' Q-37 radars joined the 11th Marines team. By the first of March, RCT-1, 1st CEB, and 1st Reconnaissance Battalion had also become combat ready with the offload and integration of their remaining gear from the FSS West ships and the arrival of ATF West. The arrival of additional FOE shipping would continue to enhance the Division's combat capabilities. Piece by piece, the Division's combat power came together at the 11th hour. The Marines could always use more time to integrate and train with fully equipped units, but were now ready to fight when the call came.

The Division's C2 plan was flexible enough to handle any number of situations that might arise. As the plan stood, the Division Forward CP would move forward and establish control immediately upon receipt of a Warning Order. The Forward CP would be in control while units pushed to their Dispersal Areas, allowing the Division Main CP to break down and displace. The Main CP would be pushed as far forward in northern Kuwait as possible, in a risk trade-off between communications coverage of the battlespace and the threat of enemy indirect fires. The ability of the Division Main to move to its battle position without drawing indirect fire would depend on the success of the desensitization plan. The Main CP would have time to 'burn-in' communications links and gain situational awareness while the Division moved from Dispersal Areas to Attack Positions. The Main CP would also coordinate actions against the *Jabal Sanam* outposts. The Division Forward CP would break down during this period and stage with RCT-7, prepared to move across the border quickly to establish the first Division CP in Iraq. If the desensitization plan had been effective, these movements would match the patterns the Iraqis had seen previously, and would not draw a large reaction from them. The Division adjusted its expected timing to the new attack dates.

Jersey Drill Two – Amphitheater in the Desert

The second Jersey drill was held just east of LSA-Matilda on 27 February. This drill was to be the capstone event for the planning efforts up to this point. The terrain model was to be expanded and the ground prepared such that it would have a greater 'amphitheater' effect. This time, no mere shovels and picks would do, and the Division's engineers prepared the ground with D-7 bulldozers. By pushing around a great deal of sand, they expertly constructed a multi-tiered amphitheater in the desert, complete with an angled sand table surface, 100 meters on a side, with plenty of seating around the perimeter. There was a 'vulture's row' for key leaders, allowing them to look down on

The second Jersey Drill was the capstone event for planning efforts. The scale of the terrain model required a major preparation effort on the part of Division engineers and intelligence Marines.

the battlefield from atop a high hill. The G-6 set up a sound system in the desert, and the G-2 and G-3 Marines prepared an even grander version of the 'mother of all terrain models' for the upcoming drill.

Participating and supporting units attended the second Jersey Drill, allowing everyone an opportunity to gain a clear understanding of the Division's intended scheme of maneuver.

Personnel from across the Division, higher, adjacent, and supporting units attended this capstone event. Lieutenant General Conway, Major General (UK) Brimms, Major General Amos, Major General Stalder, Rear Admiral Kubic, Brigadier General Natonski, and Brigadier General Usher were in attendance. The large audience included uniforms from almost every branch and service, including the UK coalition partners. The purpose of the drill was to ensure that commanders, supporting organizations, and Division battle staff had a clear understanding of the Division's intent. The task organization, battlefield geometry, and visualization of the sequence of events for the Opening Gambit were reviewed. The review allowed all units to visualize their role in the coming fight. Even with a planned sequence of events, the Division's plan remained agile enough to incorporate the expected last minute changes and timings that are always part of such a high stakes endeavor.

Lieutenant Colonel Bob Sinclair, the Division Current Operations Officer, coordinated the drill from start to finish in his usual enthusiastic and forceful manner. He and Lieutenant Colonel Groen from the G-2 conducted a tag-team briefing, covering the geographic laydown, expected enemy actions, and friendly scheme of maneuver timings for the entire Opening Gambit. Although the specific opening day timing was still unresolved, the Division used a nominal four-day separation between A-day and G-day for the drill scenario. Actions on the terrain model began in the Division's Dispersal Areas in northern Kuwait, and continued through the establishment of a bridgehead across the Euphrates River. Even though originally planned as a Division rehearsal, the drill was invaluable as it coordinated key players from throughout the MEF to the actions of the Opening Gambit.

The exercise complete, Lieutenant General Conway took the opportunity to address the assembled staffs. He remarked that the rehearsal was well done, and that he was glad all elements of the MEF had participated. Then, with an unexpected gravity in his voice, he announced that five days ago, the President had notified USCENTCOM that they were to be ready to attack into Iraq on order. The assembled crowd of professionals let the words sink in. This would be the last rehearsal. This marked the end of preparation and the beginning of execution. The MEF was going to war, and the Marines gathered here would be the ones to lead it.

Reducing the Border Obstacles

On 5 March, Kuwaiti engineers began the reduction of the border obstacles on the Kuwaiti side. Breaching commenced with the cutting of the first electrified fence, which was cut in three places by the end of the day. The work, conducted by KLF engineers and a civilian contractor, was conducted to look like routine fence maintenance scheduled for that time of year. As a precaution, however, MEF tasked the Division to provide a counterfire capability so that the engineers could withdraw if they were fired upon by Iraqi artillery. 11th Marines ordered 5/11 to displace to within five kilometers of the berm marking the UN Demilitarized Zone (DMZ), and to move the Regimental CP with two Q-46 and two Q-37 radars forward to provide a counterfire headquarters and target acquisition capability. The Division FSCC stood by to request aviation fires from 3d MAW. There was no Iraqi response the first day. The pace of work picked up, however, and soon the engineers were breaching the berm and filling in the tank ditches. The Division noted an increase in frequency of Iraqi patrols along the border, but still there was no belligerent response. The engineers worked in daylight only. Each day, Lieutenant Colonel Paul Cyr, dressed in a nondescript set of blue coveralls, went forward into the Demilitarized Zone (DMZ) to monitor the work and ensure that the engineers created the breaches where and how the Division desired. At the end of six days of work, twenty-eight breach lanes had been cut and the fence had been removed in twenty places. The engineers withdrew from the berm, but 11th Marines, now with 1/11 in position to provide counterbattery fires, remained forward.

The Enemy's Last Moves

Signs of Regime instability appeared daily as the fissures in the Regime's façade began to widen. Reports of Saddam's alleged efforts to seek asylum outside of Iraq had a demoralizing effect on the Iraqi army. The reports also raised interesting questions on the US side, as media pundits debated whether an attack would be necessary if Saddam were to step down voluntarily. The Division remained undistracted by the running political dialogue, and kept its attention on securing its military objectives quickly if the order came. The standard Regime tactics of shuffling officers between commands, threatening deserters with execution, and reinforcing conventional units with special Regime troops continued. Reportedly, any Iraqi caught with a Global Positioning System

(GPS) receiver or satellite telephone was subject to execution. In a nation-wide exercise in whistling in the dark, the Regime continued to tell its armed forces and its people that the US would not attack. Even with these false messages of assurance (written for domestic consumption) the Regime began to quietly pull forces off the 'Green Line' in the north. (The Green Line was the line of control between Kurdish and Iraqi forces in northern Iraq.) Later, after the Turkish Government had denied the US permission to stage an attack from Turkey as part of a 'Northern Option', this trend accelerated. Most notably, the Adnan Republican Guard Mechanized Infantry Division was pulled south to Tikrit and the northern outskirts of Baghdad by rail (thus gaining for itself an opportunity to meet the 1st Marine Division on a future battlefield.)

The Division's commanders and staff officers watched with disdain as the open international press began to publish some highly accurate reports of the US planned attack. Reports telegraphing the criticality of the South Rumaylah oilfields to the US plan were broadcast in the international press, and grimly debated by a series of pundits on national television. Having already observed the Iraqis reposition forces in response to media discussions of an amphibious assault, the Division waited, hopeful that the enemy would not react to these new reports by initiating the destruction of the oilfield infrastructure. From some of the forward staging areas in Kuwait, the routine glow of the Iraqi oil infrastructure from natural gas burn-off could be seen in the night sky. Division eyes watched vigilantly for any unexplained bright spots on the night horizon.

'Fire Trenches' were dug by the Iraqis. When filled with oil and set ablaze, they provided obscuration of defensive positions and artillery systems.

As if they were reading the Division's thoughts (or perhaps just reading the newspaper) signs of suspicious activity in the South Rumaylah oilfields increased. There were a number of isolated indicators of infantry units deploying to the area. Daily, the Division sent the Pioneer in a detailed search pattern of the area of the oil infrastructure. The vigilance was finally rewarded on a daylight mission in March, when artillery pieces were observed setting up near the GOSPs. Perhaps this was the long-awaited move into the oilfields. These reports were quickly followed by confirmation of conventional Battalion-sized units moving in and preparing defensive positions in the oilfield area. Large suspicious trenches were dug in scattered positions throughout, and the Iraqis were observed filling these trenches with oil, then lighting them on fire. Were these 'fire trenches' for obscuration alone? Was there some more insidious purpose for these fire trenches such as the release of chemical agents in an oil slurry? In the intelligence reports from higher, the reported size of the Iraqi unit in the oilfields quickly swelled from battalion to division size. Division intelligence personnel noted the presence of one, two, three, and eventually four artillery batteries in South Rumaylah, as well as the nearby presence of reinforcing artillery from III Corps. These reinforcing GHN-45's were particularly problematic, as they were known to be chemical-capable, and had repositioned to a point that allowed them to range all the way into Kuwait. With repeated sightings of BM-21 Multiple Rocket Launchers near Zubayr, the Iraqis finally seemed to be waking from their long slumber

Closer to Az Zubayr, a battalion-sized tank unit was observed establishing defensive positions near the Crown Jewel. Elements of the III Corps anti-tank missile battalion were observed near the Shuaybah barracks north of Az Zubayr. Reports of armored units moving south to defend closer to the border began to come in, and the Regime forces in Safwan were reportedly reinforced with heavy machine guns, and tanks. Almost daily, there was some report of major changes in the enemy disposition in southern Iraq, making it very difficult to differentiate fact from fiction. Reporting of massed armor defending near Safwan, and the reports of Republican Guard reinforcement of the southern Iraqi defenses could not be confirmed, even with repeated Pioneer and P-3 missions. These reports would continue to flow into the Division, even while crossing the Line of Departure, and were to cause several tense moments later in execution. The fog and friction of war continued to impact the Division, even at the doorstep of the much-vaunted information age. The theorized 'holy grail' of perfect clarity on the the battlefield had certainly not yet been achieved. In fact, the Division's information needs appeared to expand exponentially with each new piece of information received, in an endless and ever expanding cycle. The commanders throughout the Division took it all in stride and, armed with the CG's intent, adapted their plans as the clock wound down on the Iraqi Regime.

The Division was faced with a significant impact to the timing of the Opening Gambit. RCT-5 had planned to quickly seize the oil infrastructure against only limited resistance. Now that the enemy was present in large and growing numbers, this task might take longer than originally planned. The targeting implications of these forces occupying positions near critical oil infrastructure were also significant. The Gas Oil Separation Plants, in particular, had tremendous explosive potential that could easily be ignited by a careless artillery round. The Division sought and received the attachment of UK oilfield experts who could assist in the rapid but safe shutdown of the GOSPs. This would reduce the risk of catastrophic explosion to the Marines of RCT-5 who would seize them. Extricating the enemy from the South Rumaylah without blowing up the whole complex (and themselves) fell to RCT-5 as a primary challenge.

Additional indicators of SSMs now appeared in the South, and the Division's reaction was swift. The Division quickly nominated detected SSMs for immediate strike under the auspices of OSW. The enemy's missiles were in violation of UN restrictions that forbade certain Iraqi weapons systems south of the 32^d parallel. Specific indicators of Abibil-100 Missiles and launchers were detected in the Al Qurnah, Al Basrah, and Az Zubayr areas. The presence of these highly mobile launchers in range of the Division's

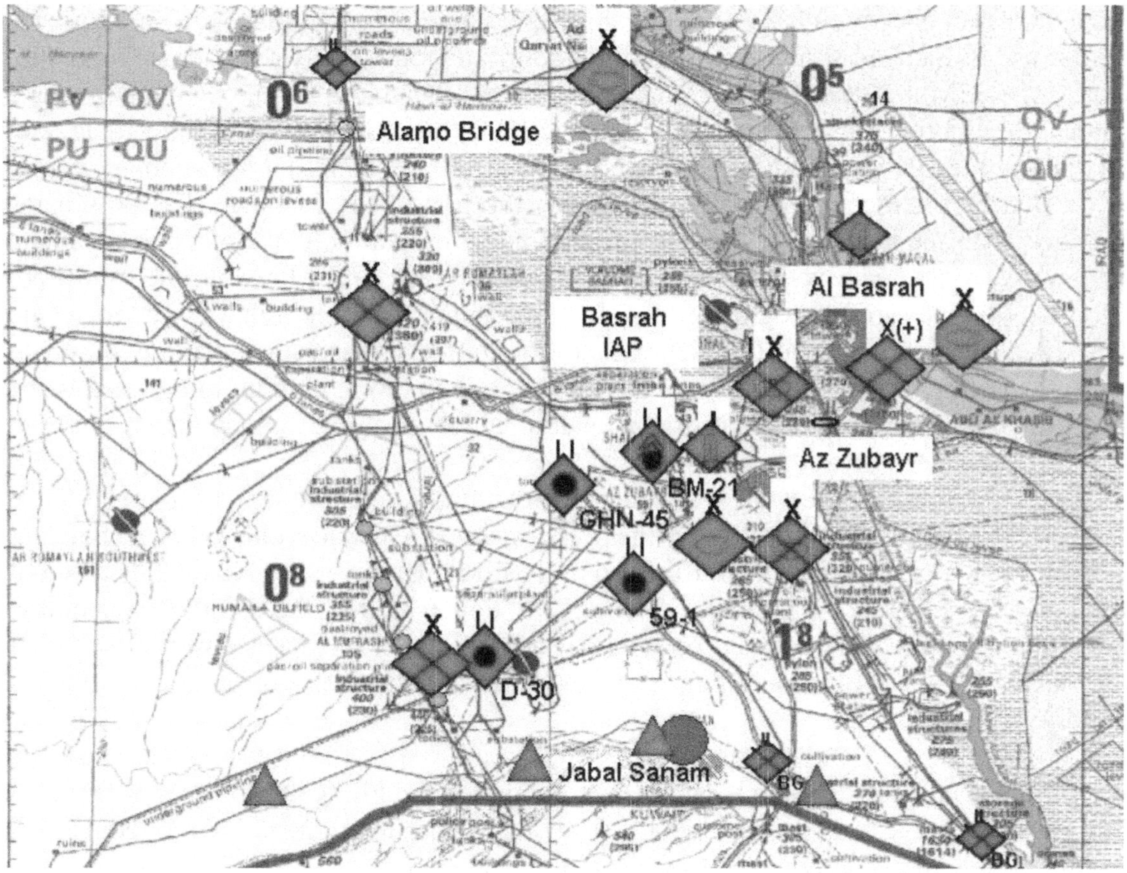

The Iraqi disposition in southern Iraq, as assessed just prior to the Division crossing the line of departure.

assembly areas marked a new threat. In a clear demonstration of the enemy's ability to outpace a ponderous deliberate targeting cycle, the detected systems usually hid or moved before Coalition aircraft were able to deliver bombs on target, and often reappeared nearby within days. This shift in targeting by OSW aircraft from anti-air defense capability to anti-offensive missile capability was a significant turning point. Balancing the elimination of threats to ground units in Kuwait with the risk of beginning a general air shaping campaign was to prove difficult. The Division waited grimly and patiently for attack orders, from within the range fan of the Iraqi missile systems. Something would have to change soon, as the Iraqis were gaining a capability that presented a threat to both Coalition forces and Kuwaiti civilians.

The Eve of Combat – The Stage is Set

As the diplomatic clock wound down, the entire world watched the Coalition forces poised to remove a ruthless dictator. Thanks to the Global Broadcast System, the Division CP was able to follow the political events, even from their advanced bases in Kuwait. UN weapons inspections had restarted in December of 2002, as Division forces were preparing to deploy. The inspectors noted the same lack of cooperation on the part of the Iraqis as previous inspectors, but had not yet been able to find any major 'smoking gun' that confirmed the presence of WMD. The Division watched as the UN inspectors visited suspect sites within its proposed zone and came up empty-handed. In February, the Iraqi's Al Samoud II missiles were declared in violation of UN restrictions, removing one of the SSM systems that the Division had been prepared to defend against. In early March, the US found itself opposed diplomatically by France, Germany, and Russia, and stymied in the UN Security Council. On the 7th of March, Chief UN weapons inspector Hans Blix reported continued intransigence on the part of the Regime. The politicians wrestled with the appropriate response to a dictator in open defiance of the UN for the past twelve years, who was known to be pursuing WMD, and who controlled the Iraqi people through a careful program of sadistic repression. On 16 March, the leaders of the US, UK, and Spain met in the Azores and committed themselves to taking action to remove the threat posed by Saddam. Diplomatic options were coming to a close, as the US sought, then abandoned, an attempt at additional UN Security Council action.

On 17 March, President Bush issued an ultimatum, giving Saddam 48 hours to leave Iraq or face his forced removal. In the wake of the President's ultimatum to Saddam Hussein, USCENTCOM declared Iraqi forces 'hostile'. The Division Main CP listened to the President's words. There was no longer any doubt. All of the months of planning, preparation, and rehearsal would now be put to the test. The Marines would be called once again to face the nation's foes and remove an armed threat, to stand and deliver a victory.

On the evening of 17 March, Blue Diamond issued the FRAGO for all units of the 1st Marine Division to move from the LSAs to their Dispersal Areas. Units confirmed receipt of the order, and prepared to move out.

The CG delivered this message to all the Marines and Sailors of the Division:

For decades, Saddam Hussein has tortured, imprisoned, raped and murdered the Iraqi people; invaded neighboring countries without provocation; and threatened the world with weapons of mass destruction. The time has come to end his reign of terror. On your young shoulders rest the hopes of mankind.

When I give you the word, together we will cross the Line of Departure, close with those forces that choose to fight, and destroy them. Our fight is not with the Iraqi people, nor is it with members of the Iraqi army who choose to surrender. While we will move swiftly and aggressively against those who resist, we will treat all others with decency, demonstrating chivalry and soldierly compassion for people who have endured a lifetime under Saddam's oppression.

Chemical attack, treachery, and use of innocent human shields can be expected, as can other unethical tactics. Take it all in stride. Be the hunter, not the hunted; never allow your unit to be caught with its guard down. Use good judgment and act in the best interests of our Nation.

You are part of the world's most feared and trusted force. Engage your brain before you engage your weapon. Share your courage with each other as we enter the uncertain terrain north of the Line of Departure. Keep faith in your comrades on your left and right and Marine Air overhead. Fight with a happy heart and a strong spirit.

For the mission's sake, our country's sake, and the sake of the men who carried the Division's colors in past battles – *who fought for life and never lost their nerve* – carry out your mission and *keep your honor clean.* Demonstrate to the world there is "No Better Friend, No Worse Enemy" than a U.S. Marine.

J.N. Mattis
Major General, U.S. Marines
Commanding

I've come to understand that in reality, war is nothing but nasty business. You can write poetry and plant flowers around it, but it's nothing but hard work and nasty business.

Chapter 4
The Opening Gambit

On the evening of 17 March, Saint Patrick's Day, the Division published a FRAGO directing the Division's units to move from their LSAs to Dispersal Areas (DAs). Early the next morning, President Bush delivered a speech that many in the Kuwaiti desert awoke early to hear. The time for diplomacy had run out, and the President gave the regime of Saddam Hussein 48 hours to depart Iraq or face military action. The LSAs came alive with activity. Final preparations were made to vehicles, equipment was remounted, additional chow was distributed, and water jugs were topped off. Because units had spent the past ten days living out of their packs, the Marines were prepared to move to the Dispersal Areas at a moment's notice.

With the 17 March order to move to DAs, the Division began a series of moves that would end in Baghdad and Tikrit. President Bush's ultimatum to the Iraqi Regime was followed by a flurry of activity on the political front, none more important than confirmation that the UK government had given its approval for its forces to participate. The execute order was gladly received by the 1st UK Armoured Division. The extraordinary partnership that the UK and Marine Divisions had established during planning was given the green light to continue into execution. US Central Command (USCENTCOM) issued the order declaring all Iraqi military and paramilitary forces to be hostile, and the wartime Rules of Engagement were put in place. The Division watched as the last of the UN Observer forces were withdrawn from the demilitarized zone along the border. As quickly as the UN Observers left, the Iraqis filled the vacated posts. There was now nothing between the two armies but a berm of sand and a ditch. The two sides eyed each other warily, each waiting for the other to make a move. Division actions near the border drew sporadic mortar fire from the Iraqis, and even the smallest action on the Iraqi side of the line drew a Coalition reaction. There was an inevitability to the coming combat now, and Blue Diamond was eager to get moving.

From the moment the Division issued the FRAGO, wheels were set in motion. A rapid tempo was important even in this first movement. The Division made a tempting target for an enemy artillery or missile attack while it moved out of its established LSAs. The Division moved quickly to establish the Main CP, get artillery fire-capable, and establish security zones ahead of the DAs. The first unit out of the LSA was the Division Forward CP, which immediately moved to its position in northern Kuwait. The 11th Marines CP, 1/11, and counterbattery radar teams were already in position. By the morning of 18 March, the Division Forward was up and running in northern Kuwait, and ready to assume 'Blue Diamond' control. (On the radio nets, the term 'Blue Diamond' served as the call-sign for the Division headquarters (Main or Forward) that currently had control of the fight. Passage of the call-sign accompanied passage of control of the Division. This way, it was transparent to subordinate elements as to whether the Division Forward or Division Main CP was running the fight.)

The Forward CP assumed control on 18 March, allowing the Main CP to tear down and displace to its battle position in northern Kuwait, just short of the range of enemy artillery units that could reach inside Kuwait. By the afternoon of 19 March, the Main CP was in its battle position far forward of the rest of the Division. From this carefully selected and practiced site, the Main CP would command and control the invasion of Iraq, sited for the best possible communications with widely dispersed maneuver elements. This positioning was fundamental to the uninterrupted tempo of the attack. 'Blue Diamond' was passed back to the Main CP, and the Marines began to track both enemy and friendly movements with a keen focus. The Forward CP, under Lieutenant Colonel Bob Sinclair, tore down and moved into its Attack Position (AP), ready to displace into Iraq behind the assault elements. For the Marines in the Division CP, the fight had begun.

Once the Division received the order, more than 20,000 Marines moved from LSA's to their DA's near the Iraqi border.

The DAs had been carefully chosen as an intermediate step that would not only keep units out of the line-of-sight of *Jabal Sanam*, but also would get them out of the fixed LSA locations, which likely had been targeted for enemy surface-to-surface missile fire. If the Iraqis were going to launch preemptive missile strikes or chemical attacks, the LSAs could easily be targeted. The DAs, located 10 kilometers or more outside the LSAs, were closer to the border but still beyond the range of Iraqi artillery systems. The Division was eager to fight, but did not want to precipitate a political event by making a provocative movement on the ground.

To ensure smooth movement of all Division units, 3^{d} AA Battalion (Reinforced by 2d and 4th AA Battalions, and a reserve company of MPs) established Traffic Management Control Points to guide units to the DAs. Strict unit movement control would enable the rapid tempo of the Division through critical chokepoints in the fight. The Division units in the LSAs began their movement to the north, continued in good order, and were in position on the afternoon of 18 March. Because of the pre-movement reconnaissance, these movements were accomplished with minimal 'friction'. The more than 20,000 Marines of the Division and their over 5,000 vehicles moved out of the LSAs and were consolidated within the DAs. Third MAW and supporting US Army Aviation helicopters were positioned with the Division and RCT CPs to provide CASEVAC and Command and Control (C2) birds.

The Division also began to move artillery into some of its intermediate Position Areas (PAs). Each direct support artillery battalion would push up to PAs that would support their respective RCTs. The positioning of artillery and counterbattery radars was key, as these weapons needed to be well forward to provide fires in support of breaching activities. Because these critical systems were so far forward, Alpha Company from 3^{d} LAR was assigned a guard mission to prevent enemy interference with the Division's artillery fire-support assets. The LAR Company also provided a measure of security for the Division Main CP, which was at risk in the unlikely event of an Iraqi attempt at a spoiling attack.

The carefully rehearsed Division scheme of maneuver continued from the Dispersal Areas with a movement to Attack Positions (APs) just short of the border obstacles. The APs were the final staging areas the assaulting elements would occupy prior to commencing the attack. The Division's scheme of maneuver had clearly identified the sequence of moving from LSAs to DAs to APs, but had left the time spent in each flexible. Each unit monitored the execution matrix containing the sequence of events, but was prepared to spend minutes, hours, or even days in the APs or DAs as battlefield conditions dictated. After leaving the APs, the attacking forces would exploit the breaches that the Kuwaitis had prepared through the border obstacles. There were a number of breach lanes across the Division's front. To the west, breach lanes 'Red' 1-6 supported RCT-5's attack. In the center, lanes 'Green' 1-5 supported 11th Marines and 3d LAR. On the eastern edge of the zone, breach lanes 'Orange' 1-5 supported RCT-7's attack near Safwan. By the plan, the Division Main CP would issue the command, and the Division would attack across the border in a simultaneous, unstoppable wave.

Endeavoring to keep the situational awareness of his commanders high, the CG conducted a final coordination/update conference to provide current operational timelines. The coordination meeting, held at a desolate, windy location, revealed no issues. The commanders were confident, and their units ready. Each passing moment now posed a risk to the oilfield infrastructure, and gave the Iraqis time to react to the attack they surely knew was coming. Although weather adversely impacted collections within the last 24 hours, the G-2 located one new Iraqi Type 59-1 artillery battery approximately 18 km north of Safwan, and an ASTROS multiple rocket launcher in the vicinity of Shuaybah barracks. Both positions were nominated and were targeted by OSW aircraft.

The CG and subordinate commanders meet one more time before crossing the Line of Departure.

The Iraqis Disposed to Defend

On the eve of battle, the Iraqis hunkered down in a largely set piece defense across the Division's zone. At the border, the Division faced screening and reconnaissance elements from the Iraqi border guards, Regime security services, and Baath militia. Human Intelligence had reported reinforcements of the defenses in the Safwan area, and a number of tanks from the 51st Mechanized Division had been pushed south down Highway 31. The border guard and Regime troops patrolled the border in isolated groups, supported by mortar fires. There were reports of hasty minefields emplaced in the wake of the withdrawing UN forces, especially to the north of

Jabal Sanam, and Iraqis were observed laying surface mines astride the few roads that led across the border from Kuwait.

To the west, the previous weeks had seen a full brigade of the Iraqi RA 18th Infantry Division move in to the previously undefended South Rumaylah oilfields, concentrated in the vicinity of GOSP 3 and the South Rumaylah intermediate pump station. A battalion of D-30 howitzers directly supported this brigade, with general support provided by at least two batteries of Type 59-1's and one battery of GHN-45's. This position appeared largely unsupported by other maneuver elements, however, and stood isolated in the oilfield infrastructure.

To the east, armored elements from the 51st Mechanized Division's 41st Brigade had been pushed south, forward of their garrison positions to defend along the Highways 8 and 31 avenues of approach in the vicinity of the Zubayr pumping station complex, or 'Crown Jewel'. Major elements of the 31st Brigade had withdrawn into the Al Basrah urban area, but at least one battalion remained in the vicinity of the Shuaybah barracks just north of Az Zubayr. Judging by its stationary equipment, major elements of the 32d Brigade likewise remained in the vicinity of garrison, with some screening elements pushed south down Highway 80. There had been a surprising lack of conventional preparation of obstacles, minefields, and integrated defensive positions on the part of this Division.

On the north side of the Shaat Al Basrah waterway, The 6th Armored Division (6th AD) remained dispersed on the approaches to Al Basrah from the east and north. The 30th Armored Brigade remained oriented toward the Iranian border and Al Faw, while the 70th Armored Brigade was poised north of the Saddam Canal in a position to counterattack to the west or south. The remaining brigade of the 6th AD, the 25th Mechanized, remained in defensive positions near Al Qurnah, guarding the Euphrates River crossing there. If the Iraqis were going to vigorously defend, the 6th AD would be a center of gravity for them. The 6th AD had older equipment, but had over 100 T-54/55 main battle tanks. The Iraqis could do little with conventional military forces to stop the 1st Marine Division's assault, but the 6th AD was their best bet to attempt to delay it.

The 11th Infantry Division (11th ID) remained spread across the marshes to the east of An Nasiriyah, and had reportedly sent one of its brigades to reinforce Al Basrah and the port of Umm Qasr. The Division G-2 had expected to see the 11th ID consolidate in the An Nasiriyah defenses, but they had not done so by the time the Division commenced its attack. By all appearances, the Iraqis had chosen to use the 11th ID to maintain control over the local Shiites, while other Regime forces defended the An Nasiriyah chokepoint. The city of An Nasiriyah was reportedly defended by at least a brigade of reinforcements from the north, composed of elements of the Republican Guard, Baath militia, and an unknown number of paramilitary fighters. Located at the waist of the 'hourglass' chokepoint, An Nasiriyah presented challenges to long term stabilization similar to the ones the Al Basrah area did. The Division hoped to avoid the enduring tasks that would accompany securing and stabilizing the An Nasiriyah urban area, leaving this mission to Task Force Tarawa.

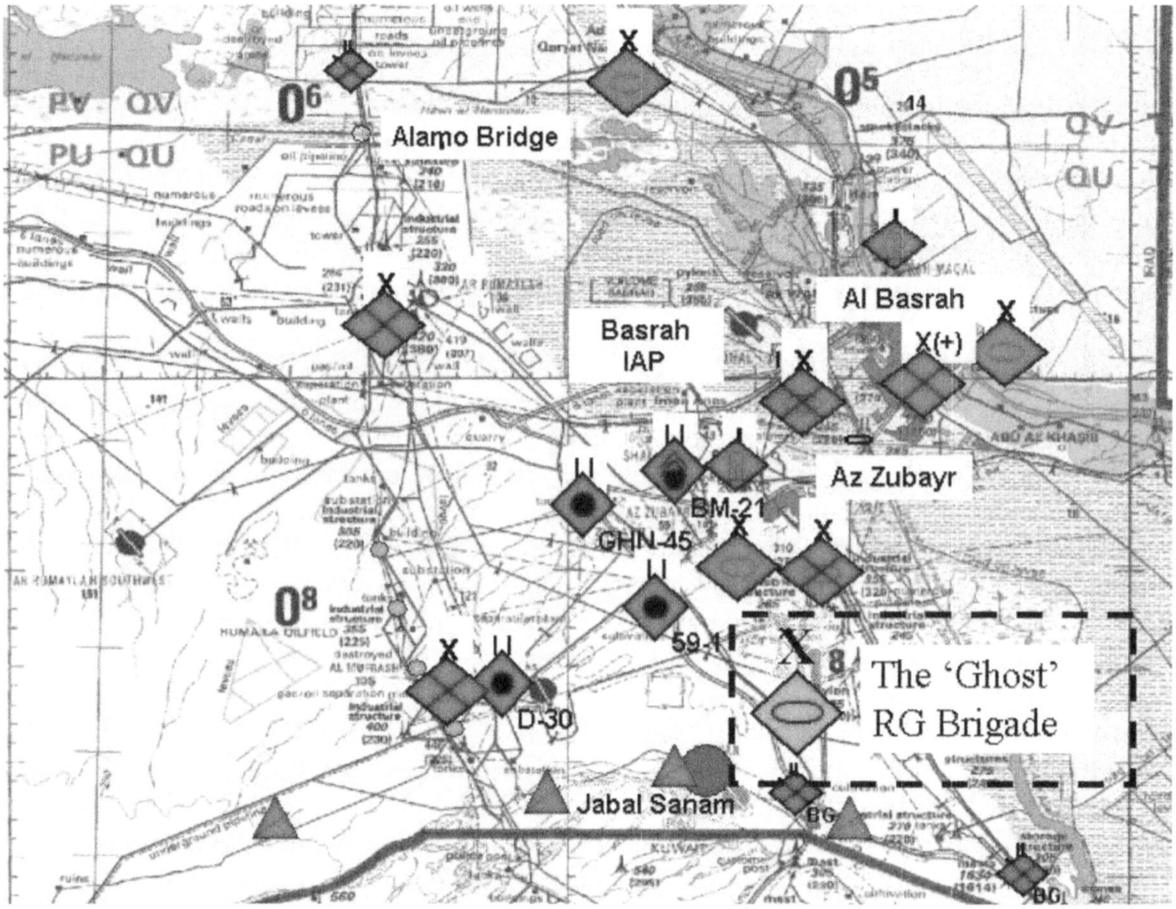

The Iraqi III Corps of the Regular Army Forces defended in the Division zone. Just before Blue Diamond commenced the attack, an Iraqi informant reported that a Republican Guard armored brigade had allegedly crept into position near the Safwan border crossing (green icon above.)

The Iraqi missile and rocket forces were still reported operating throughout the Division's zone, with at least a brigade's worth of launchers and missiles in position to range the border area and Kuwait. There had been no reduction in the number of reported sightings or imagery detections of these systems, and the words of the Regime's previous threats still rang in the Division's ears. The Regime had told the Iraqi people to be strong for 72 hours, and withstand the Coalition assault. The Regime promised this would give them time to launch a massive retaliation using its missile forces. The Iraqis had experienced some success to date in hiding these forces from coalition targeting, and all of these systems were possibly chemical-capable. The strong words were likely just more bluster on the part of the Regime hierarchy in order to buy time for their own escape, but the possibility of the employment of chemical weapons in the opening days of the war could not be dismissed out of hand or wished away.

19 March
Poised for Battle

On 19 March, as units topped off fuel and ammo in the DAs, selected elements of the Division were operating forward in security zones established to prevent enemy incursions across the border. To the west of Highway 80, 3[d] LAR Battalion prepared to tie in with the Brits on the right and with RCT-7 on the left. Alpha Company, 3[d] LAR, moved north to protect artillery and engineer elements operating along the border. Elements of 3[d] Battalion, 7th Marines (3/7), and 1st Tank Battalion took up positions near the southern Kuwaiti berm, approximately 5 km south of the border. To the west, 1st LAR Battalion established a security zone to the far western limits of the Division's battlespace. These units were to observe a number of interesting activities in the tiny space between two armies poised for battle.

That day, Major Andy Milburn, a plans officer with the RCT-7 staff, was accompanying an attached Psychological Operations (PsyOps) detachment from the US Army's 305th PSYOPS Company as they moved close to the border to broadcast a surrender message to the Iraqi armed forces. Major Milburn was interested to note the Iraqis actually grouping together to listen to the recorded message. At least some of the Iraqis were interested enough to listen, but certainly their commanders and Regime minders would not allow this sort of assembly. "I wonder why we haven't seen any indirect fires from these guys yet," he wondered aloud. There were apparently a few Iraqis who agreed and were not convinced by the PsyOps message, because he had no sooner said the words out loud than the Iraqis obliged the PsyOps team by lobbing mortar rounds at them. US Army Sergeant Thomas Stiffey, the team chief, was not amused. From that point on, Sergeant Stiffey was leery of traveling anywhere with Major Milburn.

That evening, President Bush's diplomatic ultimatum to Saddam Hussein expired. The Iraqi's had chosen to reject the ultimatum, and now faced the reality that the Regime would be removed by force. On the night of 19 March, US Navy ships in the Arabian Gulf launched over 35 cruise missiles at targets in Baghdad and other strategic sites. The successive activation of wartime Air Tasking Orders (ATOs) followed this strike. The initiation of the air campaign (A-day) had arrived, and the Division watched the firepower of 3[d] MAW begin to shape the positions of the 51st Mechanized Division and the units in the Rumaylah oilfields. In addition to the planned wartime ATOs, the Wing had also arranged eight hours of shaping which was planned to be 'cut and pasted' into the last OSW ATO. Aviation fires against ground targets commenced immediately, and took a significant toll against the enemy artillery. The G-2 and G-3 Fires sections were able to confirm the destruction of the troublesome battery of long-range GHN-45 artillery pieces by having the VMU fly the Pioneer over the position. As the Marines of the Division would later attest, these aviation fires had devastating effect on the Iraqi defenders, and began to impact the decision process of many Iraqis who were undecided about whether they should stay and fight the Americans or run away. The Division CG entered the Main CP and set MOPP 1. He approved 11th Marine's move north to the border area and their PAs. The artillery units quietly moved up under the cover of darkness.

The SCUD missile system is capable of delivering long range conventional and NBC munitions. Primary intelligence and fires tasks were to locate and destroy Iraqi missile systems.

Patriot missile batteries poised throughout northern and central Kuwait provided a multi-layer missile defense for cities, airfields, fixed facilities and maneuver forces staged in Kuwait.

As the Division sat poised waiting for the word to go, the Iraqis reiterated their intent to defy President Bush's ultimatum by launching volleys of long-range missiles at major US bases in Kuwait. The Iraqis chose to target major facilities at Camp Doha, Ali al Salem Air Base, and Al Jaber airbase, as well as Camp Commando, which housed the Division Support Area (and the MEF CP.)

The theater had established a 'SCUD-beeper' warning system to give early indication of a missile launch out of Iraq, and the beepers began to repeatedly sound off in the Division and RCT CPs. In each CP, the buzz of normal activity would cease whenever the beeper went off. Everyone would stop and look at the Senior Watch Officer (SWO) as he fumbled with the device then squinted to read the message on the tiny screen. This day, the beepers were going off at a cyclic rate, and every time a missile launch was reported, the Division upgraded to MOPP-2. The Iraqis were not going to get a 'free shot' at an unprepared Division. Almost every time, the alert was quickly followed by a report that Patriot batteries located strategically around northern Kuwait had destroyed the missiles. The initial tension soon subsided, and the Marines grew accustomed to the thought of an unseen, unheard missile battle taking place in the skies over their heads. Over the course of the day, only a few missiles made their way through the defensive umbrella, and all impacted in the open desert or splashed into the Gulf of Kuwait. With typical machismo, the Marines jeered the poor accuracy of the Iraqi missileers. One missile, a Chinese built Seersucker, brought the war close to home to

A Patriot missile races to intercept an incoming Iraqi Surface to Surface Missile in Northern Kuwait.

the Division Marines located at Camp Commando, as it impacted just outside the camp, less than a kilometer from the Division Administration Center (DAC). (This was the first of only a very few missiles that got through the shield of the Patriot system in that area. Later, others were to strike civilian targets in Kuwait City, including a pier and a shopping mall.) The Scud Alarm sounded 38 times at Commando Camp, but the Patriots successfully destroyed all other missiles. The Marines at Camp Commando were made well aware that the defensive systems were not infallible, and that they too were very much a part of this war. Fortunately, none of these missiles were armed with chemical warheads.

20 March
Clearing the Obstacles and Passage Of Lines with the Kuwaitis

The last minute Kuwaiti obstacle reduction continued. The Kuwaiti contractors hired for the job were skilled and efficient, but the Iraqis had chosen to make the task more difficult by the occasional mortar attack. Some of these mortar fires were quite well aimed and caused the Kuwaiti contractors a bit of concern. Alpha Company, 1st Tank Battalion, and Alpha Company, 3d LAR Battalion reassured the contractors by their presence, and the engineers continued to push to get all the breaching work accomplished in time. Marines rode 'shotgun' on the contractor's equipment, using their Night Vision Goggles (NVGs) to help guide the efforts. The berm reduction crew even took two enemy prisoners prior to the commencement of ground operations (G-day). Two Iraqi border guards had made their way across the border, and told a tale of beatings and executions by the Regime. The legal status of these defectors was questionable, as they had come into Kuwait prior to the commencement of ground operations, and they were turned over to the Kuwaitis. The last of the breach lanes were cleared just in time to support the Division's attack.

The Kuwaiti 26th Brigade in the east and the 35th Brigade in the west manned their positions along the border to detect and interdict any Iraqi infiltration. These units each had Reconnaissance and antitank missile elements operating well forward in the border obstacle area. In the tense pre-combat environment, conducting a smooth Passage of Lines (POL) that avoided fratricide was critical. An accident at this point could derail the early momentum of the attack, even before the Division crossed the LD. Over a two-day period, the Marines would conduct a smooth RIP with the Kuwaitis on the border. On 19 March, RCT-7 passed forward of the 61st Tank Battalion of the Kuwaiti 26th Brigade, and on the following day RCT-5 conducted a passage of the 7th Battalion of the Kuwaiti 35th Brigade. Because of the careful pre-coordination efforts of Major Clark Watson, the Division LNO to the Kuwaiti Land Forces, the POL at the border was accomplished without incident. A few days later, after the Division had passed through, the Kuwaitis occupied forward positions on the border and resumed their vigilance against Iraqi threats to the Coalition's rear areas.

Tensions along the border area remained high, especially for 11th Marines and the counterbattery fight. With the counterbattery radars in position, the 'Cannon Cockers' provided responsive fires within minutes. This aggressive, responsive counterfire was to be demonstrated repeatedly during the ensuing weeks. In one instance, 3d LAR Battalion under Lieutenant Colonel Stacy Clardy had pushed Alpha Company up to the berm to screen forward of the Division's Attack Positions and the firing PAs of 11th Marines. Captain John Custis' Alpha

Company began to receive incoming mortar fire from an Iraqi border guard post across the border, and called it in. Before Alpha Company could even get the 'snowstorm' (incoming artillery) report out over the radio, 11th Marines had detected the enemy mortar fire and had engaged with counter battery artillery. Alpha Company also returned fire, in the Division's first direct firefight cross the international border. Unwisely, the Iraqis initiated the action, and were met with a deadly response by the Marines. A combination of indirect fire, 25mm Bushmaster cannon, and a single TOW missile destroyed the border guard post, eliminating the threat.

Moving into the Attack

On 20 March, I MEF officially took control of the battlespace, and 3d MAW began vigorous preparatory fires on the Iraqi 51st Mechanized Division. The Division still anticipated a two-day separation between A-day and G-day, and waited, poised to go at anytime. The Division would get its battlespace at 1500Z on 20 March, and H-hour was set for 0300Z on 21 March. ('Zulu' time is referenced to Greenwich Mean Time. It provides a common time reference for forces operating across time zones, and is important to avoid confusion between forces that may be operating thousands of miles apart. The local time in Iraq was typically three hours ahead of 'Zulu', so 0300Z corresponded to six o'clock in the morning in most of Iraq.) Fire Support Coordination Line (FSCL) One would also be established at 1500Z, followed by FSCL Two at 1530Z. The Division CG was in constant motion as he worked with higher headquarters to adjust the timing of the opening fight to match the circumstances on the ground. Throughout the day, he was to stay in close contact with Lieutenant General Conway, Major General Brimms from the 1st UK Division, and Major General Amos from 3d MAW. The Blue Diamond team was in its 'fighting stance', and stood balanced and ready to attack.

Marines in the Division Main CP monitor the progress of the shaping efforts. Marines worked in close concert to collect and process battlefield information in order to direct fire assets and maneuver units.

As the shaping fires began to execute, and the Division prepared to attack the following day, a strange incident took place. The G-2 had confirmed the movement of additional infantry elements to the border area with the departure of the UN Observers. With the P-3 aircraft and Pioneer UAV, the battlespace just across the LD had been observed repeatedly (to the limits of these platforms) and there was no other significant activity noted. The Division was prepared to execute the attack against the base enemy laydown that had been briefed. As always, there had been a number of conflicting intelligence reports received, some indicating greater reinforcement by the Iraqis in the south. One of the challenges for the Division had been to sift through the competing and often contradictory intelligence indicators received from external sources, and patch together its own assessment of the enemy's laydown to the front. The G-2 had not been able to verify any of the more 'spectacular' claims of massive reinforcements in the south. In fact, on 20 March, NIMA had produced a comprehensive laydown of the 51st Mechanized Infantry Division, identifying the locations of most of their equipment, and not detecting any major reinforcement. On this day, however, RCT-7 received a strange visitor who was to influence the execution of the Opening Gambit.

An Iraqi informant was brought to the RCT-7 CP by a clandestine American unit. When questioned by a skeptical RCT-7 command team, this informant was presented as a credible source that had never been wrong in the past. Now he gave RCT-7 a new report that the Iraqis had stealthily positioned an entire Republican Guard Armor Brigade to the terrain just north of Safwan. To further corroborate his story, the informant made a phone call to a contact north of the border, who reported that he could see the enemy tanks being hit by air as they spoke, and the tank crewmen running for cover. This brigade had allegedly moved into their positions over the last several nights by a combination of road and rail, and had infiltrated into well-camouflaged positions under cover of darkness and bad weather. The reported presence of this brigade ran counter to the intelligence reporting from higher, and, if true, significantly altered the enemy picture to the front of RCT-7. The 'ghost brigade', if truly there, was right where one would expect it to be in order to counter the Division's attack. Complete fidelity of the Iraqi order of battle had been constrained by the lack of comprehensive imagery coverage for the last couple of days. The report seemed highly unlikely, but it was possible. There had been various reports of T-72s throughout the month of March, but none had been confirmed. The Division Commander flew out for a face-to-face meeting with the informant. An entire brigade of T-72 tanks was a justifiable concern on the part of Colonel Hummer, and the concern was voiced to the Division. The G-2 scrambled to confirm or deny this reporting, but could find nothing that would indicate it was true. As is often the case with HUMINT of this nature, however, it was not possible to prove the negative (i.e. the G-2 could not 'prove' the tanks were not present.) The tanks could not be detected by any of the assets looking at that area, but did that mean they were not there, or was this a tribute to their excellent camouflage techniques? Did the enemy tanks, as reported, have thermal covers that hid their heat signatures from infrared detection? The Division CG received all this information and contemplated his options.

The Division's reaction to this crisis was a perfect example of the aggressive and proactive spirit the CG had built into the Division. There was no way that the fog of war surrounding this incident would be lifted before the Division planned to launch its attack, and many might have recoiled from that uncertainty. Recognizing the criticality of every moment for the preservation of the strategic oil infrastructure, however, the Division was to take a different course. The CG was adamant about the Division's responsibility to facilitate the success of the RCT commanders, and this would not be an exception. Now the CG was going to simultaneously address the RCT-7 commander's concerns and accomplish the Division's operational missions by aggressive action. Far from being deterred by this information, the Division viewed it as an opportunity. Not only would the Division crush the 51st Mechanized Division, but they would also bottle up and destroy whatever elements of the Republican Guard the Regime had foolishly thrown into the path of the Marines. "If those bastards are down there, we were going to bottle them up and kill them all," was the CG's expression of a new plan that would send a shock wave through the enemy ranks. Within hours, the Division made the necessary changes to the base plan. In a tribute to the flexibility of the Marines, a plan that had been carefully worked for months was quickly adjusted to meet the realities on the battlefield within hours. No small part of the new plan was the participation of the ever-ready 1st UK forces on the eastern flank. With a quick call, the 1st Marine Division CG and the Commander of the 1st UK Division were able to coordinate the attack of 1st UK forces into a portion of the zone that would now be unused by the Marines. Without hesitation, the Brits agreed to support the revised plan, sealing the enemy's fate.

The CG and ADC discuss intelligence concerning the 'Ghost Brigade' reported north of Safwan. " If they are really down there, we are going to bottle them up and kill them all."

Under the revised plan, the Division would first launch RCT-5 to block Highways 8 and 1 north of the Rumaylah oilfields to ensure nothing escaped west once the ground fight began. RCT-5 would also take the early opportunity to seize the western elements of the critical oilfield infrastructure, preventing their destruction. RCT-7 modified its scheme of maneuver to send 1st Tank Battalion and 3d Battalion, 4th Marines (3/4) to the west of *Jabal Sanam*. This task-organized force would emerge from the north side of the rugged *Jabal Sanam* terrain into the flank of the alleged T-72 force. The other elements of RCT-7 would then attack through the smoking ruins of the alleged Republican Guard unit, and seize their original objectives. Major Ron Spears, the RCT-7 S-2, worked with the G-2 to find a new attack route for the Tank task force and continued to look for T-72s. Lieutenant Colonel Nick Vuckovich, the RCT-7 S-3, developed the new orders for the conduct of the attack. The RCT-7 staff emerged with a plan that they later assessed was even better than their original base plan at achieving the RCT objectives. The Division CG consulted with the I MEF CG and 3d MAW pushed fixed and rotary wing aircraft north of the border to locate and destroy this alleged new tank threat.

As fate would have it, one of the Division's M1A1 tank platoons from 1st Tank Battalion was in a forward position at this moment as part of the security force for the border obstacle breaching. As a condition for continuing breaching operations, the Kuwaiti contractors working on the berm had adamantly insisted upon tanks in an overwatch position. Alpha Company, 1st Tank Battalion, sent one tank platoon forward. It grew dark but the berm reduction continued, as did the enemy mortar fire. Rumors about enemy reinforcements near the border were also heard, contributing to the fog of war experienced in the tense border area. Captain Banning, the Company commander was with the forward platoon. While forward positioned, his tank traversed the turret to the rear for the gunner to pass the thermal imager to the driver. The tankers heard AH-1W Cobras overhead, and were motivated by the air support being provided on the far side of the border. Realizing that the Cobras operating in their airspace were not under

This M1A1 took a direct hit by a Hellfire missile when it was mistaken for an Iraqi tank by supporting aircraft.

their control, however, the Air Officer scrambled to find out just who was controlling them. Just then, one of the Cobras fired a Hellfire missile toward one of the friendly M1A1s on the border. The missile impacted the tank, generating a large explosion and a ball of flame. Fortunately, this was a new type of Hellfire called an AGM-114M. It was an Annular Blast Fragmentation (ABF) munition designed to produce a multi-directional blast designed to destroy targets via fragmentation. Because of the type of missile used, it did not penetrate the tank's armor and only Captain Banning was injured. He refused evacuation, as it was "too early to leave the fight." Alpha Company was later chopped to RCT-1, and the Marines would be very happy to have Captain Banning and his company later in the attack to Al Kut. The aircraft that fired the missile had come from HMLA-269, flying off one of the ships of the Amphibious Ready Group (ARG) offshore. Subsequent investigation revealed that the aircrew had misidentified the tank and its position relative to the border due to the poor visibility, and a heightened awareness of a potential Iraqi armor presence near the border. This incident served as a significant reminder to the members of the Division and 3d MAW. No one needed to be told twice to display their air panels, and aircrews across the Coalition were reminded of their positive identification responsibilities. This incident was one of a very few friendly air fire incidents that were to happen throughout the war.

Iraqis began to ignite their own oil wells as part of a 'scorched earth' campaign. Because of the rapidity of the attack, the Division was able to secure most of them intact.

Now, another threat loomed. The Iraqis were reportedly beginning to destroy oil wells near the southernmost GOSP, only three kilometers north of the Kuwaiti border. The Division's attack was still planned for the following morning, but concerns began to build as the risks to the oilfields became more acute. The Pioneer detected and intelligence reports confirmed the burning oil wells. Later in the morning, VMU reported that another wellhead in the vicinity of the GOSPs was on fire. This was an indication that the much-anticipated destruction of the oil infrastructure was beginning. Iraqi missile attacks continued, and the alert code, "Lightning, Lightning, Lightning," was repeatedly sounded. There were reports of missile strikes landing near Camp Commando, and the boom of distant explosions in the Iraqi oilfield areas could be heard (whether from American bombs or Iraqi sabotage was not known.) The risk of destruction of the oil infrastructure was becoming unacceptably high.

On the Division's eastern flank, 3 Commando of the 1st UK Division, including the attached 15th Marine Expeditionary Unit (whose ground combat element was 1st Marine Division's own 2d Battalion, 1st Marines) successfully seized the Mina-al Bakr offshore oil terminal in the Arabian Gulf, and the pipeline manifolds on the Al Faw peninsula; critical elements to restoring the oil infrastructure required to rebuild Iraq. The Division cheered the success of their comrades, and looked across to the South Rumaylah infrastructure with anticipation. It was imperative that the MEF unleash the Division soon before Iraqi sabotage made it too late to get the rest of the infrastructure intact.

The CG and MEF Commander review options for launching the attack to secure the oil infrastructure.

Marines pass the final Kuwaiti border checkpoint before entering Iraq near the South Rumaylah oilfields.

The CG was in constant communication with the MEF Commander, and together they reviewed the strategic options. Presented with the opportunity to secure the infrastructure intact and possibly destroy an RG brigade in the process, both men were eager to strike. Lieutenant General Conway communicated this to the CFLCC Commander. As the senior commanders weighed this situation, the Division CG called Colonel Dunford at RCT-5 and told him to be prepared to go earlier than the scheduled crossing of the Line of Departure (LD). "No problem," was the reply. The CG called back again and told Colonel Dunford to be ready to go the evening of 20 March. Again, "No Problem." The CG called once more to say it did not look like the Division would attack until dawn the next day. After receiving another call from the MEF Commander, however, the CG called Colonel Dunford a final time. "Grizzly 6 (Colonel Dunford) this is Chaos, how fast can you be ready to go?" "General, we can go now," came the reply. The new LD time for RCT-5 was set for 1730Z on 20 March.

Crossing the Line of Departure

In the latest of a flurry of FRAGOs, the Division issued the order for RCT-5 to initiate the attack. The main effort shifted from RCT-7 to RCT-5 to support the change. RCT-5 would attack in zone to defeat enemy forces and secure the critical oil infrastructure. This would prevent further destruction of South Rumaylah oilfields, would block the westward retreat of the alleged RG brigade, and would isolate the 51st Mechanized Division.

Artillery-delivered shaping fires on their way to silence artillery supporting the 51st Mechanized Division in the South Rumaylah oil fields.

With the accelerated timetable, the shaping fight took on new urgency. Third MAW and 11th Marines turned to the fight with a vengeance. With only a few hours before the scheduled Division LD, Aerial Interdiction strikes were focused on the pre-attack targets identified by the Division and previously cleared from the Restricted Target List. The previously scheduled preparatory fires syllabus was adjusted to the new timings, and these fires began. *Jabal Sanam* was struck repeatedly in the afternoon and into the evening by fixed wing strikes. At 0220Z, a section of Harrier jets dropped five Mk-83 one thousand pound bombs on *Jabal Sanam*. Rotary wing strikes followed closely on their heels, and the Cobras closed in to kill anyone remaining alive on the mountaintop observation post. The rotary wing assets also flew against identified fixed border guard posts, and gave them similar treatment. Elements of RCT-7, deployed in the Division's security zone, enjoyed a front row seat to this awesome demonstration of firepower, and their confidence for the coming battle continued to grow. Now it was the Cannon Cockers turn. 11th Marine Regiment preceded RCT-5's attack with a 30-minute preparation fire on the enemy's artillery positions. As the Marines of RCT-5 later passed through the devastated Iraqi positions in the South Rumaylah, Lieutenant Colonel Mark Toal expressed appreciation for their artillery brethren, "Artillery did a hell of a job on the Iraqis!"

On 20 March, RCT-5, led by Colonel Joe Dunford, became the first major ground combat force committed to the removal of the Saddam Hussein Regime. The major elements of RCT-5 included 2d Tank Battalion, commanded by Lieutenant Colonel Mike Oehl; 1st LAR Battalion, commanded by Lieutenant Colonel Duffy White; 1st Battalion, 5th Marincs, commanded by Lieutenant Colonel Fred Padilla; 2d Battalion, 5th Marines, commanded by Lieutenant Colonel Dan O'Donohue; and 3d Battalion, 5th Marines, commanded by Lieutenant Colonel Sam Mundy. The RCT was supported by 2d Battalion, 11th Marines, commanded by Lieutenant Colonel Paul O'Leary, Bravo Company, 1st Combat Engineer Battalion, commanded by Captain Brad Aiello, and CSSC-115, commanded by Captain Suzan Thompson. Elements of 1st Radio Battalion, 1st Intelligence

Captain R. Michael Ackerson describes the roll into combat with Ironhorse:

Awakening at 0100 on 20 Mar 2003 to don MOPP suits, comfort became the first, if unofficial, casualty of this newborn war. Over the course of the next 12 hours, the brevity code "Lightning, lightning, lightning" would ring out at least 3 or 4 times. Scrambling into MOPP-4 became altogether too familiar. Every Marine is well versed in "sucking it up", enduring conditions that would floor the average civilian. Day one of the war was serving up an appropriate warm-up – some time to mentally dig in for the upcoming suffer-fest. Ironically, MOPP-4 does confer one advantage: At that point, the discomfort meter is already pegged. Simmering inside my NBC mask and protective gloves, I knew I was at the maximum; "All-clear" would bring the luxurious pleasure of wearing only the MOPP suit, flak, and associated combat gear. That is, of course, until we were being shot at...

At 1530 (on 20 Mar 2003), amid the bustle of last-minute preparations and adjustments, the words abruptly cut off a thousand preoccupied thoughts: "We're going in two hours." So much for 12 hours of notice. A hundred faces, an equal number of expressions. Excitement, fear, concern, relief that the waiting was (finally) almost over. Game day had arrived. Moving along the tanks of Delta Company, I exchanged hasty words with many of the Marines, particularly those of White (2d) Platoon. Some of this group included Major Cox, Captain Lindgren, Master Sergeant Eisel, Sergeants Hughes and McDonald, Corporals White, Dolce, and Johnson, and Lance Corporals Juliana and Bowers. Major Payne jogged in, announcing our departure in 30 minutes. I gave a thumbs-up, saying "Good luck, boys, kick ass" and broke into a trot towards the COC.

A voice in the track shouted an expletive, announcing that he had just seen an incoming Scud Missile shot out of the sky. All eyes turned skyward, remaining there to watch the light show provided by outbound ATACMS and RAP rounds. Not long after that, "Lightning, lightning, lightning" was heard. Sweet. Masked up, we settled in for the inevitable – a long night in MOPP-4. We were amazed – but not complaining – when we received the order to unmask. Back in monitor-mode, the Bravo Command rumbled through the breach. Each Marine was absorbed in his own thoughts, isolated from one another by 80 decibels of mechanized racket.

Battalion, and other supporting elements were included. The RCT's assault elements successfully crossed the LD at 1730Z, after smoothly adjusting to a timing that was nine and a half hours earlier than the originally scheduled H-Hour.

20-21 March
Grizzly in the Attack

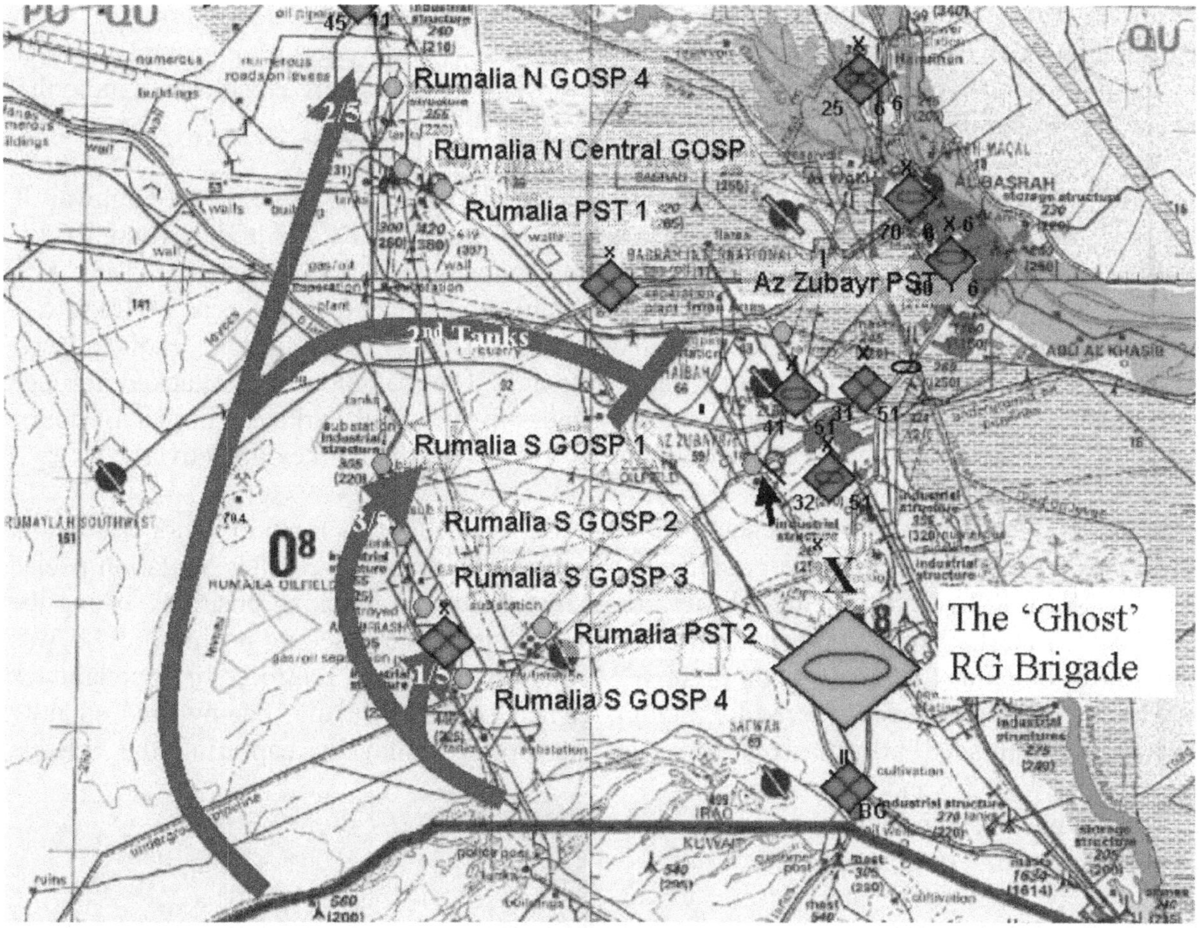

RCT-5's plan to cross into Iraq, block the alleged Republican Guard Brigade, and secure the North and South Rumaylah oil infrastructure.

RCT-5's mission was to seize the four strategically significant GOSPs of the South Rumaylah Oilfields, block the 6th Armor Division at the North Rumaylah Bridge, and block the 51st Iraqi Mechanized Infantry Division along Highway 8. RCT-7 would attack to their east at the original H-hour, less than 10 hours later. RCT-5 attacked through two breach sites, each consisting of three lanes. Just prior to the attack, the Iraqis emplaced a minefield in the middle lane of the western breach site. The lane remained fouled during the attack despite the efforts of Explosive Ordnance Disposal (EOD) teams, and a sniper who shot several of the mines causing them to explode. This obstacle had no significant effect on the attack. 1st LAR Battalion, supported by 7th Royal Horse Artillery's light gun batteries, established the RCT security zone inside Iraq, and provided overwatch across the RCT-5 zone of action as the RCT prepared to attack.

Second Tank Battalion ('Ironhorse') led RCT-5's main attack through the western breach site, moving through breach lanes 'Red 1' and 'Red 3'. Ironhorse attacked quickly through the relatively open desert on the western flank, in order to establish a blocking position that would trap the enemy in the Az Zubayr area. Over the next 10 hours, Ironhorse would push beyond the initial over-watch set by 1st LAR Battalion and advance into uncovered enemy territory. They attacked along an axis to the west of the oilfield infrastructure, uncovering Iraqi ground at a rapid rate and destroying everything hostile in their path. Twenty-seven enemy vehicles and more that 100 enemy ground forces were destroyed in their initial attack.

As dawn broke on 21 March, 2d Tanks established their blocking position on Highway 8. This was accomplished with minimal communications with the RCT-5 CP beyond RCT-5 reporting successive waypoints and significant enemy contacts. Battle Position (BP) ‘Tank’ was 25 km west of Iraq’s second largest city, (Al Basrah), and situated in the rear of Iraq's 51st Mechanized Brigade. From this position, Ironhorse was capable of disrupting the Iraqi defenses, and provided a catalyst for the Iraqi's hasty retreat. The battalion had attacked through approximately 70 km of enemy territory, entirely under the cover of darkness, in just 10 hours. Conditions were now set for adjacent elements of RCT-5 to secure their key objectives.

While Ironhorse moved to trap enemy forces to the east, RCT-5’s main effort, 2d Battalion, 5th Marines (2/5), followed in trace organized as a mechanized task force. The battalion peeled off to the North Rumaylah oilfield highway bridge, establishing a blocking position. Along the way, 2/5 secured the North Rumaylah oilfield infrastructure. The blocking position of 2/5 was important to block any attempts of the Iraqi 6th Armored Division to reinforce or counterattack south of the Saddam Canal. Second Battalion gained contact with the enemy, and soundly defeated the estimated brigade-sized force in the North Rumaylah, capturing the brigade commander. With 2/5 and 2d Tanks in their blocking positions, the enemy was held at arms length while the remainder of the RCT moved in swiftly to capture the oilfield infrastructure.

On the road north through the oil fields, columns of dense oil smoke mark oil wells the Iraqis had ignited.

First Battalion, 5th Marines (1/5), followed by 3d Battalion, 5th Marines (3/5), attacked through the eastern breach site, lanes 'Red 4 - 6'. First Battalion attacked to seize the two southernmost GOSPs and the South Rumaylah intermediate pumping station. Just before the timetable had been accelerated, the battalion staff met to conduct a last minute review of their attack (planned for the following day.) During the session, the Marine on radio watch handed the Battalion Commander an urgent message. Lieutenant Colonel Padilla read the note, then looked up and said, "Guys, we're going right now." The Iraqi actions to ignite oil wells in the South Rumaylah required 1/5 and 3/5 to cross the border immediately in order to prevent the further destruction of the oil infrastructure. Lieutenant Colonel Padilla gave some final instructions, and the meeting began to disperse. Before everyone could hurry off, he stopped everyone and said, "Hold on. Chaplain, where are you?" Lieutenant Cash, the battalion chaplain, who had been standing behind Lieutenant Colonel Padilla during the meeting, led everyone in a short prayer before they left to carry out the mission.

After gearing up, the Battalion crossed the LD and entered Iraq. One hour prior to crossing the line of departure, the battalion was notified that there would be 30 minutes of preparation fires before they crossed the LD. The battalion quickly prosecuted known targets with available assets, focusing on Air Defense Artillery (ADA) and indirect fire assets. Three km into Iraq, Alpha Company encountered an Iraqi tank platoon supported by dismounted infantry. These were quickly destroyed by javelin, tank main gun and AAV up-gun systems. The battalion simultaneously secured the southernmost GOSPs (#3 and #4) with Bravo and Charlie Companies, respectively. During Bravo Company's attack, led by Captain Jason E. Smith, they took over 200 EPWs with little resistance. Once GOSPs 3 and 4 were secure, Alpha Company conducted a forward passage of lines and turned east to seize the strategic intermediate pump station. Once committed to the attack against the pump station, the battalion's Alpha Company found the compound defended by more than 100 enemy soldiers. They enveloped the objective, and utilized artillery Dual Purpose Improved Conventional Munitions (DPICM) for suppression. It became clear that the enemy had established well-prepared trench lines and bunkers throughout the complex. The Alpha Company Commander, Captain Blair Sokol, moved forward with his company, directing a methodical sweep of the maze of enemy positions. Utilizing rockets, grenades, tank and small arms fire, his men followed his direction and rapidly moved through the complex and into adjacent positions.

AAVs transported Marines across the LD. These vehicles and their onboard weapons systems provided protection and speed for mechanized infantry units during the war.

Marines are conditioned to face danger and hardship, but nothing can adequately prepare them for the loss of one of their comrades. As the company continued its systematic clearing of the area around the pump station, they again met enemy resistance. During this fighting, the Division suffered its first Marine Killed-In-Action in Iraq. The Division mourned the loss of Second Lieutenant Therrel S. Childers of Alpha Company, 1st Battalion, 5th Marine Regiment. He was fatally wounded when Iraqi vehicle-mounted troops launched a surprise attack using civilian trucks. He died while doing what Marine Lieutenants have always done in combat, leading his Marines in the attack. Wresting control of this critical infrastructure objective would enable the rapid resumption of oil revenue for the newly liberated Iraqi people, giving hope to an entire generation of Iraqis. At the conclusion of the attack, the company had destroyed an estimated enemy battalion, and captured in excess of 100 enemy prisoners of war. Corporal Brenton P. Gross received wounds in the same action, as the Marines destroyed the remaining enemy resistance. Junior officers and NCOs leading from the front would be the hallmark of Division action from the Opening Gambit to Baghdad.

After the pumping station was secure, Charlie Company moved south to seize GOSP #6 without further resistance. Once all objectives were secure, each company's attachment of British Royal Engineers and EOD specialists utilized their specialized gear and training to enter and secure the facilities. They proceeded to shut down the facilities without incident.

LAV-25s dominate the battlefield with their Bushmaster 25mm cannon.

While 1/5 moved out in the attack on GOSP #3 and #4, 3/5 attacked to seize GOSPs #1 and #2 just to the north. Once through the breach lanes, 3/5 moved to the west of 1/5, bypassing the fight for the southernmost GOSPs, and quickly seized GOSPs #1 and #2 further north. These GOSPs were the largest of the entire Iraqi national petroleum production capability. The Iraqis defended these with conventional artillery and infantry forces, as well as a large number of fire trenches, some located immediately adjacent to the strategic infrastructure. These fire trenches caused significant obscuration of many of the key points in the oilfield infrastructure, in some cases, severely limiting rotary wing air support.

The effects of the MEF and Division fires from the previous night had obviously dampened the will to fight of the enemy at the remaining GOSPs. Kilo Company, commanded by Captain Mike Miller, called in a close air strike on a D-30 battery east of GOSP #2, then moved just beyond a burning oil trench and dismounted. Captain Miller's Marines meticulously cleared the GOSP #2 area, moving through numerous bunkers, trenches and fighting positions to round up nearly 200 Iraqi EPWs. Lima Company, commanded by Captain Scott Meredith secured GOSP #1, experiencing no enemy resistance. Later that day, 3/5 was relieved in place by 1/5 and moved to a battalion TAA west of GOSP 1. Looking back, Lieutenant Colonel Mundy (the Commander of 3/5) had this to say about the Opening Gambit attack:

"Three things stand out about that night. First, the esprit and enthusiasm of my Marines, when ordered on very short notice to conduct a mechanized night attack, 10 hours earlier than expected, was impressive ...basically they said, 'Great, let's go!' Second, the reality of this war definitely hit home when I heard Fred Padilla call in his KIA (Second Lieutenant Childers). And finally, the friction of war was, as usual, ever-present and very frustrating. We had two breaks in contact while moving the Battalion to the start point, and at one moment nearly lost our Alpha Command group when the AAV C-7 they were traveling in inadvertently blew its Halon fire extinguisher system, forcing the amtrac to stop and dismount for 10 minutes while the staff and crew threw themselves on the ground, choking and coughing to clear their lungs. We looked back the next day and laughed about it, but at the time I thought I had lost my staff principals, and could faintly hear the ghosts of my Basic School instructors asking me, 'What now, Lieutenant Colonel?'"

Despite the limitations of haze and smoke, RCT-5 enjoyed uncommon rotary wing support from the Marines of HMLA-169 (the 'Vipers') and HMM-268 (the 'Red Dragons') throughout the campaign to Baghdad. With Cobra and Huey gunships above, and the CH-46's ready to rapidly evacuate the wounded, RCT-5 was never at a loss for combat power and motivation from 3d MAW. The habitual relationships built during peacetime began to pay dividends, beginning with the first engagement.

Second Battalion, 11th Marines, CSSC-115, and the RCT-5 command elements also crossed into Iraq through the eastern breach site early on 21 March. Through the early morning hours, the RCT consolidated on its objectives and turned to processing the hundreds of EPWs they had captured. 2/11 established firing positions in the North Rumaylah, from which they could support either of the blocking positions maintained by 2/5 (oriented to the north against the enemy's 6th Armored Division), or 2d Tank Battalion (oriented to the east against any elements of the 51st Mechanized Division seeking to escape.)

As part of the Division scheme of maneuver, RCT-5 also controlled the movement of 1st Reconnaissance Battalion and RCT-1 through the western breach site to their assigned zones of action. Each in turn, 1st LAR Battalion led Recon Battalion and RCT-1 to the vicinity of the Rumaylah Airfield where they were released to screen RCT-5's western flank.

The visibility, compressed timeline, uncertain enemy sit, and C2 challenge created no shortage of fog and friction last night. That said, the battalions did a superb job of cross talking and improvising. As I write this we are in a position to accomplish assigned tasks...we've got the first one under our belt and stand ready to support RCT-7 and then continue the attack to Baghdad.

Notwithstanding the accelerated timeline for the attack and the extraordinarily poor visibility in the area of operations, the RCT had accomplished its initial combat mission with minimal loss of life or equipment. By 0956Z on 21 March, the desired end state was achieved, with all four GOSPs secured and the blocking positions in-place. Battle damage assessment, and interviews with EPWs captured during the attack, confirmed the pre H-hour estimates of brigade sized enemy forces in both the North and South Rumaylah oil fields. The enemy forces in the South Rumaylah oil fields were primarily dismounted infantry, but were supported by T-55 tanks, mechanized vehicles, surface to air missiles, air defense artillery, mortars, long range artillery, and multiple launched rockets.

With the 1st UK Division's successful operation to seize the critical oil infrastructure on the Al Faw peninsula, US Special Operations units securing the offshore export terminal, and now RCT-5's successful action to seize the South Rumaylah GOSPs; the only critical infrastructure task remaining was the Zubayr pumping station complex (the 'Crown Jewel') in RCT-7's zone. With the successful securing of this last key node, the Division would have achieved their first strategic success, and contributed greatly to setting the conditions for a free and prosperous Iraq in the future. It was RCT-7's turn.

21 March
Judgment Day for the 51st Mechanized Division

On 21 March, while RCT-5 was engaged in the South Rumaylah oil fields, the Division issued a FRAGO tasking RCT-7 to attack in zone to destroy the 51st Mechanized Infantry Division in order to prevent the retreat of this unit into Al Basrah. The timing of the attack was set for 0300Z, simultaneous with the 1st UK Division's attack on Umm Qasr on the Marine Division's eastern flank. This attack would be the 'hammer' to the 'anvil' already established by RCT-5's 2d Tanks (in their blocking position to the west, sealing off any retreat for Iraqis hoping to escape to fight another day.)

During the hours preceding the attack, the Division conducted shaping fires using the Pioneer UAV in a direct support role to the Division. From the live feed in the Division CP, Intelligence and Fires representatives were able to find targets, adjust fire, and conduct assessment of key targets throughout the zone. The Division was able to find and kill a large number of enemy in RCT-7's path, including artillery, infantry, and T-55s in defensive positions along RCT-7's intended attack routes. This was a fine example of a classic Division Fires shaping fight in advance of the RCT close fight. The close relationship the Division had established with the VMU squadrons was now paying off in observed deep fires, with rounds on target. The stated goal for the shaping fight was for the RCT to simply have to 'wade through a sea of body parts' on their way to the objective. The shaping fight had a significant impact on the willingness of the enemy to stay and fight RCT-7 only hours later.

Task Force Ripper

Harkening back to its Operation Desert Shield roots, Colonel Hummer's RCT-7 was often referred to as 'Task Force Ripper'. Built around the 7th Marine Regiment, it had nearly 5200 Marines and 1000 vehicles when it crossed into Iraq. Among the units of the Task Force were two mechanized battalion task forces, 3d Battalion, 4th Marine Regiment (3/4) commanded by Lieutenant Colonel Bryan McCoy and 3d Battalion, 7th Marine Regiment (3/7) commanded by Lieutenant Colonel Mike Belcher; and one motorized battalion, 1st Battalion, 7th Marine Regiment (1/7) commanded by Lieutenant Colonel Christopher Conlin. In addition, the RCT attached First Tank Battalion, commanded by Lieutenant Colonel Jim Chartier, Company D, 3d

LAR Battalion commanded by Captain Gordon Miller, and elements of 3d Assault Amphibian Battalion. Supporting the RCT were Marines and Sailors from Combat Service Support Company (CSSC) 117 under the command of Captain Andy Bergen. Third Battalion, 11th Marines, commanded by Lieutenant Colonel Kirk Hymes, was in direct support. Rounding out the organization for combat were units from 1st and 2d Combat Engineer Battalions, 3d Civil Affairs Group, 1st Radio Battalion (supported by a 1st UK Division Signals detachment), a VMU-1 Remote Receive Team, a 1st Intelligence Battalion detachment of Human Intelligence Exploitation Teams (HETs) and elements of a PsyOps company, added just before the order to displace to DAs was released.

The Kuwait-Iraq border was the RCT-7 Line of Departure (LD). The original plan was for a simultaneous attack with 1st Tanks and 3/7 crossing the border through breach lanes 'Orange 1-5' near the terminus of Kuwaiti Highway 80. The remainder of the RCT would follow, continuing the attack north along two axes oriented on the main thoroughfares, Highway 8 and Highway 31. Once through the breach, 1st Tanks would eliminate the tanks and APCs of the Iraqi 51st Mechanized Division and support the attack into the Basrah International Airport (BIA). The task of 3/7 was to clear the Iraqi 32d Mechanized Infantry Brigade from its garrison to the south of Az Zubayr. More importantly, 1/7 was focused on securing the 'Crown Jewel', the Zubayr

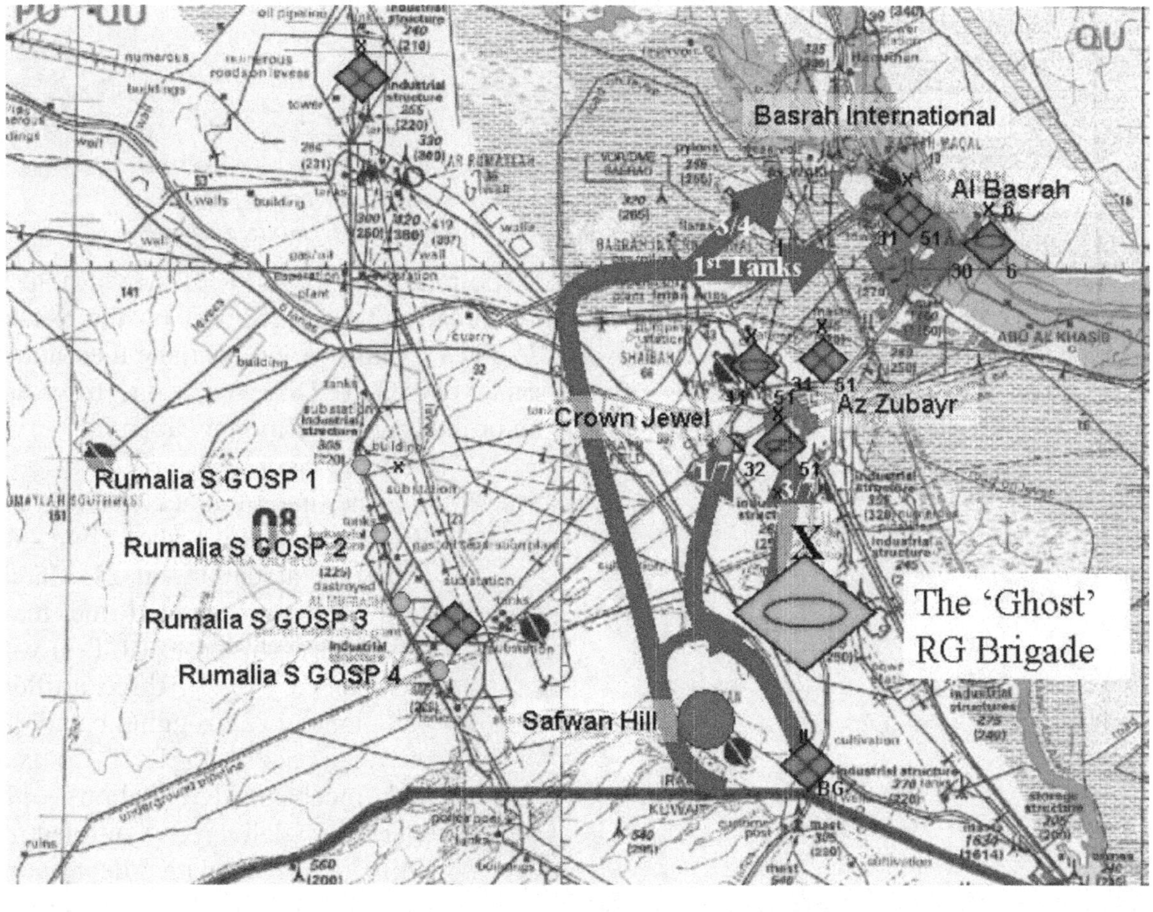

RCT-7's attack called for a rapid penetration across the border and the capture of oil facilities in Az Zubayr and the Basrah International Airport. The attack was modified with elements attacking on the west side of Jabal Sanam.

pumping station complex. This complex was the central pumping station in the South Rumaylah oil fields that received oil from the north and west, and pumped it to the offshore oil export terminals. It was the last strategic objective that had to be taken for the Marines to claim complete success in capturing the critical nodes of the South Rumaylah oil field infrastructure intact. The objectives of 3/4 included securing BIA and the northern bridges over the Shaat al Basrah waterway. These bridges, leading to the city of Al Basrah, were key to supporting the attack of the 1st UK Division's 7th Armoured Brigade into the city. They were central to blocking the enemy's movements across the waterway in either direction. It was important to the Division to set the Brits up for success in both the relief in place and their follow-on mission. RCT-7 bore the responsibility of delivering these bridges to them intact.

The guns of Eleventh Marines provide prep fires for Ripper's Attack against the 51st Mechanized Division.

The original plan was now modified because of the suspected Republican Guard brigade threat north of Safwan in RCT-7's zone. RCT-5 had already moved to a blocking position that would preclude the escape of any reinforcing RG elements in the south. Now, RCT-7 would attack to deliver the decisive blow. Instead of leading with an attack up Highway 80, two battalions (1st Tanks, followed by 3/4) were redirected to the west of *Jabal Sanam*, where they would cross the border using the Green breach lanes originally created for elements of 3d LAR Battalion and 11th Marines. The remainder of the RCT would hold in place and cross at the original sites once the enemy threat had been reduced by the assault elements. Once any RG threat had been destroyed, 3/4 and 1st Tanks would continue their attack to the Shaat al Basrah, effectively sealing off the remainder of the Division's zone from any actions on the part of paramilitary fighters in Al Basrah or the Iraqi 6th Armored Division.

Satellite image of the Crown Jewel Complex in Az Zubayr.

The time of attack was set, and RCT-7 crossed the line of departure at 0300Z on 21 March. 11th Marines weighted the main effort (now shifted to RCT-7) with the massed fires of three artillery battalions. Leapfrogging battalions to keep pace with RCT-7's tank and mechanized battalions, 11th Marines delivered unrelenting artillery fires with devastating effects, stripping the last enemy defenders' will to resist. The

RCT-7 main effort, 1st Tank Battalion ('Tiger'), crossed through the breach lane on schedule and headed north past *Jabal Sanam* and Safwan Airfield. Movement across this part of Iraq took a bit longer than expected, however, as Lieutenant Colonel Jim Chartier led Tiger between minefields and bauxite mining pits to hit its mark midway into the flank of the suspected enemy positions. Tiger took up a position from which it could strike into the flank of the RG brigade reportedly arrayed along Highway 8, but they quickly found that there were no T-72 tanks to strike. The report that had changed the entire Division scheme of maneuver was false. Once all RCT units were across the LD, and 1st Tanks and 3/4 had swung to the north of the Crown Jewel, the rest of RCT-7 units assumed their originally planned attack routes. Tiger (followed by 3/4) moved north along Highway 8, while 3/7 (followed by 1/7) moved north along Highway 31. The Ripper Tactical CP (the 'TAC') established itself 13 km north of *Jabal Sanam*, in between the two highways, and took control of the fight. Once the TAC was in place, the RCT-7 Main CP moved into Iraq to link up with the TAC.

Attacking in the wake of the earlier RCT-5 attack and a significant shaping effort, 1st Tank Battalion's movement over the next few hours was characterized by what appeared to be a collapsing conventional enemy. Some uniformed Iraqi infantrymen fought, some surrendered, and a large number of APCs, tanks, and artillery pieces were found very recently abandoned. The battalion's main body was immersed in 'civilians', many of whom had obviously just tossed aside uniforms for civilian clothes. Some appeared to be very happy with their recent release from the military, while some seemed to look as if they were watching and waiting for an opportunity to rejoin the fight. This was the RCT's introduction to the unique face of enemy resistance in Iraq, one that required discipline on the part of all the Division's Marines. Blue Diamond declined to kill hundreds of Iraqi citizen-soldiers unnecessarily, or take thousands of conscript prisoners. Throughout the campaign, the Division chose to ruthlessly destroy only those that took arms and fought against it, disarming and releasing those Iraqis who just wished an end to fighting and to return home, no longer posing a threat. Coaxial machine gun fire and tank main guns quickly destroyed the relatively few active enemy infantrymen (estimated enemy casualties for the morning were only approximately 35-40 enemy personnel) and armored vehicles encountered. By 1300Z all garrison positions of the 41st Armor Brigade had been cleared of organized enemy resistance.

A modified Iraqi self-propelled AAA piece. Destroyed equipment like this littered the routes all the way to Baghdad.

Attached to 1st Tank Battalion, Lima Company of 3/4, commanded by Captain Matt Watt, continued the attack toward the Al Basrah bridges. With darkness rapidly approaching, Lima was engaged by enemy dismounted infantry and an armored vehicle in the vicinity of the Ash Shuaybah Barracks, the headquarters of the 31st Mechanized Brigade. This enemy was quickly destroyed. As Lima pressed east, they encountered extensive civilian and military traffic on the road between Az Zubayr and Al Basrah. They continued the attack until they were in blocking positions at the major Al Basrah bridges. Now, no reinforcements could enter the Division's area of operations without fighting through the blocking position (something the enemy would try and fail to accomplish many times over the next 24 hours.)

Captain Watt and Lima Company, with a tank platoon attached, continued to fight a close quarters battle at the northern Al Basrah bridge site throughout the night with dismounted infantry on the near and far side. The ground near the bridge was marshy and restricted the movement of tanks, AAVs, and HMMWVs, making the establishment of a dispersed, integrated defense difficult. The Company had to contend with forces retreating into Al Basrah, as well as paramilitary fighters coming out of Al Basrah seeking to join the fight to the south and west. That night, Lima Company destroyed more than 12 'technicals' and troop transports, 100 dismounted infantry, and several artillery pieces at this key bridge into Al Basrah.

With the complex movement across the border complete, RCT-7 focused on seizing the Crown Jewel and defeating the remainder of the 51st Mechanized Division. The attack toward Az Zubayr and Al Basrah unfolded as planned. Third Battalion, 7th Marines moved up Highway 31, turned right, and sliced through farmlands to find a lightly defended 32d Mechanized Brigade garrison. For the next several hours, the battalion continued to clear the garrison associated with the brigade. Resistance was sporadic, and it appeared most equipment had simply been abandoned, with the Iraqi soldiers either surrendering or trying to escape dressed as civilians. Iraqi uniforms were found strewn about in fighting holes with abandoned AK-47s, Rocket Propelled Grenades (RPGs), and other gear. The frightened soldiers had rapidly donned civilian clothes and left, in just enough time to survive the direct fire attack of 3/7. Several members of the battalion's logistics trains were severely injured when they encountered an enemy minefield. HNC Leo Raya, a Corpsman with the Battalion Aid Station (BAS), recounts:

> *"After driving most of the day, we were starting to set into our position. After a couple of uneventful hours we heard a loud explosion and immediately those famous two words, 'Corpsman up!' Staff Sergeant Alva had stepped on some kind of mine, and was lying on the ground critically injured. We rushed across the minefield to stabilize his injuries and prepare him for an urgent CASEVAC. Several minutes later we heard a second loud explosion...HM3 Alaniz had also stepped on a mine as he helped set up the BAS to care for Staff Sergeant Alva. Three corpsmen had been injured from the blast in addition to the four Marines from the first explosion. We were about to be tested. After all was said and done, all Marines and Sailors survived their injuries and will live to talk about it."*

Staff Sergeant Alva and HM3 Alaniz were both evacuated to an American hospital in Germany. Each of these men lost a leg to their wounds. The remaining casualties were treated in theater by the 47th Combat Surgical Hospital, and eventually returned to their unit. These incidents reinforced the dangerous conditions that all Marines and Sailors that entered Iraq were to encounter. Before the war was over, many more members of the Division were to be wounded or killed by mines and unexploded ordnance.

First Battalion, 7th Marines, supported by combat engineers, PsyOps and civil affairs teams, attacked next to seize the Crown Jewel. As the lead Combined Arms Anti Tank (CAAT) team moved toward the objective, all indicators were that the site was undamaged. As 1/7 secured the key elements of the vast complex, the PsyOps and civil affairs teams broadcast messages to workers and residents to remain clear of the Marines. Key supervisors in the facility were identified and assisted the Marines in safeguarding the pumping station. Finally, the last of the strategic oilfield infrastructure elements was secure. After securing the pumping station complex, 1/7 sent out patrols to the outskirts of the city of Az Zubayr. CAAT 'Red', led by Lieutenant Joshua Bates, approached the city and crossed a road intersection. Suddenly, an RPG narrowly missed the lead vehicle and impacted 20 meters beyond the patrol. Immediately all CAAT vehicles and dismounts established firing positions and returned suppressive fire. One dismounted CAAT team spotted a T-55 down the block and ran into a courtyard in order to get an AT-4 shot. The team began receiving sporadic small arms fire from a nearby building. While additional CAAT dismounts joined them in suppressing the building, a TOW vehicle destroyed the T-55. Seven more T-55's were spotted around the destroyed tank. A gun truck continued to provide suppressive fire on the building with Mk-19 rounds as the rest of the team destroyed the tanks. CAAT completed the patrol without taking a casualty, leaving 10 T-55 tanks burning.

On the northern flank of 1st Tanks, 3/4 (under Lieutenant Colonel Bryan McCoy) fought through sporadic resistance past the Shaibah Airfield, the Basrah Refinery, and to the outskirts of the Basrah International Airport (BIA). The battalion was task organized, with Bravo Company, 1st Tank Battalion attached. For the attack on Shaibah Airfield, Bravo Company turned east and deployed on line against threats in the open, lining up on the flank of Charlie Company, 1st Tanks. Kilo Company, 3/4, continued toward BIA road on the left flank. Just to the west of the Basrah Oil Refinery, the tank platoon attached to Kilo made contact with enemy T-55s and MTLBs, destroying 13 MTLBs, 3 T-55s, and approximately 35 dismounted enemy. Surrendering Iraqi soldiers were simply disarmed and pointed south to make their own way to the EPW camp. Approaching the refinery, 3/4 encountered enemy resistance comprised primarily of dismounted soldiers in bunkers next to unoccupied vehicles. Several T-55s also engaged the approaching Marines. The minimum amount of fires necessary was directed at the threats in the refinery complex, leaving it largely undamaged for its return to the Iraqi people in accordance with the Division's greater mission.

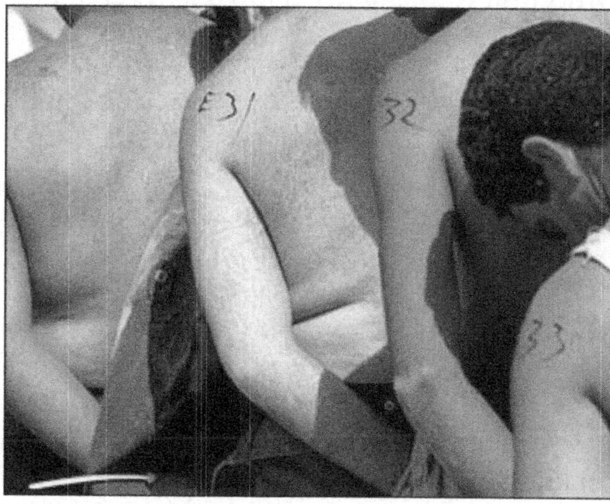

Numbered Iraqi EPWs wait in line for interrogation.

Remaining elements of 3/4 continued to push toward the east. Even though the Coalition did not intend to operate from BIA, the airport needed to be cleared of enemy units to permit the uninterrupted relief by the 1st UK Division. BIA was also an important objective from a psychological perspective. Part of the battalion including two task-organized companies (Bravo and India), supported by the battalion TAC-CP (led by Operations Officer, Major Martin Wetterauer) an Obstacle Clearing Detachment, and a section of 81mm mortars, moved toward the cloverleaf intersection just south of BIA. After a brief engagement with enemy soldiers on the southeastern corner of the airport, Bravo turned its main guns on the vicinity of the cloverleaf, quickly dispensing with 8 T-55s, 3 MTLBs, and 20 Iraqi infantry. At 2123Z, Bravo called the cloverleaf south of BIA secure. For most of the remainder of the night, Bravo remained in its hasty blocking position at the cloverleaf, overlooking the highway bridge across the Shaat al Basrah waterway.

After conducting a link-up and coordination with India Company west of the cloverleaf, Kilo moved against the airport. Its tank platoon pressed forward to the berm to scout for a breach site, moving under cover of the 81mm mortars. Upon detecting MTLBs and troops in the open, immediate suppression fires were called. As soon as the rounds from the regimental suppression mission lifted, Kilo moved through a parting in the fence, and fought its way to a support by fire position for India's movement through the breach. The attack destroyed a dozen armored vehicles, and included RCT-7's first Javelin engagement of the war, when Lance Corporal Adam Seeley destroyed a BMP-1 on the airfield.

India and Kilo Companies continued to use surface and air fires to destroy sustained enemy resistance at the airfield. India undertook the detailed clearing of the buildings that made up the international airport complex. During this operation, India Company discovered an underground military complex, and the Human Intelligence Exploitation Team (HET) led by Chief Warrant Officer Steve Blanford deployed to ascertain its purposes. One of the rooms discovered was an obvious torture chamber with a battery and cables hooked to bedsprings. Kilo Company maintained its support for the clearing by moving to new support by fire positions as India Company advanced. At 0818Z, 3/4 reported that Basrah International Airport was secure. Soon thereafter, the relief in place began with British forces.

Recon Battalion and RCT-1

First Reconnaissance Battalion followed RCT-5 through the breach lanes on the west flank with the mission of screening the Division's northern flank once west of RCT-5's zone. Once across the border, Lieutenant Colonel Ferrando and Recon Battalion ('Godfather') linked up with Lieutenant Colonel Duffy White and 1st LAR Battalion ('Highlander'), who were screening the western flank of the Division. The two commanders coordinated 1st Reconnaissance Battalion's continued movement into its screen of the Division's north flank and eventual westerly route of march toward An Nasiriyah. The LAR Battalion guided Recon Battalion through their zone to the release point. Recon Battalion then moved up to the Saddam River where they began to establish a screen to the north, slowly expanding it to the west.

Colonel Joe Dowdy's RCT-1 was referred to as 'Inchon', recalling its service in the strategic victory of the Korean War. Built around the 1st Marine Regiment, it had nearly 5,500 Marines

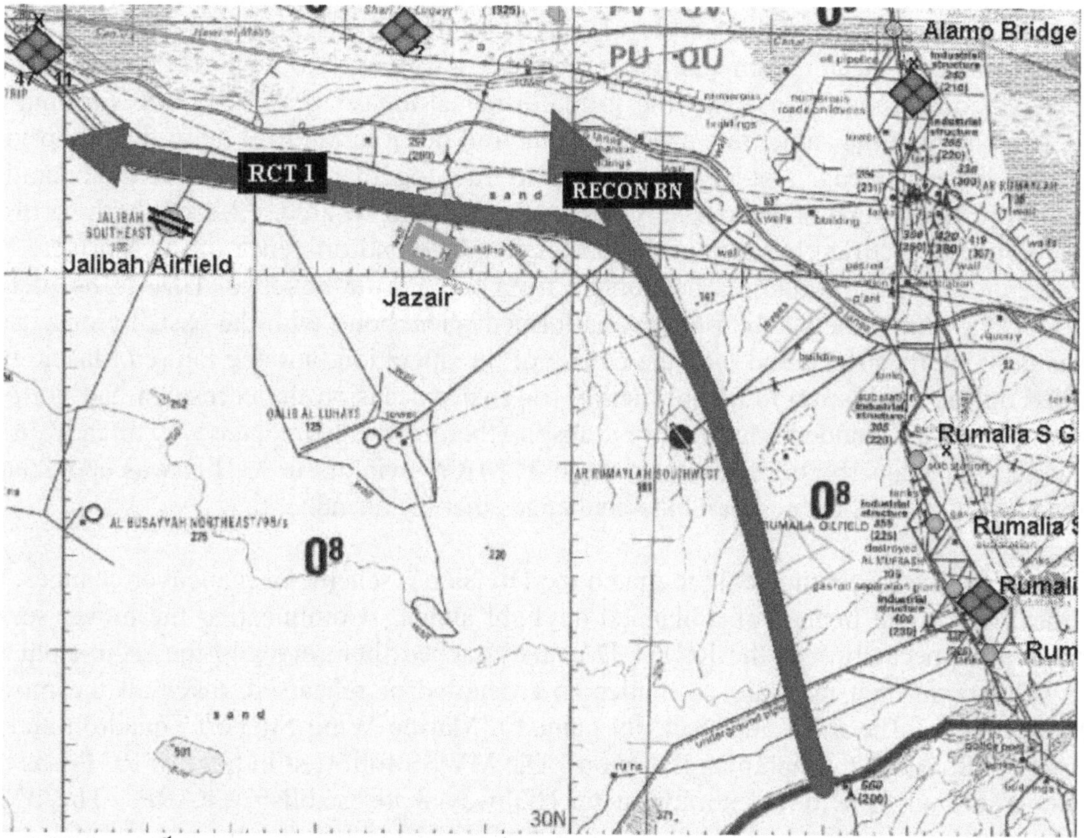

RCT-1 and 1st Reconnaissance Battalion cleared to the west of RCT-5 and RCT-7, allowing the Division to sustain the momentum of the attack to the Euphrates.

and 1,000 vehicles when it crossed into Iraq on 21 March. Among the units of the Task Force were two mechanized battalions, 1st Battalion, 4th Marine Regiment (1/4) commanded by Lieutenant Colonel John Mayer and 3^{d} Battalion, 1st Marine Regiment (3/1) commanded by Lieutenant Colonel Lew Craparotta. The third battalion, 2^{d} Battalion, 23rd Marine Regiment (2/23) commanded by Lieutenant Colonel Geoffrey Cooper, was motorized. In addition, the RCT attached 2^{d} LAR Battalion commanded by Desert Storm Navy Cross winner Lieutenant Colonel Ed Ray; and elements of both 2^{d} and 3^{d} Assault Amphibian Battalions. Alpha Company, First Tank Battalion, commanded by Captain Dave Banning, attached from RCT-7 on 23 March, and remained with RCT-1 through the end of the war. The 1st Battalion, 11th Marine Regiment (1/11), commanded by Lieutenant Colonel Jim Seaton, would provide direct support artillery fires. Also supporting the RCT were Marines and Sailors from Combat Service Support Company (CSSC) 111 under the command of Captain Grant Shottenkirk. Rounding out the organization for combat were detachments from Charlie Company, 2^{d} Combat Engineer Battalion, 1st Intelligence Battalion, 2^{d} Radio Battalion, both 3^{d} and 4th Civil Affairs Groups (CAG), and elements of the 305th PsyOps Company from the United States Army.

RCT-1 received its orders to move out of LSA-1 and to the RCT DAs at the same time as the rest of the Division. The RCT-1 task organization had come together thanks to the tremendous efforts of 3^{d} AA Battalion getting the last 2 companies of amtracks off the late-arriving ships and moving them to the LSA for RCT-1. The 'on the fly' nature of the assemblage of RCT-1 included several of their significant attachments arriving at the last minute. Fortunately, the

Regiment had an opportunity to work with some of these attachments on board the ships of ATF-West as they sailed to theater from California. Others, however, like their Air Support Liaison Team (ASLT), joined just prior to moving to the dispersal areas. CSSC-111 was forming and receiving equipment at the same time as RCT-1, and did not join the RCT until the day prior to their move to dispersal areas. Not having had time to train together, the first refueling operations and re-supply efforts were understandably less efficient than desired. All of these 'growing pains' were quickly rectified because of the professionalism and diligence of the Marines. The RCT-1 Casualty Evacuation and C2 helicopters from the 'Purple Foxes' of HMM-364 met the RCT at the DAs, and immediately a bond was formed. The bond with the casualty evacuation crews was particularly strong, and over the course of the operation, not one injured Marine from RCT-1 died during evacuation to a care facility (on many occasion the corpsmen had not even completely prepared wounded Marines for transport before the helicopter was already in the landing zone.) The superb commitment of these 3d MAW aviators to RCT-1 was appreciated, and formed a bond that held through all the challenges that lay ahead.

RCT-1 accelerated their timetable to match the Division's scheme of maneuver changes, and crossed the LD at night instead of a planned daylight attack. Complicating the move, several units arrived unexpectedly into the RCT-1 DA and attached themselves to the regimental rear elements (illustrating that no plan, no matter how detailed or rehearsed, takes all the moving parts into account.) The RCT unexpectedly gained a Marine Wing Support Squadron and the headquarters of 2d Assault Amphibian Battalion. The MWSS followed in trace of RCT-1 as they moved into Iraq and then turned northwest up Highway 8 to establish a FARP. The 2d AA Battalion headquarters was a godsend, as they brought significant maintenance capabilities with them. RCT-1 was to lean on 2d AA Battalion to produce maintenance miracle after miracle on the long march to Baghdad. RCT-1 was now an organization with enough vehicles to stretch over 40 km as they crossed into Iraq on the Division's west flank and moved up to Highway 8. RCT-1 knew it had to uncover the ground quickly, to allow the rest of 1st Marine Division and the 1st UK Division to 'uncoil' and present a less lucrative chemical target to the enemy. RCT-1 moved quickly to carry out its mission.

A Marine provides security from the hatch of his vehicle while advancing through Rumaylah.

Early on 21 March, RCT-1 reached their limit of advance just to the west of Jalibah Airfield prepared to continue their attack. The Marines and their vehicles were caked in a fine layer of dust that settled in every nook and cranny during the long overland march. Drivers and small unit leaders were already fighting the effects of two nights without sleep. Having moved approximately 80 km to the west of the Division's main attack, the RCT quickly reestablished multiple communication links with the Division headquarters. Because their movement had been so rapid, RCT-1 found itself in a position waiting to conduct an FPOL with TF Tarawa before that unit had even arrived in its battlespace. RCT-1 would have to wait, and sent patrols back to the east in order to ensure that area was sanitized of enemy.

Other Division Actions

Lieutenant Colonel Stacy Clardy's 3d LAR Battalion ('Wolfpack') adjusted to their role in the revised plan. Fighting as an independent maneuver element under Division command, they would no longer be pushed over to the east of Highway 80, as the threatened presence of a brigade of T-72s would slow their attack. Instead, the Wolfpack concentrated in the western half of their zone, opening forward PAs for artillery and seizing the historic Safwan airfield (where the first Gulf War cease fire had been signed twelve years before.) The Battalion's role in the initial attack was modified, but the Marines of the Wolfpack were confident that they would get their share of the fight later on. Bravo Company, 4th LAR (attached to 3d LAR) cleared up to *Jabal Sanam*, and linked up with the elements from 1st Force Reconnaissance Company, who had been inserted on the hill at 0500Z that morning, after the Division's attack had commenced. Wolfpack continued the attack, and seized the airfield on the other side of the hill, permitting a RIP by the 51st Royal Air Force Squadron the following morning. Almost immediately, Marines from the MWSS were there to improve the airfield and establish a rotary wing FARP, the first of many they would establish inside Iraq.

First Marine Division Forward and 1st UK Tactical CPs were co-located here near Az Zubayr. Co-locating these CPs allowed smooth control of the Relief in Place between the two Coalition Units.

From its position along Highway 80, the Division Forward CP observed the action to its front and prepared to follow 3/7 and 1/7 through the eastern breaches. The mission of the Division Forward CP was to get established as quickly as possible inside Iraq in a position from which it could take control of the fight and the RIP with the Brits. By a rapid sequence of leap-frogging Division CPs, the tempo of the Division would be maintained. The planned push of the Forward

CP into Iraq immediately on the heels of the assault regiments would allow the Division Main CP to displace to the vicinity of the Jalibah airfield and prepare the Division for the push across the Euphrates River. Given the revised RCT-7 attack routing, coupled with the earlier attack by RCT-5, the CG made the decision to reroute the Division Forward CP through the western breaches at the oilfield road rather than the eastern ones along Highway 80. This would allow the Forward CP to get into Iraq sooner, over ground that RCT-5 had already partially cleared, possibly saving time. As with many decisions taken during the Iraqi campaign, timing and speed were the critical measures of effectiveness. The Division Forward CP made the long road march across northern Kuwait from its staging area along Highway 80, and entered the Rumaylah oilfields on the heels of RCT-5. Moving in a coordinated manner in trace of the Blue Diamond CP was the Tactical CP of the 1st UK Division. The combined column maneuvered into Iraq and made its way to a position to the east of the oilfields without incident. The Division Forward CP was established and assumed 'Blue Diamond' control by 1800Z. From their shared compound near the At Tubah oil storage facility just west of Az Zubayr, the UK and US staffs made final coordination for the RIP. The collocation of the two Division headquarters elements greatly streamlined the relief in place, and effectively removed friction from the command coordination. The two staffs were quickly operating together almost as one, in yet another example of the fine comradeship between the two Coalition forces. That night was clear, and from the shared position the combined staffs could observe outgoing artillery and the trajectory of incoming rockets fired from north of Al Basrah. It looked like the Iraqis in zone had more than a little fight left in them, and the Brits would have some significant combat ahead.

Marines from the Division Forward CP speed past a British headquarters unit as they move to the Iraqi border.

22 March
Attacking to the West

On 22 March, Blue Diamond engaged and destroyed isolated pockets of the 51st Mechanized Division and paramilitary resistance in zone while consolidating, refitting, and rearming. Over the course of the day, the 1st UK Armoured Division relieved the 1st Marine Division of the sector. The relief was accomplished in-contact and under fire, as the Iraqis in zone had withdrawn into the urban areas of Al Basrah and Az Zubayr, waiting for the Coalition to come in after them. Emboldened groups of Iraqis attempted to opportunistically sortie out to attack Coalition forces, but were bloodily repulsed by the Marines. As fighting continued during the day, the Division took several hundred EPWs and passed them to UK forces. Interrogations confirmed the presence of some token Republican Guard elements among the Regular Army forces, but there had been no sign of any armored brigades.

Even though the gun barrels had hardly cooled from the seizure of the initial strategic objectives, the Division began to shift its momentum to the west in accordance with the MEF commander's intent to rapidly position Blue Diamond for its attack across the Euphrates. The Division Main CP made its way to the vicinity of the western edge of the assigned battlespace, and by the evening had reassumed control. With an eye toward a rapid advance across the Euphrates, the Division closely monitored TF Tarawa's progress in its own RIP with the US 3d Infantry Division south of An Nasiriyah. The Army's planned attack on An Nasiriyah had changed significantly since the early planning stages, and they had left the enemy combat forces in An Nasiriyah largely untouched by their rapid attack to the western bridge site. It would now be up to TF Tarawa to destroy the enemy's capability to influence the progress of 1st Marine Division across the Euphrates. Tarawa's progress in conducting a RIP with the US 3d Infantry Division, then eliminating the threat from the city, would dictate the timeline for the Division's exploitation attack across the Euphrates. The Division continued coordination with Task Force Tarawa, and prepared to support as possible.

Iraqi soldiers often ditched their uniforms and fled the fight. Few conventional Iraqi forces were willing to confront the 1st Marine Division and 3d MAW team.

Screening the northern flank of the Division along the route of march, 1st Reconnaissance Battalion operated parallel to the Saddam Canal. Their mission was to provide early warning of enemy movements and prevent the enemy from posing a counterattack threat into the westward-moving Division's right flank (from the north.) The Battalion established observation points and checkpoints near the existing rail and road bridges, and monitored activity in the former marsh areas north of the Saddam Canal. Instead of tanks attacking from the north, Recon Battalion encountered hundreds of military-age men in various states of civilian attire fleeing to the west. These were the remnants of the 51st Mechanized and 18th Infantry Divisions who had abandoned most of their uniforms and equipment and were now trying to get back to their homes. These former soldiers were more relieved than fearful when encountering US forces. They told stories of the Regime execution squads who were tasked with killing soldiers who retreated without orders or abandoned their posts. The crush of numbers precluded taking all of these unarmed men as EPWs. Instead, the Reconnaissance Battalion ensured they were no longer a threat, then let them go on their way. Over the course of the day, Recon Battalion expanded their screen to cover the area between the Saddam Canal and the Euphrates River. They continued to guard the bridges leading from the north, ensuring no organized counterattack could come across the water obstacles into the Division flank.

To the south of the Reconnaissance Battalion, RCT-1 finished clearing Highways 1 and 8, which formed Main Supply Route (MSR) Tampa. They performed a reconnaissance in force to the limits of the Division's zone. RCT-1 units reported a small number of tanks and technical vehicles moving west to east on the north side of the Saddam Canal, and AH-1W Cobra gunships were called in. While clearing MSR Tampa, RCT-1 encountered the same fleeing young men that the Recon Battalion had. Many of these former Iraqi combatants surrendered to RCT-1. Also in their zone, 1/4 secured the large Iraqi ammunition supply point at Al Jazeer (with its large walled and fenced compound), which was planned to become a major theater Enemy Prisoner of War (EPW) holding area. Later in the day, after clearing the zone and coordinating with Task Force Tarawa, RCT-1 was finally cleared to move west in order to relieve Tarawa of the Euphrates River crossing points to the west of An Nasiriyah. The Marines from RCT-1 passed through the TF Tarawa zone and took ownership of the Highway 1 bridges across the Euphrates west of the city. RCT-1 then turned the bridges over to Marines of the Division's Tactical Movement Coordination Center (TMCC), 3[d] AA Battalion. The route was now clear for the Division to continue the attack, bypassing the continuing fight in An Nasiriyah. Having accomplished their primary objective, the Marines of RCT-1 turned their attention toward passing through An Nasiriyah and beginning their supporting attack up Highway 7. There would be some delay here, as RCT-1 would have to again pass through TF Tarawa's battlespace. Task Force Tarawa was still engaged in the An Nasiriyah urban area, and there was some significant fighting required before this route would be open.

Advancing Marines encounter surrendering Iraqis
Darrin Mortenson
San Diego North County Times
March 22, 2003

Ignoring the drizzle that spattered onto the dust of the Iraq desert early Saturday morning, Lance Corporal Dustin Skaggs trained his M-16 rifle on the kneeling Iraqi soldier. 'I think if they know we'll treat 'em good, they won't want to fight us,' said Skaggs. The Iraqi soldier was one of four who abandoned a small pickup truck alongside a northbound highway and fell on their knees in front of the troops of the 3d Battalion, 1st Regiment. Four Marines from the Headquarters Company of the "Thundering Third," guided the prisoners in closer to their unit and away from the road, telling them with hand and arm signals to lay face down in the dirt with their arms out. When one of them, a captain in the Iraqi Army, complained of heart pains, Lance Corporal Roderick Fuqua ran to get a Navy corpsman. The Marines allowed the officer to get on his knees while the others lay with their cheeks pressed to the sand, which was slowly turning to sticky mud globs as a light morning drizzle trickled down. A corpsman checked the grateful prisoner and gave him some water before turning him over to an intelligence team that included an Arabic translator.

Logistics Generate Tempo

Logistics was the critical enabler for the entire operation, and the Division turned to accomplishing the tasks of rearming, refueling, and refitting the Division on this day. The focus on lightening combat loads to the smallest possible had allowed the Division to rely exclusively on organic lift to move itself. The Marines emerged from the first combat operations with a newfound appreciation for their self-sufficiency. With no requirement to shuttle trucks, operational tempo had been sustained. Essential classes of supply carried on trucks of the Division's CSSG-11 reduced reliance on the FSSG and other supporting commands for immediate re-supply. The maneuver units still needed to replenish chow, ammo, water and (most of all) sleep. Many of the Marines had not had decent rest since before crossing the LD, some 48 hours before. The Marines had much fighting yet to do, and now was a brief opportunity to catch some rest before resuming the attack.

At this point, CSSG-11 and its CSSC's demonstrated what organic, integrated logistics support meant. With precision and foresight, Colonel John Pomfret and the CSSG-11 staff established RRPs near the tactical assembly areas of RCT-7 and RCT-5 to allow them to quickly obtain the necessary supplies to sustain their attack. As the Division's objectives quickly fell, logisticians anticipated the demand for fuel forward, and redirected assets from these RRPs to

British forces from the 16th Air Assault Brigade relieve 5th Marines in the South Rumaylah oil fields.

RRP 7 closer to An Nasiriyah. Now, the entire Division made its way through these pre-staged RRPs to top off fuel tanks prior to crossing the Euphrates. By the detailed calculations of the logisticians, the refueling here would allow the Division to continue its attack all the way to the Hantush Airstrip (the next Division objective) before another RRP was required. With hardened, fully loaded vehicles consuming fuel at a greater rate than planned, this timely shift in emphasis enabled units to conduct in-stride refit, refuel, and rearm for the attack up Highway 1. CSSG-11 had met its first real battlefield challenge and overcome it handily, and the affirmation of the principle of organic, mobile CSS elements was a confidence builder for the RCTs. One concern remained, however, as repair parts and other support required from the fixed supply points were not making it to the combat forces at all. This would be a significant logistics shortfall that would plague the Division for the entire campaign.

On 22 March, RCT-5 completed a RIP in the South Rumaylah oil fields with the 16th Air Assault Brigade, 1st UK Division. Following the relief in place, the RCT conducted a road march on Highway 8 to near the Euphrates River at crossings west of An Nasiriyah. The RIP of artillery was the simplest of all. As the artillerymen had already established both US and UK firing positions in support of operations in the Rumaylah area, the US firing battalions were able

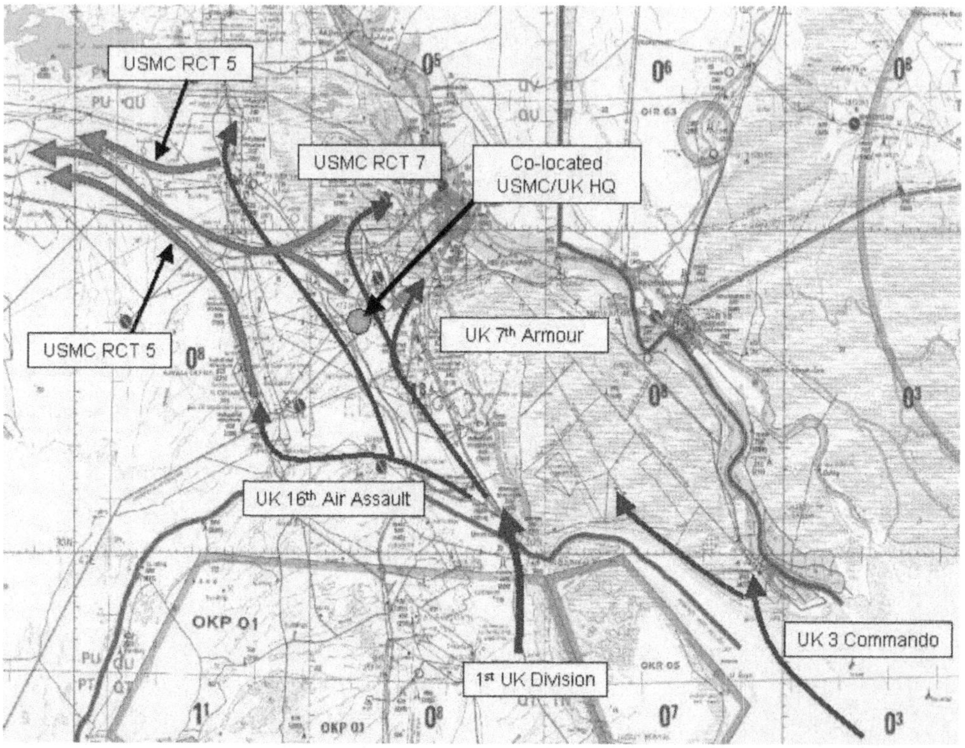

First UK Division relief in place of 1st Marine Division. They UK forces would assume control of Basrah and the Rumaylah Oil Fields, enabling Blue Diamond's rapid attack west.

to quickly disengage with no loss of fire support capability. The 7th Royal Horse Artillery, supported by its own counterbattery radar and supporting elements, seamlessly picked up the fight when 11th Marines rolled to the west.

Ripper Cleans House at the Shaat Al Basrah

Although the Division's objectives were secured, the Iraqis that had fled into the Al Basrah urban area still had fight left in them. Repeatedly, they attempted to sortie across the Shaat al Basrah bridges and counterattack the Marines.

Through the morning of 22 March, RCT-7 continued to block the bridges over the Shaat al Basrah, meeting sporadic resistance and periodic poorly organized counterattacks by uniformed and non-uniformed enemy forces. First Tank Battalion and 3/4 were continuing to use indirect and direct fire against targets on both sides of the waterway. Battle damage reports from the morning included at least two artillery pieces and one BMP destroyed. Although many of the Iraqi conscripts had fled the previous day, it was clear that many of the remaining enemy hardliners were willing to fight to the death. Technical vehicles, T-55's, and paramilitary fighters in civilian clothes continued to attack Tiger's lead elements, its headquarters, and support areas. With the fight continuing and its units spread across its zone, Tiger had not yet consolidated since crossing LD. The combat trains, command group, and TOW Platoon (-) had co-located at a crossroads near Az Zubayr. Shortly after staging at this location, the Marines in the vicinity of the command group heard the 'zip' of an anti-tank missile and an M-88 Tank retriever was rocked by a blast. Three Marines had been wounded (Master Gunnery Sergeant Guadalupe Denogean, Sergeant Eric Percy, and Lance Corporal Phillip Rugg). Captain Ruben Martinez, the 1st Tank Battalion S-4, recalls:

"As I completed shaving, I saw an RPG smoke trail out of the corner of my eye originating from about 175 meters to the south of the M-88. As I turned to look at the M88, I saw Sergeant Percy standing in the vehicle commander's cupola. The explosion rocked the M88, causing a fireball to come out of the vehicle and smoke to begin billowing from the vehicle. Corporal Rugg, a crewmember, began to exit the M88, standing just outside the hatch and turned toward Sgt Percy to ask what happened. Sergeant Percy shouted that he couldn't see. I could see Corporal Rugg's lower legs badly injured with many bleeding lacerations. Corporal Rugg then jumped down from the vehicle, obviously not realizing the extent of his wounds. He landed on his feet and immediately fell over yelling "Oh my legs!" I could hear sporadic small arms fire being fired at our position and being returned by the Marines. After washing Sergeant Percy's eyes out with water, I called in the contact report and the number of injured Marines. As I was on the radio, I observed Sergeant Jones pulling Master Gunnery Sergeant Denogean's limp body out of the burning M88, assisted by Corporal Rodriguez. We would later learn that Chief Lugo was inside the M88 passing the Master Gunnery Sergeant out of the vehicle. All three injured Marines were consolidated near the BAS vehicle and prepared for evacuation. Sergeant Jones returned and entered the burning M88 to remove AT-4 missiles that might have gone off due to the heat. Marines from the Bravo Command and combat trains secured a field on the west side of the AA for use as a landing zone."

Forward at the center Al Basrah bridge, Lima Company, 3/4, (attached to 1st Tank battalion) assessed enemy casualties had reached 100 fighters killed, but had encountered only feeble, mounted counter-attacks. This was about to change. Just as the command group's situation began to stabilize, Lima reported observing numerous BMPs and T-55's on the far side of the bridge, apparently preparing to attack. Delta Company, 1st Tanks, under Captain Greg Poland, relieved Lima Company at the bridge. A four-man British team, who had intended to cross into Al Basrah, now realized the bridge was too hot. Instead, they co-located with the Delta Company commander, providing intelligence from sources within the city. Delta Company's Red Platoon was maintaining the blocking position, awaiting the Iraqi attack, which soon came. An apparent suicide charge by a 'technical' with a mounted machine-gun, and two beat-up Sport

Utility Vehicles, was engaged and destroyed approximately 500 meters from the platoon's positions. The Red Platoon Commander, Lieutenant Vincent Hogan recalls:

"Any illusions of keeping friendlies clear of tank main gun surface danger zones during this war quickly evaporated as the enemy was mixed in throughout the Battalion's elements. The Marines of the command group, combat trains, and TOW platoon immediately engaged the enemy, and cleared buildings in the area with tank main gun fire. Captain Martinez and Major Keith completed coordination for the air CASEVAC [Note: For the injured crew of the M-88], *and the choppers were on the deck within minutes of being called. We were about a half hour into the blocking mission, and it had calmed down considerably. We first observed the three vehicles when they stopped about 1500m to our front; the lead vehicle was clearly a MG-mounted 'technical'. As we went to REDCON-1, we observed 10-12 armed personnel on the 'technical'; the gunner put on a ski mask. Suddenly they were hauling ass towards our position. Red-2 shouted, 'RPG!' seeing one passenger lift an RPG out of the vehicle bed. We engaged with coax and .50 cal. In a minute or so, two vehicles were burning, with one remaining on the road. No one was moving. A couple minutes later, people started picking up weapons scattered from the 'technical'; we continued to engage. The whole engagement lasted maybe 10 minutes. After the fight, some of the gunners were able to observe through their powered sights 'TV' sloppily taped on the sides of the SUVs. I decided to inform the S-2 about this potentially useful, enemy vehicle marking. It didn't even occur to any of us that journalists would try to participate in an Iraqi suicide charge."*

Upon completion of the engagement, it was ascertained that the two trail SUVs were, in fact, foreign journalists from the UK who had participated in the charge against the US blocking position. Whether this participation was voluntary in pursuit of a good story, or against their will at gunpoint, was never determined. Their resulting death was an unfortunate accident of war. Journalists, intermingled with the enemy combatants, were not identified as non-combatants until after the enemy attack had been repulsed. Subsequent investigations confirmed that the actions of Delta Company were consistent with established rules of engagement. This was scant solace to the families of the journalists or to the Marines who deeply regretted any noncombatant casualties.

The unfortunate action involving the journalists was just one of the many disorganized attempts by the enemy to cross the bridges. Those that tried were quickly and ruthlessly reduced. Gunnery Sergeant Steven Heath, Delta Company, 1st Tanks describes another action:

"I saw two 'technical' vehicles come across the bridge with a T-55. On the near side of the bridge, about 2700 meters away, the vehicles had formed a wedge with the tank in the lead, two 'technicals' flanking. We saw them at an angle. The tank had guys riding all over it, World War II-style. Our first shot, sabot, went through the chest of a tank rider, and destroyed the 'technical' on the far side. The tank stopped and everyone jumped off. I fired the second shot: through a berm, in front of the tank, a hull shot. Then, a guy in black jumped in the driver's hole and started it. There was just enough time to see the white puff as the engine started and the tank began to move. Red-3 put a shot square in the turret, causing immediate 'secondaries'. Good section gunnery. It was almost like the Iraqis tried to adapt Soviet tactics to 'technicals', like a Combat Recon Patrol. It wasn't working out too well for them."

Tiger used air to continue to hammer the forces on the other side of the bridges. Captain Anthony Muralt, Delta Company's FAC recalls the response of one Cobra pilot when asked for BDA as, "too much to get an accurate count on." According to the British teams working with Tiger (receiving battle damage via cell phone), eight to ten tanks, twenty to thirty trucks/SUVs, and three APCs were destroyed in these strikes.

Also on the morning of 22 March, 1st Tanks captured a group of Iraqi soldiers dressed in civilian clothes, one of whom was alleged to be the Commanding General of the 51st Mechanized Division. (This was the first of many times the 'commander' of the 51st was reported captured, as the confusing Iraqi rank structure and multiple Regime agencies on the battlefield resulted in a large number of captured Iraqi General Officers.) Among this group's possessions were weapons and uniforms in bags identifying them as Republican Guardsmen, and a white powder substance in a sealed bag with instructions in Arabic for the employment of anthrax. A Sensitive Site Exploitation team was brought in to test the contents. Although tests revealed the powder was not anthrax, RCT-7 remained vigilant and prepared to encounter WMD. As 1st Tank Battalion began its displacement to the north, mortar rounds began to fall on their position. Lieutenant Popielski and his TOW platoon, while covering the command group's move, were attacked by an RPG team. The vehicle closest to the attacker, a HMMWV (TOW-variant) responded with the only weapon system available, a TOW missile. The missile was erratic, but successfully scared the attackers off. The TOW platoon quickly attacked south and destroyed several additional weapons caches. Upon relief by the UK's 7th Armoured Brigade, Tiger moved to its Assembly Areas (AAs) in preparation for the march west. Upon arrival in the AA, all battalion elements refueled and began catching what sleep they could.

Captured Iraqi Regular Army Officers.

The battle to isolate Al Basrah was largely concluded by midday on 22 March. At BIA, the focus for 3/4 shifted to logistical operations and the RIP, which was planned to take place immediately after the Marines secured the area. The First Fusiliers, of the 7th Armour Brigade, 1st U.K. Division, wasted no time in settling into positions and making the transition. Elements of 3/4 consolidated in the vicinity of the RCT-7 Main CP and conducted resupply operations throughout the night in preparation for movement. By this time, fatigue began to play a dominating role, as exhaustion from over two days of continuous combat operations

An Iraqi traffic jam. It was not uncommon to see droves of camels or former Iraqi soldiers in civilian clothing walking alongside Marine convoys.

made the relatively simple tasks of road marches, link-ups, and refueling operations very difficult. It took Kilo Company, for example, many additional and frustrating hours in a black night to navigate out through British defenses at BIA and find the battalion's resupply point. The Marines were very ready to settle into an assembly area to rest and refit.

Many enemy vehicles were destroyed and personnel were captured during this first operation. In the largest capture of the day, First Lieutenant John Dyer accepted the surrender of 64 EPWs at a location near the Main CP where suspected rocket fire had come from. The party was surprised when one of the vehicles that was found at the location was an armored personnel carrier with the red crescent symbol painted on it (the Iraqi equivalent to the Red Cross), but with an 82mm mortar mounted and ready to fire in the back. As anticipated, Marines would encounter this tactic repeatedly in the coming weeks. Schools, mosques, and hospitals were some of the protected-status sites used to store vast quantities of ammunition and weapons. In the garrisons of the Iraqi RA, Marines found functional enemy equipment simply abandoned.

First Marine Division moves west along Route Tampa toward the Euphrates River.

RCT-7 had embedded a British Army Major Liaison Officer (LNO), Major Andy Flay, into its staff in late February to facilitate coordination with the 7th Armoured Brigade of the 1st UK Division ('Desert Rats'). The RIP began close to first light, as the 7th Brigade staff co-located its CP with the Ripper Main, to ensure coordination. As with the 7th RHA, the 3d RHA was already in their firing positions, and seamlessly picked up the fight where 11th Marines left it. Additionally, the British Signals Intelligence Units were already in-place, having gained an appreciation of the signals environment by crossing the LD with Marine Radio Battalion elements. The RIP of units in the RCT-7 zone took place from south to north and then west to east. The 'Black Watch' Regiment assumed 3/7's positions at Shuaybah barracks and 1/7's positions at Az Zubayr. The 'Royal Fusiliers' Regiment relieved 3/4 at BIA. The RIP took most of the day to complete, with RCT-7 units, one by one, pushing west to set up in a new AA. By 1948Z, the Brits and TF Ripper completed the relief in place, and RCT-7 was on the move. CSSC-117 established RRPs for each of the battalions as they withdrew toward Highway 8. Once replenished, RCT-7 consolidated in Tactical Assembly Area Seven (TAA-7), and prepared to follow the rest of the Division's movement west along Highway 8 toward An Nasiriyah.

23 March
Blue Diamond Maintains the Tempo

The RIP with the 1st UK Division was complete, and the Division now looked only to the west. The cohesive relationship cultivated between 1st Marine Division and their comrades in the 1st UK Division – developed over weeks of close coordination at all command levels – had allowed for a timely and seamless RIP. There was a tremendous lifting of a burden for staff officers in every unit, and a firm sense of confidence on the part of the Marines. For months, they had trained for the missions in South Rumaylah and near Al Basrah. They had pored over maps, reviewed imagery, and memorized every detail of the terrain and enemy situation. They had visualized how they would execute the attack, each man preparing himself to perform under fire. Now all of these issues had been settled. The tremendous success enjoyed by the Marines in seizing the strategic oil infrastructure intact was now fact. The performance of these Marines, most in combat for the first time, was proven. In dozens of different units, Marines spent quiet minutes contemplating their victory. Soon, however, Staff NCOs across the Division were coming down the line shouting that it was time to move out again.

Colonel Kelly, 1st Marine Division's ADC, is promoted to Brigadier General in the vicinity of Az Zubayr.

The Division accomplished some important tasks during this brief respite. With Secretary of Defense authority, the CG frocked Colonel John F. Kelly, the Assistant Division Commander (ADC), to the grade of Brigadier General at the Division Forward COC located in the South Rumaylah oil fields. The last known promotion of a Marine Brigadier General in an active combat zone was that of an earlier 1st Marine Division

ADC – then Colonel Lewis B. "Chesty" Puller in Korea.

Elements of 3d AA Battalion were in control of the western bridge site over the Euphrates River. The rest of the Division started to approach the crossing, passing through the Task Force Tarawa battlespace along MSR Tampa. 3d AA Battalion established traffic control points to speed the movement. The Division soon neared the 'hourglass', that key chokepoint south of An Nasiriyah that had been the subject of so many sand table exercises, Lego Drills, and Jersey Drills. Getting the 1st Marine Division through this chokepoint rapidly would become the new priority of effort.

Objects in mirror are larger than they appear. Embedded reporters moved alongside Marines as they approached the Euphrates.

Marines captured this bridge over the Euphrates, supporting the rapid movement of mechanized and armored forces across the river.

While most of the Division's combat power moved along the Highway 1 and Highway 8 corridors to the west side of An Nasiriyah, the Marines of RCT-1 began planning for their own crossing. RCT-1 was prepared to accomplish a variety of missions, and staged at the rear of the Division's column waiting its turn at the eastern bridge. The RCT was prepared, however, to move directly through the An Nasiriyah area to reach Highway 7 to the north, if the enemy situation allowed it.

At the western bridge, the Lego and Jersey drills had only partially prepared the Division for the tremendous backlog of men and equipment trying to make their way across the river. Under the eyes of the TMCC, 8th Engineer Support Battalion engineers worked diligently to improve the approaches to the two bridges at the site. The TMCC Marines tried to keep the site clear. Lieutenant Colonel Abbott described the scene that awaited his Marines at the crossing site:

"Marine columns of vehicles were heading north on both sides of highways 1 and 8. On the west side of An Nasiriyah a huge traffic jam was backing up at the river. It seems that the 6-8 paved lanes were going down to one dirt road trying to scrape across one unfinished bridge to continue up highway 1. Headquarters Battalion's MP Company was on the scene, prioritizing movement, per the Division Ground Transportation Order (GTO). Tempers were high, as every commander was hot to get into the fight across the Euphrates River. Clearly the sense of urgency and the tactical situation would not tolerate the bottleneck. The MP Company was quickly reinforced with the Tactical Movement Coordination Center (TMCC) and more Marines from the 3rd AA Battalion Combat Trains. An artillery D7 bulldozer was coaxed off its trailer and the operator made short work of a high berm that was blocking traffic. Marine engineers were already at work repairing the ramped approaches up to and down from the large heavy vehicle bridge on the right side. There was a smaller, fragile bridge to the left that was inexplicably not being used at all. The AA Battalion commander used the 'TLAR' method of bridge classification and drove across. Sergeant Danny Hamler asked, "What does 'TLAR' stand for?" He was told, "That Looks About Right. If the bridge doesn't hold the Hummer, than we won't use it anymore." It held the weight and was deemed good enough. The TMC Marines ran engineer tape and concertina wire to create two lanes up to the two bridges. All Hummers were routed left and everything else went to the right. After about two hours of frenzied activity and pandemonium, traffic really started moving smoothly without delay."

The old construction bridge over the Euphrates. Light vehicles crossed this bridge while armored vehicles crossed the new bridge.

The recently completed Highway 1 Bridge itself was an impressive structure, rising high above the river, but the approaches to the bridge were unfinished. The concrete bridge abutments dropped off steeply onto the dirt approaches. With Lieutenant Colonel Abbott's Marines looking on, Marine engineers worked rapidly to fill in the 'step' on either end of the new bridge, while light vehicles used the adjacent construction bridge. Getting the engineers to the water's edge through the tightly congested traffic was tough. From the southern bank, Division Marines watched the reduction of this last obstacle to forward progress. Over the previous months, Marines of the Division had (via imagery) 'watched' this bridge as it slowly grew across the river. The Marines had cheered the progress of the Iraqi engineers all the way from Camp Pendleton months before the attack. When the Division's plans were originally formulated, there was only a small construction bridge at this spot. Now, more than a year later, the small construction bridge was still there, but the shiny new concrete monster next to it dwarfed it. Every day that the attack had drawn nearer, Marines of the Division joked that it would sure be a shame for the Iraqis to destroy the new bridge now, after having spent to much time building it. Inexplicably, both bridges had been captured intact, and now all that remained was for the Division to work its way through the inevitable traffic delays it had foreseen in getting across the Euphrates River.

A typical mud-walled family farm in Southern Iraq. Small compounds like these housed large families or clan groups.

While the engineers worked to get the bridge site fully operational, units from across the zone began to queue up for their crossing of the Euphrates. The last of the Division's combat units had disengaged from the 1st UK Division's new zone, and the Division Forward CP was the last to leave. The Division Forward CP, however, would have to be one of the first units across the Euphrates at the western bridge to maintain the Division's operational tempo. Starting on the far eastern side of the Division as it spread out across southern Iraq, the Division Forward CP displaced from Az Zubayr, and rushed across the entirety of the Division battlespace to take its place behind 3d LAR in the column. The Forward CP would eventually set up 30 km north of the River to allow the Division Main to leapfrog forward. This preplanned repositioning of Division command and control elements forward was critical to sustaining momentum throughout the campaign, and it was a especially a tribute to the communications Marines who were able to continually support these long-distance leapfrog movements without loss of communications. Their ability to do so was the result of careful planning and preparation by the Division G-6, Colonel Nick Petronzio, who had faced similar challenges during the last Gulf War (when he had been the G-6 Operations Officer for the 2d Marine Division.)

> *"As I'm riding past a crowd of people who are waving cheering despite the fact that we have weapons pointed at them I'm struck by the fact that I am, for some reason or another trusted by peoples that have never seen me, don't share my language, customs, or religious beliefs, and know nothing about me except that I'm from the United States of America."*
>
> - Corporal Deymond Lashley
> Weapons Platoon, I 3/5

As the convoys of the Division continued their way west, dozens, and then hundreds of Iraqi young men in civilian clothes, most with only combat boots or no shoes at all, were seen walking along the main and secondary roads in the same direction. These were the conscripts of the Iraqi 51st Mechanized Division and other elements, now making their way back to their families. Division Marines realized that these men, now unarmed and no longer a threat, were not the enemy. Even though they had been reluctantly forced to bear arms against the US only days before, they were now some of the very people Blue Diamond had come to liberate.

Iraqi children wave, hoping to catch a handout from passing American convoys.

The units of the Division waited for their turn to cross with some apprehension. The fight so far had been one-sided. Highway 1 took a wide arc around the Tallil airbase on the southwest side of An Nasiriyah, skirting the urban area at a range of about 20 km. Off to the east, the Marines could see helicopters flying and smoke rising from the An Nasiriyah urban area. TF Tarawa was at that time in a tough fight with resistance fighters in the town, and it would be over a week before the city would be considered secure. The Division Marines moving along Highway 1 toward the bridge kept their focus on their next objectives. RCT-1's attack up Highway 7 would fix the enemy in-place in Al Kut. Moving up the muddy Highway 1 route would allow them to steal a march on the enemy, and would soon carry them to the Tigris River and the Baghdad Republican Guard Division.

As the Division units were now immersed among the Iraqi populace, the CG took the opportunity to remind all hands to refrain from images of triumphalism. He republished the mandate against the display of US or Marine Corps flags while in Iraq. His intent was to show the Iraqi people that the 1st Marine Division came as liberators and not as conquerors. The Marines complied with the order, not without some mixed emotions. For the Marines of the Division, these mixed emotions were easily resolved when they began to see the poor Iraqi people and the brutal conditions they were forced to live in under the repressive regime of Saddam Hussein. Very soon, the Marines would come to take great pride in their actions as liberators, and of the actions of their own Nation. What other country would act like this in victory? The Marines did not need flags to display their fierce pride, it was present in the faces that peered out from under their dusty helmets.

Handing out Humanitarian Rations was part of the "No better friend" ethos.

After spending so much time reading about the Euphrates and looking at it on the map, most of the Marines were eager for their first glimpse of the river. Heads eagerly strained out of trucks and tracks to catch a glimpse of this historic waterway as the columns crossed. The day was particularly hazy, and the muddy brown river with low banks failed to live up to expectations for this historic moment. Only an occasional date palm or mud-walled house broke up the flat muddy plains around the river. Until now, the Marines had traveled through wind-swept desert with burning oil fields, and blowing dust. The smells had been sterile, industrial - diesel smoke, oil fields, and dust. Here, the smells were organic, earthy - overpowering. Despite the historic occasion, the scene elicited a number of irreverent comments from the Marines, many referring to the smells associated with the 'cradle of civilization'.

Once across the river, it was clear why the Iraqis had not made a major defensive effort between the two Rivers. The Iraqis had certainly not expected the Americans to use this unfinished Highway 1 as an avenue of approach. The terrain on either side of the unfinished roadbed was an untrafficable patchwork of muddy fields, drainage canals, and mud berms. Beyond An Nasiriyah, there were no cities along Highway 1 until Ad Diwaniyah, 150 km to the north. The poorest of Iraq's citizens scratched a living along the banks of the canals, mostly as subsistence farmers, shepherds or camel herders. As the column advanced with 3d LAR's LAV-25s in the lead, a small number of the citizens began to appear on the sides of the road, begging for food. Once over their initial fear, the people began to tentatively wave.

Crossing the Euphrates River was anticlimactic for many Marines.

The Highway 1 roadbed was actually raised above the surrounding terrain on a 20-foot high levee. The road was not paved in this section, and numerous deep culverts cut across the road for the first 50 km, forcing traffic off the raised highway, and onto the construction road that ran parallel to it. (These culverts were bridged over the next several days, allowing the Division's re-supply convoys to move along the improved surface road on top of the levee.) The worst sections of the road were the first encountered by the Marines. As the Division pushed north toward the Tigris, the road quality improved, eventually

Armor moves across the newly constructed bridge Over the Euphrates.

becoming paved, prompting a sigh of relief from the Intelligence Marines who had recommended the route. The Division's decision to ask for the Highway 1 battlespace had been the right one. The Division had taken a measured risk, and now it was positioned on a high-speed avenue of approach leading straight to the enemy's heartland and the Division's major objectives.

The Division continued to pay a human toll in its advance toward Baghdad. Second Tank Battalion lost one Marine killed to a non-battle accident. In another action, the Division Staff Judge Advocate (SJA), Lieutenant Colonel John Ewers, and the Division's Reportable Incident Assessment Team were ambushed in the vicinity of Az Zubayr while enroute to investigate the circumstances surrounding the deaths of the UK journalists killed earlier in the Iraqi suicide attack across the Shaat al Basrah bridges. The CG had established this team to be able to immediately investigate the circumstances of any significant collateral damage incidents, friendly fire accidents, or non-combatant deaths. The team was passing through Az Zubayr on their way to the site of the incident when several RPG gunners ambushed them in the narrow urban streets. In the ensuing running firefight through the town, Lieutenant Colonel Ewers was shot in both forearms and his left foot. Lance Corporal Henry Lopez also took two rounds to the right shoulder while engaging the enemy. Both Marines were successfully evacuated and treated for their injuries. When he heard of Lieutenant Colonel Ewers' injuries, Brigadier General Kelly remarked, "Damn, they shot the only lawyer I ever liked."

A TOW gunner conducts security along route Tampa with a M-240G machine gun as his HMMWV approaches the Euphrates River during a road march.

Chapter 5
Bridgehead Over The Tigris

By securing the western bridges over the Euphrates River, the US 3[d] Infantry Division had set the stage. Expanding this bridgehead and moving rapidly across the river, Blue Diamond had successfully bypassed the entire Iraqi IV Corps, and was now in a position to exploit its success all the way to the Tigris River. Although the Division doubted that the bypassed enemy divisions had the capability to attack, it nonetheless kept a close watch on the exposed eastern flank. From 23 March until 3 April, the Division's operations focused on maintaining the tempo of the attack, and rapidly gaining a position on the north side of the Tigris River. Historically, this terrain between the rivers in south-central Iraq had been a no-man's land. Defenders had typically had tied themselves to the key water and land approaches to Baghdad, leaving attacking forces to the mercy of terrain and weather as they slogged through the swamps of Mesopotamia. The 1st Marine Division would now face the same terrain that had stymied attackers for centuries as it threaded its way between the ruins of Babylon and the Garden of Eden.

The Iraqis React to the Opening Gambit

As 1st Marine Division looked to exploit its early success, the Iraqi Regime also began to consider its next moves. The Regular Army south of the Euphrates had disappeared in the face of the overwhelming firepower of the MEF. Now, the enemy chose to show a different face as the Division resumed its attack to the Tigris River.

In An Nasiriyah, TF Tarawa still dealt with the defensive effort by paramilitary forces operating inside the urban area. The vigor of this defense surprised some, as the Shia Muslims of An Nasiriyah had shown a keen resentment of the Regime prior to the commencement of hostilities. The Iraqi RA 11th Infantry Division had been garrisoned near An Nasiriyah for this very reason. The Iraqi 11th Division was known to have had a difficult time controlling the population, and was expected to rapidly surrender when confronted by the US 3d Infantry Division's attack. Perhaps this very fact caused the enemy to reinforce the city with significant numbers of strong-willed Fedeyeen fighters, reportedly emboldened by the presence of Ali Hasan al-Majid (Saddam Hussein's cousin) in their midst. The infamous 'Chemical Ali' was one of Saddam's key advisors, and had murdered thousands of Iraqi civilians, many with chemical weapons. As this was one of the first major urban areas secured by Coalition forces, the paramilitaries had not yet tasted the defeat that would dampen their enthusiasm later. Not fully understanding the strength of resolve of the Marines who opposed them, the paramilitary fighters remained motivated to fight in the An Nasiriyah urban area. Another possible reason for their enthusiasm may have been their initial successes against American forces, including the capture of US Army Soldiers from the 507th Maintenance Company when their convoy had become lost in the city. The Fedeyeen were so confident that they did not destroy the eastern Euphrates River bridges inside the An Nasiriyah urban area. While the MEF's TF Tarawa battled for control of the city and the intact bridges, the main effort of the 1st Marine Division bypassed the city to the south and west over the Highway 1 corridor.

The events in An Nasiriyah were but a harbinger of the Regime tactics that the Division would shortly encounter. There were few conventional defenses between the Euphrates and

Tigris Rivers, but the Saddam Fedeyeen, Baath Party militia, and reconstituted remnants of other military groups launched a new defensive strategy based on human wave attacks, ambushes, and treachery. Collectively referred to as 'paramilitary fighters' or 'Regime Death Squads', these fighters demonstrated a zeal for combating Americans that their conventional brothers in the south had not. The paramilitary fighters proved treacherous and ruthless in their attacks, and were despised by the well-trained Marines. To the paramilitary fighters, violating the Law of War became a shield they could use to exploit the Rules of Engagement (ROE) of the well-disciplined Marines. They set up their command posts and operating bases in mosques, stored their weapons in schools, wore civilian clothing to hide among the civilian population, and forced the local population to fight the Americans through intimidation and murder.

The Saddam Fedeyeen was the core of this resistance. These loyalist forces, sent down from Baghdad, used intimidation to force the local populace to resist. The Fedeyeen entered the towns in the south, taking advantage of the existing Regime control structure and Baath Party offices. The Fedeyeen had been trained to resist to the death, and often demonstrated an eagerness to do so. They were joined by equally committed foreign fighters drawn from several nearby countries in a misguided *jihad*. The irony that many of these religiously motivated fighters would show such loyalty to a Regime that had systematically repressed the religious freedom of Muslims was apparently lost on these groups, not known for their enlightened intellect. This core group of fighters filled out their ranks through 'press gangs' that terrorized the local inhabitants, kidnapping young men and forcing them to fight. Stories abounded of black-pajama clad Fedeyeen bursting into local farmers' homes and putting a gun to the head of their children. The poor conscripts were given minimal training by a cadre staff, then put into fighting positions and kept there at gunpoint. The mix of dedicated loyalists and poor conscripts became readily apparent on contact. The core groups were tough fighters who had no fear of death and continued to attack beyond military reason. The conscripted component, on the other hand, looked for an early escape or surrender, as soon as their Fedeyeen 'minders' had been distracted or killed.

Marines conducting flank security along Highway 1, guarding against Fedeyeen attacks.

The treacherous tactics of the paramilitary reflected their training and misguided loyalty to the Regime. One method used by the paramilitary was to pretend to surrender to gain close approach to American forces, then quickly drop their white flags and open fire. In a variation on this tactic, trench lines of paramilitary fighters would often wave white flags when US aircraft were overhead, then quickly resume firing when the aircraft had passed by. Posing as civilians begging for food was another tactic used to gain close approach before opening fire. Still other paramilitary fighters were reported using local women and children as human shields. In some cases, the Fedeyeen took up positions in civilian homes, forcing the families to remain in the line of fire. Since these paramilitary fighters had been sent to the south from Baghdad, they had no loyalty to the local Shia populace and did not fear the collateral damage an American response might cause among the locals. In many cases, armed men would open fire then fade into a crowd of civilians, escaping the firepower of the Marines. In setting ambushes, the Fedeyeen would often assign a cadre of hard-core members to help lay the heavy weapons, mortars, and positions for the conscripts. This cadre would not only coordinate the ambush, but also would ensure the conscripts stayed at their posts. Although some labeled these dedicated fighters 'fanatics', most were well-trained soldiers fighting in a misguided cause against a liberating force whose motivations they could never understand.

The Iraqi military had once been a professional, well-trained conventional force. These soldiers had been eliminated over time, as the character of the Iraqi military had undergone a significant change. The Iraqi military was now a force of intimidation, and operated against its own populace as the greater part of its mission. The thugs and bullies that made up the new military class in Iraq had a simplistic, almost childlike, understanding of warfare. They understood intimidation and murder, but did not understand maneuver, fire support, or logistics to any great degree. When this façade of a military force was confronted with the crushing realities of integrated fires and maneuver, it crumbled to dust almost immediately. The unconventional response, however, was more complex. Many of the Fedeyeen and *Jihadi* fighters had been convinced that the Americans would not fight. Their handlers had told them repeatedly that the American warrior on the ground had no courage, and that Arab manhood would be sufficient to make the cowardly Americans run away. Many of them believed, and found out too late that this was not the case. When these overconfident fighters ran into US Marines, they were in for the biggest (and last) surprise of their lives. To their chagrin, the Fedeyeen found that Marines liked to fight, and fought with a moral and physical superiority driven not by arrogance, but by confidence, skill, and courage. The Marines were here because they wanted to be here, and there was no one better at dealing death and destruction to those who would oppose them.

The Operational Design

The Division continued its push to the Tigris, enroute to Baghdad. Not simply a movement to contact, the route and speed of the maneuver were designed to have specific effects on the enemy. The primary effect was to keep the forces on the south and east sides of Baghdad from withdrawing into the city. If Iraqis were allowed to mass in the urban area, the Coalition main effort (the US 3d Infantry Division) would be delayed. Keeping the Iraqi forces out of the urban area would be achieved by posing a credible threat to Baghdad from the east, on the north side of the Tigris River. A second desired effect was to be a catalyst for the surrender or collapse of the remaining divisions of IV Corps and the Baghdad RG Division by obtaining positional advantage astride their single lifeline to Baghdad. Getting the Division across the Tigris River, and astride Highway 6 would satisfy both of these strategic objectives.

There were a number of secondary effects that the Division sought as well. The Division had achieved operational surprise by beginning its attack up Highway 1. If the enemy reacted to this unexpected avenue of approach by repositioning forces, those forces would be exposed to air interdiction. Likewise, the effect of maneuvering against the Baghdad RG Division and turning its defenses to the north and west would expose its indirect fire assets to air interdiction. The isolation of the entire southern region (Al Kut, An Nasiriyah, Al Basrah, and beyond) from Baghdad would also greatly improve the stabilization environment. Liberating the Iraqi people and transitioning to a representative government required removing the influence of the Baghdad Regime. This was not possible as long as Regime enforcers could drive from Baghdad to Basrah through entirely Regime-controlled territory.

The Division clearly understood that the quickest way to destroy the Regime 'snake' was to sever its head, Baghdad. Thus, any Division operations south of the Tigris River were not as potentially decisive as an attack into the heart of the Regime's base of power. Because of the centralized nature of the Regime and its limited span of control, the Division G-2 assessed that defeating the enemy forces to the south in detail would be largely accomplished once 1st Marine Division was between them and Baghdad.

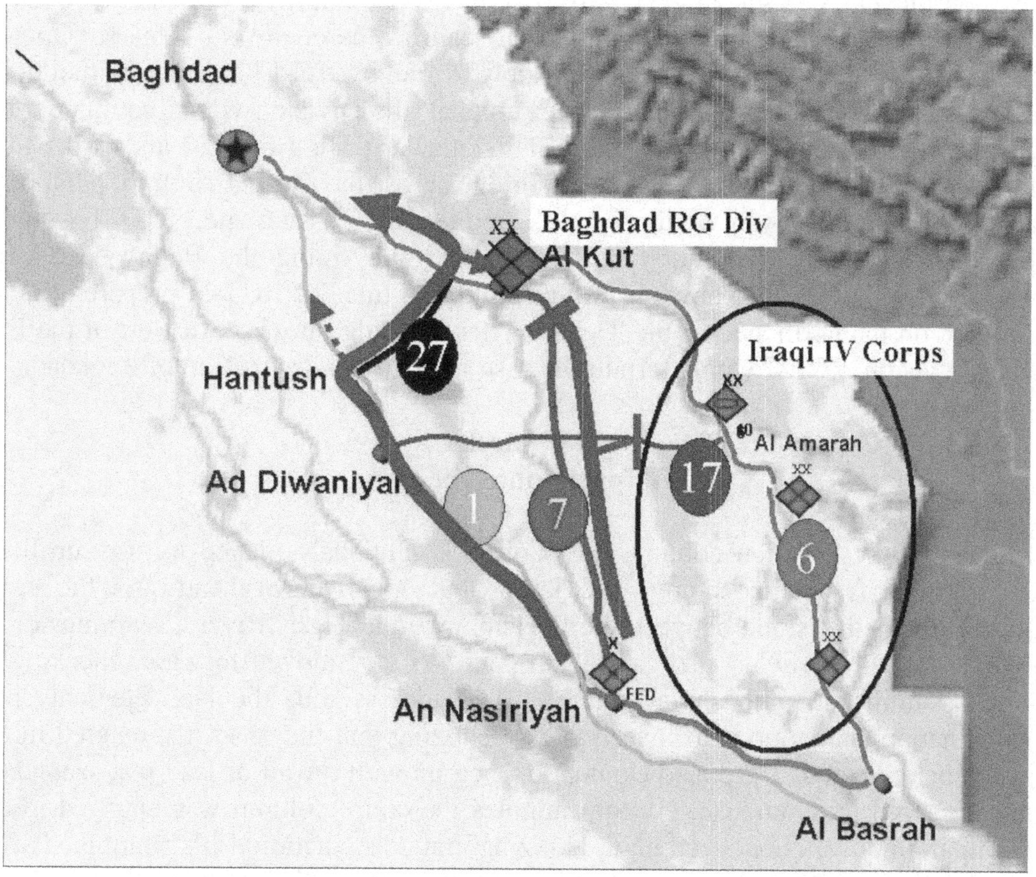

First Marine Division planned to fix the Baghdad Division in Al Kut, while bypassing them to the west in order to maintain momentum to Baghdad.

With the Highway 1 Bridge secure, the Division successfully expanded its Euphrates River bridgehead. Although not much heralded, the seizure of an uncontested Euphrates River crossing by the US 3d Infantry Division was an operational victory almost on par with the seizure of the critical oil nodes. Once crossed by 1st Marine Division elements, this bridgehead put the

Division almost a third of the way to Baghdad in a matter of less than three days. Access to Highway 1 over the western bridges, and access to Highway 7 over the eastern An Nasiriyah bridges, would enable the Division to fix and bypass the Baghdad RG Division, isolating it and the rest of the Iraqi III and IV RA Corps. The soldiers of the Iraqi 10th Armored Division may not have yet realized it, but they had already been operationally defeated when 3d LAR Battalion began to wind its way up the incomplete highway.

Leading the Division's attack up Highway 1, Wolfpack had planned to do a detailed route reconnaissance of both the Division's primary and secondary routes. When 'Chaos' contacted 'Wolfpack Six' (Lieutenant Colonel Clardy) the afternoon of 23 March, however, his message was clear. 3d LAR was to drive hard up Highway 1 and get to the Hantush Airstrip and the Highway 27 intersection as rapidly as possible. Exploitation of alternate routes would come later. The CG knew that the enemy was disoriented by the rapid pace of the Blue Diamond advance, and wanted to exploit this before the enemy realized the magnitude of the force that was coming up the unfinished Highway. RCT-5 would follow in trace of 3d LAR, ready to quickly exploit the attack to seize the Hantush Highway Airfield, then turn and go all the way to the Tigris River through the previously identified seam in the enemy's defenses. Opening the Hantush airstrip would allow for the rapid re-supply of fuel by KC-130 aircraft without clogging the single roadway with refueler trucks. Once refueled, the Division would continue a feint up Highway 1, but would change the direction of its attack at Highway 27, and strike northeast along the seam, directly for the An Numaniyah bridge across the Tigris. An Iraqi pontoon bridge 35 km upriver at Az Zubadiyah would serve as an alternate crossing site. RCT-1 would attack up Highway 7, clearing paramilitary resistance there and fixing the Baghdad RG Infantry Division in Al Kut. RCT-7, meanwhile, would follow in trace of RCT-5, prepared to continue the attack once on the north side of the Tigris. After cleaning up what was left of the Baghdad RG Infantry Division, RCT-7 would turn to the east, with a final destination of Baghdad.

23 March
Wolfpack Finds the Prey

In the lead of the Division column, the Wolfpack cautiously picked its way up the route. They had seen signs of the enemy, and knew that somewhere ahead was the screen of paramilitary fighters that stood between the Division and the Tigris River. Warnings of booby-trapped culverts and minefields also gave them pause as they moved forward, checking for the enemy and assuming an ambush lay ahead. As darkness fell, the lead element spotted a suspicious situation and stopped to investigate. Realizing that they had just entered the killing zone of a deliberate ambush, the lead element opened up with a wall of lead just seconds before the enemy triggered their ambush. Within minutes the entire column was engaged in a direct firefight with an Iraqi battalion-sized unit. Utilizing immediate action drills, combined arms, and controlled violence the Wolfpack fought its way out of the ambush. Captain Charles Blume, the Wolfpack Fire Support Coordinator describes the scene:

"Wolfpack had departed TAA Fang on the morning of 23 March. We had been ordered to begin to Recon up Highway 1 and the alternate routes called Lexus and Cobra. We had begun to move forward when the CO got a call from Chaos. The CG had told our CO to get up to Hantush and the Highway 27 Bridge over the Saddam canal as rapidly as possible, and forget

the alternate routes for now. As we pushed up, we definitely could feel we were getting well out in front of the Division. We lost comms with the DASC-A, and it was starting to get dark. We had some Cobras working in front of us earlier, and they had reported signs of enemy activity. Now, as we rolled up, we began to see abandoned weapons and equipment strewn all about along the Highway. A suspicious vehicle was seen to our front that observed us and then sped away. ROE prevented engagement of the truck, even though it was acting suspiciously. As it started to get dark, we could all feel the hair standing up on the backs of our necks. You could tell something was about to happen. Then, with Blackfoot [Bravo Company] in the lead, we began to see scattered tracers flying across the road in front of us from west to east. The scattered shots soon became a torrent of fire and the entire battalion was engaged almost immediately. They opened up on us with mortars, heavy machine guns, and RPGs. The tracers would seemingly explode when they hit the LAVs, it wasn't like the movies where they just bounce off. We later estimated the enemy at about an infantry battalion in size. There was fire coming from everywhere, and they were starting to mass and charge the column in groups. We were pouring 7.62 and 25mm on them, but they kept coming. There was no communications with the DASC, we were outside of artillery range, and there was no air on station. Unable to reach the CO on the radio and even though we were not in immediate danger of being over run the Air Officer made the decision to call 'slingshot' over the guard channel. This brought in immediate CAS. ['Slingshot' was a radio brevity code that indicated an American unit was about to be overrun, indicating all available fire support was to be vectored to its aid.] *Before we knew it, there was air coming from everywhere. There were too many sorties coming to control individually, so we established an east/west buffer of 2000m off the road and in front of our column, and turned that fight over to FAC (A) with our ground FACs directing traffic closer. We estimate we got about 20 sections of fixed wing in support of us that night."*

The Fedeyeen used unconventional tactics and vehicles to attack Marines along Highway 1.

The air support and overwhelming fires from the Wolfpack continued to effectively work over the Fedeyeen until well into the night. The devastation was tremendous. Because of the geometry of the column (linearly arrayed along Highway 1) air was able to work up to the north freely, and found a company of Iraqi tanks (10 T-55s) ready to hit the column on its eastern flank. The hunting was good that night, and 3[d] MAW effectively shut down the ambushes set along that portion of Highway 1, destroying several large enemy positions. Wolfpack received over 50 total sorties of fixed and rotary wing air support that night from Air Force and Marine aviators, making the enemy pay a heavy price for choosing to engage the Marines.

My vehicle served as the commanding officer's wingman and it was my job as gunner to watch the Battalion Commander's back. As I watched the first rounds slam into Blackfoot's flanks I recalled thinking "Holy shit!" Aside from the pucker factor, my training took over. The CO gave us the free guns order and I tried to kill as many as I could, to protect my wingman and myself. I began to engage with the 25mm main gun and watched through the thermal sights as groups of the attackers dropped. As the close air support checked in and began to attack, I had never seen air that close before. When it was all over the battalion was in a hasty defensive coil. We hit'em, and we hit'em hard.

- Sergeant David Henrikson III, of Astoria, Oregon

This attack marked the first major encounter of the Fedeyeen by 1st Marine Division. The Wolfpack fiercely responded to enemy fighters so confused or eager to die that they repeatedly rushed the column of LAVs in the face of certain death. When it was over, there were over 150 enemy casualties, 10 T-55s, and several 'technicals' with heavy machine guns destroyed. There was one Marine slightly wounded in the attack.

Lieutenant Colonel Clardy noted the complete training and technological overmatch the LAR Battalion had over the paramilitary fighters. The Marines had thermal sights, long-range direct fire capability, armor protection, and fire discipline. He recounts:

"The Iraqis obviously had someone who knew what they were doing set up the ambush position, but the guys they had executing it were really poor soldiers. We detected them and opened fire just moments

A Fedeyeen truck in an ambush lies destroyed along Highway 1. The Fedeyeen employ both military and civilian vehicles.

before they did, and that probably caused a lot of their first RPG volley to miss. After that, we had firepower dominance for the rest of the fight. The Iraqis were shocked by what they had hit. I think they opened up a whole can of whuppass that they didn't know how to deal with."

The controlled chaos within the Division's Main CP during the fight was deliberate execution at its best. Major Hank Weede, the ground watch officer, received the call of 'Slingshot' and passed the word to the 'Killing U' (as the U-shaped configuration of tables in the center of the Division CP was called.) Immediate focus from every section in the Killing U was applied to assisting 3d LAR. Division Fires immediately realized 3d LAR was out of artillery range. Radio operators cleared the TAC nets awaiting additional reports. The Division G-3 Air Officers, Major "Grams" Gramuglia and Major "Stinky" Phillips went to work coordinating 3d MAW's immediate response, which was to send (6) F-18s, (1) F-18D FAC (A), (2) F-15s, and (4) A-10s, all of which arrived on the scene within 30 minutes. As word got passed that Marines were heavily engaged on the ground, pilots of available aircraft started checking into the Division's Direct Air Support Center (DASC) at the rapid rate, looking to assist. With 3d MAW's superb immediate support, the situation was resolved in short order. It was a great exercise in confidence building, both in procedure and in the ability of aviation fires to react immediately across the battlespace.

The 'King of Battle' moves up Highway 1 to take on the Fedeyeen.

24 March
The Long March

Third LAR consolidated its position along Highway 1 during the morning hours of the 24th, and RCT-5 conducted a Forward Passage of Lines. The reported armor and increased density of fighting positions further up Highway 1 would be more appropriately cleared by tanks and dismounted infantry, and the lightly armored LAVs passed them forward. As RCT-5 moved up, intelligence reports began to trickle in that the enemy was aware of the Division's advance, and was scrambling to plug the huge gap in its defenses. Cargo trucks and busses were reportedly shuttling fighters from the Baghdad area down to the Ad Diwaniyah area, barely stopping before returning north for more troops. 'Technical' vehicles (pickup trucks with mounted heavy machine guns) and mortars were reportedly integrated into these ad-hoc defenses, as were conventional armored vehicles and tanks. Arms caches were dumped along the road in front of the Division, as the enemy raced to plug the gap before the Marines could exploit it all the way to Baghdad.

To the north of the Division, the Republican Guard still continued its preparations for the defense of Baghdad and Al Kut. In Al Kut, the Baghdad RG Infantry Division continued to dig in. They had repositioned some artillery to a position that could range the An Numaniyah airfield, but had otherwise shown no indication of changing their defensive posture. The increased use of camouflage and decoy equipment was noted in the garrison areas of the Baghdad RG Division. Perhaps unable to respond effectively to the rapid ground assault, the Baghdad RG Division was digging in deep, more concerned with surviving the on-going air strikes than the ground attack they could assume was coming soon.

Wolfpack's encounter the previous night confirmed the Division's suspicions that the nature of the fight between An Nasiriyah and Al Kut was going to be different than the conventional battles before. The late night encounter with enemy forces willing to rush into the 25mm Bushmaster chainguns of Wolfpack's LAVs was a marked contrast to the barefoot conscripts the Division had noted walking away from the oil field fight. The Division's assessment of the presence of a paramilitary screen south of the Republican Guard's main line of resistance was confirmed. The Division would have to fight its way through both the terrain and this new threat in order to reach the Tigris, and Blue Diamond looked forward to the brawl.

The Division Forward CP, moving in column behind 3d LAR, moved into position alongside Highway 1 north of the Euphrates and set up between the unfinished Highway and the parallel construction road. Master Sergeant Michael Daniels, the Operations Chief for the Forward, and Master Sergeant Jay Joder, the Intelligence Chief, had the setup of the CP down to a science. Within three hours, the CP was established and functional in an intermediate position approximately 25 km north of the Euphrates River. From this position, the Forward could control the Division's continued movement north long enough for the Main CP to displace to a position that allowed for the seizure of the Hantush Airstrip. The CP was emplaced within hours, and control was passed.

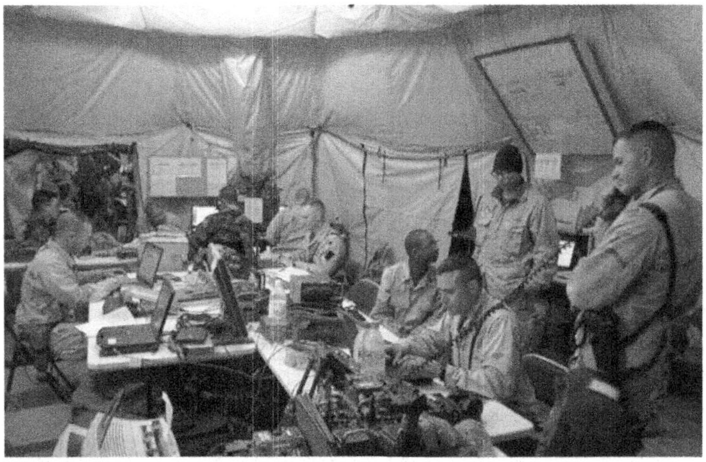

Division Forward Marines control the movement of the Division up Highway 1.

For the remainder of the night on 23 March (and continuing another 36 hours until the afternoon of 25 March) the rest of the Division moved in column behind these lead elements. For most, the movement was characterized by constant stop and go, and speeds of less than 20 miles per hour. The fatigue factor was high in all Marines, but especially the drivers. Unit leaders had to walk from vehicle to vehicle waking up drivers when it was time to move forward again. Despite constant supervision, some drivers still managed to drift off while driving. Even though they would be quickly awakened by their assistant drivers, even a few seconds caused a break in the column as many vehicles lost sight of the blacked-out vehicle in front of them. As more than two-thirds of the Division was using Highway 1 to move on, the route was filled with every type of vehicle the Division owned.

RCT-5 and the Coming Ground Attack

On the morning of 24 March, RCT-5 completed the FPOL with 3d LAR, and continued the attack along Highway 1 as the Division's main effort. They were to clear the enemy in zone and facilitate the Division's attack toward Baghdad. Each of the battalions engaged and destroyed irregular enemy forces during this period in actions characterized by initiative and aggressiveness. While the enemy along Highway 1 fought with determination, their hasty defensive positions and ambush tactics were ineffective in slowing RCT-5, whose tank/mechanized infantry teams overran them with skill and determination.

During this continued attack to the north, the Division had a negative incident with one of the non-embedded journalists that skulked around the battlefield. The incident was in direct contrast to the responsible reporting by the embedded journalists who traveled with and were trusted by the Marines. After working his way into the Marine's confidence, this irresponsible, non-embedded journalist disclosed RCT-5's position and attack plan on the air during a live phone call with CNN. His broadcast included the complete battle plan for what the RCT was about to execute. That afternoon, the I MEF CG called the Division CG to order the apprehension of the offending journalist, and within hours Division MPs escorted him to the MEF CP in the rear at Jalibah.

Second Battalion, 11th Marines in firing position. Eleventh Marines continuously provided accurate and responsive fires.

As the Division was now attacking in two widely separated columns (and the weather was deteriorating), the direct support artillery battalion for each RCT gained the gratitude of its supported unit. RCT-5 was reassured by the all weather, responsive fire support provided by 2d Battalion, 11th Marines (2/11). This battalion, and supporting/reinforcing battalions of the 11th Marines, provided continuous fire support throughout a period of very poor weather. On the 24th, for example, 2/11 continued to echelon itself forward with the progress of the RCT. Through multiple displacements forward, 2/11 always remained fire-capable in position areas along side Highway 1. During these displacements, they had fired 131 rounds in 3 missions. The Cannon Cockers provided counterbattery fire that quickly silenced the numerous enemy indirect fire attacks on the RCT.

RCT-1 Waiting for Clearance

The CG now convened his planners and asked them to assess the current situation. Did the quality of the route and our success in achieving operational surprise now make it possible for the entire Division to move up Highway 1? The Staff reviewed the requirement to send a force up the Highway 7 corridor, and decided there was still tactical advantage to be gained by sending RCT-1 along that route. The paramilitary had reportedly been using a number of the smaller cities just north of An Nasiriyah to rally their resistance fighters. Clearing the 'rat's nests' of Al Gharraf and Ash Shattrah, especially, would force the enemy fighters to focus their attention on saving their own skin, not contributing to mischief in An Nasiriyah and the other southern cities along Highway 7. In addition, the reported presence of chemical weapons in Ash Shattrah posed a significant risk to the Division and follow-on forces massed near the Highway 1 bridge crossing. The use of the eastern bridges in An Nasiriyah was recommended as a means to remove these threats and alleviate the pressure of forces moving up the Highway 1 corridor, dispersing the lucrative chemical target the Division would otherwise present. There was also a consistent tactical bias to use multiple avenues of approach where possible, to preserve flexibility and eliminate the need for a major friendly passage of lines (with the inherent friction that would entail.) Finally, the Baghdad RG Infantry Division in Al Kut would have to remain fixed if the Marines were to close rapidly on Baghdad, and RCT-1 moving up Highway 7 was planned to be this force.

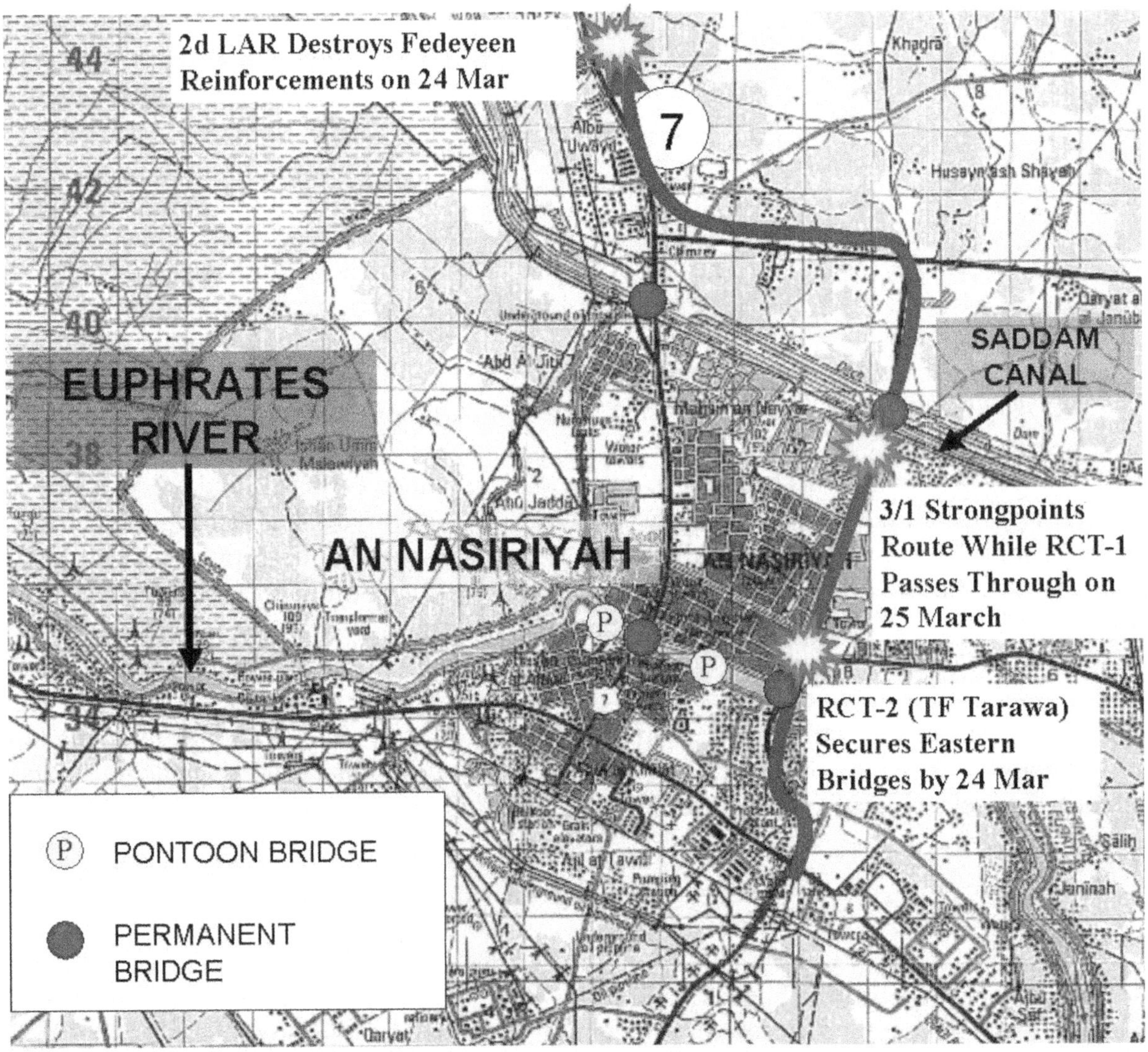

The action in and around An Nasiriyah on 24-25 March broke the back of the Fedeyeen hoping to defend along the Euphrates River line. to reach Highway 1.

The Division had long looked at the bridges in eastern An Nasiriyah as highly desirable as a route of march. There were two key bridges at the eastern edge of the city, a southern one over the Euphrates River, and a northern one over the Saddam Canal. The intact seizure of the bridges by RCT-2 (the GCE of TF-Tarawa) was welcome news. RCT-1 was issued a FRAGO sending it over the newly seized bridges and up Highway 7. The other criterion for movement across these bridges, however, was a secure route that would support RCT-1's movement north through Tarawa's lines. This was to pose a major challenge, as the An Nasiriyah urban area was anything but benign. The 4 km stretch between the bridges was subject to the fires and direct attack of Regime fighters that continued to operate in the city.

In An Nasiriyah, the battle to hold the eastern bridges continued in earnest. RCT-2 had battled the enemy elements that had chosen to stand and fight in the city. Tasked with opening the bridges on the east side of An Nasiriyah, RCT-2 had encountered stiff enemy resistance and had taken many casualties. The Fedeyeen had rallied and massed themselves against the RCT-2 assault. Air assets from across the MEF had been concentrated in support of the sharp An Nasiriyah fight. From positions along Highway 1 south of the Euphrates, the columns of 1st Marine Division waiting to cross the River could hear and see the air and artillery fires effects in the city. The Division Marines cheered their comrades, and reflected on the coming fights in Al Kut and Baghdad. As the rest of the 1st Marine Division was pushing up Highway 1 to the west, RCT-1 poised itself to push through An Nasiriyah in order to gain access to Highway 7 and conduct its fixing attack on the Baghdad RG Division in Al Kut.

On 24 March, the Division's Reconnaissance Battalion was tasked to establish a support by fire position for RCT-1's planned river crossing over the eastern bridges. Recon Battalion set in their positions and engaged in a significant firefight with enemy elements on both sides of the Euphrates. De-conflicting the battlespace was difficult, as 1st Marine Division forces were providing supporting fires in TF Tarawa's zone. The Marines of 1st Recon Battalion poured withering fire across the river in support of the RCT-2 attack. At sunset, they were ordered to fall back and make their preparation for their movement north on the next day.

That evening, 2d LAR Battalion ('Barbarian'), led by Lieutenant Colonel Ed Ray, the Division's lead element in the attack up Highway 7, passed through An Nasiriyah and over the two eastern bridges. Barbarian pushed through with blinding speed. Their impetuous attack allowed a swift passage to the northern bridge but not without some significant fighting along the way. As soon as they crossed the southern bridge, they began taking fire from the Fedeyeen fighters firing AK-47s and RPGs from windows and alleyways. Barbarian returned fire against the enemy while continuing their movement towards the northern bridge. They could see the enemy fighters on top, between, and inside the multitude of buildings and alleyways. In the darkness, Barbarian returned fire at the sight of muzzle flashes or rocket trails as they rapidly moved through the urban terrain. The aggressiveness, speed, and tenacity of the Marines was an awesome display of sheer will to accomplish the mission and take the fight directly to the enemy. Once across the northern bridge the TF pushed north up Highway 7 as sporadic fighting continued. Second LAR Battalion was able to push up as far as Al Gharraf (15 km north of An Nasiriyah) that evening. The desire was to get as far north as possible to make room for the remainder of RCT-1 to pass through An Nasiriyah and onto the Highway 7 route in the morning.

During the night, intelligence reports revealed an Iraqi counterattack force heading south from Al Kut, and Barbarian again proved their valor in the ensuing meeting engagement.

After assisting Bravo Company's move through the city Alpha Company reverted back to battalion control. Captain Monclova was reinforced with a platoon of tanks and given the mission of conducting a forward passage of lines through 2^d Battalion 8th Marines and 1st Battalion 2^d Marines and to continue the attack north up Route 7 in the lead of the battalion. As Alpha Company attacked north they encountered heavy enemy resistance and fire was taken on both sides of the road. The Forward Air Controller ran Cobra gun runs 10-15 feet from the vehicles, at times, to hit Rocket Propelled Grenade teams which were hiding in ditches along the side of the road. Captain Monclova skillfully used a tank platoon to engage and destroy a large number of technicals, ammunition vehicles, and dismounts facilitating the battalion's rapid 15 km movement north. At dusk the company established the north side of the battalion defensive perimeter astride Route 7 in the vicinity of Al Jahar.

Almost immediately after establishment of the perimeter it came under attack. Captain Monclova decisively led his company, integrating direct, indirect fire with close air support, engaging numerous vehicles carrying personnel south toward An Nasiriyah along Route 7 throughout the night. As vehicles were being engaged, the number of dismounts began to increase as the Iraqi forces discovered that Alpha Company was blocking their movement south toward the fight in the city. Throughout the entire night the company engaged the enemy. Along the entirety of Company A's lines the Iraqi's attempted to penetrate the battalion's perimeter. Due to the skillful use of supporting arms and a high volume of direct fire the Battalion was successful in defeating each attempt at penetrating its lines. As the last attack was defeated near dawn, the Battalion collected numerous EPWs and observed between 200-300 enemy Killed In Action. In addition to the Killed In Action and Enemy Prisoners of War, Alpha Company destroyed 2 busses used to carry forces south, as well as 2 trucks and several cars being used as personnel carriers.

The Division Crosses the Euphrates

Along Highway 1, RCT-7 continued to move up behind RCT-5. RCT-7 spent the entire day of 24 March and most of 25 March on the road, in a road march of over 250 km. The RCT-7 Staff worked diligently to ensure accountability up and down the line. To overcome communications challenges over this great distance, Colonel Hummer himself drove over 300 km in a single night, ensuring that all of his RCT elements continued their move from TAA-7 and into the Division file. Those in the RCT-7 TAC, spent two nights in their AAV as the column slowly advanced west then north. The Blue Force Tracker system, with its satellite instant text messaging capability became an important tool during these conditions, with vehicles spread out over long distances, mostly out of voice communications range with the RCT commander. As the weather worsened, the convoy slowed even more.

By the end of the day, most of RCT-5, RCT-7, and 11th Marines were across the Euphrates. Highway 1 would need improvement if it was to act as a major supply route, but the Division was successfully making its way along it. As long as the weather remained good and the route dry, the Division could easily handle the threat posed by the Fedeyeen paramilitaries. Ominously, the wind began to shift and pick up speed. By the time darkness fell, a smothering dust storm had begun. Marines choked on the dust and visibility was reduced to almost nothing. Soon, it was blowing so hard that it was difficult to breath outside. Tomorrow was going to be a tough day.

25 March

As the Division had more frequent contact with the Fedeyeen, a new picture of their tactics was emerging. The Fedeyeen and the other paramilitaries were a disingenuous lot. By this time, the paramilitaries had been identified using false surrenders to approach US forces in An Nasiriyah. Marines from 1/4 witnessed this first-hand, 2 km south of Ash Shattrah. Two individuals approached the battalion's perimeter with their hands in the air, offering the universal sign for surrender. Two Marines moved forward to investigate, while a scout sniper team served as their guardian angel. One Marine ordered the individuals to lay on the ground with their arms to their sides, while the other Marine served as the cover man. As the two surrendering individuals moved towards the ground, a third individual emerged from a clump of trees with an AK-47, preparing to fire on the Marines. Before the third individual could open fire, the guardian angel, Corporal Juan Vela, squeezed the trigger of his M40A1 and shot him between the eyes. The guardian angel concept was validated that day.

Even the conventional elements fighting with the paramilitaries shed their uniforms in an attempt to hide among the populace. The fighters tested the limits of the Marines' ROE, and exploited what gaps they could discover. The guerilla 'fish' needs to be able to swim in a 'sea' of popular support, however, and the Regime and foreign fighters did not have this support in southern Iraq. This was not a guerilla war. The Shia Muslims in the south resented the presence of these Regime oppressors, and were willing to work with the Marines once they knew the Marines were there to stay. The paramilitaries were able to maintain small powerbases in the urban areas, but were largely ineffectual in the countryside. As the populace gained more confidence in the presence and motivations of the Marines, they became key allies in the fight to identify the forces still fighting for the Regime. In later weeks, the population would contribute increasingly to assisting US forces remove the Fedeyeen, once the people understood the Coalition was committed to the Regime's removal. For now, the Marines would have to continue to lead with their fists, and annihilate those paramilitaries bold enough to open fire at them.

The Mother of all Sandstorms

The weather continued to severely degrade most operations the next day. As the 'Mother of all Sandstorms' raged, columns moving up Highway 1 were slowed to a snail's pace. Many vehicle accidents due to the reduced visibility contributed to the slow movement. An MTVR (7-ton truck) moving with the Division Main fell into a ditch and rolled on to its side. CSSC-115 had two vehicle accidents that wrecked two HMMWVs and another truck. Close calls, near misses, and bumps were commonplace.

The blowing sand and dust made it necessary to completely cover faces and eyes. With visibility already reduced by the blowing dust, the addition of goggles and face masks added to the burden. Penetrating as if it were liquid, the dust leaked into every opening in vehicles, tents, and uniforms, covering everything with a thick layer of talcum-like powder. That evening, freakish torrential downpours soaked it all into a sticky muck that covered everything. Loud thunderclaps accompanied the rain, and it was hard to differentiate between rolling crashes of thunder and the frequent artillery fire. Every rumble caused the Marines to pause momentarily to consider whether they were in for more sheets of water, or rain of a heavier caliber.

Visibility was drastically reduced by blowing dust , sand, and rain during the sandstorm.

The blowing dust caused brownouts that completely blinded the drivers moving over the narrow Highway 1 construction road. First Lieutenant Robert 'Akron Tough Guy' Olson, from Headquarters Battalion, recounts a typical tale of the road that day:

"In places, it was a complete brownout. You couldn't even see the road in front of the vehicle. We were making our way up the construction road when our MTVR missed one of the culverts. Before we even knew what was happening, the whole truck had flipped on its side. I jumped out into cold water up to my chest and crawled out of the mud. By some miracle, none of the Marines in the back was injured, and we got them all out of the truck unhurt. The truck rolled over until it was laying flat on its side in the muck. The column kept rolling past while we set-in a perimeter around the vehicle. After about five hours, with a tow truck and an M88 track recovery vehicle working on it, they still couldn't budge the MTVR from the mud. We jumped in an empty five-ton in the last stick and kept rolling. It was dark now and still blowing so hard that we couldn't see. We finally could go no further and had to wait for the sand to quit blowing before we could move again. I think they eventually got the truck out."

Tragedy also struck the Division while crossing the Euphrates River over the precarious night. In the raging sandstorm, unseen by the vehicles in front or behind, the M1A1 tank Charlie 2-3, the 'Harvester of Sorrows' veered off the high Highway 1 Bridge and was lost in the night, along with its crew. The tank plummeted into the Euphrates, and settled on its turret into the silt at the bottom of the Euphrates. Staff Sergeant Donald May, Corporal Robert Rodriguez, Lance Corporal Patrick O'Day, and Lance Corporal Francisco Martinez-Flores all lost their lives. It wasn't until 22 April that 1st Tank Battalion was able rejoin its detached companies and conduct a formal memorial service to mourn the loss of the crew of Charlie 2-3. On that later date, the Battalion Chaplain, USN Lieutenant(jg) Mark Redmon conducted the service with the assistance of Father Bill Devine from RCT-7. Staff Sergeant Steven Santana, platoon sergeant for the crew, spoke about each of his Marines and their loss with his fellow tankers. In a sweeping motion around him that referred to the recently liberated Iraqi people, Captain Brendan Rodden ended the service by reminding the Battalion that the crew of Charlie 2-3 had not sacrificed in vain. Even on 22 April, however, the pain was still deeply felt. As with all the men the Division lost, they will not be forgotten by those privileged enough to have served alongside them.

The weather steadily grew more intense. By midday, it looked like dusk as sand and dust darkened the afternoon sky with an eerie deep orange hue. For twelve hours, the storm pelted everything with rocks and grit, reduced visibility to less than 10 meters, and even impaired breathing. In the Division Forward CP, poles supporting the tent first bent and then snapped as Marines used everything including cammie net spreaders to keep the structure from falling down around them. Movement became increasingly difficult. Convoys inching up Highway 1 slowed movement and turned on headlights. When that proved insufficient, A-drivers dismounted and walked in front of their vehicles, navigating by dead reckoning. Vehicles moved no faster than a few kph and took hours to arrive at locations that could normally be reached in minutes. The dust particles were so fine that they penetrated the masks and bandanas Marines wore for protection and were often enough to make them choke or vomit. Venturing into the weather could mean quickly becoming lost and disoriented. One Marine left his vehicle to make a head call, and could not find it again. While trying to decide whether he should continue to look or sit down to wait out the storm, he stumbled across a UH-1 Huey parked at a FARP. The crew was kind enough to let him stay with them for the night.

Some chose to tie string to a vehicle or tent to provide a means of finding the way back to it. For others, the only option was to stop their vehicles and wait for the storm to pass. While these Marines were not getting sandblasted, they were enduring a different challenge. The air was so saturated by sand that the only way to breathe was to close all the doors and hatches. Air circulation decreased and the effects of the past hot days without showers became more pronounced with doors closed. Jammed into tracks and trucks, on top of ammo, chow and water jugs, there was little room to breathe and no way to stretch out.

Sandstorms drastically slowed the movement of vehicles.

After battling traffic and brutal weather conditions for over 48 hours, RCT-7 came to a stop to the southeast of Ad Diwaniyah, behind RCT-5. By the end of the road march, every Marine was coated in a thick layer of fine greenish dust that contrasted noticeably with both the red color of blood-shot eyes and the strangely orange, dust-filled sky. The rain had covered the vehicles with a coat of mud. Mud and dust-covered vehicles were passing endlessly along Highway 1 as Marines stomped around in ankle-deep mud and standing water. Spread out over 40 km of Highway 1, RCT-7 was tasked with protecting the MSR. The route was threatened by Fedeyeen attackers up and down its flanks, largely from the uncontrolled Highway 8 to the west in the adjacent US V Corps zone. The Marines of the RCT were just happy to have stopped, and quickly went about their mission while waiting out the rest of the storm.

Despite the weather, there were indications that the Blue Diamond advance was having the desired effect on the Medina Division. This Republican Guard Armor Division was already stretched thin by its responsibility to defend Baghdad from the south. Their primary defensive area was the 'Karbala Gap', a classic attack route into Baghdad from the south near the Shia holy city of Karbala. The US 3ᵈ Infantry Division's actions had now opened up a number of simultaneous threats to the Medina Division along the Highway 9 and Highway 8 corridors, including the Karbala Gap. With the 1ˢᵗ Marine Division astride Highway 1 and ready to continue its attack north, the Medina Division would be unable to defend everywhere at once. According to intelligence reports, the Medina Division was now being forced to commit elements of its combat power to the south to counter the threat posed by the Marines along Highway 1. This was exactly the effect the Division had hoped to achieve. A repositioning of the Medina's 10ᵗʰ Armored Brigade to the south would have the dual effect of drawing them away from Baghdad, and also removing the threat to 1ˢᵗ Marine Division's western flank once turning northeast on Highway 27. The Division was also aware that this repositioning could possibly have a more ominous overtone. With the weather situation degrading rapidly, the Division's ability to rely on air support would soon be reduced to nothing. If the Iraqis were ever to conduct a counterattack, these would be ideal conditions. Over the course of the morning, enemy prisoners and informants began to report more Republican Guard forces north of the Division along Highway 1, in areas that had been recently reported by collections platforms as being clear. Similar to the situation in Safwan days ago, there had been no positive indications of significant combat forces in this area, but the weather precluded confirmation that these forces were not present. With the weather closing down all intelligence collections, the Division again prepared for the worst case and planned to deliver heavy blows against any RG units that tried to counterattack down Highway 1.

RCT-5 Continues The Attack

As the fierce storm blew around them, RCT-5 continued the attack as the Division's main effort. The previous afternoon, 3/5 had moved north along Highway 1 for several kilometers and consolidated just before nightfall to allow the remainder of RCT-5 to close the gap. Two technical vehicles approached from the north along Highway 1 and were quickly dispatched by tank and heavy machinegun direct fires. The Marines realized that they would soon get their chance to test themselves against the stiffening Iraqi resistance. At first light on 25 March, 3/5 led an attack to clear Highway 1 in zone. As they moved forward, the Marines of 3/5 could sense their moment was approaching. Lieutenant Colonel Mundy recalls the events that day:

"Stacy Clardy told me that if we were looking for a fight, we'd soon find one, and he was spot on! As we drove up the road, things seemed to get quiet around us. Right after I noticed

this stillness, Sergeant Major Joe L. Vines and my driver, Private First Class Tyrell Victor, commented on the lack of 'locals' who had disappeared from the road sides and mud huts nearby. That's when a hail of machineguns, mortars and RPGs opened up on us."

A high volume of small arms firing broke out along most of the column as 3/5 found itself reacting to a violent near ambush by an estimated battalion-sized force. Traveling as the lead company, India Company found itself hotly engaged in a close fight with dozens of Iraqi fighters firing from behind a large berm dominating the right side of the road. Captain Ethan Bishop quickly oriented his Marines and watched as the AAV crews of Delta Company, 2d AA Battalion (attached to 3/5) skillfully turned their vehicles toward the enemy and began laying down high volumes of fire from their up-gun weapons stations. The skillful cooperation between infantry and amtrac Marines made a difference that day as the AAV crews kept up a strong volume of suppressive fire while India's Marines dismounted and went into the assault, firing at the heads of the Iraqis, and throwing hand grenades back and forth in an intense and close range firefight. Delta Company Commander, Capt Vance Sewell, reported an outstanding example of the great cooperation between his and India's Marines during that attack:

"Lance Corporal Jonathon Talbott distinguished himself while driving the lead AAV in the battalion column (India Company). When his vehicle came under heavy Rocket Propelled Grenade and automatic weapons fire, without hesitation, he turned his AAV into the fire; as he did so, he saw two enemy soldiers preparing to fire a Rocket Propelled Grenade at his vehicle in a fighting hole 15 meters away. Still under small arms fire, Lance Corporal Talbott accelerated the vehicle and crushed the enemy as they fired the Rocket Propelled Grenade, which buried itself harmlessly in the ground. He then expertly maneuvered to a defiladed position and disembarked the infantry."

Up front, 3/5's tank platoon, 'Blue 1', and First Lieutenant Brian Chontosh's CAAT platoon, remained hotly engaged with a large body of Iraqi soldiers. During this engagement 3/5 suffered its first casualties when an RPG slammed into a heavy machinegun vehicle, killing the platoon's corpsman, Petty Officer Michael Johnson, and seriously wounding the gunner, Corporal Frankie Quintera. Petty Officer (HM3) Johnson was the only corpsman Killed In Action while serving as part of the 1st Marine Division during OIF. His presence at the front of the action, sharing the hazards of his infantry brothers, was indicative of the brave and distinguished service provided by Navy corpsmen throughout the campaign. Firing multi-purpose, anti-tank (MPAT) and heavy machinegun rounds at nearly point blank range into the berms and concentrations of troops, Blue 1 and the CAAT platoon Marines soon gained the upper hand and began mowing down Iraqi defenders. Blue 1 and CAAT continued to press their attack forward, pursuing the Iraqis and engaging another well-laid ambush position further to the north. When the forward units moved beyond communication range with India Company and the battalion Forward CP, Lieutenant Colonel Mundy directed the two platoons to hold up, issued a verbal FRAGO to pass Kilo Company into the lead, and then ordered the remainder of the battalion to continue its attack north. As the rifle companies mopped up remaining resistance, passed off the enemy wounded and EPWs, the freakish sandstorm added to the friction by turning day into night. Coordination proved very difficult. 3/5 continued to press forward, uncovering multiple caches of weapons, ammunition, and equipment along the axis of advance while pursuing the Iraqi defenders.

The weather continued to impact the lead elements of the Division, even after the worst had passed. On the night of 25 March, 3/5 consolidated in a very tight defensive perimeter south of the Ad Diwaniyah cloverleaf. A few hours later, Lima Company reported to the CP that they heard what sounded like armored vehicles approaching. The weather was so terrible and it was still so dark that NVGs and thermal sights were nearly useless. Trying to maneuver the tank platoon into a position to engage, the 3/5 CP received a report that the M1A1's thermal sights could not achieve a ballistic firing solution beyond 800 meters. Captain Mario Schweizer was ordered to form a provisional reserve from available Headquarters and Service Company Marines standing security watch, while Captain Scott Meredith's Lima Company broke out AT-4s and prepared SMAW-AT rounds. Both expected to repel the enemy's assault at close quarters. The 3/5 CP reported to the RCT-5 CP that they could hear armored vehicles approaching. Enemy tanks were so close that the 3/5 Marines could feel the vibrations from the enemy tracks in their defensive positions. The Marines braced for the assault that they knew was coming soon, prepared to defend without air support. The RCT-5 CP reiterated that there was no friendly armor in that area. Through the blowing dirt and sand, 3/5 finally spotted 8-10 Iraqi tanks closing on Lima Company's lines, and called for artillery support. As the situation continued to develop, a mechanized enemy threat was also identified moving south down Highway 1 toward 3/5's positions. The all-weather capable 2/11 responded with DPICM and requested reinforcing fires from 11th Marines. The Battalion was allocated the reinforcing fires of 3/11 and 5/11. In no engagement was artillery more effective, as the combined fires of the three artillery battalions rained down on the enemy attack force, forcing them to beat a hasty retreat. Captain Spaid, the 3/5 Air Officer, also managed to direct a section of F-15s, flying above the weather, against the enemy tanks, hastening their retreat north of the cloverleaf. Throughout the night and into the early morning hours of 26 March, 2/11 processed 12 missions providing 457 rounds of support that no one else could in that environment. Overnight, 3/5 continued to experience small probes by dismounted enemy forces moving in front of their lines and engaged several technical vehicles approaching from the north along Highway 1. Dispatching both with a combination of mortar and artillery coordinated HE/illumination missions, 3/5 spent a restless and edgy night in their defensive positions.

JSTARS ground stations operated through the storm at the Division Main CP. These Army attachments added the critical capability for the Division to view the entire battlespace.

The all weather fire support provided by 11th Marines also raised the adrenaline levels of many units of the Division during segments of the fight. During the sand storm, 1st LAR Battalion ('Highlander') had unknowingly pulled up next to an artillery firing battery. When the battery responded to a call for fire, the explosions of the rounds going down range threw the Highlanders into a scramble for cover from the 'incoming'. Taken by surprise, they thought that they were once again under attack, but soon realized the true source of the noise. After a few muttered comments about the Cannon Cockers being inconsiderate neighbors, the Marines got some well-deserved rest.

Blue Diamond Weathers the Storm

Sometime that evening, the blowing dust stopped and it began to rain. Golf ball sized hail mixed in with the rain, turning the unpaved road from dust to mud. Throughout the night, periods of relative calm alternated with periods of freakish lightning and large crashes of thunder. 'Blue Diamond' passed from the Division Forward to the Division Main CP in the midst of the storm and the Forward Marines hunkered down for the night. Covered with water and mud, with the CP tent half-collapsed around them, the Marines took to any level surface they could find, glad to find at least a small bit of shelter from the storm.

The Division was to fight much more than the weather on this day. Most of the Marines of the Division found themselves enduring the sandstorm along Highway 1. But at the same time, about 70 km to the east, the Marines of RCT-1 found themselves with a more conventional opponent. On the morning of 25 March, the Commander gave the word, and the Marines of RCT-1 moved across the Euphrates River and up Highway 7.

Inchon Carries the Ball Downfield

With the attack along Highway 1 slowed down by the weather, it was RCT-1's turn to carry the ball downfield. To the east, the weather, though severe, did not have the same impact as it did to the exposed columns along Highway 1. The previous evening, 2^{d} LAR had carried the fight to the enemy north of the city, but was cut off from the rest of the RCT. The Fedeyeen still made for a dangerous passage through the eastern An Nasiriyah bypass, posing a special threat to soft-skin vehicles. The progress of the rest of the Division up the Highway 1 corridor, however, mandated immediate action to get the fixing force for the Baghdad RG Division up on the Highway 7 corridor.

The original timeline was premised on TF Tarawa's successful securing of An Nasiriyah prior to RCT-1 making this move. With Tarawa still engaged with significant enemy resistance, however, the Division decided to accept the risk of passing the entire RCT column through the edge of the urban area. RCT-1 was ordered to make its way through the eastern edge of the city, just as 2^{d} LAR had done the previous night. With RCT-2 in control of the two eastern bridges over the Euphrates River and Saddam Canal, this would require the RCT to pass through the 4km stretch of road in between. The ADC, Brigadier General Kelly, and the RCT Commander, Colonel Dowdy, decided that, even though the An Nasiriyah environment was not completely under control, the combat power of a Marine RCT would certainly be sufficient to force its own way through. Lieutenant Colonel Jim Parrington, the Regimental S-3, worked out a strategy to strongpoint the 4 km passage using 3/1 to hold the door open for the rest of the RCT to pass. The Battalion, commanded by Lieutenant Colonel Lew Craparotta, would force open the passage and remain in position for 12 hours, as the rest of the RCT convoy made its way through the restricted terrain.

The Iraqi fighters had been prepared by a great deal of propaganda regarding the willingness and ability of American forces to fight the close battle. The Marines taught them a first-hand lesson. The Marines of 3/1 and Alpha Company of 1st Tank Battalion swept into the city in the predawn hours. In the twelve-hour firefight that followed, the Marines of 3/1 literally had to stack the bodies of the Fedeyeen to clear fields of fire. The Iraqis were quickly learning a very important lesson; Marines would fight with a skill, ferociousness, and determination that they had not seen or experienced before.

The Marines of 3/1 and Alpha Company, 1st Tanks benefited from their technical overmatch during the hours of darkness. With thermal sights and night vision equipment, the Marines were able to engage squads of Fedeyeen moving under the 'cover' of darkness. The adage 'too much night vision gear is just about enough,' was true in this fight. The Fedeyeen, unable to see in the very dark night, repeatedly would walk into the same kill zone, and fall like their predecessors. As daylight began to break over the scene, the Fedeyeen gained a better appreciation for the Marines' positions, and began to change their tactics. They used the cover of walls and civilian houses to gain close approaches to the Marines. From there, they would attempt to engage the column with RPG's and grenades, but usually not before being cut down by the withering fires laid down by 3/1 and Alpha tanks.

Marines from 3/1 hold the door open for RCT-1 in An Nasiriyah.

Following 3/1 into the breach were 1/4, 1/11, 2/23, the Reconnaissance Battalion, and the rest of the units that made up the RCT. Lieutenant Colonel Steve Ferrando, CO of 1st Recon Battalion, was in the column, and recounts the story:

"As we rode up, we rolled past the three AAVs that had been shot up in the previous day's [TF Tarawa] *actions. It was a very sobering scene, as these armored vehicles lay in smoking ruin, jagged twisted metal bent in every direction. The enemy opened up with small arms, heavy machine guns, and RPGs. The Marines returned fire on everything that posed a threat as we made our way through the city. We would see paramilitary guys in civilian clothes without weapons come out and look at the column, then run back inside buildings and begin to open fire. Anything that moved was quickly suppressed. Most of these guys were hard-core, and had no fear of death. They were all dressed in civilian clothes, and would take refuge in houses along the way. It was a shooting gallery. The Marines from 3/1 and Tanks did a great job laying down suppressing fires, and every vehicle in the convoy had their pick of targets. One of the Marines from 3/1 was wounded, so we grabbed him and put him in one of our vehicles and got him to the north bridge and the aid station. When we got to the north end, we moved across the bridge and quickly got back on the Highway 7 corridor. There were weapons and ammunition caches all over, and our guys were destroying them as we went. I got a radio call from one of my*

commanders asking permission to blow up a huge truck full of mortar rounds on the north side. When we plotted it out, it was only a couple of hundred meters from our CP! We had to say no to that one."

As aerial and human intelligence reporting began to trickle in it became clear that the Iraqis were making a late effort to reinforce the An Nasiriyah defenses. The Iraqis had clearly not been prepared for the rapidity of the Division's attack, as they had to assemble reinforcements from deserters, conscripts, and Baath militiamen. These fighters augmented a cadre of Republican Guard and Regular Army soldiers that made their way south along Highway 7. Intelligence clearly revealed the busses and trucks that were making their way down Highway 7 to An Nasiriyah, and 3[d] MAW quickly swept into action. The Wing began attacking the convoys of busses and 'technical' vehicles that contained additional forces aimed at An Nasiriyah and RCT-1. The artillery fires of 1/11 were added to the devastation, as they moved in support of the forward elements of RCT-1. Although indirect fires were devastating in their attacks on the columns of vehicles, it would be Lieutenant Colonel Ray's 2[d] LAR Battalion that would spring an ambush on the survivors, finally crushing the counterattack effort. Much as they had done the previous evening, 2[d] LAR Battalion's LAV-25s and the LAR Battalion Scouts stopped the counterattack force well short of An Nasiriyah with a wall of firepower. The destructive power of the LAV-25 would earn the enemy's deep respect. Captured enemy information revealed that the Iraqi troops referred to LAV-25s as 'destroyers', and the presence of LAVs on the battlefield spread panic through their ranks.

The devastation of the enemy's attempt to reinforce by artillery, air, and direct fires was complete. In reviewing the wreckage of the counterattack force, the Marines of 2[d] LAR Battalion reported that the charred remains of Iraqi Soldiers were still holding on to brand new AK-47 assault rifles wrapped in the original wrapping paper. The counterattack was stopped abruptly and violently by the Marines and now the Iraqi's were beginning to understand that they were fighting a determined and disciplined force rather than the 'paper soldiers' they were led to believe they would face. With the reinforcements/counterattack force wiped out, the stage was set for the eventual victory and stabilization in An Nasiriyah.

A bird's eye view of the Euphrates River from a UH-1 Huey.

As soon as the Reconnaissance Battalion got to the north side of An Nasiriyah, they attempted to establish the eastern picket line to protect the Division's flank. In the planned Division scheme of maneuver, they would establish multiple outposts astride the avenues of approach the 10th Armored Division might take to counterattack or interdict the Division. The Battalion hoped to be able to pick its way to the north by route finding to the east of Highway 7. The deep

Mesopotamian mud, however, immediately began to tug at the Battalion's vehicles, and soon got the upper hand. The terrain just a few hundred meters off of Highway 7 was a mucky quagmire. The locals had built a number of canal roads on raised levees that offered some trafficability through the swamps for their light pickup trucks, but these roads were often nothing more than generations of stacked reeds on top of mud levees. When the Battalion's MTVRs and heavily laden HMMWV's traversed these roads, they compressed the material, leaving deeply rutted mud paths in their wake. After pushing several kilometers to the east, it was clear that no sustainable route of march existed to the northeast, and the Battalion would have to return to Highway 7 in order to travel north. Going back the way they had come was not possible, as the roads they had taken to this point would not support a second crossing of the Battalion's heavy vehicles. The CO decided to keep pushing to an unimproved surface road depicted on the map that led back northwest, rejoining Highway 7 at a small town called Al Gharraf.

The remainder of RCT-1 moved through An Nasiriyah and began the attack up Highway 7. Led by Lieutenant Michael Borneo's CAAT platoon, 1/4 passed through 2^d LAR's lines and continued the attack north. Approximately 5 km south of Al Gharraf, three 'technicals' emerged and engaged the CAAT platoon with ineffective small arms and RPG fire. CAAT A responded with devastating accuracy, as their .50 caliber machine guns rendered two of the vehicles destroyed. Almost immediately, the enemy revealed the rest of his ambush position and began to engage CAAT A with small arms fire from prepared positions. Alpha Company, led by Captain Chris Griffin, was ordered into the fight to conduct a flanking attack from the east while CAAT A established a base of fire to suppress the enemy. The 81mm Mortar Platoon, led by Lieutenant Andy Owings, established a mortar firing position and was responding to calls for fire from CAAT A's forward observer within minutes. The battalion's air officer, Captain Evan Wahl, directed a section of AH-1W Cobras to support Alpha Company. With audacity of action, Alpha Company secured the objective. They reported one enemy KIA and 18 EPWs, including 8 enemy WIAs. Human Intelligence Team-1 (HET-1) quickly went to work on the EPWs. They determined that the lot consisted of a combination of Republican Guard soldiers, Regular Army soldiers, Fedeyeen, and other paramilitaries.

Lead elements of the RCT now pushed all the way up to Ash Shattrah, but bypassed significant enemy resistance that had fled and sought refuge in the village of Al Gharraf. The enemy had already learned to let the tanks and LAV-25s pass, holding their ambushes for the soft-skinned vehicles that followed. RCT-1's main body began to arrive at the town of Al Gharraf by noon, and immediately became engaged with a mix of all types of Iraqi fighters. A platoon from 1/4 and the RCT-1 Alpha command group were on the southern edge of the town, while an advance party from Alpha Battery, 1/11 was also forward, scouting for new firing positions. Captain Jason Frei, the Alpha Battery commander, and Lieutenant John Carroll, his assistant, were with the advanced party when it was ambushed by the enemy.

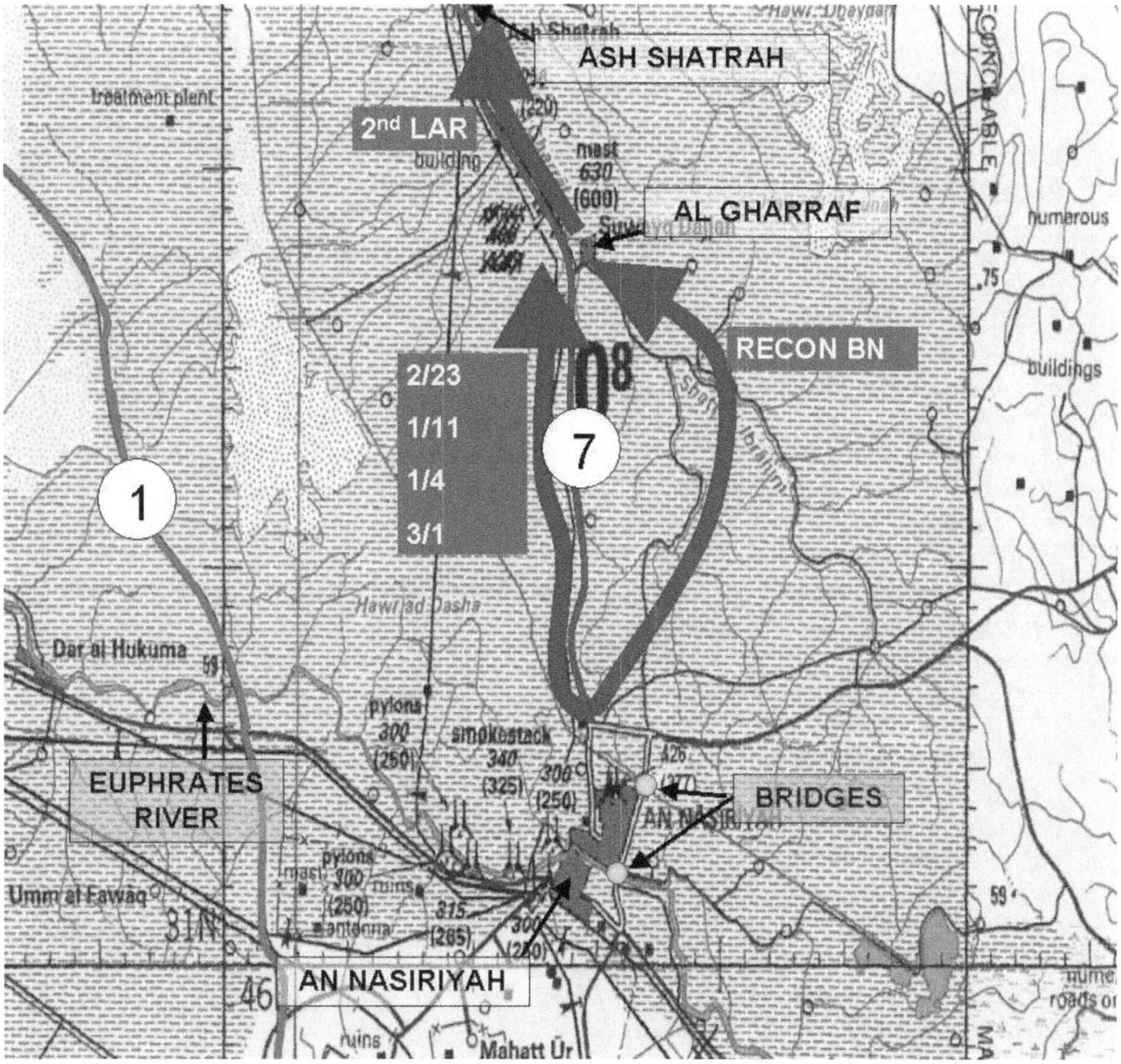

RCT-1 experienced some of its heaviest fighting just north of An Nasiriyah along Highway 7. In a meeting engagement, RCT-1 destroyed elements of the Fedeyeen attempting to reinforce the city.

While the Alpha Battery artillery advanced party was pinned down, the main body of A/1/11 also began to come under fire. The Alpha Battery main body, under the calm direction of First Lieutenant Lucas Hanback returned fires with organic crew-served weapons and small arms. The Fedeyeen continued to swarm the cannoneers, and soon India Battery (attached from 5th Battalion, 10th Marines) and the 1/11 Battalion CP also came under attack. Fedeyeen fighters were firing from rooftops and windows from multiple directions. The fire was so intense that Major Russo, the battalion Operations Officer had a 'yellow canary' (memorandum pad) shot right out of his hand. This time, the Cannon Cockers decided to conduct a fires demonstration for the Fedeyeen, as Alpha and India Batteries each lowered a number of their tubes and used direct fire from their 155mm howitzers to destroy Fedeyeen firing positions in the buildings around them.

Just south of Al Gharraf, Captain Frei and his battery advance party of six vehicles and 34 Marines was traveling to establish a forward battery position. An Iraqi force ambushed the vehicles from buildings and prepared positions with automatic weapons, rocket propelled grenades and mortars. An RPG round tore into the front passenger side door of Captain Frei's armored HMMWV, instantly severing his right hand and disabling the vehicle. Captain Frei calmly grabbed a radio handset and quickly began wrapping it around his arm for a makeshift tourniquet, and, immediately attempted to organize his Marines in a defensive perimeter to hold off the attacking dismounted troops. Over the course of the hour-long firefight that followed, he continued to lead and inspire his Marines until the rapid and significant loss of blood rendered him combat ineffective.

First Lieutenant Carroll immediately took charge of the situation, directing the members of the advance party to dismount and take cover in an irrigation ditch opposite the enemy positions. He radioed for help from the battery main body, which was traveling in trace of the advance party, and immediately reported detailed information about the situation before having to seek cover due to the high volume and accuracy of enemy fire. Once in the irrigation ditch, 1st Lieutenant Carroll quickly organized the Marines into a defensive perimeter and directed his Marines to return fire with crew served weapons and M203s in order to repulse an attack that was forming in the enemy position.

Lieutenant Carroll moved from the relative safety of the irrigation ditch and manned his vehicle-mounted radio to call for supporting artillery fires. He called in an immediate suppression fire mission within 300 meters of his position after using his GPS and the sounds of enemy fires to accurately estimate the enemy location. Then, he calmly requested medical evacuation for his commanding officer who was losing blood from his severed hand and at risk of going into shock.

When a company from 2/23 arrived to relieve the artillerymen, Lieutenant Carroll loaded the wounded Captain into a light armored vehicle, consolidated and accounted for his Marines, ensured they all loaded back onto their vehicles, and led them back to friendly positions.

While Alpha and India Batteries were moving forward, Bravo battery had established firing positions to the north of An Nasiriyah. From these firing positions, they fired in support of their fellow Cannon Cockers, the Reconnaissance Battalion, and soon the regimental headquarters as well. During the engagement in Al Gharraf, the RCT Alpha command group came under attack. The RCT headquarters had been scouted by suspicious looking 'civilians', and soon armed fighters began to appear. When it arrived, F Company, 2/23 was ordered up to the Alpha Command to aid in the extraction of A/1/11. F Company arrived at the Alpha command group, was given its marching orders, and moved out. Fox Company had only moved out about 200 meters before they began taking fire. Soon the entire company and the Regimental CP were under attack by Fedeyeen fighters. All hell broke loose, as the enemy opened up with heavy machine guns, mortars, and RPGs. The Marines in the RCT CP returned fire, and called for more forces. The headquarters Marines rallied to the fight, and were soon providing devastating return fires on the enemy. Every Marine with a weapon wanted a piece of the action. Bravo Battery, 1/11 continued to provide suppressing fires.

Meanwhile, to the east, Reconnaissance Battalion had made its way to the road they sought, and were now making their way to the northwest. This road was significantly better than the trails the Battalion had used before, and they began to make good time toward Al Gharraf. As they approached the town from the southeast, they clearly surprised the villagers by their approach. One suspicious taxicab came toward the column, and then sped away, as if it was monitoring the position of the convoy. The suspicion was confirmed a short time later, when the Battalion ran into a hasty ambush set up along the dirt road. The battalion returned fire, and was able to suppress the ambush with direct and indirect fire. Artillery support was called in, and destroyed the enemy positions with crushing accuracy. The Reconnaissance Battalion continued their advance toward Al Gharraf, and encountered additional enemy resistance there. Recon Battalion could now hear the action to the west, where F/2/23 and other elements of RCT-1 were engaged. The Reconnaissance Battalion was amazed at the accuracy and responsiveness of the artillery support provided by 1/11. If they had known the circumstances under which those fires were provided, they would have been even more impressed.

The situation for the isolated cannoneers became more desperate as the battle raged on. First Battalion, 11th Marines continued to be attacked by Fedeyeen fighters on the west side of the road. As their small security forces held off the attack, the rest of the battalion continued to provide indirect fire support in response to calls from 1st Recon Battalion. The remainder of the battalion leveled their tubes and engaged the Fedeyeen forces that were using the buildings there for cover. Providing supporting fires for one unit, while laying the guns in direct fire mode for their own defense was just another day in the war for the pros from 1/11, albeit an exciting one.

Recon Battalion was able to continue to move past the first ambush, and make its way into Al Gharraf from the southeast. Captain Brian Patterson's Alpha Company had the lead as they entered the town. They could hear the firing on the other side of town, as the RCT-1 Alpha Command, with elements from 1/4 and 2/23, were engaged by the Fedeyeen. Their approach was a fortuitous flank attack of the enemy forces attacking 1/11 and the Regimental CP; as perfectly executed as it was unplanned. As they attacked across the town, they engaged Fedeyeen fighters shooting from inside buildings and on rooftops. Captain Patterson pushed through the town and affected the linkup between the Recon Battalion and RCT-1. The low number of casualties from enemy fire was remarkable given the volume of fires received from the Fedeyeen. The absence of casualties from friendly fire between the two converging Marine forces was even more so. The unexpected flank attack had taken the wind out of the enemy's sails, and they retreated, leaving their dead on the field.

The evening of 25 March found 1/4 in a perimeter defense to allow the RCT to close the gap and continue the attack north. Throughout the evening, the battalion was subjected to probing attacks as the enemy sized up the Marines knocking on the door of Ash Shattrah. Small arms and RPGs were ineffectively directed at the battalion. The Marines could see the enemy mustering at a parking lot in Ash Shattrah, loading into trucks, and heading east into the tree line. Soon thereafter, CAAT A's Sergeant Smith, the designated TOW Sniper, spotted a 'technical', a 3-ton truck with an anti-aircraft gun loaded in the bed, at the parking lot loading men with weapons into the flatbed. He gained the proper sight picture with his TOW sight while his platoon commander confirmed the Iraqis' hostile intent. Receiving clearance to destroy the Iraqi weapons system, Sergeant Smith unleashed his missile, which traveled 2.1 km before destroying its intended target. The direct hit validated the 'TOW Sniper' concept.

At 1500Z on the evening of the 25th, 2d LAR and Alpha Tanks relieved 1/4's northern perimeter and formed a two-battalion perimeter defense in preparation for further offensive operations. To the north, 2d LAR and Alpha Tanks repulsed the enemy's efforts to drive vehicles into the established defensive line. To the south, PFC Mickey Graff, a Javelin gunner attached to Alpha 1/4, spotted a BTR-60 just over one kilometer from his position, moving with a pickup truck that was debarking armed personnel as it drove. He sighted in and fired his first missile, destroying the BTR. After reloading, he fired his second missile and destroyed the pickup truck. After these two devastating shots, all was quiet on the southern front.

For the Marines and sailors of RCT-1, the last two days had been exhausting. Fighting through An Nasiriyah, they continued the attack to Al Gharraf and then north to Ash Shattrah. As Lieutenant Colonel Parrington, the RCT S-3 put it, "Those two days were a real gut check for us, but as Marines always do, we excelled, accomplished the mission, and continued to push the fight."

26 March
The Orange Crush

Back on the Highway 1 corridor, the sun rose with an eerie orange glow. Although the winds had abated, the suspended dust in the air reduced visibility near the ground and replaced the blue sky with one of pure orange. The 'Orange Crush' the Marines called it. Under the Orange Crush, the Division worked to clear the mud from weapons, computers, and engines. The CPs that had been set up over the last two days looked like a bomb had gone off inside. Mud-smeared maps hung over dirt covered radios, manned by unshaven, red-eyed Marines sipping a mud-slurry of coffee from filthy canteen cups.

The winds had died down and Marines emerged to assess the damage. The Marine stranded at the FARP awoke inside the Huey to find the HMMWV he had lost less than 100 meters away. The fine dust that had once coated everything now accumulated in clumps of mud on every surface and in every crevice. Marines began maintenance to make vehicles and weapons operable again. The morning sky was orange with residual dust, and remained that way for the better part of the day.

The storm was not finished with the Division yet. The Division's C2 aircraft and MEDEVAC birds had been grounded at the Division Forward CP, and needed to get back up to the Main CP some distance to the north. The Division Forward Operations Officer and Intelligence Officer took the opportunity to analyze from the air the ambush location of Wolfpack and 3/5's encounters two days before. The mud had not been friendly to helicopter engines either. While circling over the ambush site, the Division Operations Officer's Huey lost an engine to the accumulated mud and dust, and crashed about 200 meters off the Highway into a suspected minefield. One crewmember of the destroyed aircraft was badly injured and was

MEDEVAC'd out by the other aircraft. The remainder of the crew and the Division staffers were left in a perimeter around the crashed aircraft until a patrol from 3/7 came along and relieved them. To the hundreds who passed the wrecked Huey just off the Highway, it was a reminder of the inherent risks of operating under these arduous conditions. 3[d] MAW was to recover the wreckage some days later.

The sandstorm took its toll on machines as well as aircraft. This UH-1N crashed because of dirt and debris blown into the engine.

RCT-5, as the Division's main effort, resumed its attack north to clear Highway 1 in zone to Highway 27, and positioned itself along the 41 Northing for its planned attack on the Hantush Airstrip the following morning. RCT-7 followed in trace of RCT-5 as a supporting effort, waiting to be unleashed on the enemy in Al Kut. The enemy continued to resist in small elements up and down Highway 1, using pre-positioned weapons caches, civilian clothing, armed sport utility vehicles/pickups, and hit-and-run tactics focused on headquarters and combat service support units. Both RCT-5 and RCT-7 engaged in several skirmishes with enemy paramilitary forces over the course of the day.

The Tragic Loss of a Leader

During the height of the horrific storm, tragedy struck 3/5. The Battalion lost their Executive Officer, Major Kevin G. Nave, to a vehicle accident, and Gunnery Sergeant Russel S. Cederburg, the battalion communications chief, was also severely injured. Major Nave was struck and killed by a tracked vehicle departing a refueling point whose driver lost sight of his ground guide in the night during the storm. Such a tragic event negatively impacted the Marines and Sailors of the Battalion immediately. The Marines mourned the loss of their leader and 3/5's Chaplain held a memorial service in remembrance of Major Nave. But even as they continued grieving, the Marines of 3/5 were forced to resume the attack north toward Baghdad.

RCT-5 continues to move up Highway 1 despite the raging storm.

That afternoon, 3^d Battalion 5th Marines Logistic Operations Center arrived at the position southeast of the town of Ad Diwaniyah near the cloverleaf interchange. There was still sporadic fighting along the MSR and we immediately began to assess the battalion's needs to determine what convoy configuration we should send to CSSC-115. It became immediately apparent that we were going to have to send a convoy to pick up critical 81 mm mortar rounds. We set up the convoy just before dark and hit the road. About a half mile from the battalion's position we were called back because the CP believed there might be some continued fighting along the MSR. We went back to the battalion's position and I went to clear our convoy mission south. Confused, I tracked down our XO, Maj Nave hoping he would have some assessment of the enemy situation along the MSR. As we spoke I could see his calm resolve that everything would be all right and that the convoy should proceed as the mortars were critical to the combat effectiveness of the battalion. The last thing he said to me was, "Make sure you don't let them take you prisoner." En route to CSSC-115 the convoy came under mortar fire. I remember as I saw the burst of a mortar shell about 30 meters away to our right, promptly telling our driver, LCpl Borders, to "Drive!" Despite the mortars, almost being run down by oncoming tanks moving north, the traffic jam coming back, and the fact that we had no idea where the battalion had displaced to in the time we were gone, we managed to get back the battalion with the mortar rounds. By the time we had arrived Major Nave had already been tragically killed. Regardless of everything that night, it was Major Nave's insight into the importance of our mission, his composure in the face of danger, and his willingness to be a leader that set that convoy in motion. The fact of the matter is that I only knew Major Nave for a very short time, but it was long enough to make an impression. He was the type of Marine who would, "Never let them take him prisoner" and should always be honored as such.

- Second Lieutenant G. J. Bayliff

Making Logistics Work up Highway 1

Through force of will, Marines now turned to making Highway 1 into the Division's Main Supply Route (MSR). Poorly maintained and damaged stretches of the incomplete Highway 1 corridor were repaired from Safwan to Ad Diwaniyah. Engineers worked around the clock to fill in culverts and reinforce existing bridges to speed the Division's advance. Everywhere along the route, dust as fine as talcum powder coated men and equipment, and provided a powerful demonstration of the term 'periods of reduced visibility.' Despite these challenges, sending more than 3500 of its vehicles up Highway 1 gained tactical surprise for the Division in its attack to Baghdad.

CSSG-11 sustained the Division's momentum along Highway 1 through intuitive placement of hasty RRPs, and the CSSG established a hasty RRP (RRP-24) along Highway 1 near the Division Main CP. Tracked vehicles were using fuel at a much greater rate than originally planned, and getting fuel forward quickly was a critical enabler for continuing the attack. The original re-supply efforts were based on a planning factor of 200 miles per full tank of fuel, but the Division's tanks and AAVs required fuel at the 120-150 mile mark. When the speed of the Division advance made pre-planned RRP locations obsolete, CSSG-11 teams showed great flexibility in establishing alternate sites by adjusting on the move. By staging refueling vehicles in convoys as 'RRP Opening Packages' logisticians created flexibility. This operational agility stood in contrast to the Logistics Support Area (LSA) sustainment plan that was not designed to accommodate the Division's rapid advance, and often left the LSAs too far behind the Division to provide timely support. The success of the logistics effort was due to those dedicated Marines of CSSG-11, operating as part of the supported unit, understanding and flexing to the needs of the commander.

A CH-46 refuels at a hasty FARP near Highway 1. Due to extremely mucky terrain, roads often offered the only decent ground for refueling operations.

The CSSG-11 Marines shared in the danger with the other elements of the Division. Captain Toby Dyer of CSSG-11 recounts his experience at RRP 24:

Logistics executed tactical refueling operations up and down the route and allowed the Division to continue its movement.

"We had just received a push of supplies from the Transportation Support Group. As they were topping off our fuel bags at the RRP, a barrage of mortars impacted at the RRP. One of our MK-19s suppressed the mortar position and we kept working. Luckily, no Marines were injured and refueling operations resumed. I remember that night as one of the darkest of the war; I couldn't see a thing as we continued sustainment operations while maintaining light discipline. There wasn't really time to stop what we were doing."

The Division recognized the requirement to shake a methodical garrison mentality in the re-supply of repair parts too. In garrison, authorization for cannibalization was granted only for combat essential equipment and only after everything had been done within the system to obtain repair parts. Given the paucity of spares, however, the CG used his authority to allow his Marines to do whatever it took to sustain the momentum of the attack. Marines went to great lengths to keep vehicles running and weapons firing with little help from an unresponsive supply system. Innovative solutions again ensured that speed equaled success, as CSS Marines worked as best they could with two incompatible logistics systems that were not very expeditionary. Mechanics became miracle-workers, using skill and imagination to make up for the dearth of repair parts. In several cases, repairs were made on vehicles as they were being towed up the Highway to rejoin their combat units.

The pause after the storm gave Marines a welcome chance to clean weapons and equipment.

Taking advantage of the momentary pause in the relentless drive to the north, the flow of vehicles continued up and down the route. The CSSG initiated Class I and III re-supply of RCT-5 and RCT-7. Convoys from CSSB-10 and CSSB-12 arrived with 128 pallets of MREs, 17,000 gallons water, 200,000+ gallons of fuel, and 520 short tons of ammunition to be delivered to RRP-24. These impressive figures made a big impact, but only a momentary one, as the Division's daily requirement of 120 pallets of MREs and 50,000 gallons of fuel continued to burn continuously. The Division looked forward to the seizure of the Hantush Airstrip, which would reduce the need for road bound logistics and provide a maintenance and repair collection point.

The Open Western Flank

The open western flank between As Samawah and Ad Diwaniyah was a continuous source of insecurity in the Division's zone. Ad Diwaniyah had been shaping-up as a hub for paramilitary activity across the south, and the Iraqis still controlled the major roads in the area to the west of the Division zone. South of Ad Diwaniyah, Highway 8 offered the Iraqi paramilitary fighters a high-speed route on which they could travel up and down the Division's western flank with impunity. This road was in the V Corps battlespace, but was not being used or controlled by the Army. The enemy found sanctuary there, and exploited the seam between the two US forces. For the combat forces of the Division, this was not really a problem. For support convoys full of ammunition trucks and refuelers, this presented a significant risk.

The Division took every opportunity to aggressively hunt down and kill the remaining paramilitary fighters in zone. The Diwaniyah area, which was a reported command and control hub for the paramilitaries, would receive the Division's special attention. In addition, the Division considered ways to close down the Highway 8 flank road and eliminate the risk coming from the seam between Marine and Army forces. The Division began to employ a number of methods to tighten the security in zone and to reduce the threat of ambush from the flanks. The stretch of Highway 1 south of the Division to An Nasiriyah, which would have to be traversed by supply convoys, was assigned to Lieutenant Colonel Chris Conlin's 1st Battalion, 7th Marines. 1/7 proved adept at the MSR security mission, and was particularly successful when they captured an Iraqi patrol:

"On morning of 26 March, as the Battalion was occupying a TAA, Bravo Company spotted four men in civilian clothing walking north-to-south west of their lines. The Battalion had recently received several Intel reports to be on the lookout for infiltrators trying to get into An Nasiriyah to help resist the US forces in that area. According to the report the men typically wore civilian clothing and carried bags with uniforms in them. Based on the report, Bravo sent out a patrol to intercept these individuals. As the patrol approached, the individuals began to run, but the Bravo Marines were able to chase them down. Immediately the individuals were turned over to the HET. Shortly after HET began the interrogation, it was determined that these men were Iraqi intelligence officers. Much valuable information was gained, none more important than the location of an American POW."

Interrogation of the Iraqis captured by the patrol led to the location of Private First Class Jessica Lynch, an Army soldier who had been captured earlier in An Nasiriyah, gaining her the infatuation of the media. Her location in the Saddam Hospital in An Nasiriyah was passed to higher headquarters, eventually resulting in her recovery by other US forces, including Division Marines from 2/1 that were part of the 15th MEU (now reassigned from the 1st UK Division to TF Tarawa.)

The Highway 1 muck claims a new 7-ton MTVR. Extricating vehicles from the muck was a full time job.

Because securing the entire Highway 1 corridor would have seriously drained the Division's combat power, the Division chose to assign combat units as escorts to mission critical re-supply convoys, and leave rear area route security to roving patrols made up of Navy Special Warfare (NSW) SEALS and Force Reconnaissance Company Marines. Fourth LAR was assigned to the TMCC to ensure the secure movement of 50 refueler trucks from south of the Euphrates up to the hastily established RRP-24. After meeting with the Division planners, the G-3 also directed the RCTs to conduct limited objective missions in their zones on either side of the Highway, killing any enemy discovered in an effort to provide a more secure LOC. The units were to clear 4km on either side of the MSR and ensure there were no Saddam Fedeyeen or other militia close enough to the road to interdict convoys. The reports of continued mortar attacks and sniping incidents along the length of Highway 1 served as a reminder of the open flanks of the Division, and pointed out the need for aggressive action. As the Division continued to 'throw elbows' at the remaining Fedeyeen along Highway 1, 3d MAW began to take advantage of the clearing skies. The Wing commenced hammering both the

Republican Guard forces to the north and the Fedeyeen opposition to the Division's effort along Highway 1. The Division would not hunker down and offer a target for the paramilitaries, it would stay on the hunt.

TF Inchon Sustains the Momentum

The lessons learned in An Nasiriyah resulted in the tactics that the Inchon Marines would invoke for the remainder of the fight to Al Kut. To counter the urban ambushes used by the paramilitary forces, RCT-1 employed the 'strongpoint tactics' proven in An Nasiriyah to pass the RCT through, thus continuing the momentum all the way up Route 7. Any area that fit the template for a Fedeyeen ambush was attacked with a robust force of armor, mechanized infantry, or light armored vehicles. Once this area was secure, the attacking infantry would remain in position as a guard force, allowing the remainder of the RCT to quickly pass through unimpeded.

On 26 March, using the strong point concept, 2d LAR and a platoon from Alpha company, 1st Tank Battalion continued their movement north in advance of the RCT. The RCT's mechanized infantry followed in trace, with 1/4 reinforced by Alpha tanks as the designated strongpoint force for the city of Ash Shattrah. This town had been a hotbed of Regime activity in the past days, and was expected to provide a challenge. The enemy had adjusted his own tactics, and now allowed the LAVs and Tanks to pass before opening up on the softer vehicles to the rear. Captain Paul Gomez's Bravo Company and a platoon from Alpha tanks picketed along the route. Captain Brian Collins' Charlie Company and Captain Dave Banning's Alpha Tanks headquarters section blocked the town proper. Captain Griffin's Alpha Company and a platoon from Alpha tanks guarded the approaches to the town from the south. As 1/4 entered Ash Shattrah they immediately began taking fire. The battalion moved to suppress the enemy fires, and quickly gained control of the situation. The battalion's forward command group drew small arms fire and several RPGs from paramilitary fighters located in the built up area. The battalion's assistant operations chief, Gunnery Sergeant Andre Askew, led the supply Marines, cooks, and scout sniper team that comprised the command group's security detail in quickly silencing the regime loyalists. Every Marine was a rifleman. As Alpha Company was eliminating final pockets of resistance, the ADC, General Kelly, approached the company commander and asked for an update to the situation. Captain Griffin stated that all enemy positions had been effectively suppressed with the exception of the southernmost building from which the company was

Convoys often had to compete with local shepherds and their flocks for road space.

receiving sporadic sniper fire. "Well, we can't let that hold up the entire regiment. What are we going to do about it?" inquired the general. Only moments before, the company commander had directed the assault section to employ a thermobaric rocket to silence the sniper. "Sir, if you wait just a minute, you'll see a thermobaric rocket shot that should eliminate the problem," the company commander replied. Seconds later, a devastating explosion rocked the target building, blowing large pieces of concrete in the air. It was a spectacular display from the first employment of a thermobaric round during the war. No further resistance came from the building.

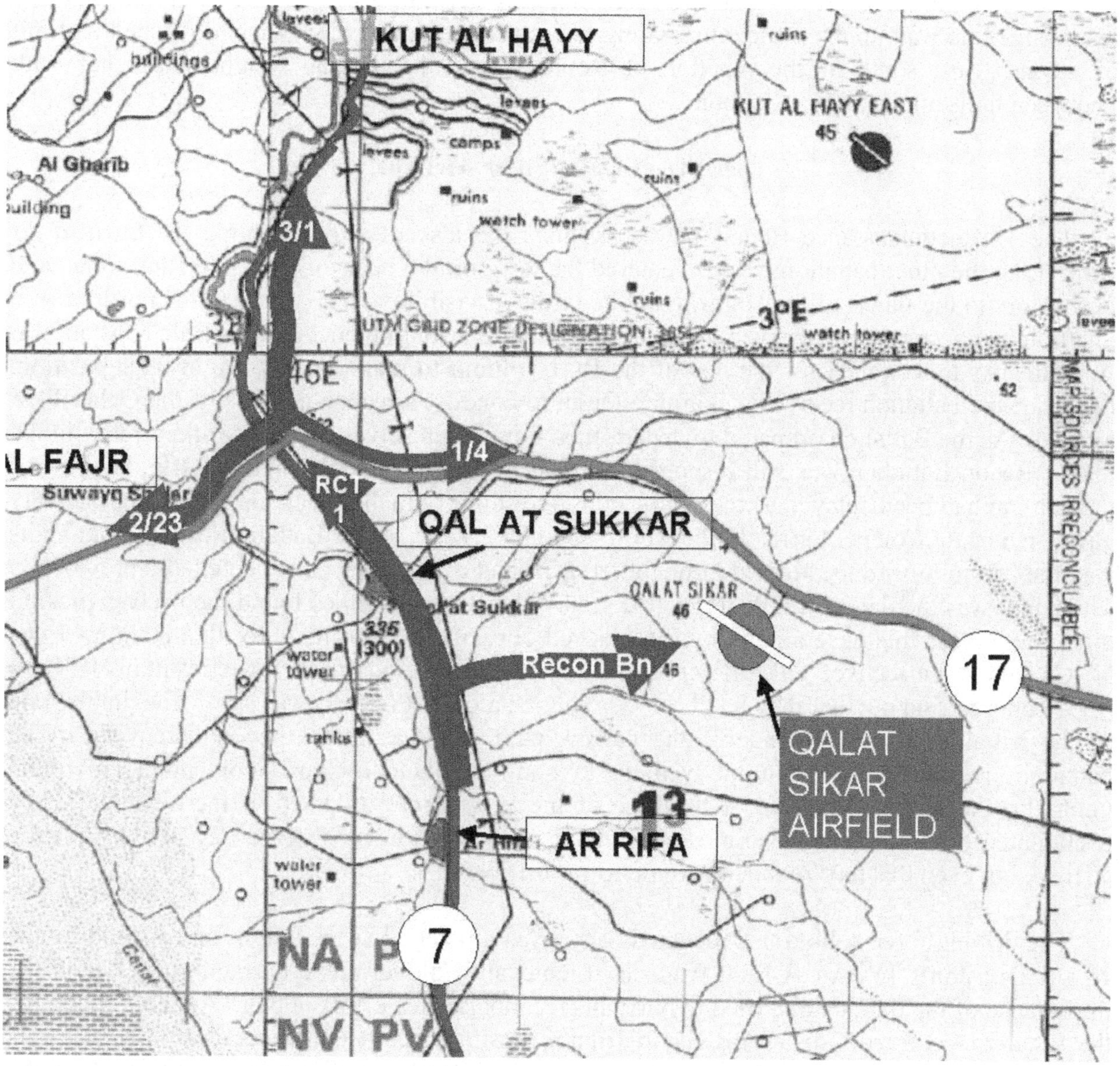

Inchon maintains the momentum, pushing north and branching out alongHighway 7.

During this encounter, the Marines also observed the enemy fighters using 'human shields'. The Baghdad Fedeyeen would grab local women and children, and force them to stand between the Marines and themselves. The superb fire discipline and tactical skill of 1/4 allowed them to destroy the enemy fighters, while causing no known casualties among the innocent civilians. First Battalion, 4th Marines had one of the highest percentages of recent Boot Camp graduates in the Division, and their actions were a testimonial to the skill and discipline of these young Marines. They held the door open for the rest of the RCT, and took their place at the end of the column once the rest had safely passed.

RCT-1 continued to use strongpoint tactics to clear Highway 7, and began to refer to their progress as the '100-mile running gunfight'. At the town of An Nasr, 2/23 held the door open while engaging enemy mortars and RPGs in the town. The Reconnaissance Battalion took its turn in Ar Rifa, followed by 3/1 in Qalat Sukkar. By this method, the RCT successfully leapfrogged its way up the route, eliminating resistance along the way. Inchon reached its limit of advance just south of the junction of Routes 7 and 17, having fought skirmishes with Fedeyeen units all day along the route.

Seizure of Qalat Sikar Airfield

The 1st Reconnaissance Battalion was assigned the task of strongpointing the town of Ar Rifa. Like the other battalions, Recon entered the town on the heels of the LAR/Tank force, and pushed out to the flanks of the Highway to the limits of visibility. The towns to the north of Ash Shattrah had become progressively more benign, and the battalion encountered little resistance in Ar Rifa. As they waited for the rest of the RCT column to make its way up to their position, however, the battalion received a Warning Order to conduct a reconnaissance of the Qalat Sikar airfield. As the Battalion prepared to depart, they found themselves waiting for their relief in the town. Recon Battalion was still responsible for securing the route through Ar Rifa, but part of the column had been delayed as they refueled to the south. Finally, near sunset, Colonel Dowdy gave Lieutenant Colonel Ferrando the word: "Go now, Steve." The Battalion began to make its way the 25 km toward the airfield through terrain as bad as any they had encountered previously. This night was the darkest the Marines had seen. The Recon Marines found themselves picking their way across this miserable terrain in blacked out conditions, with zero illumination. Even NVGs were not effective. One officer recalls his driver changing out the batteries in his PVS-14 NVGs only to find out that they had been working all along. Progress was slow. Inevitably, one of the battalion's vehicles slid off the narrow path and was mired beyond recovery. The battalion decided to abandon the vehicle to continue their forward momentum. Around midnight, the battalion arrived at the base of the access road that led to the airfield. They dismounted several teams to immediately move to a position to observe any activity on the airfield, but even this foot-mobile movement was difficult.

The driving force behind this reconnaissance was a planned air assault on the airfield by the 1st UK Division's 16th Air Assault Brigade. Lieutenant Colonel Steve Ferrando understood the importance of the mission and tried to hurry his reconnaissance efforts along. At about this time, the Battalion received a disturbing report from a passing aircraft that several armored vehicles had been spotted on the airfield. This complicated the problem significantly. Pressed by the terrain, a potential change to the enemy situation, and time pressure to get this mission accomplished, Lieutenant Colonel Ferrando was in a tight spot indeed. Only one thing could add to the pressure now. The Iridium phone rang. It was Chaos. "Steve, I've really got to know what's on that airfield. We want to support our UK allies here, and we really need this," the CG said. Sensing the urgency of providing this critical reconnaissance in support of the Division's Coalition allies, Lieutenant Colonel Ferrando huddled his Alpha and Bravo Company commanders and they crafted a plan. They would have to take a higher degree of risk to accomplish this mission. They would assault the airfield with two companies, using speed and surprise to cover their assault. After a tense two hours of waiting for dawn to break, the two companies attacked online onto the airfield, finding it deserted. The Recon Battalion spent the next two days securing the airfield and blocking Highway 17 to the east, protecting the flank of the Division from any sortie the 10th Armor Division might decide to make. As fate would have it, the UK air assault never took place, as the 16th Air Assault Brigade was assigned to another mission.

27 March
Demonstration to Hantush

The cloverleaf intersection of Highways 1 and 17 was key terrain, and fought over repeatedly during the campaign.

Artillery fires on Fedeyeen positions as Marines approach the Hantush airstrip.

At 0321Z, 27 March, RCT-5 resumed the attack up Route 1 as the Division's main effort with a mission to seize Hantush Airstrip. Confronting irregular enemy forces along the route, 2/5 engaged and destroyed an enemy company-sized unit with armor and mechanized vehicles on the objective. The enemy used scores of RPGs, mortars and heavy machine guns. Supporting the attack, 11th Marines had three battalions along Highway 1 in support of the Division's attack, and provided a crushing fires capability. RCT-5 seized the Hantush Airstrip at 0620Z, following a brief but potent firefight. During this action, Gunnery Sergeant Joseph Menusa of the 1st Combat Engineer Battalion was killed by enemy fire. As his engineer platoon was attacking up Highway 1 to seize the Hantush Airstrip (supporting 2/5), it was subjected to machinegun fire and attacked by RPGs. Aware of the danger, Gunnery Sergeant Menusa chose to replace a junior Marine in an exposed security position atop his AAV. In doing so, he gave the Marine needed relief and put himself in a better position to observe and help control his platoon. Shortly after, Gunnery Sergeant Menusa's AAV and the two M9 Armored Combat Excavators behind him came under withering small arms fire. More concerned with the mission and his Marines than for his own safety, Gunnery Sergeant Menusa chose to stay on top of his vehicle to direct the suppressive fires of his

Marines, protecting the unarmed bulldozers behind him. While in this exposed position, he was mortally wounded. Because of the training and mentorship he had provided, the Marines of his platoon were able to quickly collect themselves after this devastating loss and continue on to accomplish the mission of seizing Hantush Airstrip.

Marines advanced on Hantush with Cobra escorts. Cobras provided both CAS and intelligence on enemy threats up the road.

While 2/5 was seizing the Hantush Airstrip, 3/5 had the task of holding the door open at the 'cloverleaf', an interchange at the junction of Highways 1 and 17 just to the east of Ad Diwaniyah. The Fedeyeen continued to be concentrated in the Ad Diwaniyah area, and the Division's movement to the north had drawn several hundred of them out of the city to positions from which they could interdict the Marines' movement. Lieutenant Colonel Mark Toal, the regimental Operations Officer, recounts that while 2/5 was pushing north, 3/5 was in 'one hell of a fight' at the cloverleaf.

Third Battalion, 5th Marines holds the door open at the Hantush cloverleaf as the Division takes Hwy 27.

While watching 2/5 and the other lead elements of RCT-5 push toward Hantush Airstrip, 3/5 observed several small bands of enemy soldiers approaching from Ad Diwaniyah. These soldiers were taken under fire by Lima Company and its attached CAAT section. Tanks from the attached tank platoon moved forward from the cloverleaf toward Ad Diwaniyah and destroyed two T-62 tanks located in deep revetments. Concurrently, Kilo Company, back at the original defensive position providing security for the Battalion's Main CP and field trains, saw two BMPs and a pair of anti-aircraft guns located approximately 2 km to the west. The Company Fire Support Team (FiST) quickly worked up a combined arms package of artillery and close air support to destroy the equipment. This package did the job, as Marines throughout the battalion witnessed the secondary explosions that erupted after the air strike.

Soon afterward, however, the Battalion CP, field trains, and Kilo Company began receiving accurate enemy mortar fire. The rear elements of 3/5 quickly moved forward, out of the incoming fire, using Kilo Company as convoy security. Nearing the cloverleaf, Kilo tripped another ambush when a bypassed company-sized unit opened up with a high volume of direct fire on the AAVs. Quickly dismounting, Kilo Marines moved into the assault, returning fire and killing several Iraqi soldiers. As large numbers of Iraqis retreated into a series of mud huts and houses nearby, Captain Mike Miller pulled his men back and began to rock each target with a steady combination of rotary and fixed wing aviation. The battalion cleared at least 100 enemy troops in the trench lines lining both sides of the road, utilizing both air and surface fire missions to attack and defeat the enemy. The actions at the cloverleaf would continue for most of the day.

Iraqi life goes on as Marines advance along Highway 1. Many of the civilians in the south showed little concern for the war occurring around them.

At 0900Z, the Division received the order to abandon Hantush and withdraw to the south. RCT-5 began to withdraw back to more defensible terrain in the vicinity of the Ad Diwaniyah crossroads. After the terrible weather, the tragic loss of life along the way, and now the successful attack and gain of positive momentum, this ordered withdrawal was a hard pill to swallow for many of the Marines. The Division had its sights set on Baghdad, and any delay in getting there was not suffered gladly. The Marines from RCT-5 had a proud legacy from World War I and the actions of their forefathers in Belleau Wood, and the withdrawal drew many comments of the 'Retreat hell, we just got here' variety.

Later in the day, after the Division had pulled back to the vicinity of the Cloverleaf, the situation was clarified for the Division Marines. After the attack had been launched, the CFLCC and MEF Commanders had notified the Division that the main effort had shifted to establishing a better security environment in the rear prior to resuming the attack. The Division's rapid advance had knocked the enemy off-balance, but had left its own re-supply routes vulnerable as long as the An Nasiriyah and Al Basrah urban areas still harbored significant enemy forces. The additional battlespace taken by the move on Hantush was considered an untenable position should the Iraqis counterattack, and the Division had been ordered to withdraw back to the Diwaniyah area. During planning, the Division and CSSG had foreseen and accepted the risks to extended supply lines, but the Division was now ordered to reduce the risk by tactical action. The Division had a philosophy regarding orders from higher

Cobras provide supporting fires near the cloverleaf.

headquarters: 'Just Do It.' The Division completed its withdrawal and resumed its aggressive hunt for remaining Regime paramilitaries in zone.

After the misery of the last three days, the sun emerged with clear skies on the 27th. The Marines were excited about the opportunity to be back in the offensive, and you could feel it. The Division Forward had joined the column in trace of RCT-5. There was sporadic resistance from defenders firing from defensive prepared positions and bunkers alongside the highway. Ahead of us there was a lot of air and artillery being called against targets in a cement plant and a chicken farm. The lead elements of the convoy scattered these guys, and as we passed, the citizens of the local area had taken to the streets and were waving to us. North of Diwaniyah, the terrain began to look more like what I thought Iraq would look like. Water buffalo grazed in muddy pools at the base of Palm groves. Green fields surrounded small mud-brick villages that could have been drawn from a history book. Boys and women herded cattle, sheep, and goats alongside the roads. But the terrain was terrible. There was mud and standing water everywhere, and cross-country mobility was next to nothing. We were pretty fired-up when the airfield was secured, but then we started to see vehicles coming back the other way. While we sat along side the road, we watched in disbelief as the whole attack force came rumbling back south. When the entire tank battalion and an infantry battalion were observed heading south, we knew something was up. It was frustrating as hell to see that dominating combat force turning around and running back south. I was afraid we were losing our advantages of momentum and surprise.

- Captain Steve Dunn III

Actions in Ad Diwaniyah

The Fedeyeen presence in Ad Diwaniyah was reportedly extensive. The local people reported that these Regime enforcers were brought down from Baghdad and were allegedly terrorizing the populace, running around in their black pajamas and hoods through the streets of the town. Because of its central location, and the fact that American forces had not yet cleared it, Ad Diwaniyah became a transit point for Regime fighters brought in from elsewhere. The fighters that were bussed or trucked into town included Fedeyeen, deserters threatened back into service, conscripts, and foreigners. Allegedly, there was a cadre of Republican Guard officers who were organizing, training, and equipping these forces as they arrived in town, then sending them out for missions against the US Army to the west, or the US Marines to the east. The local populace was terrified of these fighters, and wanted them to go away. The civilians were, of course, powerless to do anything about it.

Now, it was 1st Marine Division's turn to deal with these thugs. The Division had been operating to the east of Ad Diwaniyah for several days. By the time the attack elements of RCT-5 had made their way back to the Cloverleaf just outside of town, 3/5 was busily engaged with Fedeyeen fighters on both sides of the highway. They had been fighting most of the day, and were coordinating fixed wing and rotary wing support as the columns went by. Using (by now) standard enemy tactics, the paramilitary fighters would wave white flags and signal surrender whenever helicopters appeared overhead, then would pull them down and resume firing as soon as the helicopters went off-station. The battalion kept the heat on the enemy, however, and destroyed several hundred Fedeyeen fighters over the course of the day.

Putting fires on the Fedeyeen in Ad Diwaniyah proved to be an exercise in patience, as again the seam between the Army and the Marine Corps had to be traversed. One example of the staff gymnastics that were required happened later that day. In the morning, Lieutenant Nate Boaz, and his HUMINT Exploitation Team Three (HET 3) were working with the local populace near Ad Diwaniyah in support of RCT-5. The team was advised by a fearful local citizen that the Fedeyeen had been using the Ad Diwaniyah stadium as their rally point, and that there were hundreds of fighters at this location at any given time. This information was reported up to the G-2, who quickly vectored the Pioneer from VMU-2 over the site, confirming the reported activity. Since this was not in the Division's battlespace, the request for fires on the target went to I MEF. I MEF coordinated with the CFLCC, and passed the target over to V Corps for prosecution. The Army then flew their own Hunter UAV over the stadium and confirmed the validity of the target before turning it over to joint air for prosecution. The Air Force then attacked the target, reportedly killing dozens of Fedeyeen and destroying some antiaircraft guns in the process. Such a process was painfully slow, and effective only against large stationary targets such as a troop concentration in a stadium. Fortunately, these cross-service, cross-boundary engagements were rare.

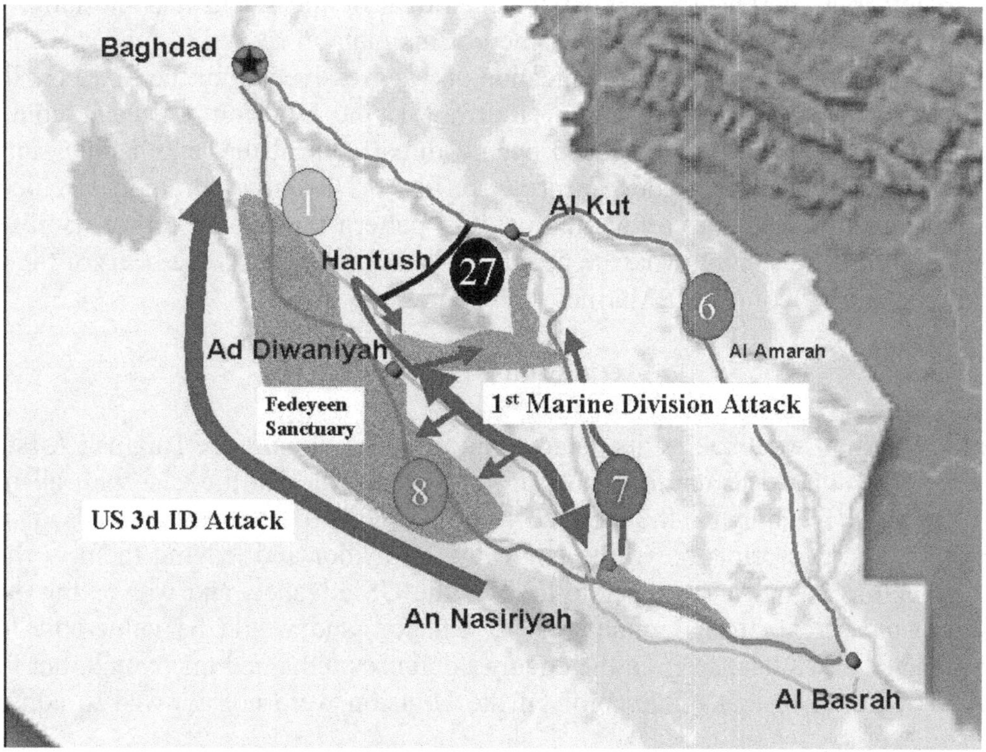

'Throwing elbows' along the Highway 1 corridor. First Marine Division maintained momentum by destroying pockets of Fedeyeen fighters remaining in the vicinity of Highway 1 while the 3d US Infantry Division continued to move to the west.

The destruction of the Fedeyeen rallying point, and the continuous fires brought to bear against any identified troop concentrations by the Division took a toll on the Regime organization in Ad Diwaniyah. The Division was not quite done with this town, however, and revisited the city with the Pioneer that night. The operators discovered a large equipment storage and repair facility housing dozens of T-55 tanks and self-propelled artillery pieces. Observing and adjusting fires with the Pioneer, the Division again demonstrated the powerful effects that could be achieved with an Unmanned Aerial Vehicle under the direct control of a Division targeteer. This had only been possible because of the superb supporting relationship the VMU squadrons had established with the Division. Because they had been placed in direct support, there was no intervening bureaucracy between the supported intelligence section and the crew operating the UAV. Both VMU-1 and VMU-2 aggressively attacked their mission, providing responsive support to the Division throughout the campaign. The VMU commanders and staff could frequently be found inside the Division CP, ensuring the effectiveness of their support. The tanks were destroyed, the concept of a directly supporting/organic UAV within the Division was proven, and the Iraqis paid the bill.

The time spent on hunter-killer operations along Highway 1 and in Ad Diwaniyah would pay off handsomely in the days and weeks ahead, but was not without cost. On 27 March, Lance Corporal Suarez del Solar, an infantryman with 1st LAR Battalion, was killed during a foot patrol conducted to find and eliminate the threat posed by remaining Fedeyeen fighters. As dusk approached, his squad was assigned the task of conducting a reconnaissance patrol in front of the screen line established by the LAVs of his platoon. Within a few minutes of beginning the patrol, an enemy mine detonated, rocking the members of the squad and mortally wounding Lance Corporal Suarez del Solar. The corpsmen of his platoon rushed to render aid, but it was too late to save his life. Because of the dedication and aggressiveness of the Marines (like Lance Corporal Suarez del Solar) clearing along Highway 1, the Division was later able to move against Baghdad with significantly reduced enemy threat against the lengthening supply lines. Over the next several days, the Fedeyeen presence in Ad Diwaniyah dramatically declined, as they were no longer able to find refuge there. In a pattern that would repeat itself across the newly liberated Iraqi cities and towns, the people took to the streets on the heels of the retreating Regime fighters, and welcomed the Marines with smiles and waves.

Actions on Highway 7

While the Division was making its demonstration attack against the Hantush Airstrip in the west, the enemy continued to reinforce the Highway 7 corridor in the east with all manner of fighters. As in Ad Diwaniyah, the Regime had reportedly been bussing loads of deserters, foreign fighters, and conscripts down the Highway 7 corridor and leaving them in the smaller towns. The Regime was clearly desperate to slow the US advance, and was taking extreme (if uncoordinated) measures. The Regime was off balance, and would be vulnerable to a rapid attack if conducted now. The seam in the enemy's defenses remained uncovered, but the enemy was bound to wake up to that vulnerability if the Division were not allowed to continue their attack soon.

RCT-1 Marines went into motion, with the mission of clearing Highway 17 near the junction with Highway 7, and continuing the pressure to the north. Second Battalion, 23d Marines (2/23) pushed a small distance west on Route 17, while 1/4 cleared to the east towards 1st Recon

Marines of RCT-1 provide cover for forces advancing on Highway 7.

Battalion at Qalat Sikar Airfield. The 'Thundering Third' of 3/1, supported by 2d LAR, attacked north to draw the attention of the Baghdad RG Division.

Later that night, a US Air Force tank-killing A-10 errantly attacked RCT-1's Alpha Command group. The pilot had received the mission request and grid coordinates from RCT-5 Marines engaged in an intense fight. Unfortunately, somewhere in the process, an error crept in to the fire mission coordinates. The grid zone designator passed to the pilot was off by one character. That one letter difference equated to 100 kilometers on the ground, and the A-10 subsequently rolled in on what appeared to him to be the enemy. The A-10 pilot strafed RCT-1 with his 30 mm tank-killing gun and returned for another attack with high explosive bombs. The unit that requested the air support was engaged in fighting exactly 100 kilometers directly to the west. Alpha Command was at the exact grid location of the requested air support, but one grid zone designator further to the east. The aircraft was successfully called off before any Marines were injured. Shaken and considerably stirred by the A-10's actions, RCT-1's Alpha Command pressed on with their attack.

This type of friendly fire incident was rare. The Marines of RCT-1 were no strangers to some awesome, on target, rotary wing air support throughout the campaign. The 'Stingers' from HMLA-267 provided superb rotary wing air support around the clock. The Marines of the RCT also noted the support of the 'Purple Foxes' of HMM-364 with considerable enthusiasm. At no time did a battle casualty from RCT-1 die while awaiting evacuation due to their rapid response times and willingness to go into harm's way to help their fellow Marines. This squadron (with one of the greatest mottos in the Marine Corps) proved themselves repeatedly.

Shaping The Al Kut Battlefield

In a final action of 27 March, the Al Kut defenders received some shaping of their own by air fires. Although the pause in the attack along Highway 1 had given the enemy a respite on the ground, this was offset by an increase in air shaping fires, permitting the MEF to remain on the offensive and to turn the heat up on the enemy. Because of the rapid execution and uncertain timing of the initial attacks into Iraq, there had been only limited shaping fires. Bad weather had interfered with flight operations and targeting in subsequent days during the storm. Now, the skies opened up and aviation assets from across the Coalition were brought to bear against the Baghdad RG Division. The Baghdad RG Division would receive hundreds of sorties over the next few days; most of it focused on artillery, tanks, and ammunition storage points. The Al Kut defenders had the bad fortune of being within a single fuel tank range of a variety of strike aircraft operating from Kuwait and the Arabian Gulf. Without the requirement for external tanker support, assets from across the coalition were piling on against the luckless Baghdad RG Division. From the limited battle damage assessment received, it was clear that the air shaping was having a crushing effect on the Republican Guard. The artillery regiments of the Baghdad Division were assessed at less than 45% remaining strength, and additional shaping fires were on the way. Division Marines eagerly awaited the attack order while applauding the destruction of the enemy by air fires.

Closer to Baghdad, CFACC shaping was having a similar effect on the artillery of the Al Nida RG Armored Division. Once 1st Marine Division was successfully across the Tigris, the Al Nida Division would be the last major force between Blue Diamond and Baghdad. The Iraqis were now making some attempt to reposition artillery and troops into the gap between the Al Nida and Baghdad Divisions, but these efforts were rewarded with precision guided munitions from Coalition aircraft. The Iraqis had lost the opportunity to reposition major elements of their ground forces now that they were under a deluge of air fires. Out of position, and unable to move their conventional forces, they found themselves unprepared for the coming ground attack of the Marines. It was too late for the enemy to close the seam. This was the classic combined arms effect the Marines had hoped to achieve, and it was working. The Iraqi military leadership that had not yet run away probably began to appreciate the power of a balanced combined-arms fight.

28 March
The Waiting Continues

As the Division waited for approval to continue the attack north, all three RCTs continued aggressive security and ambush patrols to their flanks in their respective 4 km security zones. Along Highway 1, RCT-5 conducted limited objective attacks and security patrols, and continued the pressure on the Fedeyeen remnants in Ad Diwaniyah.

RCT-5 began to experience the transition between military operations and civil operations that was to be a recurring theme throughout the campaign. During the early days of the attack, operations had been conducted in the classic manner to which the Marines had been trained. All reported mortar attacks or troop concentrations were responded to with devastating firepower. As the Regime presence declined in the area, however, the Marines had to make the mental adjustment from attacker to liberator. The recent decline of large paramilitary organizations in zone allowed the Division to become much more discriminating in their application of firepower. Counter-battery indicators, for example, now had to

Embedded media enjoy unprecedented access to Marines and commanders during the entire operation.

be confirmed by the splash of rounds or observed enemy troop concentrations prior to granting approval for return fires. Villagers started reporting to the Marine units that were being hurt by the Marine's artillery fires as the enemy would drive up next to noncombatants, fire mortar rounds, then drive away (a method termed 'shoot and scoot', designed to draw the Americans into indiscriminately firing on civilians.) Positive identification of targets had always been the rule, but now the security situation allowed the Marines to more deliberately assess the risks to local civilians before they opened fire. This extension of the positive identification ROE saved many noncombatants from injury or death. Making this adjustment required a disciplined force with a strong moral compass, one of the Division CG's emphases from the earliest days of preparation. From the beginning, the members of the 1st Marine Division had been taught to be 'No better friend, no worse enemy.' It was a sliding mental adjustment the Marines would be required to make in both directions until the entirety of southern Iraq was secure.

Tanks from 2d Tank Battalion provide security along Highway 1. In the flat open terrain, tanks were always near the front of the Division's advance.

The Division continued its security patrols to the south and west of the zone, and closed up its positions along Highway 1. Two platoons from the I MEF Force Reconnaissance Company, temporarily assigned to the Division, were given a zone south of the Division's current positions along Highway 1, and carried out their task with zeal and professionalism. The Division's intent was quite clear; to kill those who had been attacking the Division and Coalition forces along the LOCs, using innovative and aggressive techniques. No enemy attacks took place along either of the Division's columns this day, another clear indicator that the time to resume the attack had come. The MEF continued to press CFLCC hard to let them unleash the Division, and the Marines were more than ready.

The Division continued to rely heavily on the P-3 and Pioneer platforms that were flying in direct support. As the only direct support imagery intelligence collections assets assigned to the Division, these were the only ones that provided continuous reliable support for the entire operation. The Division was to have little success in navigating a cumbersome collections bureaucracy to obtain support from non-directly supporting platforms, but aggressive use of these two assets provided excellent coverage for what they could range. Between the efforts of the VMU squadrons and the Sailors of Combined Task Force 57 (CTF-57), the Division Marines had eyes that would have been the envy of their predecessors. As the Division moved north into Iraq, the Joint Forces Air Component Commander (JFACC) gave its approval for the P-3 to begin operations north of the Iraqi border. P-3s, began to fly missions in 'safe' killboxes further north, remaining in step with the Division's movements. The P-3 was invaluable in supporting the Division's efforts up Route 7, especially in the vicinity of the town of Qalat Sukkar. Similarly, the Pioneer Unmanned Aerial Vehicle continually demonstrated the value of a direct support or organic UAV in prosecuting the Division's deep fight and targeting efforts. During the maneuver pause, intelligence collections continued apace, and a large number of Iraqi armor and artillery units were located and struck. On this day, the Pioneer detected a G-5 155mm howitzer battery just north of the Highway 27 Bridge, enabling the Division to direct precision guided munitions against the position.

28 March
2/23 Strikes a Blow Against the Baath Party

The town of Al Fajr had been a hotbed of Regime activity. This town anchored the eastern end of the Highway 17 cross route between Highway 7 and Highway 1, the route of march for any forces seeking to travel cross-corridor within the Division's battlespace. On the morning of 27 March, 2/23 was tasked to secure the west side of the intersection of Highways 7 and 17.

Marines used hand-written signs to tell Iraqis to stop at checkpoints.

The account of the guarded cooperation on the part of the local citizens is telling, as it is representative of the encounters the Marines had with the Iraqi people:

> *After Fox Company killed a driver attempting to run the road block we had established, one of the battalion's own Arab speaking Marines, Sergeant Dory Habachy, identified several citizens willing to discuss enemy activities in the town. These informers seemed fearful of talking to U.S. forces in the presence of other Iraqis so the battalion arranged to meet with them several hours later. With the help of these Iraqis and others in the surrounding hamlets, information suggesting a Baath Party presence in Al Fajr began to come to light. The informants described the structure of the local Baath party identifying four distinct groups and their leaders.*
>
> *This information helped the battalion to further develop the situation and plan for actions into Al Fajr. To locate the reported Baath party houses, the regiment's Dragon Eye was launched to fly over named areas of interest within the town and its findings were reported back to 2/23. At the same time, the scout-sniper platoon began planning for a night insert of three teams to confirm the presence of armed individuals and air defense weapons at sites on the west side of the Al Gharraf River.*

On 28 March, 2/23 was sent to Al Fajr to clear out the Baath party members that had been reported in the city. They initially met no resistance, and turned to Civil-military operations in the town instead. The Baath party members had wisely chosen to stay out of sight while the Marines were in town. The local populace was friendly and willingly received the Marines into the town. From information received from friendly local citizens.

The battalion concluded that approximately 60 Baath Party loyalists controlled the town using local schools as their headquarters and meeting places. The battalion learned the precise locations of the four top Baath Party leaders and their weapons caches in the town. As the rest of the battalion pulled back just outside of the town, the S-2 assigned objectives for scout snipers that would allow them to observe these suspected sites, and perhaps catch the Baath militia (who were avoiding contact with the Marines) in the open.

Day and night, Marines keep a watchful eye for the enemy.

As night fell, two scout-sniper teams conducted their insertion, and by daybreak they were observing and reporting targets. One of the sniper teams observed enemy forces carrying Kalashnikov rifles in a crowd of people. The third scout-sniper team began its insertion into a building when they ran into a group of armed Iraqis. The team immediately broke contact and fought their way out of the building, bounding back to the link-up point. During the emergency extraction, a HMMWV hit two members of the response team, tragically killing Staff Sergeant James Cawley and seriously injuring Captain Harry Porter. This tragedy occurred incident to the liberation of the fearful citizens of Al Fajr. The Battalion mourned the loss of Staff Sergeant Cawley, recognizing that the brave Marines who participated in this action played a key role in removing the Baath militia as a force of oppression on these people, and set the conditions for a successful attack the next day.

Arab linguist interrogates an Iraqi detainee near the Tigris.

On 29 March, based on this new information, the battalion staff quickly completed planning and issued a raid order on the key sites inside the town. At 1300Z, a platoon of M1A1s from RCT-1 approached the town from the east. Citizens of Al Fajr, recognizing that an attack was imminent, gathered to see the Baath members receive their due justice. As the tank platoon entered the town, the 120 mm guns fired into the building where the scout-sniper team had been attacked the previous night. To the rising cheers of the town's civilians, the tanks continued west over the bridge, remaining on the main road to provide over-watch. An Army PsyOps team followed the tanks into town and broadcast a message for the Baath Party members to surrender. A division of Cobra helicopters engaged an open truck carrying armed Iraqis and provided forward air support during the raid.

The battalion's CAAT sections led each of the companies through the town to their objectives and provided local security as the companies cleared their objectives. Fox Company secured the suspected Baath Party residences while Echo Company assaulted the Baath Party Headquarters buildings. Each company uncovered large caches of weapons and numerous documents, which confirmed the extent of Baath Party activities in the vicinity of Al Fajr. Golf Company moved rapidly through Al Fajr and cleared the school on the north side of town. Iraqi civilians, eager to help rid themselves of the former Regime thugs, pointed out the location of the Baath Party members. Golf Company cleared a nearby built-up area where they located another large weapons cache and documents. Marines spotted armed men burning documents in an adjacent building and provided covering fire as another platoon cleared the building, which yielded an even larger cache of documents, information, and weapons.

Liberated Iraqis everywhere expressed their appreciation for Marines.

Meanwhile, Weapons Platoon Sergeant, Staff Sergeant Charles Spence, and the Mortar Section led by Sergeant Antonio Dominguez, confronted 12 well-dressed men wearing traditional Arab robes. As these men were searched, a patrol found military uniforms and equipment in nearby bushes beside the road. These men were detained, as was another group of men traveling in a tractor-trailer. The driver discreetly pointed out eight of these passengers for further search. Underneath their gowns, they were fully dressed in Iraqi military uniforms.

RCT-1 Commander pushes north while simultaneously transitioning to stabilization operations in part of the Regimental zone.

As the raid continued, Marines began looking for opportunities to assist the local citizens. The town's inhabitants had been instrumental in capturing the Baathist thugs and the Marines wanted to show their appreciation. Soon after, the battalion began conducting humanitarian operations in full earnest. Medical supplies were issued and medical assistance was provided to both children and adults while other 2/23 Marines distributed humanitarian rations to members of the town.

29 March
Plans and Re-plans

The Division awaited orders to resume the attack while stabilization operations continued in the 1st UK Division and TF Tarawa zones in the rear. The Division remained convinced that the quickest path to stability in the south was a Marine Division astride the single LOC leading there from Baghdad. The Division noted that it was still possible to drive from Baghdad to Basrah on major highways controlled by the Regime, a continuous source of weapons, fighters, and Regime control. The remaining enemy forces to the south of Al Kut had been effectively shaped, and were powerless to take major offensive action.

The operational pause also gave dangerous time for the planners at higher headquarters to think, and a number of schemes had been pushed down to the MEF for consideration. A number of supporting plans were considered and rejected, including a scheme to mount up an RCT from 1st Marine Division on Heavy Equipment Transports, and ship it to the east in order to attack the 10th Armored Division in Al Amarah. Most of the proposals would have served to shift the momentum away from the successful advance toward Baghdad, and back to the south and east. On 29 March, the Division sent Operations and Intelligence planners to the I MEF Headquarters at Jalibah Airfield in southern Iraq in order to participate in the planning efforts, and to drive home the Division's perspective that chopping the head from the snake would be the most effective way of reducing its influence in the south.

Marines keep a watchful eye on prisoners.

The media reporting of the day was an interesting study in perceptions. Some media reported, almost gleefully, that the Coalition attack was falling apart. They implied that the highest decision makers in Washington were eager to continue the attack, while others were sounding notes of caution. The media cited 'commanders in the field' as those most opposed to the continued attack. To the Division poised on the leading edge of the attack, this all came as somewhat of a surprise. The Division had full ammunition bays, full fuel tanks, and was eager to get back into the attack. The MEF planners had a similar appreciation of the situation. Some, outside the operating forces, took the counsel of their fears, playing into the hands of a sensationalist media that was anxious to exploit any potential negative news. At all echelons, Marines on the ground were itching to resume the attack.

Afak Operations

As the Division pressed its advance along two separate axes, the divided RCTs were not immediately mutually supporting on the tactical level. Although the RCTs were mutually supporting at an operational level, regaining the ability to provide mutual tactical support and gain additional maneuver flexibility became a priority for the Division. On 28 March, 3/4 was ordered to conduct a reconnaissance in force along Highway 17 in order to link up with RCT-1

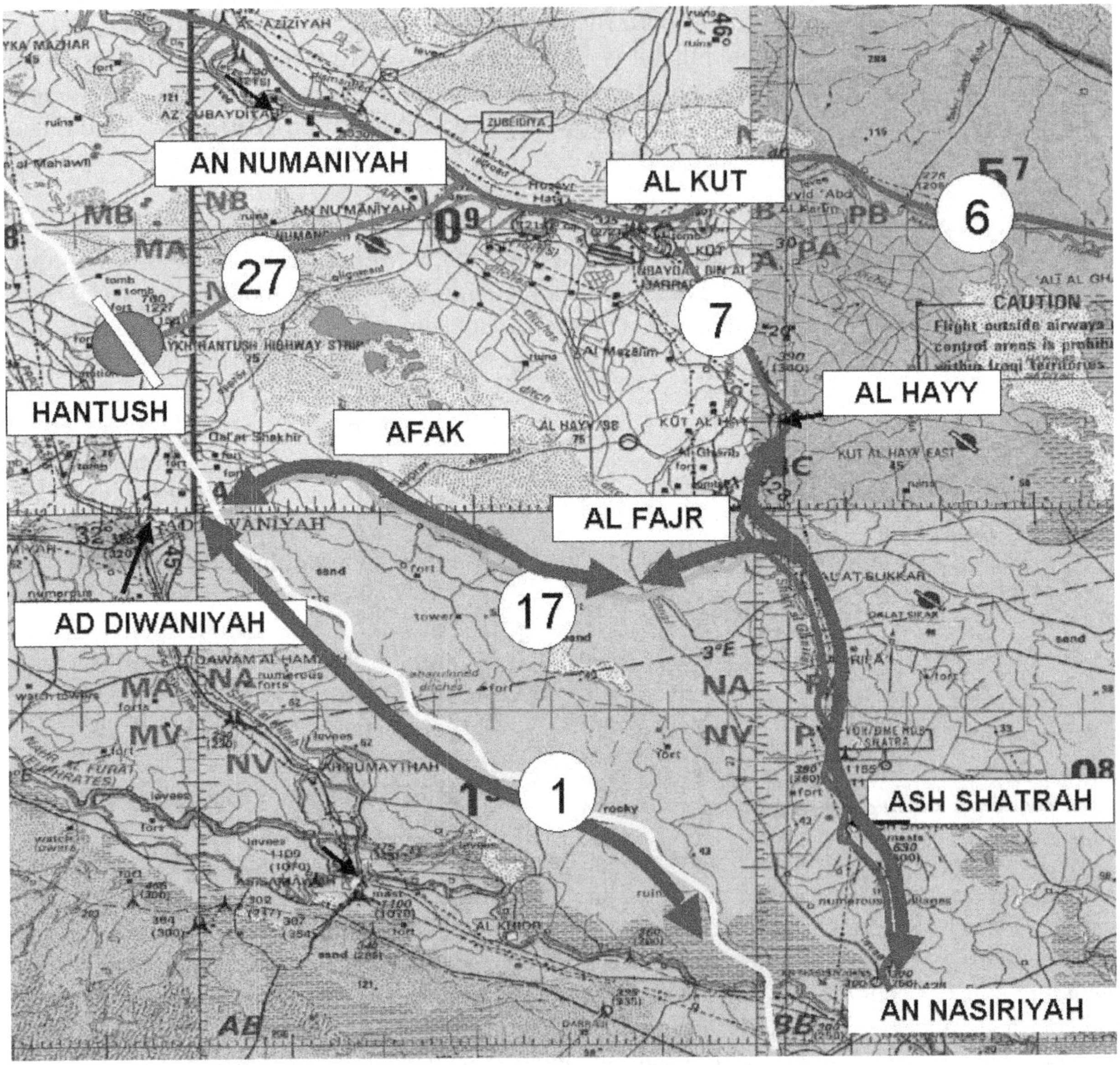

Afak and Al Fajr operations called for elements of RCT-7 and RCT-1 to link-up on Hwy 17 in order to provide mutual support and a logistics route that connected the two corridors..

east of the Saddam Canal. That evening, Lieutenant Colonel Bryan McCoy issued the order for a reconnaissance in force, east along Highway 17. "Go in like you own the place," he charged, "We're gonna kick the beehive and see what comes out." Once the connecting route was open,

A Battalion huddle gives the commanders of 3/4 a chance to talk through the plan of attack for Afak.

the two attack corridors of the Division would again be mutually supporting. This route would become a crucial west-to-east re-supply route for the Division, and serve as the connecting file between RCT-1 and RCT-7.

Operations in the town of Afak turned out to be representative of the actions that occurred in each of the towns 3/4 encountered along the route. The pattern had become predictable in the smaller towns of southern Iraq. In each town, the Iraqis used dismounted paramilitaries supported by technicals in hastily arranged ambushes and defensive positions. In response, 3/4 used shock action and combined arms. In Afak, CAAT and supporting air blocked the town's ingress and egress routes. Tanks led into the center of the urban area with rifle companies following in trace conducting the reconnaissance in force. Once in the city, infantry dismounted to clear buildings and gain the rooftops. The result was the same in every case; accurate supporting and direct fires destroyed the paramilitaries and their equipment. Follow-on Marines swept into the town on the heels of the attack, giving medical attention to the locals, dispensing humanitarian supplies, and using the HETs to gain intelligence.

By early afternoon, 3/4 approached the last town in the planned zone of action, Al Budayr. A section of AH-1W Cobras, uncovered three separate Anti-Aircraft Artillery (AAA) pieces and engaged two, under Kilo Company's control. The aircraft dropped smoke on all three sites to aid follow-on aircraft in locating and destroying the AAA.

An 81mm mortar platoon emplaces its mortars to support the attack on Afak.

Bravo Company (1st Tanks, attached) established a support by fire position while CAAT platoons once again surrounded the town. Kilo and India Companies, led by the Battalion Commander, pressed into the town center. Eerily quiet, the expectation was that this would be a much stiffer engagement than the previous towns. As the units progressed, however, the town citizens began to come out in force to watch the invading forces. Lieutenant Colonel McCoy called to report that he was surrounded by townspeople, and called up the HET and security. Kilo had cleared through the Baath Party headquarters and penetrated deeper into the town. India did likewise, both companies pressing in for a distance, and then dismounting to clear the nearby city blocks. In all, 3/4 captured 23 detainees, of whom 16 were ultimately designated EPWs. The logistics effort was significantly helped by the discovery of a fuel depot, which served to refuel all the tanks in the task-organized battalion. Having completed its efforts, 3/4 pulled out of the town and consolidated along the road northwest of Al Budayr.

30 March
Regaining the Momentum

The Division remained ready to continue the attack to the north, and had repeated the mantra in its daily Intentions Message:

> *The area around the Division is increasingly secure. Offensive operations on Route 7 and North on Route 1 will help set conditions to turn over our rear area to TF Tarawa when we go across the Tigris. Our current vulnerability to chemical attack and the potential for the enemy to realize they have a gap in their defenses indicate we should conduct a limited objective attack soonest to cross the Tigris.*

The Division Commander flew to Jalibah to join the Division planners and to meet with the MEF Commander. The MEF planners had agreed that the security situation allowed a limited objective attack by the Division across the Tigris, and were as eager as the Division to get moving to Baghdad. The MEF and Division Commanders agreed, and the Division was given authority to proceed with a limited objective attack across the Tigris to isolate Al Kut. Once

Blue Diamond was positioned across the River, the MEF and Division would be in position for a variety of further actions. The CFLCC Commander had mandated a shaping reduction of the Baghdad outer cordon by 50% before launching an assault on the capitol, but this fight was several days away. A continuous flow of joint air strikes against the Al Nida RG Division persisted over the next several days. The Division, who planned to meet the Al Nida RG face-to-face very soon, greeted the strikes with enthusiasm.

The commanders and staff of Division HQ gather to receive the plan for crossing the Tigris.

The CG and his planners flew back to the Division's positions along Highway 1, and straight to the RCT-5 CP where Colonel Dunford and his staff were briefed on the plan. The attack would take place tomorrow, and RCT-5 would lead it. A feint by 3d LAR would press north along Highway 1 (the avenue of approach expected by the Iraqis), while RCT-5 would abruptly attack northeast along Highway 27 through the seam in the enemy's defenses. Crossing the Saddam Canal and seizing the Tigris River bridge at An Numaniyah, the Division would gain a bridgehead and sever southern Iraq from Baghdad's control. The Division was finally on the move again. The CG made the rounds and briefed the plan to all of his RCT commanders. In every position, the life went back into the faces of the Marines as the CG explained his concept. Blue Diamond had waited long enough, and the Marines were more than ready to go.

Linkup Along Highway 17

Having accomplished the task of reconnaissance in force along Highway 17, 3/4 looked to complete the link-up with RCT-1 to the southeast. Moving west from Al Fajr, 2/23 had uncovered Highway 17 from RCT-1's positions to the Saddam Canal. Early on the 30th, the link-up between 3/4 and 2/23 was completed when both units made physical contact at the bridge over the Saddam Canal. The Division was no longer attacking up two isolated corridors, and now had a viable connecting route. This route would see a great deal of Division traffic over the next four days, and the welcome flow of supplies through this corridor enabled RCT-1 to continue its fixing attack north.

One of the remarkable aspects of the continued success of RCT-1 up Highway 7 was the ability of the Marines to continue the attack with minimal resupply. Unsecured LOCs and the great distance from fixed logistics nodes had precluded ground resupply, and a perceived Surface to Air Missile (SAM) threat limited the use of aircraft for a few days. Highway 7 through An Nasiriyah continued to be impassable for resupply convoys. With the linkup on Highway 17 accomplished, Delta Company, 1st Tanks escorted a CSSG convoy up Highway 1 and across Highway 17 to link up with RCT-1. After remaining overnight with 3/4, the convoy was escorted to 2/23 and 1st Marines. After several days without resupply, and subsisting on only one MRE per day, the Marines of RCT-1 were glad to see the convoy roll up. The uninterrupted flow of logistics from RRP-24 to RCT-1 via surface transport greatly alleviated the critical shortfalls the RCT had been operating under.

From Stabilization to Combat Operation

EOD Marines prepare to destroy a cache of enemy mortar rounds.

The battalions of the Division were experiencing significant success in bringing peace to the locals in the Division's zone. Local civilians in Ad Diwaniyah were actively cooperating with the Marines, and had pointed out a sizable paramilitary stronghold in a local schoolhouse located in a grove of palm trees. The locals were eager to have the Marines of the Division conduct an operation against this position, but with Blue Diamond moving back into the attack north, this would have to be passed down from 2/5 to 3/4. Acting on another local tip, the Marines of 2/5 identified a sizable weapons cache 3 km to the east of Ad Diwaniyah. "This thing is like Fallbrook," quoted the excited EOD team members referring to the extensive naval ammunition supply station at their home base, a Camp Pendleton landmark. The weapons storage facility included over 40 bunkers containing everything from RPGs to thousands of brand new 155mm artillery rounds. The Division expressed its appreciation to the Iraqi General Staff for assisting with its artillery ammunition re-supply.

Inchon Stretches to Cover All of Highway 7

With elements spread along the length of Highway 7, RCT-1 was now conducting missions in Al Hayy, Qalat Sikar, Ash Shattrah, and Al Fajr. The RCT span of control was over 100 kilometers along the route. In addition to straddling a large section of battlespace, the RCT also straddled a range of missions that ranged from wresting control away from Baath party thugs, to civil-military operations.

The sun sets as Marines continue to patrol along Highway 7.

The Reconnaissance Battalion had been relieved at the Qalat Sikar airfield by 1/4, and now established a 360-degree perimeter around the intersection of Highway 7 and 17, guarding the multiple avenues of approach. On 30 March, 1/4 (less the units at the airfield) conducted a relief in place with 3/1 and defended the Highway 7 avenue of approach from the south. That night saw a spike in enemy activity as 1/4 was probed by a squad-sized unit and the Recon Battalion came under intense enemy rocket fire from the direction of Al Hayy. With extremely limited visibility, 1/4 relied heavily on the Javelin gunners and their Command Launch Unit (CLU) thermal sights to observe the activities outside their perimeter. As Corporal Michael Reyna scanned his sector with the CLU, he picked up the unmistakable thermal signature of individuals with AK-47s and RPGs moving along a ditch approximately 150 meters north of the battalion's defensive line. The Marines engaged the enemy, spoiling his plan to launch a surprise attack. Caught by surprise, the enemy hastily and inaccurately launched two RPGs that sailed harmlessly above the forward command group's AAVC-7. Unable to effectively direct their fire against the wall of steel that honed in on their position, the would-be ambushers beat a hasty retreat and left their boots in the mud in the process. Shortly thereafter, the distinct sounds of a multiple rocket launcher releasing its load were heard. The enemy had pre-registered fires against this intersection, and was now executing a well-planned artillery raid. The first rounds impacted very close to the Recon perimeter, and Major Rich Whitmer, the Battalion S-3 was sure the next rounds would be adjusted right into their position. The fires were unobserved, however, and the rounds continued to impact just outside the perimeter. Rudely awakened, but unharmed, the battalion command group tried to get a few hours of sleep before their mission the next day.

31 March
1st Marine Division on the Move Again

The Division attack plan for its push across the Tigris River would entail all three RCTs acting in a coordinated action of fire, maneuver, and exploitation. RCT-1 would serve as the fixing force for the attack, fixing the Baghdad RG Division by fires from a position south of Al Kut on Highway 7. The RCT, with the Recon Battalion attached, would establish blocking positions on both sides of the Gharraf River south of Al Kut, and commence its fires from these positions. RCT-5 would attack to seize the Hantush Airstrip, seize a crossing over the Saddam canal, then finally to seize crossings over the Tigris River in the vicinity of An Numaniyah. The RCT would lead with 3d LAR Battalion to Hantush, who would continue to feint north along Highway 1 in order to fix the Medina Division from the south. First LAR would attack in trace of 3d LAR and set up blocking positions south of the small town of Ash Shumali. They would allow the lead elements of RCT-5 to pass through, then follow in support. First Battalion, 5th Marines would attack to seize the Highway 27 Bridge over the Saddam canal, while 2d Tanks and 3/5 would make the attack into An Numaniyah to seize the 600 meter long bridge over the Tigris located there. Second Battalion, 5th Marines would provide security to the RCT's western flank, and guide the 8th Engineer Support Battalion to an alternate crossing site to the west of An

An LAV-25 from 3d LAR Battalion moves forward to participate in a feint against the Medina Division

Numaniyah. The alternate crossing site was upwind of An Numaniyah, and gave the Division a good location to construct another bridge it could use if the An Numaniyah bridge was disabled or closed down by persistent chemicals. The exploitation would come from RCT-7, who would follow in trace of RCT-5 to the Tigris, seizing the An Numaniyah Airfield in the process. Third Battalion, 7th Marines would relieve RCT-5 at the bridge site, and secure the town of An Numaniyah itself.

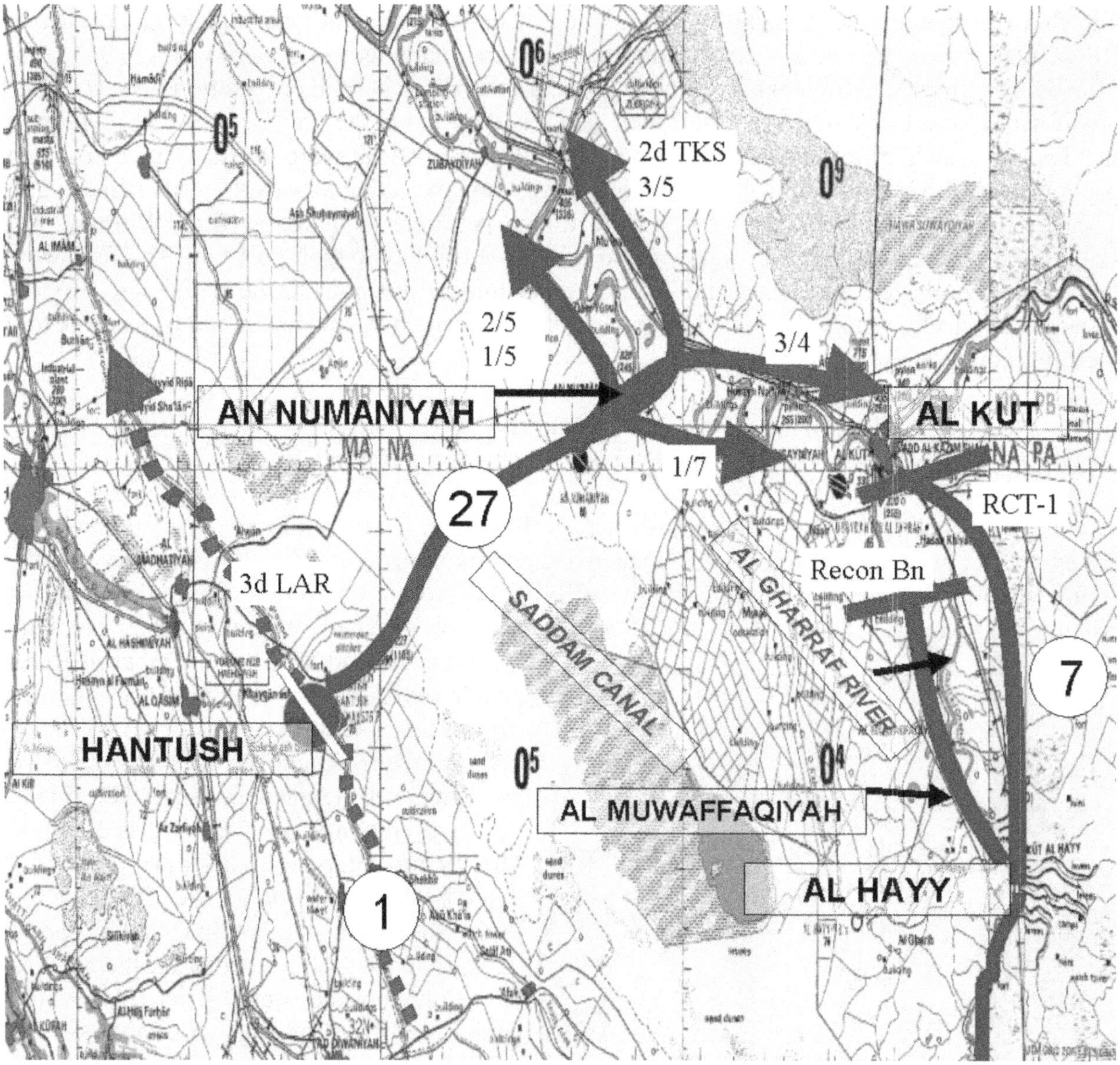

Details of the attack plan. RCT-1 would fix forces in Al Kut while RCT-7 and RCT-5 crossed the Tigris at An Numaniyah. RCT-7 would attack east toward Al Kut, while RCT-5 would continue to roll to Baghdad. RCT-1 and the Reconnaissance Battalion would conduct a roundabout march to link-up with the Division.

All of these attacks facilitated the movement of the exploitation force, as 3/4 and 1/7 would conduct a coordinated attack to the east on both sides of the river, attacking the Baghdad RG Division in its defensive positions on the outskirts of the Al Kut urban area.

Back to Hantush Airstrip

With the Wolfpack back in the lead, RCT-5 re-attacked to Hantush. This time, there was little resistance until they reached the airstrip itself. There, an Iraqi AAA unit, firing in direct fire mode, engaged them. Wolfpack continued the attack through the airfield, taking 15 EPWs and destroying 12 AAA pieces. They opened the critical FARP for the Division and continued to push up Highway 1. Their continued advance up Highway 1 was meant to give the enemy reason to believe that the Division's attack would continue up the main route to Baghdad. Over the next several days, 3d LAR aggressively patrolled north of the Hantush airstrip to the limit of the Division's zone. In the process, they discovered two Al Samoud Surface-to-Surface missiles (recently declared in violation of UN resolutions) and tagged them for destruction. In one engagement, Bravo Company, 4th LAR Battalion, engaged and defeated an infantry counterattack in the vicinity of Al Madhatiyah.

Scouts from 1st LAR Battalion search buildings in Ash Shumali.

While the Wolfpack attacked through Hantush, 1st LAR turned northeast just prior to the airfield and established a blocking position near the small town of Ash Shumali to prevent any potential enemy forces from attacking south. Highlander conducted reconnaissance of potential routes around the town and sent security patrols inside the town center, where they discovered a Baath party headquarters. The inhabitants of Shumali informed the patrol that the Baath members had already fled. With the town secure, the lead elements of RCT-5 continued up Highway 27.

The MWSS Engineers followed closely on the heels of RCT-5 as they cleared the airfield. Within hours, they had cleared and improved it. The MWSS and CSSG Marines immediately laid out the bladders for a fuel farm. The CG of 3d MAW, Major General Amos, arrived on scene for a personal inspection, and declared it fit for KC-130 operations. The plan became a reality, and within 24 hours the field was in full operation, with critical fuel and supplies pouring in. When the first KC-130 landed at Hantush, it marked the logistics turning point in the attack to Baghdad. Hantush was the first KC-130 serviced hard-surface airfield north of the Euphrates, and was to enable the delivery of the fuel that the RCTs used to continue the attack from Al Kut to Baghdad.

The seizure of Hantush demonstrated the criticality of 3d MAW support as Division units stretched the 1st FSSG's arc of sustainment. CASEVAC helicopters positioned with the RCTs and Division CPs shortened the distance and time from point of injury to treatment. CH-53s brought pre-staged ammunition packages to units beyond the end of secure roads. MWSS

Marines dispensed fuel to the Division's vehicles and many others as they moved up Highway 1. In addition to providing fuel and ordnance to rotary wing CAS aircraft, FARPs became supply points for ground units. The Division worked with 3d MAW to get ten KC-130 sorties and six CH-53E sorties each day to bring 50,000 gallons of fuel, 88 pallets of MREs, petroleum oil and lubricants and miscellaneous parts into Hantush and Wrigley (an expeditionary airstrip further south, near the Division Main CP on Highway 1.)

The Division also requested that 3d MAW lift 20,000 gallons of fuel with two CH-53s daily to Qalat Sikar to sustain the Division's movement along Highway 7, its eastern line of communication. The Wing went so far as to bring pallets of MREs from its own stocks to dispense to Marines at this critical FARP (although the Division Marines kidded their Wing comrades that they would rather have them bring the steak and lobster from their chow hall.) Shortly thereafter, the Marine Logistics Command began to take advantage of KC-130-capable airstrips as major logistics nodes. They staged MREs at Ali As Saleem Air Base and Joe Foss Airfield (in Camp Matilda, Kuwait) to shorten supply lines from theater stocks to front-line Marines. The strength of the relationship between Division and Wing staffs paid dividends from the LD to Baghdad. CSSG-11 appreciated the reinforcing capability of the Wing, as both organizations committed body and soul to keeping the Division on the move.

Marines of 3d MAW establish a forward operating base at Hantush Airstrip to support actions near Al Kut.

With the successful seizure of Hantush Airstrip, the Division surged ahead to stage itself for the continued attack. The Division Forward CP jumped and established itself several kilometers east of Hantush, along Highway 27. The roadway was jammed with bridge companies, MWSS vehicles, combat forces, and logistics trains from CSSC-115. The limited area around the airfield became a parking lot. The TMC pushed forward to take charge of the intersection, and soon had traffic flowing to the northeast. RCT-5's lead elements gained some maneuver space by pushing toward the Saddam canal Bridge, which lay about 23 km further up the road.

There were a number of contacts that day as units moved to the northeast. MPs from the TMCC were ambushed by 82 mm mortars and direct fire from a cluster of mud buildings along Highway 27. They immediately attacked through the ambush and discovered a squad of paramilitary fighters using women and children as human shields. The MPs killed or chased off the paramilitary fighters without causing any civilian casualties, much to the relief of the families that lived in the compound. The Iraqi women and children had been forced to stand in a row at the intervisibility line between the paramilitary fighters and the Marines.

That evening, the Division continued to shape the battlespace in detail for the next day's attack. Using the Pioneer, the Division was able to identify a small defensive position at the Highway 27 Bridge over the Saddam canal. The Iraqis had prepared what appeared to be a small ambush on the southwest side of the canal, while creating a conventional defensive line along the northeast side. This position was attacked by air and reduced. The targeteers turned to several large military facilities to the northwest of Highway 27 between the Saddam Canal and Tigris

River, and found vehicle activity in a large ammunition storage area. This target, far from any populated areas, was subjected to sustained aviation fires and later, a B-52 strike. The Division remained like a coiled spring, ready to push up Highway 27 the next morning.

Attack on Ash Shattrah

With the forward elements of RCT-1 pushing north of Al Hayy to fix the Baghdad RG Division, the RCT was also required to support an additional mission in Ash Shattrah. This small town astride Highway 7 remained a troublesome source of Baath activity, and a haven for Regime fighters. Several days earlier, a 3d MAW support convoy had passed through the town, attempting to reach the Qalat Sikar airfield. The convoy had come under fire, and in the confusion, there was a vehicle crash. The convoy withdrew from the city, and later discovered that an MWSS Marine had been left behind in one of the crashed vehicles. On 30 March, 3/1 was ordered to move to isolate the town in support of clandestine forces that were planning to attack Baath party facilities and search for evidence of the missing Marine. The battalion went into attack positions, isolating Ash Shattrah. When the other forces (Iraqi irregulars) that were to conduct the attacks did not arrive at the linkup point, the RCT took matters into their own hands. Working with information from informants in the town, the RCT S-2, Major Tom Sharp, and his crew, were able to precisely identify the buildings housing the Baath members. They also identified a hospital that was said to hold the body of the Marine (who reportedly had been killed in the earlier vehicle crash.) The RCT nominated the targets (minus the hospital, of course) for precision strike, and planned to send 3/1 into the town to search the hospital for evidence of the Marine. The next day, the nominated targets were struck, and 7 of 9 were destroyed. The precision of the strikes marveled the local populace, who turned out into the streets in droves to witness the demonstration of both firepower and restraint. Into the confusion, 3/1 raced into the town. They quickly searched for signs of remaining Baathists and the missing Marine. At the hospital, they confirmed that the Marine had been killed in the initial crash, and that his body had been buried by the local citizens outside of the town. The recovery task was passed to TF Tarawa for exploitation, and 3/1 hurried north to go into the attack on Al Kut.

An Iraqi civilian tries to keep up with his transportation on Highway 7.

Inchon Attacks to Al Kut

RCT-1 elements continued to receive indirect fire from artillery raids being conducted out of Al Hayy. These enemy artillery units had been moving from the cover of the urban area to firing positions south of the city. In fact, some of these systems had been spotted displacing during the previous night. Aerial intelligence from the Pioneer UAV and two P-3 missions (with Division observers Colonel Jim Lukeman and Major Mike Rodgers aboard) discovered the movement and efforts were started to take out the weapon systems. As attack aircraft checked on station, the frustration began to build as none of the aircraft were configured with a sensor capable of positively identifying the mobile artillery pieces. Pilots felt the frustration of the RCT as they could make out movement through their night vision goggles, but were unable to discriminate the target well enough to be certain it was enemy artillery. With the friendly fire close calls that had occurred in the recent past, Marine pilots wanted to be certain this was foe and not friend. Unfortunately, the positive identification was never achieved. RCT-1 began planning for a ground takedown of Al Hayy and the source of the indirect fire. The RCT planned to attack with the Reconnaissance Battalion moving along the west side of the Gharraf River to block the city from the west and north. Alpha Company of 1st Tanks, and 1/4 would attack north along Highway 7, and act as the hammer to Recon Battalion's anvil. Their mission was to attack into the city, destroy any enemy, and put a stop to the artillery attacks.

The Cannon Cockers deliver shaping fires on the Baghdad Infantry Division.

On the western flank of the RCT-1 attack, the Reconnaissance Battalion made its way up Route 7A on the west side of the Gharraf River. The Battalion planned to get into blocking positions on the west side of the river, keeping enemy forces from withdrawing to the west out of town. The Battalion made good progress up the poor road. As before, the terrain limited the Battalion's movement, and they found themselves in a column of companies, on top of the raised levee road for much of its length. As the Battalion column approached a sharp bend in the river, paramilitary forces ambushed them. This was a well-placed ambush, supported by mortars and AAA guns being used in direct fire mode. The Battalion quickly executed its reaction plan, pulled back out of the ambush kill-zone, and called in supporting arms. Once again, the Battalion began to receive incoming mortar fire, and watched its approach. The first rounds impacted about 800 meters from the column, but the second series were much closer. The third series began to land in the water next to the Battalion, and they waited for the next set to be right on target. Miraculously, the mortar fire ceased. The air support soon arrived, and cleared the ambush positions.

Life continues as it did thousands of years ago, even as modern armies battle nearby.

The Reconnaissance Battalion moved up the remaining distance to the bridges at Al Hayy, and approached the northernmost. Despite their earlier scrap, their arrival at the town from the west was an absolute surprise for the enemy in Al Hayy. The Battalion could see Iraqis running through the streets on the far side of the river as they reacted to the presence of the Marines. The enemy managed to man fighting positions and began to fire. The enemy positions were well prepared, and they were able to pour heavy fire on the Marines. Recon Battalion was able to gain fire superiority with its heavy machine guns, and fought its way across the Al Hayy Bridge. RCT-1 was still south of Al Hayy on Highway 7, so Recon Battalion continued their attack all the way to Highway 7. The Battalion was ordered to conduct patrols and establish a blocking position to the north of the city, on the east side of the river. That night, the Recon Battalion engaged hundreds of targets as enemy forces attempted to flee to the north up Highway 7. Bravo Company under Captain Craig Schwetjewas patrolling to the west of Highway 7 when they detected hundreds of paramilitaries attempting to bypass the battalion's blocking position on the highway. Once again, 1/11 provided devastating fires, and the enemy forces trying to bypass the Recon Battalion's positions were destroyed.

As the Recon Battalion was pushing up the west bank of the river, 1/4 conducted a simultaneous attack north along Highway 7 on the east side of the river towards Al Hayy. As 1/4 continued its attack north, it uncovered numerous weapons caches along the way. Just south of the city, the battalion received mortar fire, and Captain Coby Moran, the battalion's Weapons Company commander and fire support coordinator, called in the 'snowstorm' report and requested close air support. The RCT-1 Air Officer vectored a section of AH-1Ws to the battalion, and they promptly identified and destroyed an Iraqi tank, artillery piece, and other Iraqi weapons systems. As RCT-1 kept pressing the enemy, they began to find the route littered with abandoned equipment. As the Blue Diamond attack began to break their will, the Iraqi fighters continued to abandon their positions and weapons. The assaulting companies found abandoned uniforms, small arms, RPGs, and hundreds of 120mm mortar rounds.

A suspicious Iraqi vehicle is disabled and searched at a checkpoint.

1 April
The Saddam Canal

Destroyed Iraqi armor near the Saddam Canal

By the morning of 1 April, the enemy's resistance had been severely weakened. The intelligence assessment of the Baghdad RG Division was that 80% of its major weapons systems had been destroyed. The status of the infantry in the Division's three brigades was harder to ascertain, but without their supporting arms they lacked much of their former combat power. The Baghdad RG Division had always had mobility problems due to a shortage of transport vehicles, but now it was truly fixed. To the west, the enemy position on Highway 27 near the Saddam Canal bridge remained, as did the positions near the An Numaniyah Bridge. The conventional defense of Al Kut had been significantly disrupted, but the paramilitary threat remained unclear. The paramilitary forces had rallied in the defense of An Nasiriyah even when conventional forces had failed, and had given TF Tarawa a tough time of it in the city. Here in Al Kut and An Numaniyah, closer to the Regime center, would the reception be any different? The Division continued its attack, with the full expectation of hard fighting on the way to the key bridges. This threat caused the CG to issue additional guidance to the attacking commanders. This attack would be conducted with deliberation and maximize the usage of shaping and preparatory fires. Positions confronted were to be reduced by fires prior to assault. The enemy showed every indication of a weakened posture, and the risk of unnecessary casualties precluded a rush into the enemy's prepared defenses. In keeping with the CG's intent, the Division Forward CP and their imbedded DASC coordinated a number of B-52 strikes on the ammunition storage area along the west flank of RCT-5's attack. RCT-5 was locked and loaded for their attack, aiming to seize the Tigris River crossings.

Marines from Charlie Company, 1/5 in the attack.

In the morning, RCT-5 launched their attack up Route 27 to seize the bridge over the Saddam Canal. Lieutenant Colonel Padilla's 1/5 led the attack as the RCT's main effort, followed closely by 3[d] Battalion. An enemy company-sized element put up strong resistance from defensive positions both north and south of the canal. From positions on both banks, the enemy engaged Bravo Company, led by Capt Jason E. Smith, with mortars and heavy machineguns. Captain Smith quickly deployed his company on line, and boldly began crossing the bridge after effectively suppressing the enemy. Once Bravo had secured the far side of the bridge, Charlie Company continued the attack, pursing the enemy with elements of Bravo Company. They quickly overran the enemy 120mm mortar platoon and remaining defensive positions that were suppressing the Battalion's anti-armor platoon on the near side. The enemy company guarding the existing bridge was destroyed, and 59 EPWs were captured in the process. Because of concerns with the structural integrity of the existing bridge, 8th Engineer Support Battalion constructed an additional bridge over the Saddam Canal. Although not used by the assault forces, this new bridge provided an alternative crossing for the large number of heavy logistics vehicles that would later traverse the route.

As 1/5 was mopping up the far side and collecting prisoners, 3/5 conducted final coordination for the forward passage, then quickly moved across the canal and through the leading elements of 1/5, approximately 15 kilometers northeast along Route 27. While supporting the crossing, 3/5 soon encountered small groups of soldiers fleeing the area. As twilight approached, India Company dismounted and cleared a suspected enemy trench line while Kilo Company and CAAT platoon pursued an estimated squad of Iraqi soldiers by fire using the Battalion's organic 81mm mortar platoon. India Company also prosecuted a rotary wing air strike on a D-30 artillery battery five kilometers further east, successfully destroying the enemy guns in the process.

First Marine Division builds a pontoon bridge over the Saddam Canal. This bridge would serve major support convoys that followed the Division along the major roads.

The artillery performed magnificently during the attacks up route 27. The Cannon Cockers from 2/11, reinforced by 5/11, set a standard of impressive performance (here and throughout the campaign.) Terrain was extremely constricted, but somehow the artillerymen were always able to get into firing positions.

Elements of RCT-7 moved up to their attack positions on the south side of the Saddam Canal. They relieved RCT-5 of all of their remaining battlespace, and were poised to participate in the next day's attack across the Tigris River. With RCT-1 fixing the enemy, RCT-5 establishing the crossing points over the River, and 11th Marines poised to provide devastating fires, it was as if the rest of the Division were cocking back the hammer and preparing to shoot RCT-7 out the end of the barrel. What remained of the Baghdad RG Division was coming into Blue Diamond's sights.

Marines take up positions near An Numaniyah to cover their fellow Marines as they near the Tigris River.

The Division learned that the Army V Corps had launched an attack on the Medina RG Division at the same time as the Blue Diamond attack. After the significant staff coordination that had been required for the MEF to gain reluctant approval for an attack over the Tigris, the Division was somewhat surprised that CFLCC had all-along been planning for a US Army attack on the Medina Division on all fronts. The Division was launching a major supporting attack, uncoordinated with the main effort.

That evening, the Division again used the Pioneer to shape the Division's fight for the next day. In An Numaniyah, the Division identified a large number of paramilitary defenders bivouacking in a large date palm grove at the foot of the An Numaniyah Bridge, and a number of armored vehicles on the far side. The Division targeteers in the Forward CP were able to guide multiple air and surface fire missions against these positions, eliminating many of these fighters. The night also saw the use of napalm on the assembly areas of the Fedeyeen fighters in town, destroying many, and scattering the rest. A number of enemy fighters remained in the vicinity of the An Numaniyah Bridge, but they had received a significant amount of shaping fires. This had been a textbook integration of intelligence and fires in support of the Division deep fight. The ability of the Division to tightly coordinate the actions of the UAV collections and the delivery of fires allowed for a very precise delivery of shaping fires to only those specific targets that required them. Because of the timeliness and precision enabled by Division control of the UAV, fires were brought to bear only on the enemy positions, largely sparing the adjacent town from a massive preparatory bombardment. In the final hours of darkness before the resumption of the attack, the G-2 used the Pioneer to search the streets of the city for any indications of a paramilitary defense and found nothing other than at the bridge.

The evening of 1 April also graphically demonstrated the viability of another intelligence tool, the JSTARS Common Ground Station. This night, the sensor detected large numbers of vehicles evacuating the city of Al Kut to the north. The Division directed aviation assets to ascertain the nature of the targets and found them to be mostly civilian vehicles. A large portion of the population was choosing to flee the fighting going in the only safe direction remaining, north to the Iranian border.

Darkside Moves on Diwaniyah

Although much of the Division's attention was focused on the bridge at An Numaniyah, the Fedeyeen in Ad Diwaniyah had not heard the last of Blue Diamond. Third Battalion, 4th Marines ('Darkside') conducted a limited objective attack against paramilitary forces in the vicinity of the city in order to ensure no interference with the Division's main attack. Having conducted a relief in place with 3/5 on 31 March, 3/4 remained in a defensive position on the cloverleaf. The aim of the attack was to aggressively advance on the town, attacking all hostile forces encountered, forcing the last remaining paramilitary forces to scatter from their hiding places.

The attacking force that day was leaner, with India Company serving as rear security and the battalion reserve near the Combat Trains. Two companies formed the bulk of the combat power. Bravo tanks led the Darkside, quickly rolling up to an enemy observation post (OP), registering one enemy KIA as he tried to run back to warn his unit. Pressing on, Bravo and CAAT 2 cleared a series of trenches and bunkers, destroying them with heavy machine guns, TOWs, and tank main guns. Between these two elements, 55 enemy KIA were counted, as well as numerous WIA and 24 EPWs.

The pace quickening, the Battalion Commander ordered a more deliberate approach, allowing the situation to develop. While the lead elements continued their movement west into the city, Kilo Company dismounted to clear a cluster of mud huts on the south side of the road. Approaching the town, Bravo established an attack by fire position on the eastern outskirts, allowing Kilo Company to maneuver around and aim for the

A door gunner provides cover with his minigun from a UH-1N.

southeastern area of the town.

As Kilo and CAAT 1 advanced south, Bravo reoriented on a date palm grove, clearing through it and destroying two technicals and three motorcycles. Now at the edge of the city, the tanks began to receive numerous RPG volleys, some of which appeared to be lobbed from behind the first row of buildings inside the city. Air Force strike aircraft were aborted by order of the Battalion Commander due to collateral damage concerns. Instead, the Dragon Eye UAV was deployed to reconnoiter the area and identify the location of the RPG shooters. Cross talk between the Battalion Commander, Bravo, and CAAT 2 allowed the tank company to vector in on the RPG shooters and engage with a platoon volley of main gun rounds, eliminating the threat.

Kilo Company reoriented northeast and began a deliberate movement in an attempt to push any resistance from the outskirts of the city into Bravo Company's sector. This advance uncovered numerous mortar tubes and infantry fighting positions, but saw no enemy contact. The numerous weapons caches discovered were blown in place and units began their retrograde to the cloverleaf. 'Longshot', the battalion's 81mm Mortar platoon, fired disengagement fires and all units returned to previous positions in vicinity of the cloverleaf. Enemy casualties were reported as 92 enemy KIAs and 56 EPWs following the engagement.

Barbarian and Godfather Take out Al Muwaffiqiyah

RCT-1 continued the attack through Al Hayy toward Al Kut in order to fix the Baghdad Division. Second LAR, occupying the Kut al Hayy East Airfield, would attack to clear northwest along the Kut al Hayy road, and effect a link up with 1/4 near Ar Rashidiyah. Captain Banning's Alpha tanks led the way into Al Hayy, destroying fleeing military trucks loaded with ammunition in the process. Alpha 1/4 followed in trace, and located a cluster of bunkers and trenches with men armed with small arms and RPGs. The enemy had positioned their defenses in extremely close proximity to the city. Alpha Tanks established a support by fire position to suppress the enemy while Alpha 1/4 maneuvered to clear the strongpoint. Alpha Company, 1/4, closed with the objective supported by precision direct fire provided by the tanks and accurate close air support provided by the Cobras, their guardian angels. With the bunkers destroyed and two enemy KIA, the enemy at the former strongpoint withdrew in a 'technical'. Seeing the devastating display of combined arms, the remaining enemy chose to withdraw out of Al Hayy and proceed northwest.

The townspeople soon came out to greet the Marines that had liberated their town, and began to lead the Marines to numerous weapons caches. Spotting a Baath Party building, the Marines cleared the building, discovering numerous RPGs, mortars, and small arms ammunition. Deliberately clearing this building of the ordnance would have taken the better part of one day. Chief Warrant Officer Parks and Lieutenant Gerry Roeder, the Combat Engineer Platoon Commander, rigged the building with explosives, cleared the townspeople from the area, and leveled the building. The Gunner would be extremely busy that day, destroying nine artillery pieces, destroying thousands of artillery, mortar, and small arms rounds. While the Marines of 1/4 began the task of clearing as much of the town and ordnance as time would allow, 2d LAR and Alpha tanks passed north through their lines to continue the attack north.

In Al Hayy, conversations with many of the local citizens indicated that the Regime operated a terrorist training camp to the northwest in a town called Al Muwaffiqiyah. The locals reported that the enemy who were able to escape Al Hayy had also holed up in this Regime stronghold. As the Reconnaissance Battalion (Godfather) continued to move to the north, it approached the town of Al Muwaffiqiyah from the east side of the river. The battalion had been ordered to attack over a bridge and through the town the next morning to resume their march north on the west side of the river. As Recon Battalion prepared for their attack in their assembly area well east of the river, Lieutenant Colonel Ferrando walked over to Lieutenant Colonel Ed Ray of 2d LAR (Barbarian) and told him of an Iraqi ambush that had been established at the Al Muwaffiqiyah bridge. Lieutenant Colonel Ray turned to his Alpha Company commander, and told him, "Go take care of it." On their approach, Barbarian encountered another well-laid ambush on the east bank of the river. The enemy's base of fire on the west side of the river opened up on the LAVs, followed by the rest of the 'L' shaped ambush in a cluster of buildings on the east bank, to the north of the company. The Recon Battalion, back in the assembly area, watched as additional LAVs rapidly drove past their position to the west. Volley after volley of artillery, and repeated sections of air appeared overhead, all pouring fire into the ambush site down the road to the west. Crossing the river promised to be an interesting experience. After a period of hours, 2d LAR had reduced the threat from the ambush and then withdrew to the assembly area. Recon Battalion waited until 2000Z, when it had air available, and commenced its own movement to the bridge. Under cover of darkness, Captain Craig Schwetje moved up to an obstacle that lay across the bridge. The Marines immediately came under fire from a prepared position on the eastern bank of the river. The enemy in this position was uncharacteristically tough, and fought with a tenacity and skill not seen prior to this ambush. They continued to fight until the Battalion had taken them out individually. These fighters were dressed in full combat dress, with chest-rigs, antitank grenades, heavy machine guns, and RPGs. It was clear that these were not ordinary Iraqi fighters. When the Marines inspected the bodies, paperwork was found indicating these fighters had come from Syria only weeks before. They even had tattoos in English between their thumb and forefinger. These Syrian hardcore fighters were possibly linked to the terrorist training camp near the town.

Pontoon bridges stage near Highway 1, ready to push to the Tigris.

Back on the bridge, the Recon Marines tried to reduce the bridge obstacle by fire. AH-1Ws tried to dislodge the obstacle with TOWs and Zuni rockets without success. Bravo Company had two tanks attached from Alpha Company, 1st Tanks. The tankers could not get on the rickety bridge, and could not dislodge the obstacle either. The tanks reported however, that a HMMWV might be able to squeeze past. As the sun rose, Bravo Company tried to bypass the obstacle, and found they could just squeeze a HMMWV past it on the bridge. They proceeded to squeeze a platoon through, moving a single vehicle at a time past the obstacle. The trail HMMWV in the first platoon across was towing a trailer, which foundered in a hole in a damaged section of the bridge. The HMMWV was now wedged in the hole, and could not be budged. Through some creative use of motorcycle ramps, the Marines were able to get the rest of the platoon back across the bridge, but the damaged vehicle remained stuck. Stymied in the attempt to get across the damaged bridge, the Recon Battalion and 3/1 were sent south back down to Al Hayy to cross the river there. They would resume the attack on Al Muwaffiqiyah from the west side of the river. The initiative and aggressiveness displayed by the Recon Marines was another example of the spirits of the Division Marines that would find a way to accomplish the mission, or make one themselves.

Division Marines in An Nasiriyah

As part of the 15th MEU, the Marines from 2/1 continued combat operations in support of TF Tarawa, as the rest of 1st Marine Division attacked further north. On 1 April, 2/1 played a key supporting role in the efforts to rescue a US Army POW being held in a hospital in An Nasiriyah. The battalion conducted a diversionary attack on multiple objectives throughout the city, opening the way for a Special Forces team to enter the hospital unopposed and rescue the American POW.

Third Battalion, 7th Marines, relieve 3/5 at the An Numaniyah Bridge.

The Marines from 2/1 would continue to play a significant role in stabilizing the An Nasiriyah battlespace over the next several days, conducting raids on Baath party buildings, Fedeyeen strongholds, military facilities, and other areas of potential resistance identified by intelligence. During combat operations and patrols, the battalion destroyed over thirty arms caches. Even though not operating under direct 1st Marine Division command, these Marines were instrumental in executing a 'No better friend, no worse enemy' policy on the streets of An Nasiriyah. The battalion conducted a number of humanitarian missions, and supported the distribution of fresh water to needy civilians. On 8 April, the Marines of 2/1 supported an operation in Suk As Shyuk (southeast of An Nasiriyah) to recover and transport wounded civilians to Talil Army Airfield.

The Marines of 2/1 would eventually be responsible for a number of towns north of An Nasiriyah, providing security through the same Highway 7 zone that their parent regiment had earlier maneuvered through. The battalion continued the stabilization operations begun by RCT-1, handing out over 6000 humanitarian rations to hungry civilians along Highway 7. The battalion eventually retrograded to Kuwait, and backloaded onto amphibious ships to complete their MEU deployment.

2 April
Crossing the Tigris River

As the sun rose, the much-anticipated attack across the Tigris commenced. RCT-5, the main effort, crossed the line of departure at 0055Z astride Route 27 to maneuver forces across the Tigris River and cut Highway 6. During the night hours, the RCT moved off Highway 27 in order to give 2d Tanks an unencumbered route straight to the An Numaniyah Bridge. At 0355Z, 2d Tank Battalion successfully passed through An Numaniyah. The Scout Platoon, led by 1st Lieutenant Matthew Zummo led the way, identifying enemy positions and returning fire through significant enemy resistance, identifying uniformed Republican Guardsmen defending the approaches to the bridge. When the resistance stiffened, the rest of the Battalion moved up and brought their main guns to bear. Ironhorse attacked through An Numaniyah and crossed the bridge while battling a hailstorm of RPG and machinegun fire. They continued the attack to the far side of the bridge, destroying an estimated enemy battalion-sized force, while experiencing one M1A1 immobilized by enemy fire.

Moving up behind the tanks, 3/5 was ordered to clear Route 27 through the city and secure the bridge for follow on forces. For the next several hours, 3/5 fought dismounted through the northern half of the city. India Company led the attack, moving on foot through the charred and smoking remains left behind by a section of Cobras providing outstanding close air support to the company. Encountering only light resistance, India Company deliberately cleared both sides of Route 27 through the town, and then began methodically destroying Iraqi equipment and rounding up dozens of EPWs. Capt Ethan Bishop recalls that his Marines, especially his Assaultmen (trained in demolitions), had a "field day" blowing up two T-72 tanks, numerous anti-aircraft guns, and killing a couple of Iraqi snipers.

Lima Company moved through India, seized the far side of the Numaniyah Bridge and chased off the remnants of the enemy battalion. In the process they destroyed several ammunition trucks, and rounded up approximately 50 EPWs. The battalion remained in control of the bridge, while 2d Tanks moved up to the intersection of Highway 27 and Highway 6. For the first time, Highway 6 was blocked by US forces, severing the connection between the Regime leadership in Baghdad and Al Basrah, physically cutting off the Iraqi III and IV Corps and the remnants of the Baghdad Division. The Division was now north of the Tigris, astride the primary avenue of approach into Baghdad.

Weapons Company, commanded by Captain David Hudspeth, was the main effort for 1st LAR Battalion on the morning of 2 April. As they moved north on Highway 27, they passed through 1/5 at the Saddam Canal and joined 2d Tanks and 3/5 in the attack on An Numaniyah. They continued to the 'Y' just to the southwest of the city when they began taking fire. The company immediately returned fire and deployed on line. The overwhelming firepower of the 25 mm guns and M240G machine guns was too much for the Fedeyeen fighters and they began to flee. After the fighting at the edge of the city had died down, 3/5 passed through and began clearing the city, while 2d Tanks continued to the Tigris bridge. Weapons Company counted nine enemy KIAs as a result of their actions that morning. Guarding the Division's flank, 1st LAR continued its movement to the west on 'powerline' road paralleling the Tigris River on the south side.

To the west of An Numaniyah, another major bridging effort was underway. Leading the effort to the west of Highway 27, 2/5 advanced on the Division's western flank, and cleared a path for the 8th Engineer Support Battalion (commanded by Lieutenant Colonel Rick Nelson.) The engineers had determined that the long An Numaniyah Bridge was capable of supporting 70 ton vehicles, but the Division required an alternate crossing point in case the primary bridge was destroyed or made impassable through enemy action. The principle concern was the enemy using persistent chemical agents now that the Division had crossed the Tigris River. The engineers of 8th ESB began emplacing assault bridging at crossing site 'Pearl 3' at 1426Z, and anticipated completion of construction at 2100Z. When it was done, the Pearl 3 bridge would be 145 meters long, another historic achievement by the engineers. Simultaneously, the engineers ferried tanks and AAVs across the Tigris during the night.

Ripper Moves up to An Numaniyah

RCT-7 moved up to relieve 3/5 of the bridgehead, seize An Numaniyah Airfield, and establish blocking positions on both banks of the Tigris River. Entering the An Numaniyah Airfield, 1/7 seized it without incident. Third Battalion, 7th Marines, followed by 1st Tank Battalion, relieved 3/5 of the bridge. RCT-7 took responsibility for the Tigris River Bridge and the town of An Numaniyah, freeing RCT-5 to focus on the attack to the west. With all of its forces prepared to press the attack into Al Kut the following day, RCT-7 looked to securing the city of An Numaniyah itself. The Division Main CP also moved up to a location in the vicinity of the airfield, and began to set up.

The CG provides guidance to his staff near An Numaniyah.

Cordoning off the city, 3/7 established positions to the north, west, and south of An Numaniyah during the relief with 3/5. The cordon formed by the Marines on three sides and the Tigris on the fourth cut off enemy forces in the town. Late on the evening of 2 April, 3/7 attacked. They cleared an army training camp on the northwest corner of the urban area, finding abandoned positions and discarded weapons. During a night urban penetration, they advanced against token resistance to secure the central market, soccer stadium, and a key road intersection at the north edge of the city. A local informant gave them valuable information on locations of weapons caches and safe houses containing regime leadership and sympathizers. Maneuver companies exploited this information by conducting raids and removing the ammunition caches. By the morning of 3 April, they had eliminated organized resistance.

Lieutenant Colonel Belcher and select staff met with local community leaders to discuss critical civil issues including stabilization and security operations. These operations would challenge the battalion in ways combat had not. The capture of a local Imam with reported ties to Uday Hussein caused a crowd to form in protest. After assessing the situation, 3/7 released the Imam, who made a speech supporting the Marines and denouncing the Regime. In another instance, Marines at a checkpoint risked themselves in order to protect An Numaniyah citizens:

About fifteen minutes into it, Lance Corporal Reuben Muniz noticed a truck approaching at a fast pace. Realizing the danger the speeding vehicle posed to the Marines and civilians, Muniz decided to stop it. He proceeded to stand in the middle of the road while waving at the car and yelling "Stop! Stop!" The driver apparently paid no attention and kept on driving. As the vehicle kept approaching, Muniz continued his efforts by putting his weapon at the alert, followed by the ready. Seeing that the driver did not respond, Muniz decided to double-tap the truck on the grill. The truck came to a halt about five feet in from of Muniz and the people in the back of the truck jumped out. Muniz got everyone on line and the passengers all started making signals with their hands and saying "No stop, No stop!" After that, we knew that the vehicle could not stop because it had bad brakes.

Marines from 3/7 establish a firing position inside an Iraqi army camp near An Numaniyah

Tiger relieved 3/7 of the bridge and crossed the Tigris River, while 3/4 spent the night south of the bridge in a position to go into the attack against the Baghdad RG Division on the north side of the Tigris River. In trace of RCT-7, 1/7 cleared An Numaniyah Airfield and prepared for their attack the next day against the 4th Infantry Brigade of the Baghdad RG Division on the south bank of the Tigris. Intelligence was developed for both of these attacks, identifying tanks that would be destroyed by 3/4, and company-sized defensive positions that 1/7 would be engaged by 1/7.

Fixing the Baghdad RG Division

Meanwhile, the Inchon Marines were setting the conditions for the Division's attack into Al Kut by drawing the enemy's attention to the south through limited objective attacks up Highway 7. This effectively fixed the Baghdad Division in place, allowing RCT-7 to attack into their rear and complete their destruction. The Marines of 1/11 established firing positions and continued the destruction of the enemy's indirect fire assets through counterbattery and observed fires. In order to observe the steel rain, RCT-1's S-2 sent the supporting VMU-1 Pioneer to Al Kut to provide eyes on the D-30s and Type 59-1 artillery pieces that remained in and around the city. By the evening of 2 April, the enemy was effectively surrounded and the conditions were set to complete their destruction.

The Recon Battalion and 3/1 attacked through Al Muwaffiqiyah on the west side of the river was successful. The AAVs of 3/1 made their way through the town without resistance, and the Reconnaissance Battalion passed through. Recon continued the attack all the way up the west side of the river, and established a blocking position just south of Al Kut. The blocking position in a small village called Al Qaryah was positioned just south of Al Kut's Al Jarrah airfield. Meanwhile, 3/1 relieved 1/4 in Al Hayy and blocked Highway 7 to the south. Second LAR established a screen line and passed 1/4 north to clear a village and seize a bridge on the approaches to Al Kut. Securing its objective without enemy contact, 1/4 learned from a village elder that the Republican Guard soldiers had withdrawn the previous night towards Al Kut. The RG had reportedly emplaced mines 10 to 12 km north of the bridge along Highway 7.

> *Company E (Easy Company), resumed their patrol of the area the following day, and again the Iraqi's came out to fight. "The RPGs (Rocket-propelled Grenades) sounded like really loud bottle rockets," stated Corporal James Santoro, 2d Platoon (White Platoon), of Martinsburg, West Virginia. One rocket passed between Lance Corporal Kevin Corrigan Salsberry, Maryland, and his LAV-25, "Then another RPG missed my drivers hatch by about four inches." As Sergeant Michael Carista of Plum Point, Maryland, 1st Platoon (Red Platoon) raced on to the scene, his vehicle began receiving fire from the direction of an adobe hut. "I ordered Gorman to blow that house down!" With a smile on his face, Lance Corporal Daniel Gorman fired a single M203 high explosive grenade at the hut and knocked it down. When he viewed the result of his gunnery, the lance corporal exclaimed with a resounding "WUUUUUUHOOOOOO!" When the fight was over the company had taken six enemy prisoners of war, one of which was the unit commander. Sergeant Eric Miller of Baltimore, Maryland, summed it up best, "It felt good to take the fight to them and push them out."*

Back at Hantush Airstrip, the 3d LAR Battalion remained aggressively postured on Highway 1. Northwest of the airstrip, they came into contact with a probing enemy unit:

The EPWs from that unit possessed new gas masks and Atropine injectors of Turkish manufacture, similar to these found on Iraqi dead along Highway 27 outside the An Numaniyah airfield. This information was passed to the G-2. The Division now had to consider the likelihood that a successful crossing of the Tigris River may be the trigger for the use of chemical weapons. The Marines remained in a heightened MOPP posture, and continued the attack. Third LAR Battalion continued to block Highway 1, run patrols, and protect the growing logistics footprint at the Hantush Airstrip.

To the south, the rapid movement of the Division was facilitated by pulling up its battlespace behind it. The relief in place of the Division's southern battlespace by TF Tarawa commenced at 1400Z as elements of 3^d^ Battalion, 2^d^ Marines coordinated with 3^d^ AA Battalion along Highway 1. To the east, elements of the 24th MEU made preparations to relieve the Division of its last battlespace along the Highway 7 corridor. The Division already had its eyes on the next prize, and was building momentum for its attack to the west. The Division Forward CP rolled across the Pearl 3 bridge late on the evening of 2 April, and conducted a road march in trace of RCT-5 to Al Aziziyah on Highway 6.

3 April - Final Assault on Al Kut

The Division was now poised to complete the destruction of the Baghdad Republican Guard Division in Al Kut. Third MAW, along with 11th Marines, had delivered a steady flow of shaping fires against the Baghdad Division, and most of their armor and artillery were destroyed. The remnants had shown a willingness to continue fighting in An Numaniyah, and it was likely that there would be significant bloodshed yet to come in the attack on Al Kut. In keeping with the Division's motto, 'No better friend, no worse enemy,' the Republican Guard Commander would have one last chance to avoid the destruction of his command.

> *"To the commander of the Iraqi forces in Al Kut: You are surrounded. There is no hope for your forces to be reinforced or re-supplied. We will continue to attack unless you choose to stop this unnecessary killing. If you choose not to surrender, all the killing will be your responsibility and yours only. At 0700 Greenwich Mean Time, we will stop the attack and if you choose to come forward to the western side of Highway 6, we will arrange a safe passage for you to surrender. I suggest you consider your choices carefully.*
>
> *From the commander of the United States Marines surrounding Al Kut."*

The Information Operations campaign had included a set of radio frequencies that enemy commanders could use to reach US forces should they decide to surrender. The CG constructed a message to the Baghdad Division's Commander and at 0700Z, Corporal Ali Abdelgawd, the CG's talented and conscientious interpreter, passed the message over multiple radio frequencies from the intersection of Highway 27 and Highway 6.

There was no response from the Baghdad RG Division commander. It was later assessed that he had been killed in the shaping fires or had fled the city in the civilian convoys seen leaving to the north.

The attack plan called for 3/4 to attack east along the north side of the Tigris while 1/7 attacked on a parallel axis south of the river. On 3 April, 3/4 and 1/7 advanced with little resistance until they closed within 1000 meters of their objectives. As expected, enemy small arms and RPG fire increased as each battalion grew closer to the outskirts of Al Kut, and the battalions responded with overwhelming force. A limit of advance (LOA) had been put on the west side of the city of Al Kut as it was anticipated that the enemy wished to draw the Marines into a costly urban fight. Neither RCT-7 nor the Division were going to fall for that. To defeat this enemy, it would not be necessary to kill them all. Leaving them bottled up in Al Kut would be just as effective. Numerous close air support sorties, including a B-52 strike, had devastating effects. White phosphorous rockets marked targets. GBU-16's destroyed T-62 tanks. Mk 83s destroyed APCs and support vehicles. On the ground, 3/4's Marines fought with similar effectiveness. Fire from Iraqi infantry in a bunker complex was answered with heavy machine guns, TOWs, and tank main gun rounds. During the fight, Corporal Mark Evnin, a Scout Sniper serving with 3/4, was killed as he provided suppressive fire for units engaging an enemy strongpoint. While under continuous enemy fire himself, Corporal Evnin provided suppressive fire with his M203 against enemy bunkers that were in a position to surprise an advancing platoon from 3/4 with deadly grazing fire. Corporal Evnin first fired from a covered position, located behind an AAV, then moved from the covered position of the AAV to a better firing position 15 meters away in order to better support his fellow Marines. It was then that he was mortally wounded by enemy machine gun fire. Corporal Evnin's selfless dedication to his fellow Marines and courage under fire were above and beyond that expected, and met the highest standard of Marine Corps NCO leadership and aggressiveness. In a desperate last attempt, Iraqi infantry armed only with small arms attempted to rush the battalion's M1A1 tanks before being cut down. Darkside continued the attack through the garrisons of the Baghdad RG Division, but stopped short of clearing the Al Kut urban area. There were no longer offensive forces in the city that could influence the Division's continued attack into Baghdad. Without a credible offensive threat, remaining paramilitary holdouts could be left isolated in Al Kut. To clear these remnants would have drawn off significant Division combat power and would likely have caused extensive damage to the city, resulting in significant numbers of civilian casualties.

First Battalion, 7th Marines enjoyed similar success across the river to the south. Preparatory fires from Marine air and artillery led the way. The Battalion's CAAT teams engaged and destroyed Iraqi forces attempting to ram their vehicles near Al Akhrar. Bravo Company cleared a date grove and destroyed several artillery and AAA pieces. Marines found buildings full of ammunition that were simply too big for the Battalion to destroy. They marked the sites and pressed on. At the outskirts of Al Jarrah airfield southwest of Al Kut, they cleared a bunker and took eight enemy prisoners. They also uncovered a site containing aircraft ordnance, fuel, and missiles.

RCT-1 continued its attack on the Baghdad RG Division from the south. 2d LAR along with the 'Horse Marines' of 1/4 attacked towards the southern Al Kut bridges. The Reconnaissance Battalion attacked along a parallel road, dubbed Route 7A, to establish an overwatch and attack by fire targets of opportunity. After completing the clearing of Al Hayy, 3/1 blocked south along Highway 7 to support the RCT's eventual movement south on Highway 7 towards Highway 17, the planned route to link back up with the Division. Second LAR with Alpha Tanks led the attack to Al Kut, supported by 1/4's Obstacle Clearing Detachment (OCD). At 0306Z, as the attack commenced, 2d LAR encountered a hastily employed obstacle not covered by enemy fire. As the Armored Combat Earthmovers (ACEs) went to work to clear the mines, one of them struck a mine and was rendered ineffective. Barbarian's OCD hastily reduced the minefield, clearing a route for the attacking force by 0401Z. Inchon continued the fixing attack on Al Kut.

RCT-1 planned to uncover a safe route through the city to affect a quick link up with the Division in order to maintain the momentum all the way to Baghdad. As Barbarian and 1/4 approached the bridges, however, they were engaged with heavy enemy direct and indirect fires. Immediate returned fire suppressed the Iraqi direct fires, while the enemy artillery was silenced by accurate counterbattery fires from 1/11. Barbarian was at the southern bridge in contact and requested additional infantry support. Colonel Dowdy came up and positioned himself with Barbarian 'six', Lieutenant Colonel Ray. He tasked Lieutenant Colonel Mayer ('Pale Rider-six'), to pass through 2d LAR, continue the attack to the Highway 7 bridge into Al Kut, and detach a rifle company to support Barbarian. Captain Collins and Charlie Company, 1/4, passed through 2d LAR's lines. As they passed between the LAVs, an enemy combatant (who was pretending to be among those killed by 2d LAR) pushed a dead enemy's body aside, got up, and began to shoulder his weapon. Charlie 1/4's Fire Support Team shot and killed the individual before he could fire a round. Lieutenant Colonel Mayer led 1/4's forward command group through 2d LAR's lines, followed by the 81mm mortar platoon and Alpha Company. Bravo Company, 1/4, linked up with Barbarian and supported 2d LAR's actions on Al Kut's southern bridge.

At 0545Z, Charlie Company, 1/4, entered Al Kut from the south. As they approached the road intersection prior to the northern bridge, the remaining enemy forces unleashed small arms, RPGs, and mortar fire on the Marines from the north and west side of the road. Captain Collins ordered his 60mm mortars to suppress the enemy while his assault platoons returned fire. While the 60mm mortars suppressed the objective, Captain Richard "Shaky" Parkinson controlled a section of Cobras as it delivered its lethal load on target.

With fires now coming from both sides, and Charlie 1/4 having to fight the enemy from three directions, Major Martin Casado, 1/4's Operations Officer, ordered Captain Griffin to attack west, past the forward command group, to relieve the pressure on Charlie Company. As Captain Collins acknowledged the orders, an RPG struck his command vehicle, but failed to detonate because the enemy had failed to pull the safety pin. Unfazed, the company commander located the source of the attack, and focused direct and indirect fires at the target. Moments later, secondary explosions were observed as the Marines destroyed a building housing an enemy weapons cache.

As the firefight continued, a family of five emerged from a building and began to move south along Highway 7 to escape the fighting. Recognizing the dangerous predicament for this family, Colonel Joe Dowdy, Gunner Parks, and Private First Class Xavier Cobb rushed towards the family, picked up the children, and shielded them with their bodies as they led them to a defilade position. The embedded journalists with 1/4 (Dr. Bob Arnot and Jim Bruton, both from MSNBC) captured the two-hour battle for Al Kut live via satellite phone. Riding with the forward command group into the heart of the action, the journalists showcased the Marines' tenacity and compassion all at once.

By early afternoon, the attack was complete. To support the Division through this complex battlefield maneuver, 11th Marines had adroitly changed the missions of firing battalions to match the shifts in direction and status of the main effort. To support RCT-5, 2/11 was in DS, with 5/11 reinforcing and 3/11 in GS. To support RCT-7 when it became the main effort, 3/11 was shifted to DS, with 5/11 reinforcing. The 11th Marines fires professionals smoothly made the transitions in fires control without disrupting supporting fires to either attack. Finally, 11th Marines fired disengagement fires and all three prongs of the Al Kut attack disengaged. RCT-1 was ordered to withdraw and affect a link up at An Numaniyah by the most expeditious means. The Division hurried west to strike at the enemy's heart in Baghdad. Would-be Regime fighters in Al Kut were cut-off from Regime support, and could only watch helplessly as the Americans continued their unrelenting march to Baghdad.

The Inchon Marines pulled out and began the second of their 'Mother of all Movements' retracing their steps south on Highway 7, turning west on Highway 17, moving north on Highway 1, and northeast on Highway 27 to An Numaniyah; covering 200 km in 24 hours. The RCT crossed the Tigris River at the An Numaniyah Bridge and raced to join the Division in the continued attack to Baghdad.

First Recon Battalion reverted to Division control and remained on Route 7 south of Al Kut to block until relieved by the 24th MEU. The Battalion moved up and established a blocking position to cover the withdrawal of RCT-1. The Battalion expected to remain in place for at least 24 hours until elements of TF Tarawa could come up to relieve them.

The Division had accomplished its second strategic victory. The deliberate planning that had identified the routes, seams, and scheme of maneuver for the Division's actions had all paid off. The combined arms team had set the conditions, and the blood and sweat of the infantry battalions had made it a reality. The Division had successfully weathered the 'Mother of all Sandstorms', and had taken the time to secure its LOCs coming up from the south. The Division cared for its casualties, topped-off fuel and ammunition, and marched to the northwest, heading for the final battle to bring down Saddam Hussein's Regime.

The sun sets on the 1st Marine Division's Antenna Farm.

"We're the Marines. We took Iwo Jima. Baghdad won't be shit."
- Brigadier General John Kelly
while under fire along Highway 7

Chapter 6
Attack to Secure Eastern Baghdad

The Division was now poised on the north bank of the historic Tigris River. The attacks of RCT-7 and RCT-1 against the Baghdad Republican Guard Division had destroyed the last remnants of the conventional capability around the city, leaving a small number of Fedeyeen and paramilitary fighters holed up in Al Kut. The Regime no longer had unfettered access from Baghdad to Basrah, and the survivors of III and IV Corps were cut off from the capitol. More importantly, the Fedeyeen fighters in the south were now cut off from any hope of reinforcement or support. Behind Blue Diamond, the door from Baghdad was closed. The Regime's loss of control in southern Iraq was now merely a matter of time.

To the Marines this meant that they now had a straight shot at the enemy's capitol, but it also meant that they were nearing the final confrontation with the Regime. If the Iraqis had the capability and intention to use chemical weapons against the Division, this would have to be the time. The terrain around Highway 6 offered little maneuver room, and the Division would be stretched in column, generally along the downwind hazard line. Having spent the last 15 days in chemical suits, the Marines had gotten used to the constant awareness of the risk of chemical attack. The Division had already crossed over several alleged 'triggers' for Saddam to employ chemical weapons without incident. Each alleged trigger line crossed increased the potential that the next could be the one that would trigger a WMD response.

The Iraqis Prepare to Defend Baghdad

The Iraqi conventional resistance in the south had seemingly melted away in the face of the Division's advance. The Fedeyeen and paramilitary threat that had replaced the conventional defense had been surprisingly vigorous, but had proven ineffective. The Iraqis had not made good use of their significant numbers of tanks and artillery systems, and had not launched the much-anticipated barrage of surface-to-surface missiles. The weakness of this response was not entirely unexpected from the two southern corps of the Iraqi Regular Army. Although the resistance in the south had been less than anticipated, even the Regime had not counted on the southern forces to put up a rigorous defense. Baghdad was expected to be a different story. Baghdad was the Regime's strategic center of gravity, and critical to Saddam's survival. Baghdad was also at the end of all lines of retreat for Iraqi forces that had withdrawn in the face of the US 3^{d} Infantry Division or the 1st Marine Division advances. Whatever forces the Regime had left were now reinforcing the capitol defenses.

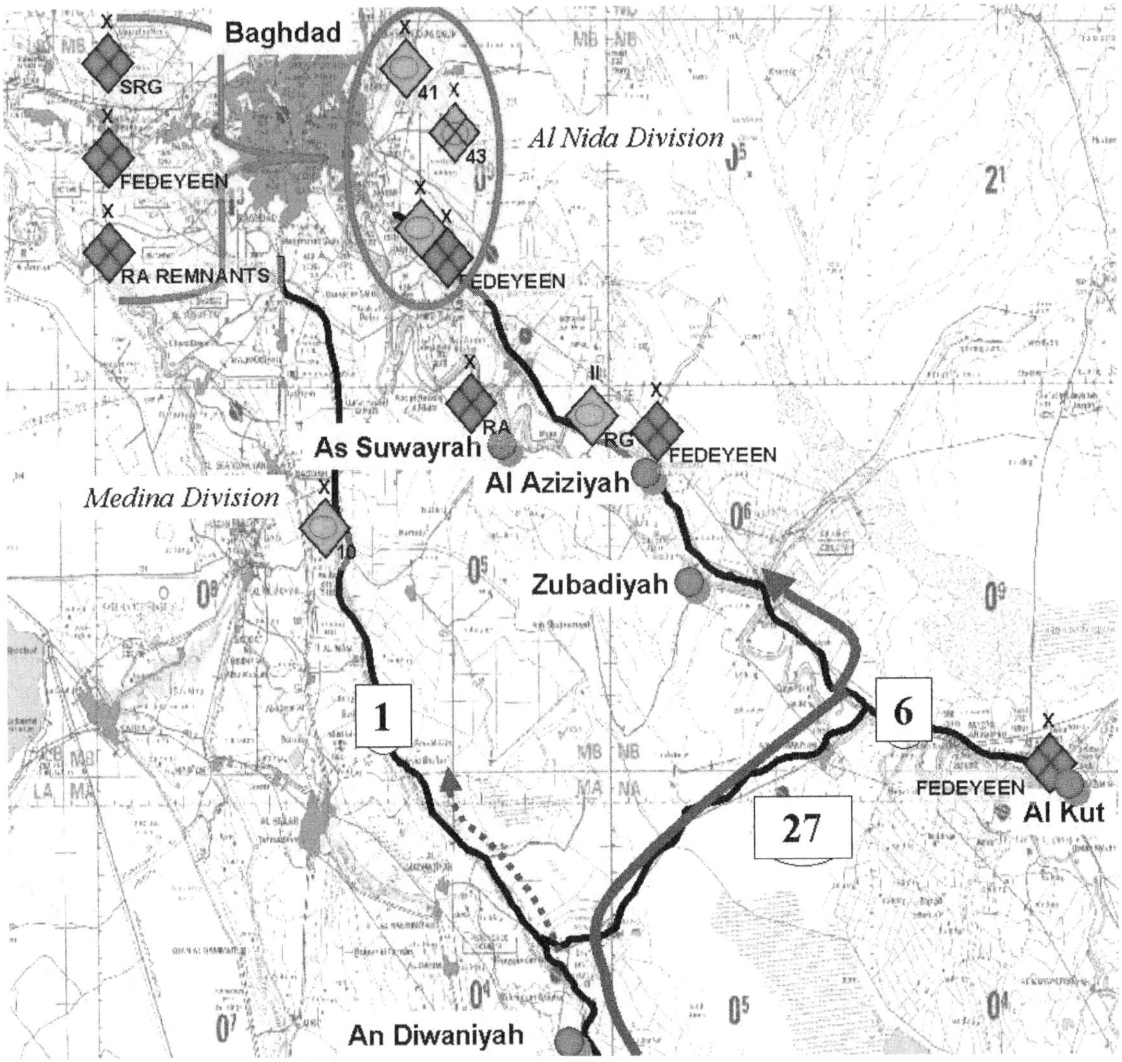

The Republican Guard, Regular Army remnants and paramilitary groups establish an outer cordon around Baghdad. The defense of Baghdad itself was left in the hands of the Special Republican Guard and the Fedeyeen.

The intelligence reporting continued to assess that the Iraqi strategy was to 'circle the wagons' around Baghdad and Tikrit. The Republican Guard (RG) was expected to defend the capitol in a series of concentric defensive rings. The outer ring would be established by the RG, defending in depth in a series of supplementary positions 25-60 km from the city. Inside that ring, the Fedeyeen and paramilitary forces were expected to establish an urban-centric defense in the capitol itself. Inside this defense would be a third ring of defense, composed of the Special Republican Guard and Iraqi Intelligence Service troops. These would be the last holdouts, defending Saddam Hussein and his sons to their final deaths in the palaces of Baghdad. In the international media, the Iraqis claimed that they would draw the Americans into the city, where the US technological advantage would be diminished. The Iraqis had visions of posing a 'Chechnya' or 'Stalingrad' problem to the Coalition in the suburbs and streets of Baghdad.

Allegedly, the defense of Baghdad would also entail the free and liberal use of chemical agents on the battlefield. If the Iraqis had this capability, there did not seem to be a deterrent to their use at this late stage in the fight. There was no more diplomatic leverage to be gained, and the presence of 'infidel' invaders on the holy ground of Iraq might actually gain some sympathy for the Iraqis in the Arab press. The use of chemicals, moreover, would probably not be restricted to use against attacking US forces. Missile strikes on Israel, the Gulf Cooperation Council countries, and southern Iraqi cities under Coalition control were considered possible.

The Division began to question this assessment as it crossed the Tigris River. The Baghdad Republican Guard Division had certainly not posed the significant challenge it had been expected to. Many of the defenders had adopted the same civilian-clothes strategy of their brothers in the south, making it easy for them to fade into the civilian populace as their military situation became untenable. The Baghdad RG Division had also been the subject of hundreds of sorties by Coalition aircraft, and had been largely disabled by air fires. Most identifiable armor or artillery targets had been destroyed. Without their combined arms, the infantry did not possess the combat power required to defend the city. The only strategy left to the Baghdad RG Division had been to fade into the urban area, and defend as irregulars. Would the same model hold true for the Baghdad defenders?

The Al Nida Division, to the east of Baghdad, had been on the receiving end of a significant volume of shaping fires already. Curiously, the Al Nida Division's equipment was still reported holed-up in large numbers in the revetment fields to which they had moved early in the war. The continued presence of these systems in their revetments might indicate a crafty strategy of patience on the part of the Iraqis, waiting until the last minute to mass for a counterattack or move to a final defensive line. The Division began to arrive at an alternative conclusion. The Al Nida Division may have abandoned most of its conventional equipment, and had either taken up positions in the city as infantrymen or had abandoned the fight altogether. The 'weapons system' focus of the US intelligence collection apparatus could count the tanks in the revetments, but could not determine their intent. The Division would attack toward Baghdad keeping a wary eye on the equipment of the Al Nida Division, and would continue to shape it as if it were a viable threat. If the Al Nida remained a viable fighting force, destroying them in their current positions would be much easier than if they were allowed to withdraw into urban areas where they would possess additional advantages, and the potential for innocent civilian casualties would skyrocket.

The Regime had also made attempts to bring selected forces down from the north. Although the threat posed by the Kurds had actually intensified, the Regime had decided to pull elements piecemeal to reinforce the southern defenses. The Adnan RG Mechanized Division was the most significant unit to be recalled to the defense of the capitol. In addition, there were reportedly additional brigades from the Regular Army's northern corps that made their way to Baghdad. Whether any of these reinforcing units retained their full combat power by the time they reached their defensive positions was questionable, but their reported presence colored the Division's perceptions of the assault on Baghdad. The significant Coalition shaping efforts on the Medina and Hammurabi Divisions had left Baghdad's approaches vulnerable from the south and west. The forces pulled from the north were now plugged piecemeal into the gaps in the Baghdad defenses until unit integrity was meaningless. It became nearly impossible to assess the position or strength of any conventional unit. The Iraqis were trying to preserve enough combat power to put up stiff resistance inside the Baghdad urban area, using the remnants of their faltering tactical efforts outside the capitol.

The remaining wildcards for the Iraqi defense were the Fedeyeen and the foreign *Jihadis*. As the Division had been told on multiple occasions, there were reasons to fight other than loyalty to Saddam. It was likely that some fighters with a nationalistic, religious, or pan-Arab motivation would continue to fight even if the Iraqi soldiers of the Regime would not. Intelligence reports were received indicating that terrorist extremists from surrounding countries were rushing to the Baghdad area. The Regime had aggressively recruited this extremist element, and believed that thousands of these aggressive fighters would remain to fight the Americans. These fighters could fight conventionally, and could also serve as a cadre for popular resistance. The Regime had already dispersed surprisingly large amounts of ordnance in weapons caches and equipment across the countryside (undetected by US intelligence efforts) stored principally in schools and mosques, to arm the population in an uprising against the Americans. These weapons could also arm thousands of foreign volunteers if they were to come to the aid of Saddam's regime. Because of the rapidity of the US attack and the lack of popular support, they never had a chance. The Iraqi people were not about to take up arms in defense of the Regime that had so brutally repressed them for decades. When the Fedeyeen and paramilitary cadre were dead or had faded into the population, their dreams of jihad or popular rebellion faded with them.

The Operational Design

The Iraqi's Baghdad defenses were in disarray, and maintaining the Coalition's momentum through relentless attack would prevent them from reorganizing or gaining the support of additional fighters. The original operational design was to establish an outer cordon around Baghdad that would prevent both the escape of high-profile Regime figures and keep reinforcements from adding to an urban-centric defense. By the plan, CFLCC forces from the outer cordon would then conduct raids into the city against selected targets. This pattern would continue until conditions in the urban area would permit a more deliberate and long-term presence in the city without prohibitive casualties.

The Blue Diamond concept was slightly different. To the Division, destroying the Al Nida Division would eliminate the last remaining obstacle to operations inside Baghdad itself. The heavy concentration of Shia Muslims in east Baghdad, especially in the Saddam City area, offered a potentially sympathetic area of the city that might offer a less dangerous entry into the urban area. Additionally, once the Division had fought to seize objectives in the urban area, it did not intend to give them up and have to fight for them all over again. Withdrawals from portions of the city after seizing raid objectives would embolden the enemy and lessen the 'dominating effect' the Division wanted to portray to the enemy and to the international media. This was a lesson learned through bitter experience by Marines past, as discussed in the Division's MOUT Seminar in Camp Pendleton the previous December. If the Iraqis contested the city, identifying important targets by raiding and then abandoning them would give the Iraqi fighters the opportunity to reoccupy, mine, booby trap, or preplan fires. This would likely increase casualty figures from an outright seizure and occupation strategy. On these points, the Division shared the same vision as the US 3rd Infantry Division.

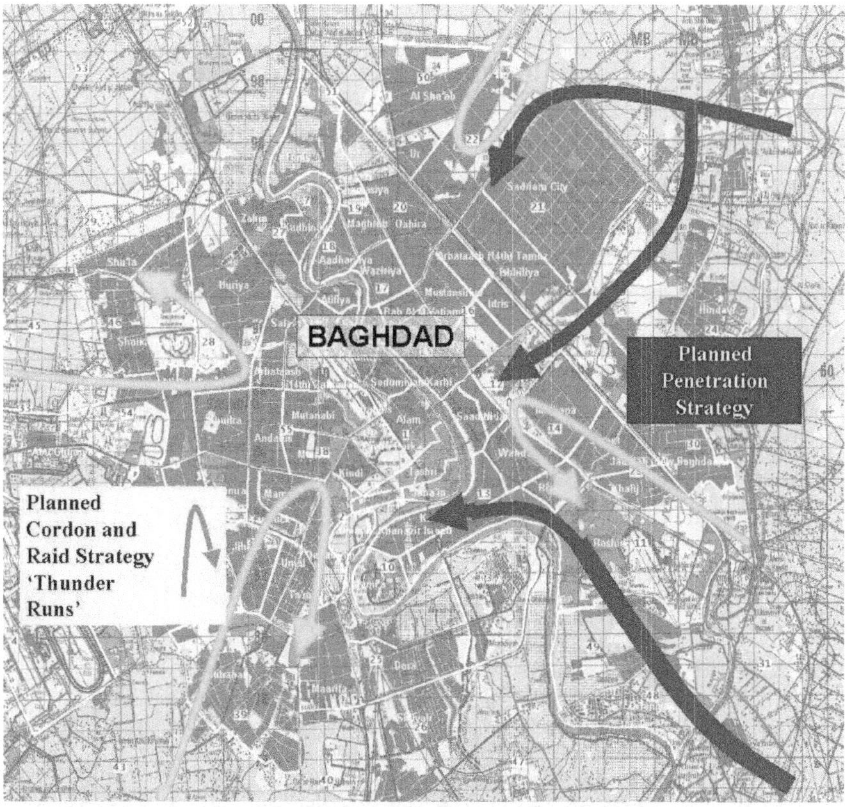

Two strategies emerged for dealing with the Baghdad urban area. One strategy involved cordoning the city and conducting raids into the urban center. Blue diamond preferred a more conventional penetration strategy.

If the Iraqis mounted a credible urban-centric defense, the Division was prepared to respond. The payoff from months of hard urban training had resulted in a core element of the Division that was skilled in urban operations. The tactics, techniques, and procedures for intelligence, fires, and maneuver inside the city were at least developed, if not yet combat tested.

The Division's operational scheme was to attack rapidly to cut off the city by establishing the Baghdad outer cordon, concurrently assaulting into this urban terrain to hunt down and destroy the last defending elements of the Regime. The Division would secure neighborhoods and Regime areas in the city, isolating pockets of resistance for piecemeal destruction. Penetrating the enemy's outer cordon was mainly a function of destroying the Al Nida Division and Fedeyeen fighters along the way. After punching through these outer defenses, the Division would attempt to seize the two existing bridges over the Diyala River intact, using them to support an attack to fix the remaining eastern Baghdad defenders to the southeast. The Division's main effort would then shift to an attempt to cross the Diyala River at a point north of the city, and continue the attack into Baghdad from the northeast, in the vicinity of Saddam City. The G-2 developed a number of target sets that represented the key terrain sites that had Regime or WMD significance. Securing these sites would serve to orient lines of operation once inside the city, but the focus would be on hunting down and destroying any remaining enemy fighters. After combat operations were complete, the Division planned to rapidly disengage from the city, prepared for the anticipated order to continue the attack north to Kirkuk or Tikrit.

3 April
Grizzly Continues the Attack to the Northwest

The Marines and sailors of RCT-5 occupied their blocking position north of the Tigris River on the afternoon of 3 April. Having seized the Tigris River crossing sites in an 18-hour action, the RCT now was positioned on both sides of the river. Second Tanks and 3/5 had turned over An Numaniyah to RCT-7 and were blocking to the northwest in positions on Highway 6 north of the river. On the south side of the Tigris, the remainder of RCT-5 was located between Highway 27 and the Pearl 3 crossing site, where 8th ESB had continued to ferry tanks and trucks across the river while the bridge was being constructed. In the morning, the Division's planned main effort would be RCT-7's attack on the Baghdad Division, while RCT-5 continued to block to the west. The Marines of RCT-5 were eager for their own contest with a Republican Guard Division, but a day spent in a blocking position would have advantages as well. For one thing, a day that was not spent in the back of an AAV would present an opportunity to tend to feet that were already suffering in the airless confines of rubber chemical protective overboots. The Division had been in MOPP 2 since crossing the Saddam River, and would remain in this posture for as long as the Iraqi Regime was considered capable of employing WMD.

The CG, Colonel Dunford, and Lieutenant Colonels White and Lethin discuss the details of RCT-5's assault into Baghdad.

Overnight, RCT-5's mission changed. While RCT-7 dispatched the Baghdad Division in Al Kut, RCT-5 would immediately conduct a limited objective attack to the west along Highway 6 in order to open up battlespace north of the Tigris River crossings. There would be no day spent recuperating; Blue Diamond was on the move again. As well as being suited to the temperament of the Marines, attacking in two directions would allow the Division to position forces and supplies north and west along the Tigris in preparation for the drive to Baghdad. As if throwing a basketball head fake, the Division's eyes were moving east to Al Kut, but the momentum of the body was already moving north and west toward Baghdad. This swift transition from one phase of the operation to the next was a consistent theme of the scheme of maneuver, and a significant contributor to the Division's successful tempo. Not pausing to consolidate or count its gains, the Division would continue a relentless tempo until the final objectives were achieved. Colonel Dunford was instructed to choose his own intermediate objectives and to get moving.

The RCT-5 staff focused on the enemy to their northwest. The previous day, the Pioneer UAV had detected a platoon of T-72 tanks about 15 km northwest along Highway 6. The Division G-2 assessed that the platoon was from the Al Nida Division and that the presence of conventional combat power that far forward meant that the Iraqis were going to make a more vigorous attempt to delay along Highway 6. They had not experienced good success in delaying in a similar fashion along Highways 1 or 7, but the addition of RG forces and foreign fighters could make a significant difference. The tank platoon's destruction became RCT Objective Alpha.

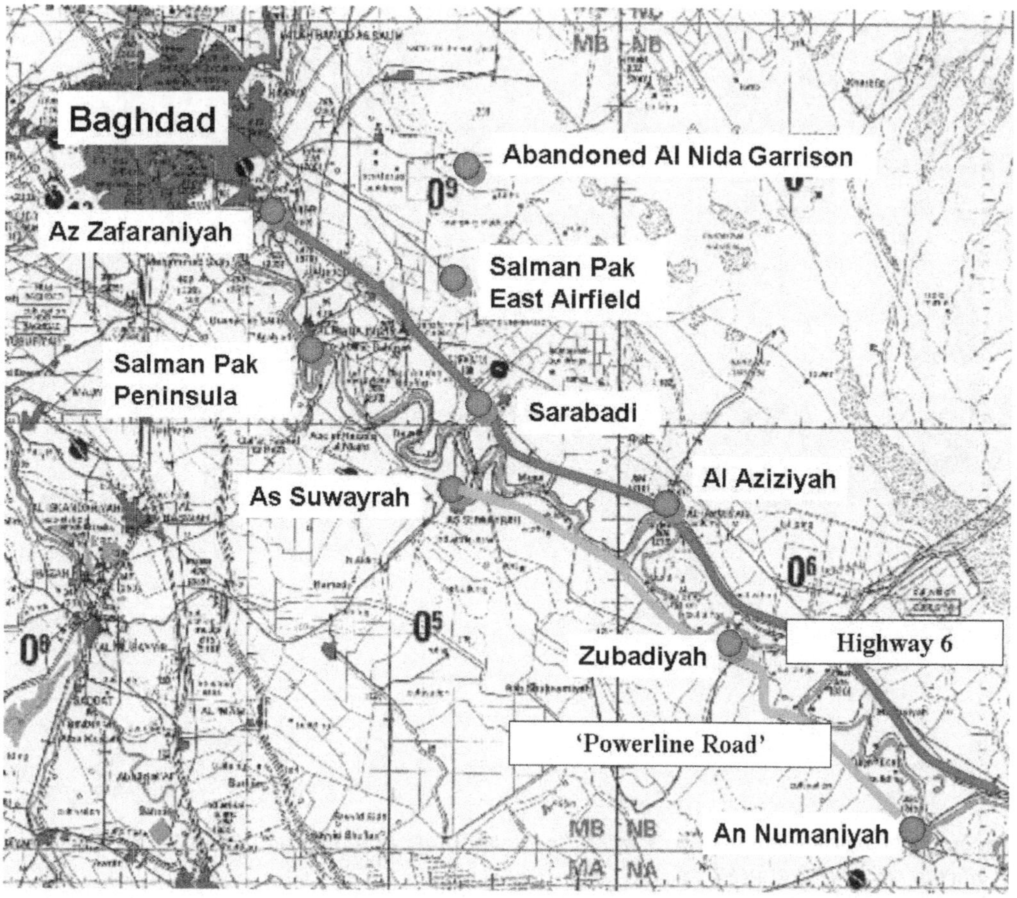

First Marine Division initially utilized both sides of the Tigris as it pushed northwest into Baghdad. Later, the entire Division would shift to the northern Highway 6

A pontoon bridge crossed the Tigris River at the small town of Az Zubadiyah, about 35 km west of An Numaniyah. The Division had ordered the pontoon bridge to be destroyed earlier that morning. Busloads of Fedeyeen reinforcements had been observed traveling south from Baghdad, and the decision had been made to remove the bridge to keep them from crossing the Tigris and posing a threat to the Division's southern flank. Twenty km further to the west, a second RCT objective was a II Republican Guard Corps Anti-Armor Regiment near Al Aziziyah. Imagery showed a company of armored vehicles mounting anti-tank missiles defending along Highway 6 just east of the city.

The final RCT-5 objective for the day was a second pontoon bridge at Al Aziziyah, located 70 km upriver from An Numaniyah. Advancing on either side of the Tigris River, the attacking columns would converge at this pontoon bridge, the next intact bridge over the river. Even further west (25 km west of Al Aziziyah), the town of As Suwayrah was the site of a third key bridge. As Suwayrah was also believed to be a logistics and transshipment point for Iraqi fighters and equipment. There were indications that a brigade of RA troops and a large number of Fedeyeen fighters were near As Suwayrah, concealed in the palm groves lining the river. RCT-5 would be prepared to continue the attack to As Suwayrah the following day.

Destroyed enemy armor on Highway 6. Where the enemy chose to stand and fight, they were quickly defeated by combined arms attacks.

In the early morning of 3 April, RCT-5 moved into the attack. To the rear of the westward-facing columns, artillery made a dull racket as two battalions of the 11th Marines sent volley after volley east toward Al Kut in support of RCT-7's main effort attack. Leading the attack the opposite way, the Scout Platoon of 2d Tanks turned onto Highway 6 and moved to the head of the column. The Ironhorse began their move to the west with Scouts forcing civilian traffic to the side of the 4-lane highway and vigilantly scanning for enemy positions.

Second Tanks met no resistance to the west as they passed Az Zubadiyah on the north side of the river. Along the highway they encountered tanks, armored personnel carriers, air defense artillery, and howitzers. All were abandoned. Some had been destroyed by air strikes; the Marines destroyed others as they passed. Instead of the conventional defense they expected, the Marines encountered vehicle after civilian vehicle traveling south. Cars, trucks, and minibuses, loaded with men, women, and children of all ages displayed white flags, tee shirts, or scraps of cloth as they fled southeast out of Baghdad.

As the battalion approached Al Aziziyah, the first manned enemy positions were spotted, and the Scouts in their armored HMMWVs began to take sporadic small arms fire. The Scouts returned fire and continued to push west and north. The sporadic enemy fire quickly turned into sustained fire, and the Marines began to spot the flash of RPG rounds or saw the grenades as they skipped off the pavement next to their vehicles. Unable to pull off the highway to pass the lead tank platoon forward, the Scout vehicles pulled into the highway median so the tanks could move forward. Alpha Company moved up, and the tanks began to engage the enemy. Moving along the median, the Scout Platoon Commander's HMMWV was struck by two RPGs. The platoon commander, First Lieutenant Matthew Zummo, was seriously wounded, but remained with his Marines and continued to fight through the barrage of gunfire:

> *"Leading from the front, he uncovered a company-reinforced ambush, forcing the enemy to engage prior to the arrival of the Battalion's main body and foiling his attempt at a successful ambush. Personally engaging enemy rocket-propelled grenade and machine-gun teams from an exposed position, he courageously continued to destroy enemy targets, lead his platoon, and provide accurate reporting to the Battalion command group. After receiving wounds of an urgent nature from an enemy RPG, he refused to be evacuated until the enemy had been completely defeated."*

With the tanks moving up on both sides of the highway, and the Scout vehicles in the center, 2d Tank Battalion pushed through the city. Enemy resistance was fierce. With T-55 and T-62 tanks, mechanized vehicles, and air defense artillery employed in a direct-fire role, the enemy battled in reinforced battalion strength. Iraqi tanks and armored vehicles fought from positions along Highway 6, supported by mortars and artillery, while two companies of dismounted Iraqi infantry defended from within the confines of the city. The Marines fought them with combined arms, using tanks, dismounted infantry, and close air support. It was one of the more significant battles the Division would fight against a conventional force during the war.

RCT-5 had trained long and hard for such a fight, and the Marines were ready. While 2d Tanks engaged the enemy defensive positions along the highway, 3/5 (who had been tasked to follow and support 2d Tanks attack) moved up close behind them. The speed with which 2d Tanks moved was daunting, and 3/5 fought to keep the gap between battalions closed. Hearing reports of stiff fighting on the RCT-5 tactical radio nets, Marines from 3/5 could tell they would encounter resistance this day. As the Battalion approached the southernmost edge of the town, several groups of Iraqi soldiers from well-hidden positions suddenly opened up on the 3/5 column. Kilo Company, in the lead, quickly dismounted and maneuvered platoons to clear a compound on the right side of the road. Marines used fire and movement to close with the enemy, while the company FiST team boldly called in 60 and 81mm mortar fire at danger close ranges. Aided by the close and steady fires of the tank platoon and CAAT section, Kilo's lead platoon, led by Second Lieutenant Michael Prato, moved into a series of trenches, bounding from position to position, firing and killing 30 Iraqi soldiers at close range. The other platoons maneuvered to clear the flanks of the company's position, killing and capturing a score of Iraqi soldiers.

Marines of 3/5 pursue Iraqi defenders into the city.

Lima Company, 3/5, and the Battalion's 81mm mortar platoon began receiving accurate machinegun and small arms fire from groups of Iraqis occupying positions along their flanks. Dismounting quickly, Lima Company aggressively cleared the Iraqi soldiers, killing a good number using direct fire and devastating sorties of rotary and fixed wing CAS. The 81mm mortar platoon paused long enough to dispatch a ground assault, pursuing the Iraqi soldiers through a wood line before returning to their guns to pour indirect fire on the enemy.

India Company was ordered to take the lead and resume 3/5's original mission of securing the Al Aziziyah Bridge. It became apparent, however, that many Iraqi defenders had slipped into the city and were attempting to engage in an urban fight with the Marines. India moved forward quickly, dismounted, and began to clear the town. As the infantrymen scrambled from their AAVs and lined up behind a low sand berm, the AAV crewmen kept up a lively suppressive fire with their .50-caliber and Mk-19 up-gun turrets. In the heroic fashion of Marine Corps Non-Commissioned Officer leadership, individual squad leaders would stand, and with a wave of his arm, lead his squad over the berm, dashing across the open space to the nearest alleyway or building. As the Marines vanished into the urban maze, the next squad was already filing in behind the berm and making ready to enter the city. While India made its approach to seize a foothold, the CAAT platoon, which had moved up on the right side to provide suppressive fires, greatly aided the assault. The effects of these fires were devastating, and quickly defeated Iraqis attempting to ambush the infantry with RPGs.

House to house, the fighting continued for several hours, although by the time 3/5 entered the town in overwhelming force, it became apparent that much of the stamina of the Iraqi fight was gone. As the fight progressed through the city and into surrounding farmland, the enemy resistance dissipated. Second Tank Battalion eventually halted their move when the lead element was about 10 km west of Al Aziziyah, and was joined by 3/5.

South of the Tigris, the rest of RCT-5 moved on a track parallel to 2d Tanks and 3/5. The numerous irrigation ditches and canals confined movement to a single paved road dubbed 'Powerline Road' because it ran alongside a line of powerline pylons. This road extended from

An Numaniyah all the way to As Suwayrah. First LAR Battalion ('Highlander') led the RCT (-) attack along Powerline Road. First Battalion, 5th Marines (1/5) followed in trace of Highlander, with orders to seize the pontoon bridge at the town of Az Zubadiyah. The battalion cleared through the town with two companies on-line, meeting no enemy resistance and confirming that the bridge located in the vicinity had been destroyed by aviation fires. The Marines quickly assessed the bridge, evaluating its potential for repair and use as an alternate crossing site. Following 1/5, 2/5 continued the advance to Al Aziziyah, where the battalions on the south side of the Tigris intended to cross and link up with the battalions on the north side of the river.

The Iraqis fired rocket-propelled grenades from behind a taxi parked along a distant canal. One grenade zipped across the nose of an armored amphibious vehicle and exploded in the dirt. That angered Major Andrew Bianca, executive officer of the Marines' 2d Tank Battalion. Sheathed in aluminum plate, the tracked amphibious vehicles known as AAVs can withstand rifle fire, but not rocket grenades. And Major Bianca's support team was in AAVs. He ordered his tank crew to fire a round at the Iraqis. The 120 mm cannon barrel dropped slightly, then erupted with smoke and flame. The noise ripped the air so violently that Marines standing in an AAV behind the tank were knocked off balance. A cloud of dirt appeared behind the taxi. The shell had gone through the taxi's open windows. But shrapnel from the round finished the Iraqis, and a finger of black smoke and flame soon rose from the taxi.

The tank column resumed its march to Baghdad. For Marines and Army soldiers fighting throughout southern Iraq, this was their war: armored columns blasting through urban ambushes. The 2d Tank Battalion started several fights with these urban opponents. Speed mattered more than body counts or seized ground. Using tanks to punch through enemy ambushes put the Marines closer to Baghdad and Saddam Hussein's regime. Lieutenant Colonel Mike Oehl, the battalion's commanding officer, put it to his officers this way: "Speed is the essence of this endeavor." He was talking about a planned raid, but the remark held true for the battalion's mission in the war. "It's hard to know what our part was in the overall war, but I'd like to think we made it a shorter war because we got here so quickly," Colonel Oehl said when his unit reached Baghdad.

Yet several analysts agreed that the battalion exemplified the strategy and tactics that toppled Mr. Hussein in just three weeks of warfare. Colonel Oehl's crew named its tank "Deadly Mariah" and animated the name with an angry cloud blowing swords from its mouth. Major Bianca's tank crew reached back to Greek mythology for the name "Two Furies" - anger and vengeance, minus the third fury, jealousy, which seemed out of place in Iraq. The 2d battalion spent most of the war at the tip of the Marines' spear. It came within three miles of Baghdad on April 4, after three days that destroyed what was left of the Al Nida Division of the Republican Guards. They cleared the way to Baghdad for the 1st Marine Division and killed a large number of Arab Muslim volunteers who heeded Osama bin Laden's call to come to Iraq to kill Americans.

By Jim Landers in the *Dallas Morning News,* 18 May 2003.

On Powerline Road, RCT-5's advance proceeded without challenge, at first. Passing through dilapidated farming villages, the Marines were amazed to find themselves welcomed with surprising hospitality. Despite living in what was obviously desperate poverty, the smiling villagers offered sweet tea, flat bread, and dates to the Marines. Meanwhile, inquisitive children touched the Marines' camouflage uniforms without fear. These farmers, members of an oppressed Shiite Muslim religious community, had no love for Saddam and the Iraqi Regime. This response from the Iraqi people would soon become routine, characterizing the population's relief at their liberation.

First contact was reported by Highlander as its lead company advanced on the south bank of the river. Observing a pickup truck speeding toward them, a platoon of Delta Company identified a heavy machinegun mounted on the vehicle as it turned south along a canal. One section engaged the pickup truck with 25mm high explosive rounds, destroying the truck and killing the occupants. Soon after, 'Highlander Three' called for Delta Company to form a screen line along an aqueduct leading south from the river for about 10 km . Two batteries of 2/11 moved up behind the screen line and were soon providing artillery fires in support of 2d Tanks and 3/5 in their fight on the north side of the Tigris. By then, 1/5 had moved up as well, and the near side of the aqueduct had become crowded with tanks and AAVs.

Eleventh Marines units cross the Tigris River at the Pearl 3bridge west of An Numaniyah.

As the fight to the north in Al Aziziyah drew to a close, 1st LAR resumed its reconnaissance in zone along the south bank of the river. In the lead of the battalion, Lieutenant 'Lucky' Cullins' Third Platoon, Delta Company, rounded a bend where Powerline Road ran west right along the reedy riverbank. All at once the platoon found itself in an ambush. With heavy small arms and RPG fire emanating from an orchard and from around a bridge about a thousand meters down the road, the platoon returned fire using all of its organic weapons. Lieutenant Cullins was ordered to pull back so that the enemy position could be engaged with supporting arms. Destruction of the ambush force was a perfect choreography of supporting arms that demonstrated the combined arms killing power of the MAGTF, and ended in the certain destruction of the Iraqi defenders. The battalion's 81mm mortars went into action first. Fox Battery, 2/11, shifted trails and was soon ready to fire in support. Captain Reuel Pietz, the company FAC, arrived and air began to check on station. Meanwhile, the 1st and 2d Platoons moved up on line and the company attack by fire position was set. As the LAR company began to engage the enemy with direct fire, Lieutenant Cullins directed mortars and artillery into the orchard and a nearby walled compound into which some of the ambushers fled. Captain Pietz controlled a division of AH-1 Cobra gunships and a section of F/A-18 Hornet jets, which worked over the ambush area. Fox Battery recorded Lieutenant Cullins' surveillance as he ended the artillery mission, "Everything is destroyed, everything is destroyed. The buildings are destroyed. Everything is destroyed."

As 1st LAR was battling the enemy, 2/5 reached the bridge at Al Aziziyah. Engineers from the Combat Engineer Battalion evaluated the captured pontoon bridge and determined that it would not support the armor and heavy equipment of the elements of the RCT advancing on the south side of the River. After a discussion about continuing the attack to a potential crossing site further west, the Division CG ordered the RCT's elements on the south bank of the Tigris to turn around and cross the river at An Numaniyah. With 1st LAR and Battery F covering the withdrawal, 1/5 and 2/5 began the long movement back east down the powerline road, across the Tigris at An Numaniyah, and then back west on the north bank. There, they prepared for the planned attack to the outskirts of Baghdad the following day. Initial movement was slowed for 2/5 by difficult terrain near the Tigris, where the marshy ground swallowed up three M1A1s and an M88, which were left behind for recovery by follow on forces. In the dark while moving to link-up with the rest of the RCT, three Marines were severely injured when their LAV ran into a dump truck abandoned on the road. The artillery convoy of 2/11 also came under enemy indirect fire, causing them to mask up and drive for several additional km in MOPP 4. As night fell, the RCT-5 units on the south side of the Tigris were spread out on a 90 km road march to the north side of the Tigris River. Enroute, they re-armed and re-fueled. They joined the rest of the RCT early the next morning, physically tired, but prepared to press the attack to Baghdad.

Marines get some much-needed sleep near Al Aziziyah. The Division would soon launch the assault on Baghdad.

The Division noted a number of intelligence indicators that the Regime's control inside Baghdad was beginning to fracture. Although the military forces remained at least partially under control, it appeared that the civilian populace was beginning to take to the streets against the Regime. There were reports of Regime leadership figures abandoning Baghdad and attempting to head to Syria. Additional intelligence hits indicated that the large number of tanks and BMPs in the vicinity of the Al Nida Division's garrisons had been abandoned. The Division CG was eager to exploit this situation, and reiterated the importance of speed in getting to Baghdad. There, it was likely that conditions would allow a continued attack right into the Baghdad urban area. The Division Forward CP was ordered to Al Aziziyah to collocate with RCT-5 and be in a position to let the Division Main CP leapfrog to the outskirts of Baghdad.

Logistics would provide the fuel, ammo, water, and food that would allow the Division to move into the city. The importance of KC-130 strips in keeping the Division's momentum forward was already proven. By the end of the day, An Numaniyah Airfield (RRP-19) was clear of unexploded ordnance and made operational. To enable speed against Baghdad, Al Kut could not be allowed to draw off combat power in the defense of this airfield and the very important RRP. RCT-7's attack to Al Kut along the southern bank of the Tigris ensured no enemy threat existed from this direction. The G-4 and CSSG-11 continued to work with 3[d] MAW and MWSS as they expanded aerial re-supply capability in support of the Division's high tempo combat operations. In addition to serving as a FARP, RRP-19 would serve as the Division's maintenance collection point for damaged vehicles and other gear, and eventually became home to the MEF Forward headquarters.

The CG and his staff plot the Baghdad strategy at the Blue Diamond Forward CP near Al Aziziyah.

4 April
Pushing to the West

On the morning of 4 April, Ironhorse led the Division's Main Effort, RCT-5's attack to Baghdad. With signs that the enemy's ability to conduct a conventional defense in Baghdad was collapsing, rapid progress was expected. Commencing their attack from approximately 10 km to the west of Al Aziziyah, Ironhorse ran into stiff resistance not far after crossing the LD. It was clear that the enemy along Highway 6 had not gotten the word that they were losing the war. The combined arms power of the MAGTF was brought to bear. Eleventh Marines brought up 2/11 to support the attack, and the Battalion opened up to devastating effect. The Division Forward CP and the RCT-5 CP were rocked by volley after volley of outgoing artillery, and the sky was filled with friendly aircraft overhead as 3[d] MAW and JFACC air entered the fray, demoralizing and destroying the enemy.

The MEG emplaced Maybey-Johnson bridges over water obstacles where a more permanent structure was required.

RCT-5 began the attack as it had the previous day. Ironhorse (with attached infantry from 2/5) led the movement with a strong push up Highway 6 through numerous small villages and hamlets. A battery of 2/11 providing close and responsive fire support followed them closely. Lieutenant Colonel Sam Mundy's well-oiled 3/5 machine came next, with the remainder of 2/11 and 5/11 in hot pursuit.

As 2[d] Tanks moved through a stretch of Highway 6 north of Salman Pak, a hail of rifle, machinegun and RPG fire ripped into the battalion from both sides of the road. The fiercest resistance came along Highway 6 between the 61 and 66 east grid lines (absent significant terrain features, battles were often noted simply by map grid references.) The battalion saw small arms, machine gun fire and RPG teams near the highway, launching attacks against the entire column. Numerous enemy fighters were spotted throughout the area along Highway 6 in abandoned buildings, man-made bunkers, fighting holes, and behind vehicles. The enemy was a mix of regular Al Nida Division units as well as irregulars clad in black uniforms and non-military attire. These were not just Iraqi soldiers, but Fedeyeen fighters and *jihad* extremists imported from as far away as Egypt, Yemen, Syria, Lebanon, and the Sudan. Their sole motive may have been to kill Americans, but these *jihadis* seemed committed to die for the Saddam Hussein regime. The Marines of Ironhorse efficiently complied with their wishes, killing many with air strikes and tank main guns. As with the day prior, fixed and rotary wing aircraft provided accurate and effective close air support in conjunction with the RCT's maneuver.

Iraqi Artillery (like this D-30) was found abandoned all along Highway 6. Artillery pieces were often found in schoolyards, mosques, and urban areas.

An M1A1 destroyed by a fire during a Fedeyeen ambush on Highway 6. This burned out hulk was a sharp reminder of the dangers that remained in the fight for Baghdad.

Second Tanks continued the attack along Highway 6 with the Scout Platoon in the lead. In the vicinity of the 65 easting, the scouts uncovered another enemy ambush. In the process of engaging the enemy, the new Scout Platoon Commander, First Lieutenant Brian M. McPhillips, was killed by an enemy sniper. Lieutenant McPhillips had taken over the platoon from the seriously wounded Lieutenant Zummo when he had been evacuated. Knowing the dangers of his position but still leading from the front, Lieutenant McPhillips led the Division's attack in the face of a determined enemy. Despite the loss of their Lieutenant, the scouts accomplished their mission of identifying the enemy, then fell back to allow Company C to continue the attack west and north. As the lead elements of RCT-5 approached to within a few km of Baghdad, the enemy set fire trenches adjacent to the road ablaze, making visibility extremely difficult. At this point in the battle the fog of war began to thicken for the battalion. The intersection of Highway 6 and Route Green (the route leading north to the battalion's objective) was known as the 'Diyala Crossroads'. The intersection was to play a significant role throughout the coming fight. Upon approaching this key crossroads, the lead tank was hit with multiple RPGs, instantly killing Corporal Bernard G. Gooden, the lead tank's loader. Corporal Gooden's loss was indicative of the ferocity of the enemy's defense, and the bravery of the Marines, like him, that were willing to meet the enemy in combat.

Damaged, but moving, the lead tank continued through heavy fire, but soon realized they had missed the turn on Route Green. Under fire, the battalion executed a counter-column as rear elements waited just prior to the turn north. Not far behind, Charlie Company's Commanding Officer had taken fire and his tank was immobilized. Captain Jeffrey S. Houston dismounted his tank in order take over another tank within his Company. While moving to the other tank, an enemy sniper found his mark and Captain Houston was severely injured with a gunshot wound to the jaw. Captain Houston's immobilized tank became an instant target for Iraqi gunners. In the fusillade of fire that followed, an RPG pierced the rubber *flexcell* fuel bladder strapped to the side of the tank, and the leaking fuel set the entire tank ablaze. This burned out M1A1 hulk was a sobering sight to follow-on units and drove home the bitter fighting in the area. There were dangers beyond bullets and RPGs too. A truck bomb mixed in with the mass of civilian traffic on Highway 6 maneuvered next to another tank of Charlie Company and was detonated. Fortunately, the injuries to the crew were minor and the tank continued in the attack.

With plenty of aggressive enemy still anxious to fight (2d Tanks had passed), Colonel Dunford ordered 3/5 to clear both roadsides in the vicinity of the 61 easting and provide security for the burning M1A1. The battalion's lead tank and CAAT section moved to support the damaged tank. Lima Company also halted, dismounted, and prepared to conduct a hasty attack. Ordered to clear a large, fenced military complex on the north side of the road, Captain Scott Meredith quickly organized his platoons and sought to lay down some indirect fire suppression. Unfortunately, due to the density of the urban structures and vegetation lining the road, he was unable to observe the fires. Unwilling to cause civilian casualties, Lima resorted to entering the compound using direct fire suppression only. As Lima moved into the assault, Lieutenant Colonel Mundy ordered India Company to move forward, dismount and clear the south side of the road:

"Because 2d Tanks had already moved through and we were unable to conduct face-to-face coordination, I wasn't exactly sure where the enemy was located. Looking at the map, I guessed that they must have been defending from the apparent military compound on our right flank, and so I ordered Lima to clear it. I pulled India Company up mainly to protect Lima's flank and to provide some flexibility during the attack. The left side of the road was dominated by a large canal giving way to a huge open field running south and southwest. The ground was broken and had some scrub vegetation on it. My Battalion Gunner and I were discussing the fact that it

didn't look or feel right to leave that flank unguarded. I'm glad he planted that seed because it turned out the open ground was hiding a hornet's nest of jihadis."

India Company began its movement along the left side of the road and all seemed well. Without much warning several Marines flushed pockets of well-camouflaged *jihadis*. These clustered groups maintained a steady and willing defense of their respective areas. The attitude they displayed was fatalistic, *"kill me or I'll kill you."* The Marines of India willingly obliged them. During the next several hours, a raging, close quarters battle ensued where India Marines, operating by platoons and squads, closed in on scores of defending *jihadis*. Though not particularly good marksmen, these *jihadis* fought tenaciously. During this intense battle, Corporal Erik Silva was shot and killed by a Syrian fighter hiding underwater in a canal. The presence of these foreign fighters, eager to fight to the death in their misguided cause, reinforced the larger purpose of the Division's mission. The Marines had come to Iraq to liberate the innocent Iraqi people from just this sort of brutal repression, and the loss of fine Marines like Corporal Silva was a painful reminder of the lengths these terrorists would go to retain their grip on power. Silva's squad members reacted violently to the loss of their friend and comrade, bayoneting and shooting the attacker before he fell dead. After the fight, several Marines mentioned that the small caliber 5.56mm round failed to stop the Iraqi fighters even when hit, spawning conjecture of possible *jihadi* drug use and causing many discussions among the Marines about having a larger caliber bullet to shoot. Initially, the fighting was too close for artillery. India's artillery FO called in several missions but the enemy were in too close to have much effect. India Company resorted to employing a division of AH-1W Cobras to root out the defenders. As usual, and even in clear danger from ground fire, 3[d] MAW's killers closed in on the enemy alongside their infantry brothers. In one case Staff Sergeant Gonzalez, commanding a

Marines advance near the Diyala Crossroads, while Iraqis begin to return to their homes.

rifle platoon, popped smoke in front of his position so the Cobras could run a 20mm strafing run only 30 feet off the ground into a stubborn group of 15-20 terrorists. It was a spectacular display of combined arms fighting and when Gonzalez swept through the trench shortly thereafter he found nothing but body parts.

While India continued to pursue the *jihadis* deep into the open field, the rest of 3/5 experienced sporadic, small arms attacks up and down the battalion column. Most were beaten back by aggressive small unit actions on the parts of sergeants and lieutenants. Lima continued to clear the compound, killing an estimated 25 Iraqi soldiers and capturing approximately 40 more who lacked the courage of their convictions. When the lead platoon of India Company encountered stiffening resistance 500 meters deeper in the open field, Lieutenant Colonel Mundy ordered the platoon to return, then directed the FSC, Captain Robert Piddock, to unleash a fury of artillery fire. The display that followed was a sight to behold. Back to back battalion volleys that lasted for 10 minutes or more caused the earth to shake and lifted a pall of smoke that lasted for several hours. Because the Marines had pulled back to a covered position, 3/5 was able to direct this devastating fire within danger close ranges and right on top of the enemy positions. Afterward, India Company reported that the effects of the fires had left the *jihadis* shattered and bloody. In this fighting, 3/5 had sustained fallen comrades, as well. Along with the loss of Corporal Erik Silva, the battalion also experienced four seriously wounded Marines, including Staff Sergeant Anderson, a platoon sergeant in Lima Company, leading from the front in the finest tradition of infantry platoon sergeants.

As 3/5 fought furiously along Highway 6, 2d Tanks turned north on a narrow hard surface road to seize the intersection of Route Green and an east-west hard surface road that provided access into and out of Baghdad. Ironhorse, still led by their Scout and TOW platoons, rumbled to the intersection under sporadic gunfire. The last block of buildings near the intersection proved to be another ambush site, and the column came under intense machinegun and RPG fire. Trucks full of enemy ammunition began to explode, causing numerous secondary explosions. The sector designated for Fox Company, 2/5 (attached to 2d Tank Battalion), turned out to be an enemy ammunition supply point (ASP), now exploding out of control, sending shrapnel, debris, and RPG rounds into the air. During the fighting, First Sergeant Edward Smith, the Fox Company First Sergeant, was wounded in the head. Several Marines from the company heroically tried to evacuate their beloved leader under enemy fire, but he died of his wounds shortly thereafter. His loss was keenly felt throughout the entire battalion.

Once the intersection had been secured and the sun began to set, the Marines of Ironhorse faced the challenge of trying to sort panicked civilian traffic from fleeing or attacking military personnel approaching this critical intersection. At one point, shortly after 2d Tanks arrived on the objective, a small white car raced towards the intersection from the east, headed towards Baghdad. Staff Sergeant Efrain Torres' TOW section was blocking the road and engaged the vehicle after it ignored warning shots, killing both occupants. Upon searching the vehicle it was obvious one of the occupants was a very high-ranking general officer. Upon closer inspection, it was determined that the general was the Chief of Staff of the Special Republican Guard, and his vehicle was filled with important documents and large amounts of cash. The Division had moved so quickly, and the Iraqi's command and control systems had been so completely disrupted, that even this senior general was unaware that RCT-5 was between him and his capitol city. From that time, the intersection was referred to as 'Dead General's Crossroads.'

As 2[d] Tank Battalion and 3/5 led the attack, the rest of RCT-5 moved up behind them on Highway 6, facing sporadic enemy resistance from remnants of enemy forces who had been bypassed by the lead assault battalions. Highway 6 remained a dangerous place on the battlefield, as the *jihadis* who had escaped the wrath of the initial Blue Diamond assault occasionally returned to the Highway to reengage the Americans. One such attack took the life of Sergeant Duane Rios, who was killed by an enemy sniper as he stood exposed in the turret of his advancing AAV. Sergeant Rios, an engineer squad leader supporting 1/5, had molded his Marines into a tight, cohesive unit, and had been a defining influence on many of his junior Marine's lives. He had led his squad through battles in the Rumaylah oilfields, Al Aziziyah, and An Numaniyah. His loss weighed heavily on his Marines, who had enjoyed a special bond with a leader they respected and admired.

LAVs roll past an abandoned T-72 near the Diyala Crossroads.

Having successfully completed their attack on the Baghdad RG Division the day before, RCT-7 conducted a movement from Al Kut to the Diyala crossroads, a distance of over 100 km. They moved up to relieve 3/5's position on the night of 4 April. The following morning, 3/5 moved up to the 'Dead General's Crossroads' to join the rest of RCT-5. RCT-5 immediately began to aggressively patrol toward their objectives along the Diyala River, searching for crossing sites.

As the Division worked its way up Highway 6, Lieutenant Colonel Stacy Clardy's 3[d] LAR Battalion ('Wolfpack'), as an independent maneuver element, conducted a moving flank screen to the north. After being relieved of their Hantush Airstrip security duties, Wolfpack had crossed the Tigris at An Numaniyah and was now operating to the north of Highway 6, paralleling the Division's attack route. Wolfpack picked its way toward the airfields at Sarabadi and Salman Pak over a dense network of canal roads and trails. The terrain was extremely limited, giving even the highly mobile LAR battalion pause. As the Battalion closed on the Sarabadi Airfield, it engaged with and destroyed several armored vehicles and tanks. Third LAR reports of the limited trafficability from the flank gave the Division confidence that no sizeable armor force from the Al Nida Division could attack from this direction. Having transited some of the worst terrain imaginable, 3[d] LAR was ordered back to Highway 6 for follow on tasking in the vicinity of Baghdad.

5 April
Pulling the Division Up to the Diyala

The Division was reasonably confident that the enemy had abandoned most of the equipment belonging to the Al Nida RG Division's 43[d] Mechanized Brigade in the vicinity of its garrison east of the Diyala River. A number of intelligence sources reported tanks, artillery, and other vehicles in the revetment fields around the garrison. Blue Diamond had received multiple indications that this equipment was abandoned, and agreed with this assessment. In a scene that was often repeated during the war, however, every time an unfamiliar Coalition pilot or new theater imagery analyst would observe the site, the Division would receive the alarm that there was an armored brigade on the Division's eastern flank. Each iteration of this erroneous reporting would require the commitment of additional intelligence collection assets to confirm that these were the same abandoned vehicles seen before. Most times, additional fires were also directed against the potential threat. During the night of 4 April, the Division DASC had once again scored a series of B-52 strikes, which the Division vectored to the Al Nida garrison (the garrison was located 20 km to the east of the Diyala River, and had the added benefit (misfortune) of being away from populated areas, affording a low risk of collateral damage.) After tiring of these repeated false alarms, the Division dispatched 3[d] LAR Battalion to establish a security zone to the east, and to sweep through the area to confirm this equipment it no longer posed a threat. Echo Company drew the mission of reducing a brigade's worth of abandoned equipment. After a cautious approach, they came upon the large revetment fields containing intact and operational T-72 tanks, BMPs, and other vehicles. The Marines happily turned to rendering this equipment inoperable. Echo Company spent the next 12 hours destroying, exploding, or calling in air on the gear of the former 43[d] Mechanized Brigade. For some, it was the most enjoyable 12 hours of the war, as they reduced one of the most-vaunted enemy units to scrap. This action removed any possibility of a threat to the Division's rear as the Marines oriented west into Baghdad.

The north-south running Diyala River now stood between the Division and Baghdad proper. The ineffective outer defensive cordon of Baghdad had been defeated. Compared to the crossing of the Euphrates or the Tigris, the Diyala crossing had received less planning attention. Both of the bridges over the Diyala in the south of the Division's zone remained intact. By the plan, these southern bridges would support an RCT-7 fixing attack, while RCT-5 would swing wide and attack into the vicinity of Saddam City. RCT-5 began its reconnaissance up the Diyala to the north in order to find a suitable crossing site that would support its attack. Crossing the river there would allow RCT-5 to extend the cordon around Baghdad to the north. That morning, the Division and 8th ESB planners reviewed the data on the Diyala River, and 'flew' over sections of it on the Falconview 3D terrain visualization system. A number of sites appeared to offer relatively easy crossing of the river, and the Engineers worked their way north to check each of them first hand. By this time, the Iraqis had also figured out that the Americans were looking to envelop Baghdad to the north, and the Marines soon found themselves receiving mortar fire along the banks of the river.

The companies of Highlander moved out to reconnoiter their zones. None would locate a suitable crossing site that day, but all found plenty of trouble. Around noon, a platoon from Delta Company was the first to make contact when it received fire from a compound on the west bank of the Diyala. The remainder of the company arrived and came under small arms and mortar fire from enemy forces across the narrow river. Delta Company returned fire and called for support from artillery and its own organic mortars, destroying the enemy positions. An hour later, two platoons from Alpha Company also received fire from across the Diyala and returned it. Shortly after midnight, Charlie Company's screen line encountered a dismounted force of platoon size, which it engaged with artillery and rotary wing CAS as well as 25-millimeter and 7.62-millimeter fire, destroying the enemy force.

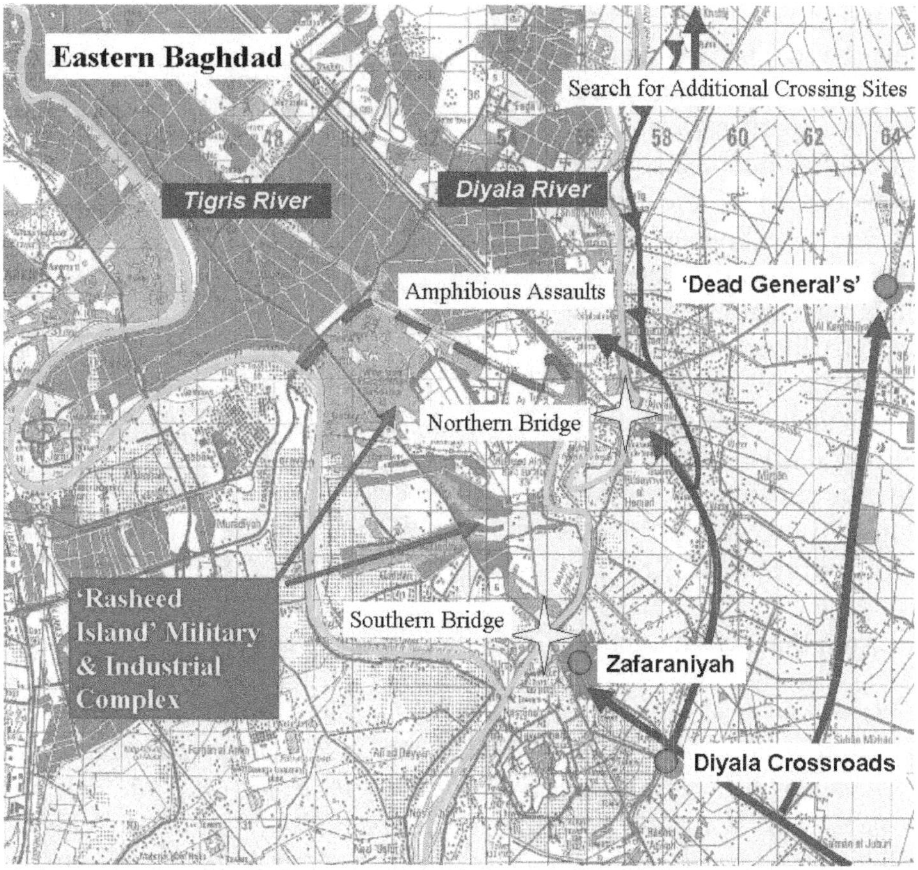

The Division planned to cross the Diyala River in multiple places, including the two intact bridges near the Rasheed military complex. Attempts to find a crossing further north were unsuccessful.

Weapons Company was the busiest that day. After searching throughout the morning, a promising crossing site was uncovered near a place called Ummal Abid. It appeared that the Iraqis had prepared this site for the same purpose, and had staged pontoon-bridging trucks on the west side of the river. The approach on both sides of the river had been prepared and was suitable for the emplacement of standard bridging and would support the huge amount of traffic that the Division planned to put across it. The fording report was filled out and sent to higher. The Company Commander, Captain Dave Hudspeth left one of his LAV-25 platoons to provide security of the site. Included in this security element was the company's Command and Control LAV that housed First Lieutenant Jonathan St. John, the company's Fires Support Team (FiST) leader, and the rest of his crew. As Captain Hudspeth continued north to look for other sites, the security element arrayed itself in a screen line along the east side of the river among sparse trees and vegetation. Opposite them separated by the steep banks of the Diyala River were several thick palm groves and a road running parallel to the river and an approach to the potential crossing site.

At about 1300Z, the company's FAC, Major Randy Nash, noticed a squad-sized element of Iraqi Republican Guardsmen conducting a dismounted patrol just across on the west side of the river. Although he was perched on top of the LAV-C2, the trees were providing concealment and the squad was oblivious to his presence. He alerted Lieutenant St. John, who directed his team to call on the company's organic mortars for a fire mission to take out the enemy soldiers. Second Lieutenant Mark Reinhart, the Fire Support Team (FiST's) artillery Forward Observer (FO), did a quick map study and passed a grid to Corporal Clifford Walker, the mortar FO. Corporal Walker called in the fire mission and the accurate fires immediately dropped five of the Iraqi soldiers, while the remainder retreated toward the palm grove returning fire as they ran. Crewmen on the LAV-C2 quickly formed a fire team and opened fire with their rifles.

Rotary Wing CAS was critical to success throughout.

Sergeant Clayton Blankenship, the Vehicle Commander, picks up the story from there:

"With the XO's orders I moved to the ledge of the bluff over looking the enemy troops. I was given an order to only shoot once the mortar mission was complete, and if there was anything left moving. The troops had moved toward the bend of the river when the mortar rounds fell exactly where they stood. I thought to myself 'Corporal Walker, you just did some damn good shooting'. Soon after, several enemy soldiers ran to the top of the bend setting up perfect silhouettes. I estimated the range at about 320 meters. As I saw them advancing to the main tree line I didn't waste anytime before putting rounds down range. I noticed the first three impacts a bit low and adjusted my aim. As I fired again I watched one fall down immediately, then I waited for the next one to pop up. Meanwhile, Corporal Walker had called in another fire mission and joined me with his M-16 on the bluff. I noticed a few of the soldiers trying to get up; it looked as though they were injured. We observed fire coming back at us, so we put some more rounds down range on the enemy soldiers. It seemed we had taken 'em all out when a vehicle approached on the parallel road across the river. I heard the other LAV-25s light up the three individuals who had dismounted as well as the truck. While all that went on, we lay there watching the tree line across the river, but there was no movement. Then the last mortars fell making sure that no one was left alive."

At nearly the same time, a section of LAV-ATs (Anti-Tank variants), equipped with TOW missiles, was moving to the south in order to reinforce the engaged security element. The lead vehicle commanded by Sergeant Brian Siebert, observed a T-72 moving south on the opposite side of the river. Sergeant Siebert took aim with his thermal sight and destroyed the tank with a single TOW shot. The tank erupted in a huge fireball that continued to grow as the ammo inside cooked off. The AT section immediately stowed their turrets and continued south along the road as they continued to search for enemy across the Diyala. Suddenly, an F-14 Tomcat appeared overhead and flew right over the LAV section. First Lieutenant Cameron Albin, the platoon leader, took notice of the aircraft and inquired if the FiST was controlling CAS. The answer came back negative, and Major Nash tried desperately to contact the Tomcat's pilot. It appeared that the aircraft was posturing for an attack on friendly positions, and the jet aircraft roared by another time. Suddenly a tremendous explosion erupted directly between the two LAV-AT s, obscuring them from the rest of the company. Luckily, the vehicles had been traveling along a raised road with a berm off to the right side. The Navy F-14 had dropped a 500-pound bomb directly in between the vehicles, but on the other side of the berm. The section escaped with only a minor injury to Sergeant Siebert, who and immediately called back to the company alerting them of the incident and advising everyone to display their air panels. Major Nash finally reached a section of Marine F/A-18Ds on the TAD net and instructed them to announce on the emergency 'guard' frequency that a fratricide incident had just occurred between 1st LAR and an F-14. The Navy Tomcat disappeared and never returned.

Other units in RCT-5 also saw action that day. At first light, the battery advance parties of 2/11, led by the battalion survey section and with a platoon from Charlie Company, 1/5, moved through the "hornet's nest" to prepare PAs near Hatif Haiyawi. The party started to take fire from a date grove to the west of their positions. Hearing the small arms fire, the platoon from Charlie Company reoriented and engaged a BMP-2 as well as dismounted infantry threatening the artillery position areas. The security platoon requested artillery support to engage the Iraqis. The 2/11 forward CP received the call for fire, and issued a fire order to 5/11 to engage the BMP and dismounts that were only 800 meters away. Once the artillery started falling, an AT-4 was used to destroy the BMP and the resistance was quickly mopped up.

As would be the case for the next five days, most units on the east side of the Diyala received sporadic mortar and artillery fire in and around their positions. The enemy indirect fire was inaccurate, and Marines quickly assumed a degree of nonchalance on hearing the sound of explosions. In fact, Marines were more likely to jump at the sound of the Division's own guns, as 11th Marines attacked enemy firing units located by counter-battery radar with a vengeance. Sometimes, however, the Iraqi cannoneers and mortarmen got lucky:

On 5 April, the Corpsmen in the Field Train were called into action when two mortar rounds impacted, injuring two Marines. Chief Kelly Richardson was treating Lance Corporal Frederick Evans for shrapnel wounds to his foot and leg when Lance Corporal Evans began to go into shock. Chief Richardson maintained his composure and provided some comforting words to the wounded Marine: "I'll see you at the next Marine Corps Ball. You'll be easy to spot because you'll have all the ribbons and you'll be surrounded by all the women." Lance Corporal Evans and the other wounded Marine were successfully treated and evacuated.

Preparing to Enter the City

Selection of maneuver objectives was made easier because of the preparatory planning the Division had done in Kuwait.

In the southern portion of the zone, RCT-7 had a number of actions ongoing. Its main effort was reconnoitering the two crossing sites for the fight into Baghdad. Third Battalion, 4th Marines was in the lead, and cleared in zone to the southeastern outskirts of Baghdad. The Fedeyeen fighters continued to use ambush techniques against the Marines' advance. The Battalion continued to prosecute detected targets with CAS. The air officer for 3/4 got some 'non-routine' close air support while calling in on some vehicles across the river. A B-52 had just checked in to support the advance, and was pushed to the India Company FAC. The India FAC took control of 'Diecast 32' (the B-52) and set up a Mk-82 string. The total preparation time from check in of aircraft to ready for drop was approximately 17 minutes, and the B-52 dropped 17 Mk-82s on target. The ironies abounded in an aged platform built for a now-defunct strategic mission dropping modern weapons in support of an infantry company's close fight. Later that day, 3/4 was able to use the battalion's Dragon Eye UAV to reconnoiter forward to the two bridges crossing the Diyala in RCT-7's zone. Both bridges were, surprisingly, still intact. Lieutenant Colonel McCoy briefly considered pushing forward to seize the bridges there and then, but the RCT had been given a limit of advance short of the canal. Additionally, Bravo Company, 1st Tanks (attached to 3/4) came upon a compound with many high-tech machines covered in tarps and sandbags located outside. Suspicious of having discovered a WMD sensitive site, the attached infantry platoon cleared the compound. The research facility was completely abandoned, but had a large number of computers and precision machines. The battalion focused on securing this site, and had its hands full clearing out enemy resistance on the east bank of the river. It was late in the day, and attacking across the Diyala would have to be left until the next morning.

Marines from 1st Platoon, India Company, 3d Battalion, 4th Marines clear the town of Az Za'Faraniyah in preparation for seizing the southern Diyala bridge.

Salman Pak

The remainder of RCT-7 continued to conduct operations along Highway 6 to search out and destroy enemy resistance and ammunition caches. South of Highway 6 was an isolated peninsula formed by a bend of the Tigris River. The Regime had long used this peninsula, near the town of Salman Pak, to conduct a variety of activities in an isolated setting. There had been reports a few days earlier of a large concentration of Saddam Fedeyeen at the Salman Pak military facility. Lieutenant Colonel Mike Belcher's 3/7 was tasked to conduct a raid on the facility in order to eliminate the threat to the Division's southern flank. The facility housed a terrorist training camp, the II Republican Guard Forces Corps (RGFC) Headquarters, and a presidential palace. The battalion commander's plan called for India Company to secure the training camp and headquarters, while Lima Company secured the palace. Kilo Company would provide the reserve and Tanks would provide a blocking position to the north. Two combined teams of operatives from other US clandestine forces would support the Battalion's attack and intelligence collection efforts.

Because of the potential to catch a large number of enemy fighters in this cul-de-sac, 3/7's timeline was accelerated by five hours. The attack that was planned and briefed for 1800Z now had to commence no later than 1300Z due to the rapidly developing situation. Major Mike Samarov, the 3/7 Operations Officer, quickly contacted each of the battalion's elements to adjust the attack. Departing on the new timetable, Third platoon from Delta Company, 3[d] LAR Battalion, encountered a significant enemy ambush by well-entrenched Iraqi paramilitary soldiers. The Marines defeated the ambush through a combination of suppressive fires and maneuver. Several LAVs became mired in the muddy Tigris terrain, but were able to extract themselves while under enemy fire.

By 1430Z, with the battalion poised to strike, they were informed that the pre-planned unguided air strikes were cancelled. It was decided to strike the lengthy target list with artillery and observed CAS only, keeping effects of unguided fires as far as possible from the critical sites to be exploited for their intelligence value. Working feverishly during a 30-minute artillery preparation fire, the Battalion Fire Support Coordinator, Captain Daniel Schmitt, and the Air Officer, Captain David "Digger" Kulik, assembled six fixed wing and one rotary wing section from the available close air support stack. Briefing desired munitions points of impact in rapid succession, the FiST addressed all necessary targets to support the Battalion's scheme of maneuver. Supporting the efforts of the Battalion FiST cell, Captain Philip "PB" Smith and Major William "Wagon Burner" Dunn, forward air controllers

Marines detonate an arms cache containing Iraqi landmines and artillery ammunition.

for Companies India and Lima respectively, controlled fixed and rotary wing aviation on additional key targets. Captain Smith recounts:

"After the Hornets dropped their JDAMs, I had them run multiple passes with rockets. Good thing they came loaded with rocket pods, because it enabled us to hit targets of opportunity without the collateral damage to the surrounding buildings. It also afforded them a real good look at what we were about to go into, and I had a very comfortable feeling that we had destroyed everything that moved."

An enemy surface to air missile battery fired at the Cobras during one strike on their objective. The AH-1W helicopters pulled back and the artillery forward observer for Company L, First Lieutenant Walter Maessen, called in a battalion concentration of artillery. Once again, the effects of combined arms were devastating to the Iraqi defenders. The efficacy of surface fires, air fires, and maneuver all working together was repeatedly demonstrated, as the enemy SAM battery was knocked out and the cobras returned to the fight, unhindered by the anti-air threat.

The battalion's first elements broke through the compound gates at 1720Z that night. Mechanical difficulties hindered Lima's advance, but they eventually completed the attack. By 2000Z they secured the presidential palace. Little more than an hour later, the suspected terrorist training camp and headquarters were secured. Buses, trains, scuba equipment, and even an airplane mockup confirmed the area as a terrorist training camp. Intelligence exploitation of the sites was conducted, while the battalion concluded operations to secure remaining objectives. By 0300Z the next morning, 3/7 reported all objectives secure. Even though the reported enemy force had disappeared before the attack, the uncovering of the terrorist camp and a significant amount of Regime documentation marked the operation as a resounding success. Intelligence exploitation teams used trucks to haul away the voluminous documentation uncovered.

Division Shaping Efforts

Pushing fires across to the west side of the Diyala River, the Division worked to reduce the risk to RCT-7 in a shaping effort against Regime forces in the Rasheed military complex. The Rasheed area had been a suspected chemical weapons storage site, and was known to house a number of Regime helicopters at its airfield. Division Fires and the 11th Marines took it as their special mission to reduce this pocket of capability, and to eliminate the threat potentially posed by a Regime helicopter loaded with chemical munitions. From both the Division Forward and then the Division Main CPs, Division drove B-52, F/A-18, AH-1W, DPICM, and HE artillery missions against the enemy equipment and facilities observed at the airfield and scattered throughout the huge Rasheed military complex. The shaping fires were ultimately effective, and the Rasheed complex offered little resistance when uncovered two days later.

In an economy of force measure, the Division detached 2d LAR Battalion from RCT-1, and assigned it the security mission for the critical 150 km stretch of Highway 6 back to An Numaniyah and the Tigris River crossing. The fact that Barbarian was able to cover such an extended battlespace was testimony to the agility and flexibility of an LAV mounted unit. Imposing their will on lingering fighters all along this route, they also built trust and endeared themselves to the Iraqi population along the highway. During the accomplishment of this mission, Barbarian was to reduce the number of Regime holdouts along Highway 6, and intercept a number of terrorist bombers who had been enroute to southern Iraq.

Murals of Saddam were on nearly every street corner in Baghdad. The local Iraqis had defaced most of these prior to the arrival of the Marines.

The Division Main CP was established along Highway 6, just to the east of the Diyala Crossroads. The personnel of the Main immediately regretted this choice of CP locations, as it was in the middle of a slaughterhouse-rendering field. The blowing dust and dirt were mixed with animal hair and skin; and the ground was littered with animal bones. The environment motivated the Marines in the Division CP to bring swift support to the attack, as it would get them out of there quickly! The shaping and counterbattery fires went on for many hours. From the field, the Division watched the effects of the Division shaping fires first-hand. On those clear nights, the steady glow of the western horizon showed the way to Baghdad.

On the Division's far eastern flank, the Reconnaissance Battalion maintained a blocking position near the intersection of Highway 6 and Highway 27 oriented toward Al Kut. Only two days prior, the Reconnaissance Battalion had blocked the Baghdad Division from the south. Now, they had the same mission, only oriented a different direction, ensuring no threat to the Division's rear along Highway 6. The 200-km jaunt in-between was already receding into memory. By now, the Marines of the Reconnaissance Battalion had their fill of Al Kut, and were eager to get back to the fight at the front of the Division. Their efforts as an economy of force mission here, however, permitted the Division's flow of combat power forward. Having bought time for the Division to close Baghdad with its main assault elements, the battalion sent Alpha and Bravo companies west to link up with the Division Main CP near the Diyala River in anticipation of additional tasking.

The Division's push up Highway 6 continued to stress extended LOCs. The Hantush Highway Airstrip was the nearest KC-130 capable strip and the transfer point for much-needed supplies. The Hantush FARP was workable but required a 150 km round trip that included two crossings of the Tigris to support the Division movement into the suburbs of Baghdad. I MEF was busy establishing An Numaniyah Airfield as a logistics node but it was not yet fully operational. The Division had outraced the supporting logistics effort, and as a result, CSSG-11 had to reach back to 'pull' support forward. In one instance, 11th Marines downloaded their own logistics trains and made the drive south to retrieve critically needed artillery ammunition when it could not be pushed to them. In another instance, CSSG-11 searched unsuccessfully for MREs, ammunition, and repair parts. The G-4 made repeated requests to forego establishment of intermediate support nodes in order to push support as far forward as possible, but the support plan continued on its established timetable unaltered. The Division continued to live on 'pull-logistics.'

Assessing the situation, both CSSG-11 and 3d MAW responded magnificently to support the Division. On 4 April, CSSG-11 established a hasty Rapid Refueling Point near Sarabadi Airfield to provide fuel, food, and ammunition. Located along the Division's route of march on Highway 6, it provided easy access and ample storage space for the CSSG. For their part, the MAW evaluated the landing strips at both Sarabadi and Salman Pak East – an airstrip 10 km closer to Baghdad. The Wing chose to use Salman Pak East Airfield as their FARP site, and the MWSG rapidly closed in and expanded the field's capabilities in short order. These two sites satisfied the differing needs of elements of the MAGTF, and together greatly increased the support to the Division.

6 April
To the Banks of the Diyala

RCT-1 arrived along the Diyala eastern corridor and conducted a RIP with RCT-5 at the 'Dead General's Crossroads'. RCT-1 had made its original 200 km march from Highway 7 to An Numaniyah in 24 hours. They had made a second arduous march of 120 km from An Numaniyah to the Diyala crossroads. After covering all that distance, RCT-1 was eager to get back into the fight, and began planning for its own assault crossing of the Diyala. The RCT took its new battlespace south of RCT-5, and assumed responsibility for the cordon along the eastern side of the Diyala River, north of RCT-7. This move set the Division up for its planned course of action in Baghdad. RCT-7 and RCT-1 were poised to attack into Baghdad from the south and east, while RCT-5 had been relieved of its battlespace in the east so it could force a crossing further north and enter the city from the northeast. The entire Division was now in position for the coordinated attack into Baghdad.

As the Division worked to establish its cordon around the city, the 11th Marines continued to provide extensive counterbattery fires against Iraqi artillery firing from within the open areas of the city (stadiums, racetracks, roadways, military complexes) often co-located near noncombatants. In order to minimize collateral damage to noncombatants and civilian infrastructure, many of the radar-acquired targets were passed exclusively to aviation for engagement with precision munitions using the 'quickfire' procedures the Division had practiced back in Camp Pendleton. All targets were vetted with high-resolution imagery to check for potential collateral damage prior to conducting counterfire missions. Artillery targets became fewer and fewer as the 11th Marines gained fires dominance over any would-be Iraqi indirect fire capability, and infantry units overran enemy firing positions.

RCT-5 continued its exhaustive reconnaissance along the banks of the Diyala north of the existing bridges, but found no viable alternatives, except for the Highway Bridge at Baqubah, 50 km to the north. Even the promising site found the previous day by 1st LAR Battalion proved unsupportable. The difficulty was not the river itself, but the miserable terrain on both sides of it. The area along the Diyala River was very muddy, and was crossed with drainage ditches, canals, and muddy fields. Individual small vehicles in small numbers could make their way to the river's edge using the network of muddy canal roads, but any larger vehicles would not make it. The bridging could be done, but building a road to service the bridge on both sides of the river would eliminate it as a timely option. The maneuver elements might get across, but the RCT commander did not want to get stranded on the far side of the river with no combat trains or logistics support in what could be a logistically intensive fight. Using the existing bridge in Baqubah would require a round trip distance detour of over 100 km, and would depend on securing the bridge intact. It appeared as if RCT-5 would have to use one of the bridges in the south to get across the Diyala.

The Division began to focus intently on the two southern bridges over the Diyala, as these appeared to offer the only intact crossing points into the city. Both bridges were in RCT-7's zone, and 3/4 continued their reconnaissance in force along the eastern banks of the river. The resistance in the suburban area on the nearside approaches to both bridges would prove significant.

Scoping the Diyala Bridges

The town of Az Zafaraniyah was the scene of intensive small arms and RPG fire against the advancing Marines of 3/4. Initially the movement was clear, as CAAT 1 and Kilo Company moved along the road on the southwestern side of the town. With Kilo in place to provide

After the Fedeyeen defenders were driven out, Iraqi citizens welcome the Marines.

overwatch, India Company moved up, dismounted, and began a limited clearing operation in the urban area. It was then that the paramilitary forces in the town initiated their ambush. Numerous RPGs flew at the Marines, but no one suffered a direct hit. The intense exchange of fire lasted for some time, until a well-aimed tank main gun round into the building sheltering the enemy fighters turned the tide. By pushing into the town approximately three blocks on the flanks of the route, India established a buffer zone between urban threats and the road.

As the battalion secured the east side of the canal just short of the bridge, enemy infantry and mortars were sighted across the river, and gunfire was exchanged across the canal. The battalion's Dragon Eye UAV was used to confirm an intelligence report of possible damage to the southern bridge, allowing the battalion's Obstacle Clearing Detachment (OCD) to conduct a more detailed assessment. They found that both of the Diyala bridges had been damaged overnight. The Iraqis had taken advantage of the brief respite, attempting to destroy the two existing bridges, successfully damaging both. The Division would now have to conduct an amphibious assault across the river and build its own bridges for follow-on forces.

The Dragon Eye UAV provided a view of the damaged Southern Bridge over the Diyala River.

As Marines assessed the damage sustained at the southernmost Diyala Bridge, Lieutenant Colonel McCoy decided to move to the river and view the bridge himself. He needed to assess if the gap created by the Iraqi's demolition attempts was narrow enough to support the use of an Armored Vehicle Launched Bridge (AVLB) attached to Bravo Company. Throughout the day, the battalion had received intense but inaccurate AK-47 and RPG fire from across the river. Advancing alone towards the bank of the river, Lieutenant Colonel McCoy quickly began his assessment of the gap and ran through the capabilities he had to address the problem. As he waited in the prone he was surprised to look towards the river to find a French photographer in bright civilian clothes right next to him taking his picture. Initially perturbed about having visitors in civilian clothes drawing fire towards his position, he revised his assessment when he realized that the photographer was between him and the incoming fires. After that, the photographer's bright clothes were a welcome addition to the Diyala River landscape as the commander conducted his leader's reconnaissance.

Later that afternoon, the Division engineer was sent forward to assess the supportability of a Division crossing. The Iraqis had made an attempt to destroy both of the bridges crossing the Diyala Canal. The engineers discovered that the southern bridge had an entire span blown out.

The bridge was damaged beyond repair, and the damage was too wide to span with an AVLB. At the northern site, the Iraqis had blown a hole in the far span of the bridge, and the entire western end of the bridge was sagging. It would require a detailed engineer assessment, but neither crossing looked good. With the engineer reconnaissance that RCT-5 had conducted also coming up empty, the Division had gone from three planned crossings to zero in 24 hours. The Division had crossed oceans and deserts to get to Iraq, had come nearly 500 km into Iraq over two 'big name' rivers, and was now challenged by a ribbon of water that most barely knew the name of.

After being in a heightened chemical defensive posture since before crossing the LD three weeks prior, the Division directed all elements to transition to MOPP Level 0 on 6 April. For the first time in three weeks, Marines would be able to wear something other than their now filthy MOPP suits.

An abandoned Iraqi T-72 hidden under the palm trees near the Diyala River.

Lieutenant Colonel Chris Conlin's 1/7 moved into a suspected sensitive site about 8 km southeast of Baghdad, the Tuwaitha nuclear facility. It was completely unoccupied and had been partially looted, but was obviously extremely important to the Regime. The facility was encircled by sixty-foot high berms, with bunkers evenly spaced, surrounding the facility. Tuwaitha was the site of the Iraqi atomic energy facility. UN weapons inspectors had recently inspected the site and there was some residual concern about exposure to chemical agents and radiation. The battalion's NBC officer, Chief Warrant Officer Carl Hinson, ran a series of tests to determine that all levels were normal. Elements of 1/7 stayed in the facility overnight waiting for the order to move north. The battalion was relieved of the facility in the morning by Marines from the Combat Engineer Battalion.

A Nighttime Assessment of the Diyala Footbridge

In the dark, 3/4 tasked Kilo Company with clearing and supporting the southern bridge approaches. Luckily, there was a damaged footbridge next to the main bridge, that the engineers of Charlie Company, 1st Combat Engineer Battalion decided could be used with some improvisation. Second Lieutenant Daniel Francis and Sergeant Mauricio Guevaracano led the engineer platoon attached to 3/4. As the Marines approached the river, the enemy began fiercely defending the bridges with a high volume of machine gun, RPG, mortar, and artillery fires. Sergeant Guevaracano's squad was one of the first units at the bridge, and they immediately established fighting positions and returned fire. Upon securing the near bank, Sergeant Guevaracano organized a small bridge reconnaissance team and moved out onto the footbridge to conduct a night reconnaissance. With no cover and being highlighted by flares, he carefully gathered as much information as possible on the bridges before returning to his defensive position. He discovered that the 20-foot gap in the footbridge could be covered with improvised materials from the riverbank. This would allow dismounted infantry to cross the river at the southern bridge site.

Abandoned Iraqi surface-to-air-missiles were found throughout the battlespace. The poor condition of most of these weapons made them unusable.

Crossing dismounted infantry was one thing, but the RCT staff huddled to plan how to project the rest of the RCT's combat capability across the Diyala. RCT-7 had been reinforced with two Bridge Companies, and the engineers assisted in developing a plan that would get the most combat power across the river quickly. It was decided that 1st Tanks would conduct a feint at the northern bridge site, located seven km up-river from the southern bridge. Engineers would conduct an explosive breach of a mine field that lay at the base of the bridge, and the Tank Battalion would drive an AVLB forward to span the damaged section. Meanwhile, 3/4 would continue to flow dismounted forces across the southern footbridge, while the engineers constructed a ribbon bridge a few hundred meters to the north. Once armor crossed the ribbon bridge to reinforce the dismounted Marines, a more permanent Medium Girder Bridge (MGB) would be built over the southern bridge's damaged span. To enable the emplacement of the MGB, 3/4 would conduct an amphibious assault crossing to seize the far bank. LD was set for 0300Z, 7 April. RCT-7 would be conducting the first combat assault crossing operation in this fight.

Intelligence from multiple sources indicated the enemy was trying to concentrate reinforcements and indirect fire assets opposite RCT-7's planned crossing points. To support Colonel Hummer's attack, Colonel John Toolan of RCT-1 proposed a supporting attack that included an assault crossing of the Diyala on RCT-7's northern flank. Newly arrived, short on food, but full of motivation, RCT-1 would use their AAV's to swim across. They would thrust into the city on the northern flank and draw pressure off the Division's main effort, RCT-7. The Division CG quickly approved the plan.

In for a Penny

The Division expanded its planning for the operation in Baghdad. If there was to be an assault river crossing operation into the City, then there no longer appeared to be a severable line between establishing a cordon and maneuvering forces into the city. The CG tasked the G-2 with coming up with the objectives for a Division attack into the city, and the G-2 produced a list of 27 Regime targets, gleaned from the lists of SSE sites, Regime strongpoints, government offices, media centers, and recent intelligence reporting. Securing each of these sites would demonstrate the Coalition's intent to disarm the Regime and liberate the populace. In preparation for the attack, map packages on these 27 targets were printed and disseminated. If the Division was going into the city, it would at least be going in with a sense of what it needed to accomplish. A large number of these objectives were located in the Rasheed military complex just over the river from RCT-7. As a 'middle ground' between establishing the cordon and actually securing eastern Baghdad, seizing the Rasheed complex could be considered part of an outer cordon that offered a superb launch platform for continued attacks or raids into the city. The Division established objectives in the Rasheed complex for the initial attack by RCT-7 in the south, and defined the perimeter road around the complex as a limit of advance. Forcing the door open a crack at Rasheed would allow the rest of the Division to flow through and complete the cordon. The Division effected boundary adjustments to afford all 3 RCTs battlespace east of the Diyala so that they could effectively employ indirect fires in the coming assault.

The Division's flanks were secured by the LAR battalions, with 3^d LAR Battalion screening the northeastern flank of the Division battlespace to protect against counterattacks by remnants of the Al Nida Division. Third LAR conducted a RIP with RCT-1, enabling the RCT to continue the attack into Baghdad with all of its combat power. The Wolfpack zone now extended for nearly 30 km, as it established its screen to the east. The Battalion continued to screen to the east until ordered to join Task Force Tripoli on 12 April. Second LAR Battalion continued to provide LOC security from An Numaniyah to the Diyala crossroads, aggressively keeping the MSR open for transportation of critically needed supplies.

Far to the south and east, members of the Division were fighting another battle against the terrain. On 3 April, four M1A1 tanks and one M88 tank retriever had become mired in the muck on the south bank of the Tigris across from Al Aziziyah. Two tanks had pulled into what they thought was an open field to set in for the night. When they went to move the next morning, they found themselves mired in mud, completely unable to move. The M88 retriever sent in to pull them out suffered the same fate. A third M1A1 was a mobility kill due to enemy fire and the last inoperable due to transmission failure. RCT-5 had evacuated the crews, passed grids for recovery to CSSC-115, and moved up Highway 6 to continue to press the enemy. Major Greg Frich, the Division Ordnance Readiness Officer, led a recovery team and security force to the site on 6 April. By the time they arrived, the tanks and retriever that were stuck in the field had slowly sunk to the level of their turrets, and the mobility kill tank had been pushed off the road into the Tigris. Master Sergeant Mike Hamlett, Staff Sergeant Shane Perry, and Staff Sergeant Kevin Jones drew on years of experience to solve the problem. Trees felled from the surrounding area were laid across the field to support the CSSB-10 retrievers. Sergeant Michael Zutz and Corporal Jesse Pogodzinski used excavators to clear away dirt and build earthen ramps for additional support. The light enemy activity in the northwest tree line was quickly resolved with some medium machinegun fire. The first retriever attached a cable to the tank while the second retriever attached a cable to the first in daisy chain fashion. As the first retriever began to sink pulling out the tank, the second pulled it out. At the end of the first day, the first M1A1 was successfully extracted. Using tow bars on the South River Road, the team returned to the security of Captain Erich Krumrei's 8th ESB perimeter near the Pearl 3 crossing at An Numaniyah. The second and third days would see the same operation – starting well before first light, continuing through the day, and returning 45 miles to the secure perimeter at night – under the watchful eyes of the security team led by Sergeant William Roper from CSSB-10's MP detachment. In five days, the 16-Marine team salvaged all four M1A1s and the M88 that had been reported as unrecoverable. Even the Mesopotamian marshes that had denied attacking armies for millennia could not stand up to the resourcefulness and stubbornness of the United States Marines.

Recovering these M-88 and M1A1s from the unforgiving Iraqi muck south of the Tigris proved a tremendous challenge.

7 April
One River, Three Attacks

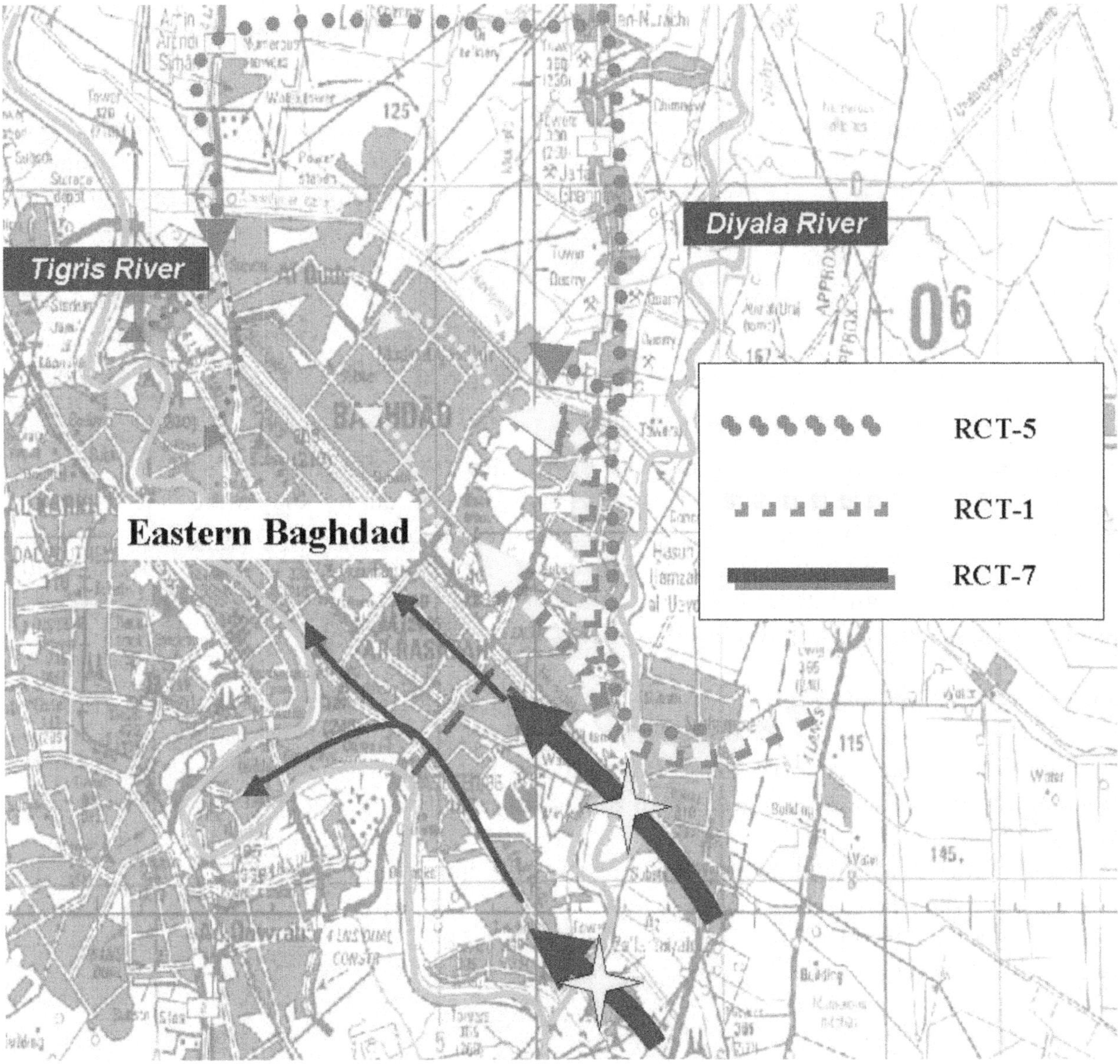

RCT-1 and RCT-7 would attack straight into Baghdad while RCT-5 completed the cordon around the north side of the city. Once the cordon was complete, all three RCTs would advance into the city.

At 0300Z, the Division attacked to seize the bridgeheads at the north and south bridges over the Diyala. RCT-7 would lead the attack as the Division's main effort. Within RCT-7, 8th Engineer Support Battalion was the main effort for bridging operations. RCT-1 would conduct its own assault river crossing later in the day on the northern flank of RCT-7. RCT-5 would follow in trace on 8 April, and move to establish the northern cordon around Baghdad. The combat bridging operation incorporated a full fire support plan that used combined arms to permit the crossing at minimal risk to the assault elements.

A BMP- 2 burns on the east bank of the Diyala River.

Near the southern bridge, the first tasks included putting infantry across the river to secure the far side, and identifying the specific site where the ribbon bridge would be put in. The Division conducted the southern crossing operation under both direct and indirect fire from the enemy. Intelligence reports indicated that the enemy commander continually demanded reinforcements, and was continually denied. The enemy had the bridges under observation of their Forward Observers, however, and brought artillery and mortar fire to bear on the crossing sites from firing positions in the Rasheed complex. Although the Iraqi indirect fire was generally inaccurate, they were able to hit one Marine AAV by massing their fires at the crossing site. Two Marines, Corporal Jesus Medellin and Lance Corporal Andrew Aviles were killed by the massed artillery fires, both by an enemy artillery round that struck their AAV. Twelve other Marines were wounded as 3/4 moved into position to attack across the river into Baghdad.

The bravery of Marines like Corporal Medellin and Lance Corporal Aviles carried the Division across the Diyala River and into the enemy's capitol. The Marines gained fire superiority over the enemy, enabling the engineers to make an attempt at crossing the still-standing footbridge. Sergeant Guevaracano again led his squad out onto the southern footbridge, with no cover and amidst a massive volume of friendly and enemy fire. He moved his squad to the gap in the bridge and began spanning it with improvised materials, coordinating the movement of his Marines without regard to his own safety. In less than two minutes, the gap was spanned and prepared for use. It didn't look pretty, but it accomplished its purpose and allowed two companies of 3/4 to initiate their assault across the river. Crossing the two dismounted companies via the footbridge, 3/4 quickly secured the areas suitable for an improvised crossing site and dug in not far from the bridge on the enemy side.

A medium girder bridge is eventually used to span the damaged Southern bridge over the Diyala River.

It had been reported that the banks of the river just to the north of the main bridge looked like they would support this operation, and the engineers from the Army's 65th Multi-Role Bridge Company (MRBC) quickly confirmed this. By 0630Z, the ribbon bridge operation was beginning. The engineers and heavy equipment operators of CEB did the near side bank preparation, while the 65th MRBC constructed the first raft. Once the raft was completed, the first vehicle sent across was a D-7 dozer. The operator, Private Benjamin Krueger, heroically continued to develop the far shore while under mortar fire. CEB engineers continued with the preparation of the near side approach, while the MRBC finished assembling the actual pontoon bridge. The engineers were under the eyes and guns of any remaining enemy throughout their operation. The bridge engineers smoothly executed the operation, shrugging off sporadic enemy fire as they worked. The first tanks from 3/4 crossed at approximately 1200Z.

The Iraqis had caused more problems for the Marines at the damaged southern bridge because they had done such a poor job trying to destroy it. While they had successfully dropped the middle span, they had damaged the adjacent span in such a way that the engineers would have to drop that span as well before they could start work on the MGB. The engineers' first attempt to complete the job with explosives did not work. The span dropped down only another couple of feet. The engineers from Charlie Company ventured out into the river in small boats and tried again. On the second attempt, they blew the stanchions down closer to the water. This attempt was successful and the span dropped cleanly out of the way. The Marines of 8th ESB quickly went to work building the MGB over the gap. The ribbon bridge would support the assault forces, but the heavier MGB was required to support long-term sustainment and civilian traffic into the city. By 1430Z, the ribbon bridge was complete, and 3/4 spent the night pushing across its armor to provide support for the two dismounted companies on the far bank. It would not be until the next day that the hard-working Marines and soldiers would have the MGB complete, but it soon became the best bridge into eastern Baghdad.

The footbridge (left) and medium girder bridge (right) facilitate the crossing of Marines and equipment over the Diyala River shortly after RCT-7's assault.

Seven kilometers away on the northern bridge, RCT-7 had a different set of challenges and the Combat Engineers had a chance to earn their hazardous duty pay as well. Just to get to the bridge, they had to breach two surface laid minefields on the approaches. They accomplished this using Mine Clearing Line Charges (MCLCs). Under the suppressive fires by elements of 1st Tank Battalion, The AAV with the line-charge pushed forward through enemy fire to get into position. On the first attempt, the line charge did not detonate (a common occurrence.) An engineer ran out under fire to re-set the detonator and manually blow the charge. Even after these heroics, the engineers' work was not complete. Once the way had been opened to the bridge, engineers were required to determine if the bridge was wired with explosives. Although a cursory examination of the exterior had been done, a detailed analysis of the interior portions of the bridge was required before tanks could be risked across. This was accomplished by sending one engineer through each of the service tunnels that ran along the sides of the bridge. Crawling through the pipes on their hands and knees, Corporal Cobian and Lance Corporal Kinney from Charlie Company, Combat Engineer Battalion, cleared up to the point of the damaged westernmost span. They crawled back and reported that there were no explosives remaining. Next, the engineer officer, Captain Duncan 'Buck' Buchanan, proceeded to the damaged portion and 'stuck his head in the hole' to determine the extent of the structural damage. Using an M1A1 for cover, the engineers crept their way up to the damaged portion. Upon reaching it and getting a closer look, Captain Buchanan and Lieutenant Paul Bock saw that the Iraqis again had done a poor job. They had blown the stringers along the sides of the bridge, but those in the middle (the most important ones) had been left intact. The engineers informed Lieutenant Colonel Chartier, the 1st Tank Battalion commander, that an AVLB could be used to breach the gap.

The AVLB was emplaced, and the first tanks crossed the river, only to discover another minefield 200 meters on the far side of the bridge. Again, the engineers were called up and fired a MCLC. When the line charge misfired, Lance Corporal Bianchimano got out of his AAV and ran up under heavy enemy fire to manually ignite it. While the detonation cleared many mines, there was still nearly 100 meters of minefield in front of them. Having only two remaining MCLCs that might be needed later, they began destroying the mines with bangalore torpedoes. Corporal Cobian and Lance Corporals Noonan, Kirkpatrick, and Warpness, each rushed forward into the minefield with a bangalore torpedo. While taking enemy fire, they emplaced the explosives then returned to the AAV. A lane was cleared through the minefield, but it was still too narrow to pass tanks. Seeing this, Sergeant Lauritzen dismounted and led his squad in picking up and moving the mines by hand. Once all the mines were cleared, 1st Tank Battalion continued the attack into Baghdad.

Two days later, when Army engineers came to complete repairs to the bridge, they found that the entire bridge had indeed been wired for destruction. The Iraqis had wired it in such a way, however, that when the first span blew, it had cut the wires to the remainder of the bridge. When the wires were traced back to the far side of the river, a dead Iraqi with a plunger was found. It had been a close call.

Back at Ripper's Tactical CP, events seemed to be unfolding as planned. After news was received that the northern bridge could be made functional with the use of an AVLB, it was decided to convert the feint into an attack, and force a crossing on that bridge as well. Throughout the night of 7 April, and into the next morning, 1st Tanks and 3/4 continued to flow across the river and into the Rasheed area of eastern Baghdad over both bridges. Both battalions began expanding their bridgeheads on the far side of the river as the forces became available. RCT-7 found that once the initial defenses had been breached, enemy resistance began to decline. Resistance was stiff in a few isolated areas, but it was beginning to collapse.

A steady stream of supporting aircraft had enabled the successful crossing at the two sites. On this day, RCT-7 received fantastic rotary wing air support, just as it had throughout the campaign. The support of the outstanding Marines of HMLA-169, the 'Vipers', was noted frequently in Regimental accounts. The Vipers were a combat multiplier for the RCT and regularly went above and beyond what was asked of them, earning the respect of their comrades on the ground. Likewise, the CASEVAC helicopters provided by the Purple Foxes of HMM-364 were indispensable.

An AAV crosses into Baghdad over the northern bridge over the Diyala. An AVLB was just able to span the damaged section of the bridge. Marines cleared minefields on both sides of the bridge before using it.

Marines from RCT-1 demonstrate their amphibious capability over 600 km from the sea.

As RCT-7 crossed and entered the Rasheed complex, RCT-1 pushed to the Diyala River. RCT-1's battlespace was to the northeast of the RCT-7 battlespace, and Colonel Toolan was eager to get moving. The Pioneer UAV identified a partially completed bridge only 300 meters further north of the northern-most RCT-7 bridge, and the site appeared suitable for an assault crossing and an additional ribbon bridge. CAAT B Platoon, led by First Lieutenant Oscar Rodriguez, scouted a route through the farmers' fields and led Alpha Company and the forward command group to the road that ran parallel to the river along the east side. Once at the road, CAAT B and Alpha Company cleared the length required to conduct a reconnaissance of the crossing site. First Lieutenant Gerry Roeder led an engineer reconnaissance detachment augmented by two CAAT vehicles, and led his team to the site. As the reconnaissance was underway, Lieutenant Rodriguez's Marines spotted armed personnel across the river. When the enemy suddenly appeared and manned a recoilless rifle positioned across the river, CAAT B engaged them. CAAT B's forward observer began to call indirect fire on the objective, while the team provided suppression with .50 caliber and Mk-19 machineguns.

As fires were exchanged across the river 500 meters north of their position, Lieutenant Roeder and Corporal Brian Goff stripped down to their 'green on green' shorts and tee shirts. Armed with only their pistols, the two combat engineers swam 150 meters across the river. Iraqis farmers tended to their water buffalo only 200 meters away. As the mortars landed on CAAT B's intended targets to their north, First Lieutenant Roeder and Corporal Goff identified fighting positions and bunkers that had been recently abandoned by the enemy. The engineers made quick work of the far side survey and swam the span of the river to rejoin the waiting CAAT Marines. These two stalwart Marines had confirmed the only viable crossing site along miles of river line. This site would soon be exploited by RCT-1 to open a second advance on Baghdad.

AAVs cross the Diyala River during RCT-1's historic amphibious assault into the heart of Baghdad.

Colonel Toolan informed the CG of the crossing site and was given approval to get RCT-1 across the river to press Baghdad from two axes of advance. As a section of CAAT B guided the amphibious vehicles to the crossing site, a squad from CAAT B was sent to link up with Lima Company, 3/1, south of the crossing site to guide 3/1's tracks to the crossing site. As the link-up was affected, a BMP with uniformed enemy personnel around the vehicle was spotted through a grove of palm trees approximately one kilometer west of the company's position. Sergeant David Kaufer, CAAT B's TOW sniper, was ordered to destroy the target. Sergeant Kaufer skillfully threaded the missile through the grove and destroyed the BMP; another victory for the TOW snipers.

As the mechanized vehicles were moving towards the crossing site, the AAV company commander, Major Nichols, informed 1/4's commander, Lieutenant Colonel John Mayer, that the sea-worthiness of the AAVs was unknown. Having received the vehicles three days before moving to the dispersal area, the Marines had not had time to confirm the integrity of the seals and bilge pumps. Additionally, 1/4's combat trains and CAAT vehicles would not be able to swim the river, meaning the assault elements might be separated from their combat support for an unknown amount of time against a desperate, cornered enemy. Colonel Toolan made the decision to swim the AAVs across the river with the top hatches open (to allow a quick escape if they foundered before reaching the other side of the river.) At 1100Z, the first vehicle from Alpha Company was 'feet wet'. The Marines cheered as each vehicle swam across the river. As 1/4 continued inland to expand the beachhead line, it encountered sporadic small arms fire. The enemy began to pull back in the face of the assault, and now the Division had two fronts from which to pressure the capitol. Third Battalion, 1st Marines, crossed the river and passed through 1/4's lines, heading south to secure the far side of the northern Diyala Bridge.

At the northern bridge, 8th ESB had been reinforced with an Army Multi-Role Bridge Company (MRBC) from the West Virginia National Guard, and they quickly emplaced the new assault bridge under enemy fire. The Division coordinated this support, and soon RCT-1 was moving its wheeled equipment across using both the northern Diyala Bridge (now vacated by RCT-7) and the new assault bridge. RCT-1's forces consolidated along the river road that ran along the west bank of the River. There were extensive palm groves along the banks of the river, which the Iraqi defenders had used for cover and concealment. After completing their assault and routing the last of these defenders, RCT-1 set up their own bivouacs in the same groves. There were abandoned weapons and equipment everywhere, ranging from bayonets to SA-3 missiles loaded on trailers. The Iraqis that had chosen to fight had been destroyed, and the amount of wreckage lying across the battlefield on the west side of the river was an indication of the effectiveness of US shaping fires. As the AAVs splashed across the river, they captured a prisoner who gave an indication why resistance had faltered, *"when we saw the 'tanks' floating across the river, we knew we could not win against the Americans."* The amphibious capabilities of the Marines were breaking the enemy's will and winning battles almost 600 km from the sea.

In a hastily arranged conference, commanders of RCT-1, RCT-5, and 1/11 map out their plan for their continued attack into the Baghdad urban area.

8 April
Closing the Door on Eastern Baghdad

On the morning of 8 April, the CG crossed over the river to assess the situation. Meeting with Colonel Dunford and Colonel Toolan, the CG reviewed the situation on the ground. The CG spread a map over the hood of Colonel Dunford's HMMWV. Behind him, an Iraqi herdsman and his water buffalo waded in the Diyala River, while a few curious civilians observed from the high banks. Iraqi flags still flew over most neighborhoods, and there was still significant sniper fire coming from the houses and buildings to the northwest. RCT-7 had pushed into the Rasheed military complex, and was preparing to seize the initial objectives they had been given. RCT-7's limit of advance had been set in compliance with the CFLCC order to establish the Baghdad cordon, but not to attack into the city itself. There were enough military objectives on Rasheed to keep RCT-7 occupied for most of the day, while the rest of the Division worked to complete the cordon mandated by higher. The Rasheed complex also offered a secure area from which to launch further operations, as it was a large military facility bounded by water on three sides, and a major highway on the fourth. 'Rasheed Island' the staff had called it, and it jutted like a dagger toward the jugular of downtown Baghdad. Bisecting east Baghdad just to the north of Rasheed was the Army Canal. This waterway now served as a regimental boundary between RCT-1 and RCT-7. On the north side of the Army Canal, RCT-1 was prepared to push into the city on the northeast flank of RCT-7. If RCT-1 were to push forward for about 4 km, they would uncover Highway 5, the major route into and out of the city from the Baqubah area. Beyond this terrain feature, there was an area of former Regime administration buildings, including several military and intelligence service compounds. Since the envelopment of Baghdad by crossing the Diyala north of the city was no longer feasible, the CG decided to bring RCT-5 up over the same bridges that RCT-1 had used. The CG told Colonel Toolan to take his boys and push to the Highway first. "You need to push up to Highway 5 so that Grizzly can get behind you and get around to the north of the city." The results of the RCT-7 and RCT-1 attacks would put the two forces roughly on line in the vicinity of Highway 5, and Phase Line Andrew was established there as a limit of advance. RCT-5 would pass behind RCT-1 on the west bank, and move to cordon off the northern escape routes from eastern Baghdad. CFLCC did not want to push into the city until the cordon was complete, including a physical link-up between Army and Marine Corps forces at both the north and south sides of the city. Forces were already on the move to accomplish this task, and the CG wanted to maintain the momentum for an attack into eastern Baghdad.

The ever-present 3d MAW conducts air patrols over Baghdad.

Until RCT-5 could maneuver north of Baghdad on the ground, 3[d] MAW would play a key role in cordoning off the city. Between the Tigris and the Diyala Rivers, the two main highways leading out of Baghdad were Highway 2, which ran along the Tigris, and Highway 5, which ran along the Diyala. While the Division maneuvered into position to block both of these routes on the ground, 3[d] MAW patrolled them from the air. Their actions would ensure that no military reinforcements would be able to enter the city from northern Iraq, and that no military hardware would be able to leave the city to establish a defense in the Tikrit area. Third MAW would be relieved of their post only on 9 April, when RCT-5 was able to complete its maneuver into position across Highway 2.

Ripper Takes Rasheed Island

A hasty conference with RCT-7 near the Tuwaitha nuclear research facility.

RCT-7 began to prosecute the objectives inside the Rasheed complex. 'Rasheed Island' was a sprawling military and industrial complex approximately 6 km on a side. It was bounded by the Tigris River on the west, the Diyala River on the east and south, and isolated by Highway 5 and a perimeter wall/fence to the north. The complex included an airfield, which had contained the Iraqi's alleged suicide helicopter fleet (and potentially chemical weapons.) In addition to the airfield and the nearby military garrison, there were munitions factories, military storage facilities, a Special Security Office (SSO) prison, a medical complex, and several military prisons which had reportedly held US POWs at one time or another. The Rasheed complex had been the subject of a tremendous shaping effort, and now little conventional resistance remained. Identifying and securing these various sites on the ground would prove a time consuming task.

Throughout the day, RCT-7 continued to flow across the river, with 1/7 moving across the southern bridge and 3/7 following 1st Tanks over the northern bridge. After crossing the bridge, 1/7 oriented and attacked to the southwest, clearing that corner of the Rasheed complex, while 3/4 continued to drive northwest to the limits of the zone. First Tanks attacked and seized the Al Rasheed airfield, and 3/7 passed through them to attack along the northern flank of the RCT's zone to the edge of the complex. There, the battalion cleared the Rasheed Medical Facility, which reportedly was used to house American POWs. Upon arrival at the facility, 3/7 found no American prisoners, but found uniform items that belonged to US soldiers captured by the Iraqis early in the war. Continuing the attack, 3/7 cleared another suspected SSO facility on Rasheed. The battalion found no American POWs here either, but did find a semi-trailer full of decomposing Iraqi bodies. The apparent victims of the Regime each had at least one bullet hole in the head.

Horses from the Rasheed Military Stables run through the streets of Baghdad.

By the evening, RCT-7 had secured the entire Rasheed complex, and was prepared to continue the attack into the rest of Baghdad. The RCT-7 Main and TAC were now collocated at the Rasheed Airfield. Upon further inspection of its area upon occupation, RCT-7 discovered hundreds of crates of 122mm rocket ammunition buried under 15 foot high dirt mounds. The S2, Major Ron Spears, found top secret communications documents in the back of a disabled communications truck. At the south bridge, 8th ESB completed the Medium Girder Bridge, and the Division Forward CP moved up to Rasheed, setting up in the stadium of a Fedeyeen training facility, and preparing to take control of the attack into Baghdad proper.

Engineers emplace a ribbon bridge over the Diyala River to speed the attack.

Near the Tuwaitha nuclear facility, recently vacated by 1/7 and turned over to the Combat Engineer Battalion, Marines investigated an Iraqi 'yellowcake' site and found it wide open, with extremely high, unsafe radiation readings. Yellowcake is the raw form of uranium, and a great deal of it had been stored at this facility. In areas not yet under the control of the RCT, local civilians had looted the storage facilities and taken barrels of unenriched uranium away (just for the metal barrels themselves.) Recognizing the obvious health hazard to local citizens who did not understand the radiation threat, the Combat Engineers closed off the site and marked it with NATO hazardous area signs. A team was called forward to conduct detailed site exploitation and to handle the risk posed. Both the yellowcake and Tuwaitha facilities were secured until they were turned over to the relieving force, the US 2d Armored Cavalry Regiment.

Inchon Is In

While RCT-7 attacked through the sparsely populated military-industrial complex, RCT-1 began the difficult task of pushing through the dense urban neighborhoods of eastern Baghdad. RCT-1 continued to make slow progress through the course of the day, meeting moderate resistance from snipers and mortar attacks. The remaining resistance in zone was concentrated in schools and mosques, complicating the targeting process. The amount of mortar fire remained high, as 11th Marines registered 39 counter-battery radar acquisitions. Of these, most were passed to the DASC for aviation attack, or were not fired at all. At this point in the fight, returning unobserved indirect fire into the urban area was no longer necessary, and the risk of Iraqi civilians becoming injured or killed was too great. Once again, the Marines quickly adjusted their engagement criteria, sparing innocent lives in the process. The enemy's reckless, but generally ineffective, indirect fires out of and into civilian neighborhoods did not goad the Marines into returning fire where it could be avoided. Division Marines, operating in accordance with their 'No better friend, no worse enemy' credo, gave the Iraqi people every reason to welcome Coalition forces, not fight them. The superior fire discipline practiced by the Marines carefully discriminated between enemy and innocents, demonstrating to the Iraqi people that the real enemy was the Regime and those loyal to it.

Marines press the attack into the urban areas of Baghdad.

Once both RCT-1 and RCT-7 were across the river, the main effort switched to the Inchon Marines as they attacked to create enough maneuver space for both themselves and RCT-5. The RCT attacked with two battalions abreast, with 1/4 attacking northwest along the Army Canal and 3/1 attacking on 1/4's right flank. The two-battalion front extended the RCT zone all the way to the abandoned canal on the outskirts of Baghdad. First Battalion was tasked to seize the Highway 5 bridge across the Abandoned Canal. Bravo Company, 1/4, under Captain Paul Gomez, was ordered to secure the bridge to deny access into the zone. Alpha Company, under Captain Chris Griffin, attacked abreast of Bravo as a supporting effort to clear in zone to the Abandoned Canal. Charlie Company, 1/4, maintained security at the river-crossing site.

As Bravo Company approached the bridge, they started to receive sporadic small arms fire. The lead platoon crossed the bridge and soon realized that the company had overshot its objective area and was in Grizzly's zone. The enemy waited until almost the entire company had crossed the bridge before initiating a coordinated ambush with machinegun and RPG fire. At least five RPGs were fired, along with a heavy volume of machinegun fire. First Lieutenant Dallas Shaw, Bravo Company Executive Officer, made a critical decision to turn his AAV into the rear of the ambush instead of back across the bridge. There, he found several Fedeyeen preparing to assault the rear of the company column. He immediately charged, leading the Marines in his AAV against this enemy force, killing at least three and disrupting the would-be ambush, and the company was able to cross back over the bridge into its proper objective area. During this contact, Corporal Christopher Linck, a sniper attached to Bravo Company, was struck in the neck by gunfire as he was suppressing the enemy located in the buildings. Alpha Company dispatched a platoon to locate a suitable LZ to support a CASEVAC, and helicopters made their way into the zone.

Meanwhile, the platoon blocking the bridge continued to receive a heavy volume of fire from across the bridge. As Captain Moran was working to receive approval for cross-boundary fires into Grizzly's zone, Private First Class Juan Garza Jr. was mortally wounded by an Iraqi sniper hiding in a civilian house. Private First Class Garza had been part of the RCT's attack to liberate these oppressed Shia neighborhoods from the last remnants of the Regime. Bitter at the loss of one of their own, RCT-1 worked against the sniper, finding that many of these last remaining fighters were well trained and very capable. It was a sobering first-hand introduction to the difficulties in fighting in MOUT. Had the Regime and foreign fighters been able to rally popular support using the ammunition supply points and weapons caches in the schools and mosques, there may have been significantly more Coalition casualties.

As the Marines from Bravo Company continued to gain ground in the town, Major Casado (the 1/4 Operations Officer) ordered a platoon from Alpha Tanks to reinforce Bravo Company. Bravo Company's fire support team called-in accurate artillery missions, while the tanks unleashed devastatingly precise main gunfire. This deadly combination destroyed two enemy bunkers, and effectively ended enemy fires directed at Bravo Company. The Marines of RCT-1 cleared the town, killing several enemy fighters and capturing a significant amount of weapons and ordnance and EPWs, to include a two-star general. RCT-1 stood poised to continue the attack towards Saddam City.

Blue Diamond Marines continue to roll into Baghdad.

Grizzly Goes Long

RCT-5's columns wound their way through the narrow avenues of approach to the river crossings, and crossed the river. RCT-5 used the same combination of bridges and fording sites that RCT-1 did to get across. Once on the western side of the Diyala, RCT-5 continued its advance north along the bank of the river. After passing north behind RCT-1's front lines all the way to the old abandoned canal on the northeast edge of Baghdad, RCT-5 turned west and began its attack in its own zone on the right flank of RCT-1. Third Battalion, 5th Marines, led the attack across the canal, pushing to the east to uncover the ground to the west bank of the Diyala River. Lieutenant Colonel Fred Padilla's 1/5 followed, clearing the ground to the west. In this suburban area, the RCT immediately experienced significant resistance from Fedeyeen snipers, and triggered a number of RPG ambushes. RCT-5 successfully fought through each enemy attack, eliminating the threat. The RCT was to come under a significant number of mortar attacks, but these were poorly aimed and ineffective. Like in previous areas, once the Fedeyeen had been cleared out, the local populace returned to the streets, welcoming the Marines with cheers and waves. For this and other areas of the capitol, the 'Three Block War' was often fought on a single block, only over three time periods in a single day. The Marines found themselves in intense combat in the morning, followed by deterrence of factional fighting at noon, and ending with the smiles and waves of the citizens as they entered the streets without fear, many for the first time in their young lives.

With 3/5 and 1/5 consolidating north of the Abandoned Canal, RCT-5's plan called for Lieutenant Colonel Dan O'Donohue's 2/5 to attack across the north side of Baghdad to establish a blocking position along Highway 2 to complete the cordon of the city. Just north of Saddam City, however, the battalion encountered stiff resistance from Fedeyeen and other irregular forces, fighting from prepared positions. The enemy seemed surprised as to the direction and timing of the attack and 2/5 skillfully employed 3[d] MAW aviation and direct fires to exact a heavy toll on them. With large numbers of enemy reinforcements moving from the built up areas of Saddam City into the restrictive terrain ahead of the battalion, the decision was made to seek an alternate route to the Highway 2 objective. The battalion returned to Highway 5 and sought a bypass of the poor terrain at the north end of the city. By the evening of 8 April, 2/5 had made its way on a roundabout 55 km detour around the north side of the city, and established a blocking position astride Highway 2. Once 2/5 was in position, the Regimental CP followed, and shifted the headquarters of RCT-5 to the far northwestern edge of the Division's zone. The city was now surrounded, with Marines to the east and north, and Soldiers of the US 3[d] Infantry Division (on the west side of the Tigris River) generally west and south.

Adventures on the Screen

On the Division's flanks, 2[d] and 3[d] LAR Battalions continued their security zone operations. Wolfpack Battalion extended their screen line into the zones vacated by RCT-1 and RCT-5. The Wolfpack now patrolled the entire eastern edge of the city all the way to the Tigris (a zone 45 km in length) while Barbarian continued to secure the critical Highway 6 route all the way back to the Tigris River crossings. This area included the critical airfields at Salman Pak East and An Numaniyah, where much needed food and ammunition were being flown by 3[d] MAW to sustain the Division's attack.

Over the course of days, there had been numerous reports of ongoing enemy activity in Baqubah to the north of Baghdad. This had been the site of the Al Nida Division headquarters and one of its combat brigades. Although the garrisons were thought to have been abandoned, it was necessary to send a ground force to assess the threat to the Division's northeast flank. The Reconnaissance Battalion, having collocated with the Division Main CP on the east side of the Diyala River, received the order to attack to Baqubah, to establish a security zone to the north and east. The Al Nida Division's 41st Brigade had been located in Baqubah, and it was not clear what level of resistance remained in the city from conventional or paramilitary forces. The Battalion was to conduct an advance on the eastern road (5 km to the east of the Diyala River) to Baqubah and began to roll at sunset on 8 April.

Reconnaissance Battalion's Attack to Baqubah

In the late evening hours, the Reconnaissance Battalion, with Charlie Company of 1st LAR Battalion attached, approached the Baqubah area. When they were still 17 km south of the city, the lead LAVs began to receive fires from either side of the road. The enemy was arrayed in depth to make the entire roadway one continuous engagement area. Air was called, but the visibility was very poor due to fog and well-concealed positions occupied by the defenders. Recon Battalion pushed flanking patrols on either side of the ambush, attempting to turn the defenders out of their fixed positions, but the enemy had pre-registered mortars on these routes to their flank, and successfully kept the flanking patrols suppressed. The enemy began to adjust their mortar fire on to the LAVs who were by this time returning fire from the road, and forced them to pull back out of the kill zone. These skilled resistors were likely Republican Guardsmen of the Al Nida Division, but were all dressed in civilian clothes. Recon Battalion continued to suppress the defenders with 81mm mortars while they waited for daylight to continue the attack. When the battalion decided to engage again at first light, it found these positions vacated.

With the Division's successful attack across the Diyala, the Division found itself in possession of a significant portion of eastern Baghdad already. By establishing a cordon, the Division had closed the door for the enemy's escape, and opened the door to operations inside the city. With RCT-7 poised at the edge of downtown, and RCT-1 prepared to push into Saddam City, all of eastern Baghdad was within a few km of the Division's positions. The Division worked quickly to generate and refine a target set in the Baghdad urban area that would meet the MEF Commander's intent. With resistance crumbling, the Division was positioned to continue the attack into Baghdad now, and this list would at least give the attack some specific objectives.

On 8 April, Charlie Company, 1st LAR was attached to 1st Reconnaissance Battalion and led the attack towards Baqubah. White platoon commanded by Second Lieutenant Frank McCabe was moving up route green (Highway 5) to establish a blocking position at the 12 Northing. A light colored pickup truck was traveling towards them when it stopped 500 meters away and immediately made a u-turn upon seeing the platoon and sped away. The platoon thought they saw a heavy machine gun on the back of the vehicle but couldn't make a positive identification. As the truck continued to retreat it stopped about 2000 meters away and began to open fire on the platoon with its 12.7mm Heavy Machine Gun. White platoon immediately gave chase to the truck and closed within 750 meters of the vehicle. Second Lieutenant Paul Webber, the commander of the LAV Anti-Tank section, noticed several red streaks skimming over the ground directly toward his two lead LAVs. Second Lieutenant Webber called out the RPG contact over the radio just as the projectiles began exploding to the side of the road. At this point Second Lieutenant McCabe realized they had driven into an ambush and ordered his platoon to halt. Mortar rounds began impacting directly off to the right side of the road as small arms and light machine gun fire began erupting from the tall grass to their right flank. An RPG round impacted dangerously close to the lead LAV-25, blowing out four of its tires, and temporarily blinding Sergeant Steele, the vehicle commander. The concussion knocked him sideways tearing off the cord to his communications helmet and severing communications with the rest of the platoon. While Lieutenant McCabe watched his wingman lose control and swerve to the side of the road, he directed the rest of the platoon to pull back. He then called for an immediate suppression mission from the company's organic 81mm mortars. Sergeant Steele had regained control of his vehicle at this time and pulled back out of the engagement area. Once his wingman was clear, Lieutenant McCabe withdrew south to a safe distance and directed a Close Air Support (CAS) mission with a section of UH-1N Huey gunships. Using their rockets and guns on the enemy ambush position, the aircraft completed three attacks and fires from the enemy ceased. Charlie Company remained in their original defensive blocking position for the night.

These objectives would serve as guideposts for operations, but the real objective remained to destroy enemy fighters wherever they were holed up. There were indications that Regime forces had already vacated large portions of the city, and now these areas were being looted by civilian mobs. It was clear to all on the scene that the chaos in the city mandated an immediate attack to crush the remaining resistance and restore order. The attack could not wait any longer. Although significant pockets of resistance remained in the city, the beginning of the end of the Baghdad fight was at hand. The first objective of CFLCC had remained the completion of the cordon, and by the end of 8 April it was finally complete. The Division had now met the cordon requirement, positioned itself for a simultaneous attack into eastern Baghdad from multiple directions, and developed its set of maneuver objectives in the city. By the morning, the Division would begin the takedown of eastern Baghdad.

The MEF Commander reviews Baghdad options with the Division Commander.

9 April
Getting Permission

From the successes of all three RCTs on 8 April, the Division assessed that the conventional military threat in Baghdad was nearly eliminated. The US 3rd Infantry Division had conducted repeated raids to the gates of the presidential palace on the west side of the Tigris, and conventional resistance had been greatly reduced there as well. Only paramilitary forces and terrorists remained in eastern Baghdad, notably in the northern parts, where they still operated in company size. In the southeastern part of the city, they maintained an uncoordinated presence, operating in small groups from strongholds in mosques, schools, and safe houses. The enemy still had the ability to mass these groups in areas of continued Regime influence, but as the attack continued, the resistance from these disparate groups gradually faded away. The changing nature of the fight was apparent, even as early as 8 April. The transition from liberation to stabilization in Baghdad would be a gradual process, but had obviously begun.

During this time, the Division received notice that there was consideration at higher headquarters for instituting an operational pause, with US units positioned on the Baghdad cordon. The premise behind the pause was that there was still an organized Iraqi conventional resistance inside the city, and that there was a centralized authority that still controlled this resistance. The plan under consideration would offer this centralized authority a 72-hour period to turn over the city peacefully. From one perspective, this appeared to be a reasonable way to gain the city of Baghdad without forcing a lengthy and costly urban attack. The reality of the situation on the ground had been changing rapidly, however, and MEF and Division assessed it quite differently. The enemy defenses had already collapsed, and the new enemy was the chaos that was beginning to permeate the city as the developing power vacuum permitted looters, vigilantes, and criminal gangs to run rampant. It was clear to the MEF and Division that the situation in Baghdad would be made more difficult by delay, not less so. The Division made its recommendations accordingly. In the early hours of 9 April, Blue Diamond gained approval for continued operations in the urban area, and launched its attack. The Division battlespace in the city was sectored and assigned to the four Regimental headquarters (including 11th Marines). Under the Division's plan, the regimental commander would exercise authority in his zone, and quickly move to secure the public infrastructure, root out any remaining resistance, and provide a security presence on the streets. The 11th Marines commander would also have the task of providing artillery or counter-battery support as required as units cleared the city. Within 24 hours, Blue Diamond would control large segments of eastern Baghdad, and would have already begun to conduct Civil-Military Operations.

Shaking Hands Across the Tigris

There was still a significant amount of fight left in the Fedeyeen and paramilitaries of northern Baghdad. The Division was required to conduct a physical linkup with US Army forces on the other side of the Tigris, and assigned the task to RCT-5. The northern neighborhoods of Baghdad had not received the same attention by US forces that the southern neighborhoods had. Fedeyeen and other fighters remained in strong isolated pockets, making the northern side of town a dangerous place to operate. On 9 April, 2/5 fought its way through heavy resistance to the Tigris River bridges on Baghdad's north side, accomplishing the linkup with the US 3d Infantry Division forces across the river.

While 2/5 fought its way to the linkup, the rest of RCT-5 strengthened the cordon by conducting aggressive patrols in their zone. First Battalion, 5th Marines sent a patrol to investigate a prison. As the patrol approached the compound, the gate was opened. Out spilled dozens of children wearing threadbare clothing and looking undernourished. Crossing their hands as if to mimic handcuffs, they approached the Marines. Then they gave the 'thumbs up' sign and shouted their thanks. According to some of the Iraqi citizens who thronged to the site of the prison, including many overjoyed parents, the prison was for Iraqi children who had refused to join the youth branch of the Baath Party. More than one hundred children, some of whom had been imprisoned for years, spilled out of the jail when the gates were opened. The satisfaction that the Marines would experience that day left no Marine untouched, and would carry them through some hard fighting that remained in the city.

Entering the City

RCT-7 had been impatiently held-up at the Rasheed limit of advance for most of a day pending approval to enter the city. They now broke out of the bridgehead with enthusiasm. Their first goal was to secure their designated zone, nearly one-third of eastern Baghdad. Although the Marines had received a generally warm welcome by Iraqis up to this point in the campaign, it was nothing compared to the reception they began to receive in Baghdad. Crowds of civilians came out to cheer on the Marines as they drove by. Chants of "Good, good!" along with a thumbs up were most common. Every Marine felt like a hero as he waved to the crowds. The stories started to quickly come out about how happy the people were to be rid of Saddam Hussein. In a letter given to the Marines, one Iraqi told his story, "...My father was an officer in the Iraqi army in the war between Iraq and Iran. After that, Saddam killed him. After that I want revenge and I want to go to the USA and to be an officer in the US Army.

Marines secure a Regime Palace east of the Tigris.

Late on the afternoon of 9 April, 1/7 moved into Baghdad. Its objectives were all located on a distinctive peninsula in central Baghdad. As the Battalion moved onto the peninsula, it was greeted by thousands of cheering Iraqis. They blew kisses, threw flowers and one man carried a sign supporting George Bush. While it was reassuring to see the people's support, there was still work to be done. Company objectives were assigned, with Bravo Company tasked to seize Baghdad University. Bravo entered the University compound and took small arms fire that they quickly suppressed. Shortly after receiving fire, a large ammo cache of anti-aircraft rounds began to go off. The Battalion's 81mm mortar position was receiving so much fragmentation from the cook offs that the Battalion Gunner sought to move them from their position. The mortar men refused to leave their guns, out of concern for not being able to support their fellow Marines and answer a fire mission if called. When the mortar men also received fire from a building next to their position, they formed a provisional rifle squad to clear it. This bias for action and aggressiveness characterized Marines from across the Division. The University had posed a threat to Army units in their zone across the Tigris River as well, and the Army had poured volley after volley of DPICM munitions against targets on the university grounds. The high 'dud rate' of these types of munitions made the grounds of the university an extreme UXO hazard to the Marines and civilians. (Colonel Mike Marletto, CO of 11th Marines and the Division's Fire Support Coordinator had wisely shifted to 'VT'-fused artillery ammunition when the Marines closed on Baghdad, limiting the DPICM dud threat in most of the Division's zone.)

A statue of Saddam moments before it was pulled down under the glare of the international media.

Charlie Company was ordered to seize a small Regime palace in the Division zone, across the Tigris from the main Presidential Palace. Charlie conducted a deliberate clearing of the Regime palace, finding evidence of the lavish fraud the dictator had perpetrated on the poor citizens of Iraq, but nothing of military significance. The battalion was also tasked with searching the residence of Tariq Aziz. The former Deputy Prime Minister of Iraq had his residence on the peninsula near the University. The residence was searched for intelligence documents, and many boxes of documents were retrieved for later intelligence exploitation. By that afternoon, the commander of 1/7, Lieutenant Colonel Chris Conlin, reported, "the peninsula is in my hands, and I think I can hold it."

During the 'statue incident', this US Flag was quickly replaced by an Iraqi Flag, demonstrating the Marines' commitment to restoring freedom to Iraq.

Initially being tasked to make contact and clear the enemy in zone, Lieutenant Colonel Brian McCoy's 3/4 moved out late in the day on 9 April expecting to run into the teeth of the Baghdad defensive fortifications. Operations conducted that day were planned to be slow and deliberate, as the battalion cleared the remaining enemy defenders. Meeting far less resistance than had been expected, however, 3/4 found that it was available to execute several new orders to secure key buildings in downtown Baghdad. Among these were the Palestine Hotel, Sheraton Hotel, Baghdad Hotel, and the Embassies of Japan, Germany, the Vatican, Indonesia, and Poland (which housed the US interests section.) The battalion commander ordered India Company to secure the Palestine Hotel. The scene that greeted the advancing Marines at the traffic circle in front of the Palestine and Sheraton Hotels was crowded and chaotic, as the streets quickly filled with civilians. Members of the press, until now restricted to the Palestine Hotel by the Regime, poured out to record the moment. In the Firdos Square traffic circle, a large statue of Saddam Hussein dominated the scene. Just beyond it, the cameras of the world's press covering the war in Baghdad went live to record the moment American Marines entered downtown Baghdad. As the Marines of 3/4 approached the statue, the crowd asked the Marines to help them tear it down. It was not long before an M88 Tank Retriever was called up to help tear down the symbol of the freshly deposed dictator. Corporal Edward Chin of Bravo Company, First Tanks, draped an American flag over the statue's head in a proud display of what the Marine Corps had accomplished (causing some considerable consternation at higher echelons.) Just as quickly, the Marines replaced it with an Iraqi flag, demonstrating their commitment to the real objective of restoring Iraq to its people. To the jubilant shouts of the Iraqis, and the whir of cameras from the international press, the statue came tumbling down. The famous 'statue incident' received the attention of every major news outlet, and signaled all that the Marines had liberated eastern Baghdad. The symbolism was lost on no one as the image was flashed around the world.

Lieutenant Colonel Mike Belcher's 3/7 also immediately set out to secure high value targets across their zone. These targets included a suspected Saddam Fedeyeen training compound, the

Ministry of Information building, the Fedeyeen Headquarters, Uday Hussein's Headquarters, and the Ministry of Oil building. After the battalion crossed the LD, it engaged a charging vehicle, destroying the vehicle and killing its two occupants. By 0930Z, they had secured the Fedeyeen compound. Elements of the battalion also engaged suspected Fedeyeen who fired on them from the tenth floor of the Ministry of Information building. Two TOW missiles silenced the enemy weapons. Fedeyeen sniping continued against all elements of the Battalion, including the Alpha and Bravo command groups, the 81mm mortar platoon, and the logistics train. Under the direction of the battalion Executive Officer, Major Anthony Henderson, elements of the battalion returned fire, killing a number of enemy fighters. Kilo Company attacked to clear the Ministry of Information, a twelve-story tower. One platoon secured the perimeter while two platoons alternated clearing floors. The training the Marines had received in urban warfare paid off in confidence and flawless execution. Lance Corporal Mark Enos remembers:

> *The thing that impacted me the most was when we began to clear the twelve-story building. Knowing in the back of my mind that there was a really big possibility that there was enemy in the buildings, we used our MOUT tactics and depended on one another for our own safety. Our squad leader knew exactly what to do and what had to be done.*

The RCT-7 TAC moved in on the heels of 3/7 and set-up just below the Ministry of Information building, while the building was still being cleared. A firefight raged in front as 3/7 security elements blocked the entrances to the compound.

Lima Company, 3/7, attacked to seize the Ministry of Oil building after finding that its first two targets already burning. Uday Hussein's office on the top floor of a nearby building was engulfed in flames, likely started by irate civilians seeking revenge on the Regime after years of brutal repression. The building burned for several days. One platoon secured the sprawling Oil Ministry structure, while the remaining two platoons cleared two floors at a time. By 1410Z both the Ministries of Information and Oil were secured. Companies begin to back-clear the buildings top to bottom.

Iraqi children welcome the Marines with shouts and waves.

Late that afternoon 3/7 also engaged two enemy "technicals" and one civilian vehicle that attempted to drive through a checkpoint. All three vehicles were destroyed. While the enemy soldiers were all killed, the Marines managed to pull two civilians from their car and affect an immediate CASEVAC. Shortly afterward, two Marines were injured when an Iraqi POV filled with gasoline drums attempted to ram a Marine vehicle. Fortunately, the gasoline did not ignite. These last acts of resistance, however, were overshadowed by the swell of popular support received by the Marines. Everywhere they went, the Marines continued to be bombarded with waves, shouts, and kisses. No Marine went through that 'fire' without being effected.

As RCT-7 pushed into southeast Baghdad from the Southeast, RCT-1 operated on their right flank, to the northeast of the Army Canal. Inchon had pushed past Phase Line Andrew at Highway 5 the previous day, and was at the edge of the government center and the outskirts of Saddam City. They faced light resistance all morning, as 1/4 and 2/23 commenced the attack into the RCT's portion of Baghdad.

The DGS Headquarters received a cruise missile on the first night of the war. This facility would soon house the Division Headquarters.

Lieutenant Colonel Geffrey Cooper's 2/23 pushed forward through the center of the zone, looking for the headquarters of the Iraqi Directorate for General Security. While pushing toward the suspected location, the RCT's Alpha command group found the headquarters thanks to its Iraqi defenders. The otherwise non-descript government buildings were not at the location described on the map, but were readily identified when the Alpha command began to take intense enemy fire from their vicinity. Fox Company, 2/23, led by Major Jonathan E. Kirkpatrick, attacked in order to clear the high-walled compound that housed approximately 100 defenders. The defense increased in intensity upon the Company's approach, and Highway 5 soon was a maelstrom of RPG, sniper, and machinegun fire.

Civilian vehicles loaded with armed men in civilian attire began assailing Fox Company's lines. One such vehicle came barreling into the position drawing fire from all friendly positions. The vehicle came to a halt at Sergeant Scott Montoya's position on the firing line. Sergeant Montoya saw that the enemy driver was reaching under the seat (for what was later found to be hidden explosives.) Sergeant Montoya dashed out into the street under fire and eliminated the threat. From his vantage point in the street, Sergeant Montoya saw Lance Corporal Segovia, a rifleman from 3^{d} platoon, lying in the open on Highway 5 and bleeding profusely from the inner thigh. Rushing to the Marine's side under the crossfire of friendly and enemy bullets, Sergeant Montoya stripped the Marine of his gear, hoisted him upon his shoulders, and carried him 500 meters to safety. As the engagement continued, Sergeant Montoya repeatedly left the relative safety of his covered position to assist three more wounded Marines and an Iraqi civilian to a casualty collection point.

Second Platoon received a barrage of hostile fire including rockets, machineguns, and small arms the preponderance of which was directed at Corporal Robert Tomczak's machinegun team. Realizing that his position was critical to the platoon's security, Corporal Tomczak remained in place and countered the heavy volume of fire with his own highly accurate and effective fire, disrupting an attempted flank assault. Under the barrage of fire, Corporal Tomczak maneuvered from his sand bag fortifications and deployed an AT-4 to silence an RPG position. When an armed gunman in a vehicle attempted to ram his position, Corporal Tomczak steadily returned fire, dodging the closing vehicle by inches as it crossed the curb and impacted his bunker.

During the height of the battle, several civilian vehicles were caught in the deadly crossfire between the Marines and the enemy positions. Observing several wounded civilians in disabled vehicles, Chief Warrant Officer Frank Castiglia organized a squad size element to rescue the wounded. During a lull in the engagement, he led the rescue team from covered positions into the exposed intersection to bring the civilians into friendly lines. They came under intense machinegun and rocket propelled grenade fire. Chief Warrant Officer Castiglia advanced his Marines under the suppressive fires of the company, and recovered a wounded Iraqi civilian. During the engagement, nine Marines were wounded. The DGS compound had been the primary facility for Saddam Hussein's secret internal police. The records captured inside this facility were a significant intelligence find and would help document the crimes Saddam had committed against his own people.

Marines from RCT-1 apprehend an Iraqi man who stole a UN vehicle.

First Battalion, 4th Marines, attacked in zone, clearing the northern sector up to the immediate outskirts of Saddam City. Enemy contact was sporadic and light. Those who chose to stay, put up minimal resistance and were quickly overwhelmed by the assault units. Killing several enemy fighters and capturing over 25 EPWs, 1/4 secured all objectives, to include the most dominant building in its sector, a cigarette factory. The battalion established its CP in this building, and employed scout snipers and a Naval Special Warfare (NSW) counter-sniper team with a commanding view from the roof. Responding to indications that there had been significant rioting in the streets of Saddam City, RCT-1 also tasked 1/4 to enter this vast Shia neighborhood to restore order. The surveillance platforms that detected the rioting could not ascertain whether the mobs of civilians on the streets were celebrating, looting, or exacting revenge on those who had cooperated with the Regime before. Moving into this Shia hotbed, 1/4 was unsure if they would be welcomed or taken under fire. The Marines' dominance of the zone was made clear to the inhabitants by a program of saturation patrolling; swarming the streets for short intervals, then moving to another area to repeat the process. On their first combat patrol into the city, thousands of well-wishers greeted the Marines with flowers and cheers of joy.

Showing that they were 'No better friend' to the Iraqi people, RCT-1 provided assistance and support to an International Committee of the Red Cross hospital site in their zone. With all of the destruction and misery in the city, it was important to get as many hospitals functioning as quickly as possible. At the Red Cross hospital, there had been a significant threat from looters. This threatened the patients currently in the hospitals, as well as the ability of the Iraqis to restore medical services to the population. Leaving the facility unsecured was not an option for the Marines, and they took action:

"On 10 April, the 'Jump' CP of 3d Assault Amphibian Battalion was coordinating with 1st Marines, when the battalion civil affairs officer, Lieutenant Colonel Mike Van Nordhiem asked to drive by the HQs for the International Committee of the Red Cross. They found the location and had a great, very informative meeting with the ICRC director. The director's greatest concern was the safety of the Medical City complex. With the collapse of organized resistance, looters and Iraqi Fedeyeen fighters were threatening to destroy the facility. The greatest concern was all the hundreds of patients on life support throughout the three hospitals. If power was lost, or equipment was not safeguarded, those Iraqis would die. Lieutenant Colonel Abbott and his interpreter, Sergeant Danny Hamler found and met with one of the Iraqi doctors, a Dr Ali and explained that the Marines were coming to protect the hospital complex. The doctor was visibly shaking with fear and was absolutely convinced that we were going to murder the patients in their beds. That's what the government had told them. After nearly hyperventilating and chain smoking an entire pack of cigarettes, he calmed down and explained that the looters had gotten away with ambulances and some medical equipment. The appearance of the Marines had frightened away the boldest looters, but they would be back.

The Jump CP only consisted of two soft skinned HMMWVs and eight Marines with no heavy weapons. As the security force had not arrived yet, the Jump Marines began aggressively patrolling the complex to give the appearance of a much larger force. Amongst the tall hospital buildings, external communications were limited to chance, as one in forty tries to get through on an Iridium phone. As the sun went down and there was still no infantry, the desperately exposed nature of the situation became quite apparent to everyone. Backlit by burning cars, shadowed figures moved along the perimeter. Medical City was on a bluff looking west across the Tigris River. The Marines watched as the 3d ID battled Iraqi forces on the west side with tracers, explosions and fireballs arching through the night. Around 2300, another Iraqi doctor came by and announced that more Americans had arrived at one of the hospitals. The AAV Marines linked up with the task-organized platoon of tanks, AAVs and mechanized infantry from 3d Battalion, 4th Marines. Lieutenant Colonel Abbott, Sergeant Hamler and Doctor Ali led the infantry team leader, First Lieutenant David M. Bann of 2d platoon, India Company, 3/4 on the recon of Medical City in order to rapidly effect the

Marines patrol the narrow streets of Saddam City.

relief in place. It was surreal as the small group walked the perimeter, discussing security priorities while the battle went on around them. 3/4 remained in place and continued to safeguard the complex. The next day, we heard on the short wave radio how the military was supposedly negligent for not protecting hospitals, but we knew there were at least three large hospitals where Marines had risked their lives to protect innocent Iraqi civilians."

The Sailors and Marines of all of the RCTs helped to the extent they were able, with supplies, security, and even medical help. Where medical supplies stored by the former Regime could be located, they were distributed to the civilian hospitals. Until widespread medical aid could be flown in, however, the few operating medical facilities did the best they could under difficult circumstances.

Marines assist in humanitarian missions from the moment they arrive.

During one of their patrols, RCT-1 discovered a food storage warehouse, and set up a food distribution site to get it to those who needed it the most. The food distribution operation aired as a story on CBS news. RCT-1's actions this day continued in the fashion of the 'Three Block War'. While 2/23 was engaged in an intense MOUT fight, other Marines of the RCT were deterring looters, while yet a third group were handing out food and supporting the local hospital.

The experience of RCT-1 this day and others also served as a visible demonstration of the balanced and flexible capability of the Marine Air Ground Task Force. Since crossing the LD, nearly every function of the MAGTF had been committed to the fight, and every one had proved itself indispensable at one point or another. Balance in capability is what brought the Division success. All capabilities were used in concert, but each, in turn, came to the fore. Engaging in the Three Block War required large numbers of trained infantrymen. The tanks and artillery that had been central to victory on the southern desert ceded their dominant position to dismounted riflemen in the urban area. Aviation, which had been critical to the shaping effort, had ceded their precedence to tanks on the ground near Al Basrah. Tanks had been supplanted as the most critical capability by engineers building and improving key bridges in An Nasiriyah. The engineers had enabled the LAVs to conduct a rapid advance along a route of unknown quality until the enemy's strength was located. The LAVs stepped aside as mechanized infantry pushed through the enemy's defenses. Artillery covered fire support tasks as fierce sandstorms grounded aviation fires. Logistics brought by KC-130 and new MTVR trucks enabled expeditionary logistics to fill-in for a slower-moving logistics support system. Modern communications equipment had enabled a swift command and control capability that allowed rapid tempo throughout, without requiring an operational pause while waiting for the headquarters to catch-up. Even with all of these technological capabilities, however, the primary weapon for the fight for Baghdad, stabilization operations and civil military interaction, was the Marine Corps' most basic core competency, the individual US Marine.

Recon's Security Zone Operations in Baqubah

On the morning of 9 April, while the Division was conducting its attacks into the heart of Eastern Baghdad, Recon Battalion continued its attack to Baqubah back on the east side of the Diyala River north of Baghdad. Captain Gil Juarez's Charlie Company, 1st LAR was able to push forward through the (now abandoned) positions that Lieutenant McCabe's platoon had attacked the previous evening, and proceeded up the road. Charlie Company again moved north along the road to Baqubah. The earlier positions had been abandoned, but the Recon Marines were on the hunt and the enemy was soon located a short distance up the road.

Saddam's forces are unable to hide from US artillery and air fires.

As the battalion continued north, Iraqi civilians informed them of a company-size enemy force set in another ambush in tall grass just ahead. The battalion attacked to this area and once again mortar fire erupted just off the road. At the same time, the Marines began receiving direct fire from uniformed Republican Guard forces who were firing from both sides of the road and from nearby buildings. LAR's White platoon returned fire with their 25mm cannon and mounted M240G machine guns, engaging the enemy sometimes as close as 10 meters away. The LAV-AT s launched several missiles into the nearby houses where the most concentrated fires were coming from.

The company's FiST went into action, led by the FAC, Captain William "Chocka" Bartolomea. The team dismounted their LAV under fire and obtained a 10-digit grid utilizing the LASER Viper. The team returned to their LAV and called in a JDAM strike and then finished the enemy strongpoint off with Hellfire missiles from AH-1 Cobra gunships. As Charlie LAR, Bravo Recon, and Delta Recon Companies continued to push up the main road, Alpha Recon and Charlie Recon companies pushed up a flanking route to the east. While Captain Brian Gilman's Charlie Company protected the east flank of the movement, Major Brian Patterson (he had been promoted since the Al Gharraf fight) led his Alpha Company to a berm just to the north of the enemy defenders. As the flanking force got into place, and more Cobras appeared overhead, the enemy decided he had had enough. Uniformed enemy soldiers disappeared into culverts and buildings, 'miraculously' reappearing as unarmed young men in civilian clothes mixed with local civilians. This was the same tactic used by Fedeyeen in Ad Diwaniyah days before. As Captain Gilman describes it, "Whenever rotary wing air showed up these guys would just bug out. It was one of the most frustrating experiences I've ever had watching these guys walk away, but they'd disappear and suddenly reappear with no weapons, wearing civilian clothes. It was tough, but the Marines showed incredible restraint. We didn't want more civilians getting killed." The Recon Marines' discipline was sufficient to prevent harm to the innocent by not continuing to engage even when they had good reason to believe that enemy troops were in civilian disguise amongst the locals.

On 9 April, one of the strangest things I saw during the war occurred. We were ambushed at the same spot as the night before, and were called to a halt. I was the lead vehicle at the time. As mortar rounds impacted all around us, I saw a family of one mother, a daughter, and two men in their late teens walk right by my vehicle straight towards the enemy positions. My platoon began firing on troops in the open, enemy vehicles, and emplacements to our north on the right side of the road. As the 25 mm fire flew over the family's heads, they just continued walking as if impervious to all that was occurring around them. The enemy mortars continued to fall all around the family and my platoon continued to shoot over the top of them. They only broke stride once to move to the other side of the road because the fire obviously was too intense to stay on the side they were walking. I curiously watched them walk all the way to the enemy positions and out of view. They must have been the luckiest Iraqis I've ever seen. Or maybe they got killed once they were near the enemy positions where all the impacts were landing. I'll never know.

Alpha Company Recon arrived at the berm and observed a BMP firing at them from the other side. The Marines were able to get a section of F-15s and had them drop two JDAMs on the target. The first landed only 300 meters from the Marines' position (due to a weapons malfunction) but the second obliterated the BMP. The attack continued. The Reconnaissance Battalion destroyed a total of 3 BMPs and 2 abandoned T-72s as they continued to advance to the east of Baqubah.

As the last resistance between the advancing reconnaissance force and the city fell, the Recon Battalion commander called back to the Division CP over the secure Iridium phone. Over the phone, the Division G-2 and the Recon Battalion Commander were able to react to the changing situation on the ground. The G-2 recommended direct observation of the garrison of the Al Nida Division headquarters, and the headquarters of the 41st Brigade. Charlie Company Recon conducted a daring daylight operation into both of these facilities, confirming the enemy had abandoned them. Charlie Company cleared both the Division and Brigade Headquarters while the remainder of the force conducted roadblocks along Highway 5. As no further enemy resistance or forces were identified, the battalion returned to Baghdad and was integrated into the Division's operations there.

10 April
On to the Next Fight

In Baghdad, the Division continued its attacks into the eastern half of the city. The list of Regime targets was slowly whittled down as the day proceeded. All four regiments were able to conduct a physical linkup at the edges of their zones, and the enemy resistance existed only in pockets around the city. With the Division in a 35 km cordon around the city, the Division Main CP found itself out of position to adequately communicate with its widely spaced elements. The Main CP moved to a site farther north along the Diyala from which it could better communicate with the widely dispersed elements of the Division. The site, between the banks of the Tigris and the Dead General Crossroads, was certainly better than the rendering field they had been in previously. The Division Forward CP in the Fedeyeen Stadium briefly took control, while the Marines of the Division Main conducted the teardown and displaced away from the animal bones, hair, and teeth in record time.

The Division Forward CP at Rasheed Fedeyeen Stadium. This was some of the first green grass the Marines had seen since leaving the US.

The Division had marked every mission success with a new assignment, and the securing of eastern Baghdad was no exception. As quickly as it began to position itself for successful completion of combat tasks in eastern Baghdad, the Division was tasked to plan a mission for securing the oilfields in Kirkuk, 200 km to the north. These key oilfields of the north, which contributed approximately one-third of Iraq's oil production, originally fell into the battlespace of US forces attacking from the north. Because the Coalition's northern option had been greatly reduced by Turkey's reluctance to support the war effort, US forces had not yet secured these fields. Blue Diamond had already secured almost two-thirds of Iraq's oil production capability during the Opening Gambit in the south, and was now given a warning order to attack to secure another third.

The town of Kirkuk was not only a considerable distance away, but also was defended by at least a brigade of RG and elements of three Iraqi RA Divisions. Given the Division's recent experience, it was known that the enemy's strength on the ground was probably significantly lower than the strict order of battle published on paper. The planners agreed it was an acceptable risk to send a light force to attack this remaining threat, and the Division immediately turned to the task. As the warning order contained no supporting intelligence, the Division scrambled to assemble a study of the enemy, routes, objectives, and oilfield infrastructure to support the attack. Obtaining maps for an objective this far away was an immediate challenge. The G-2 planners quickly designed and began to print specialized maps using its organic Topographic capability, and dispatched an officer to Kuwait to quickly obtain standard maps from the MLC. The Division expected to have about 48 hours to prepare before having to cross the LD. Undeterred by the challenges, the Division began to assemble a light task force that could rapidly close the distance to Kirkuk, would have enough combat power when it got there, would communicate by satellite, and could be re-supplied by air. The Division's infantry battalions were largely consumed in seizing and controlling the major eastern Baghdad metropolitan area, with its several million inhabitants. Balancing the various Division missions and the forces available, Warning Orders went out to the three LAR Battalions as the principal units that would make up the task force, and Brigadier General John Kelly was assigned to command what would be a de-facto Light Armored Reconnaissance Regiment (Reinforced). Even as much remained to be done in Baghdad, the Division planners rapidly began to look to Kirkuk.

Downtown Objectives

Saddam City was made up almost exclusively of poverty stricken Shia Muslims and had been a hot bed of anti-Saddam fervor. It was key for the 1st Marine Division to provide a presence not only to show Iraqi support for the Coalition forces but also to mitigate the looting that was rampant in the city. RCT-1 planned to form zones by splitting Saddam City in half. The urban area was split between 1/4 and 3/1, and each occupied its respective zone. The attack met little resistance and the Shia populace was extremely happy to see the Marines. They openly waved and cheered the Marines during patrols that were conducted throughout the city. People lined the streets to see their liberators first-hand. The sound of gunfire was replaced by the same repeated phrases offered by the young and old alike, "Good, mister, good," and "Bush good; Saddam donkey."

Above and Below: ***Ammo caches in Baghdad***

Later that day, RCT-1 conducted an attack to the north side of Saddam City in response to enemy indirect fires received earlier and located by 11th Marine's counter-battery radar. The local population was found to be very supportive and aided in uncovering weapons caches and identifying locations of the last holdouts among the Regime forces. The Shia populace had no love for Saddam and the Baathist thugs who had terrorized them for so long. This was especially evident during one of 1/4's counter-mortar patrols in Saddam City. After uncovering and disposing numerous weapons caches, the rifle companies were ordered back to their outposts. As the forward command group turned a corner,

The Division Main CP operates in the abandoned DGS facility in Baghdad.

a group of Iraqis began to gesture and point down the street, tipping off a potential enemy ambush. Gunner fname Parks sprung from his vehicle and began to fire-on the would-be assailants. The enemy returned fire, broke contact, and ran down an alley. The assailants ducked into a house and locked the door behind them. The Gunner attached a grenade to the doorknob and forced an entry into the building with Lieutenant Colonel Mayer and Private First Class Cobb following close behind. They were greeted by the screams of women and children who were being used as shields to hide the assailants, who were captured without firing a shot.

During multiple counter-mortar patrols, RCT-1 uncovered four FROG missiles, and a number of other strategic missiles and weapons caches. Because of the on-going search for chemical and biological weapons, one of the missile systems was tested for the presence of WMD agents. RCT-1 also found what appeared to be a mobile, chemical weapons lab. The RCT called in an exploitation team to have it tested. The tests came out negative, but the RCT and all Division elements remained vigilant in their search for evidence of WMD. The ubiquity of weapons caches showed that the Iraqi Regime had prepared to fight for the city block by block. It was obvious that the battle for the city could have been a lot worse if the Iraqi Army had chosen to stay and fight or the Iraqi people had fought against their liberators. The rapidity and violence of the Division attack on the city, coupled with the firepower brought to bear by 3[d] MAW and 11th Marines, had crushed the enemy's ability to resist. The seeds of popular uprising
againstth e Americans had been crushed by the Regime itself during their years of brutal repression of the Iraqi people. The recently liberated Iraqis were not eager to fight to regain their shackles.

Marines stop weapons smugglers at checkpoints throughout the city.

After the operations that morning, RCT-1 controlled an area from the Diyala River through Saddam City. RCT-5 controlled the area from northwest of Saddam City to central Baghdad north of the Tigris. RCT-7 controlled the downtown area along the Tigris in central Baghdad. The Cannon Cockers controlled the Rasheed complex and all of the surrounding neighborhoods.

RCT-7 made adjustments to the battalion sectors within its zone. In the first days after the arrival of the Marines, widespread looting in the city was all too common. Third battalion, 4th Marines was called on to provide security for a Red Cross warehouse, and also got involved in stopping bank robberies. One such 'bust' resulted in the securing of $3.6 million US dollars that Lieutenant Colonel McCoy personally delivered to RCT-7 for protection and forwarding to

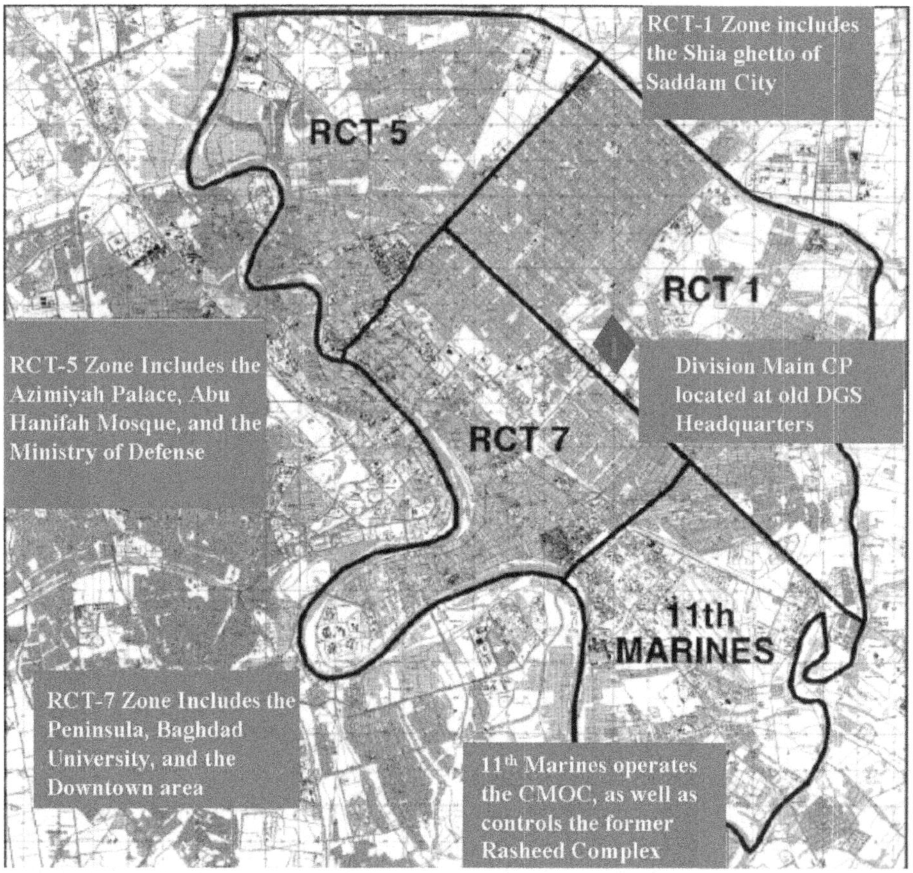

Baghdad was divided into Areas of Responsibility for all four of the Division's regiments. Eleventh Marines also operated the CMOC.

higher headquarters. First Tanks secured a suspected Regime SSO site in a school and uploaded three MTVR trucks full of documents and communications equipment to be transported to the rear for document exploitation. Still, the Marines' focus remained cleaning out all remaining pockets of resistance spread across the city.

The people of the city took the opportunity to take their vengeance out on the buildings and property of the former Iraqi government. Looters streamed out to pick clean these areas. Too often, as they would finish clearing out a building, they would set fire to it. Smoke pillars were common sites throughout the city from numerous buildings on fire. The Marines saw this growing problem, and had many complaints from civilians, who repeatedly asked for security for their neighborhoods. In response, Colonel Hummer established an embryonic Civil-Military Operations Center (CMOC) at the centrally located Palestine Hotel. The CMOC immediately began to grow with the reinforcement of the Division Civil Military Operations (CMO) team, led by Lieutenant Colonel Pete Zarconi (a reservist whose civilian duties as a Los Angeles police sergeant made him singularly qualified for restoring order to Eastern Baghdad.) The Division's Public Affairs Officer (PAO), Captain Joe Plenzler, also arrived to take advantage of the significant media presence at the Palestine Hotel.

In sector, all Regiments began setting up meetings with local leaders, establishing local patrols, and gaining the trust of local contacts in the neighborhoods. The Marines found that local civilians were more than willing to point them in the direction of any Iraqi force holdouts or Fedeyeen fighters. Even though the needs of a city with millions of residents in a tightly packed urban center were well beyond the abilities of any single division to meet, places like the CMOC offered a place where some of the problems and solutions could start to find each other in a systematic way. Because of its central location, the CMOC at the Palestine Hotel was soon expanded and placed under the control of 11th Marines, using the manpower and systems of the Division Fire Support Coordination network as the 'backbone'. Representatives of 11th Marines, the 3^{d} and 4th Civil Affairs Groups participated. The CMOC soon became the hub of civil military interaction, and a magnet for Iraqis wishing to practice their newly discovered right of free speech.

Through constant vehicle and foot patrols, the regiments cleared objectives throughout the city. Government Ministry Buildings, key electrical facilities, water distribution facilities, and key road intersections all required coordinated effort, but soon all of eastern Baghdad was under the control of the Marines. In addition to these objectives, it seemed that virtually every block in Baghdad had a weapons cache. Some of these caches were monstrous, containing every conceivable type of weapon and ammunition, to include tanks, mortars, artillery pieces, and even FROG surface-to-surface rockets. For the remainder of its time in Baghdad, the Division had its hands full trying to dispose of these huge stockpiles.

The liberated people also had a million stories to tell, and the Marines were the closest sympathetic ears. Major Mike Samarov related one of the more interesting cases from 3/7:

A man of Iraqi descent approached Weapons Company. He was accompanied by two children, a thirteen-year old girl, and a nine-year old boy. He presented a new US Passport as proof of his identity. He claimed that he and his brother, both American citizens, were spies working with the Central Intelligence Agency. ***Nine years ago****, they were betrayed, his brother killed, and he thrown in jail. He escaped yesterday, went to his old neighborhood, gathered up some few possessions and his children, and found the first US unit that he could. Captain Schmitt escorts the three civilians to the Battalion firm base. HET interviewed the gentleman and crosschecked his information. The information was legitimate. Headquarters and Services Company provided the family with a tent and sleeping bags. The following morning, a Military Police escort arrived at Battalion to escort the Iraqi man and his children to a flight that took them to the United States.*

10 April
Battle of the Mosque

The fight for Baghdad was far from over. In one of the last but most fierce battles to secure objectives in Baghdad, RCT-5 attacked into the northwestern side of east Baghdad as the Division's main effort. First Battalion, 5th Marines was tasked with the seizure of the Azimiyah Palace, one of Saddam's palaces in eastern Baghdad.

International press vehicles staged near the Grand Mosque in downtown Baghdad.

Lieutenant Colonel Padilla's 1/5 began moving from their positions near the Diyala River, heading north out of Baghdad on Highway 5 late on the evening of 9 April. As the evening progressed into night, the battalion headed west and then finally turned back south on the east side of the Tigris River along Highway 2. In the pre-dawn hours of 10 April, 1/5 conducted a passage of lines with 2/5. The night was suddenly brightened with red tracers fired in the direction of the column as 1/5 pressed into the heart of the city. Every street brought RPG, heavy machinegun, and small arms fire. The tanks and tracks responded with .50-caliber, Mk-19, and 7.62mm machinegun fire as they continued to press toward the palace as quickly as possible.

The rate of incoming small arms and RPG fire continued to increase as 1/5 progressed further into the narrow streets and alleys of Baghdad. Navigating along the maze of Baghdad streets was extremely challenging. The maps that existed for the area did not show the level of detail required to navigate through the tightly constricted alleyways of an older section of the city. After making an exit off the main route and onto another highway, Alpha Company, in the lead, found itself on the wrong road, made a U-turn, and headed back along its original route. In the dark streets, the battalion pressed on to the objective, at times passing parallel to friendly units. "Don't fire to the left!" yelled Marines in the AAV's as they warned against possible friendly fire. Under the hailstorm of machinegun and RPG fire, Alpha Company's command AAV threw a track. The company halted and crewmembers hastily put the track back on. Bravo Company, led by Captain Jason E. Smith, seized the initiative, took the lead, and pushed on toward the palace. Visibility increased with the rising sun, and enemy fire became more desperate as they attempted to halt the Marines' advance.

While the battalion was still making its way toward the palace, they received an additional FragO, tasking them with the search of two possible American POW holding sites. This portion of the mission was given to Charlie Company. Near simultaneously, the Battalion was given yet a third mission, to proceed to the Imam Abu Hanifah Mosque where Saddam Hussein had reportedly been sighted only hours before. This portion of the mission was given to Alpha Company.

On the tactical radio nets, reports could be heard listing Marines wounded from incoming enemy fire, as well as reports of enemy locations and activity. The fires continued to hit the entire battalion convoy; including headquarters and logistics vehicles to the rear of the attacking armor. Marines stayed the course, calmly continuing the attack and returning fire. The calm disciplined leadership of Lieutenant Colonel Fred Padilla and the other leaders of the Battalion kept the Marines on an even keel in the confusing and dangerous firefight.

Bravo Company finally reached the palace and attacked to seize the entire compound. They were supported by the 81mm mortar platoon and the counter-mechanized platoon. They met fierce resistance surrounding the palace, but quickly secured the eleven buildings inside the wall. Though the palace was secured, Bravo Company continued to receive fire from enemy forces outside the complex. They quickly established defensive positions around the compound and cleared an LZ to evacuate casualties and to receive needed resupply of ammunition. Under enemy fire, Bravo Company Marines were able to evacuate 12 casualties wounded early in the fighting. Charlie Company, simultaneously, moved towards the two suspected enemy POW sites they had been ordered to investigate, meeting moderate resistance but finding no signs of POWs.

The Azimiyah Palace after it was secured by 1/5 in a fiercely contested action. This battle marked the end of conventional combat operations for 1st Marine Division. The Division was to find that security and stabilization operations could be just as deadly.

Second Lieutenant Joshua Glover of C/1/5 recalled his attack to the POW sites:

"Our company was sent two km out into the downtown area to search for POWs. The stiff resistance that met us in the streets included small arms and many RPG shots. As my platoon dismounted the tracks and hastily began preparing to move out for the search, we were taking fire from several directions. As we moved out, fire teams placed suppression down the streets to cover movement across these danger areas. My radio operator, Private First Class Hahn, and I approached one of the cross streets and one at a time bumped across behind the Marines providing suppression. As we each darted across, three or four enemy fired AK-47s at us from down the street. The rounds they fired at us as we crossed impacted uncomfortably close to each of us, sending chunks of walls flying. After we both got across the cross street, I turned and saw PFC Hahn standing there with wide eyes. He looked at me solemnly and said, "Sir, that hundred fifty bucks a month of combat pay sure don't make me feel too much better right now!"

Upon reaching and setting a perimeter outside the palace, 1/5's tank platoon and AAVs continued to fire on enemy targets in houses, down streets, and even inside mosques after receiving fire from them. Alpha Company arrived at the palace and prepared to attack to the mosque. While Marines began resupplying the company, word was passed that the Alpha Company Gunnery Sergeant, Jeffrey E. Bohr, had been shot and killed. The expression of numbness could be seen on the faces of the Marines as they fought to maintain their composure. Gunnery Sergeant Bohr had been a battalion stalwart, a fine example of the highest standards of Staff Non-Commissioned Officer leadership. How could this seemingly 'untouchable' Marine be killed? The shock and grief had to be momentarily put aside, however, as the embattled force made ready to reenter the maelstrom. The earlier report that Saddam Hussein was hiding out in a mosque just to the north of the palace was time-sensitive information, and time was short. Gunnery Sergeant Bohr was killed just prior to entering the palace:

"Approximately an hour prior to crossing the line of departure, the battalion commander directed that all non-critical vehicles and soft-skin vehicles be moved to the logistics trains to link up with the battalion at a later date. The precedence during the operation to date was there could be up to several days before we could link-up with our company logistical train. The Company Commander discussed the issue with the Company Gunnery Sergeant over the radio and said that going on the mission would be on a voluntary basis. The company's water and ammunition supply could become critical if cut off from the log trains for an extended period of time. Shortly after the offer, Gunnery Sergeant Bohr responded that he would be moving with the mechanized column in his two high-back HMMWV's.

During the movement, the battalion was attacked from both sides of the roads, at all intersections and overpasses. Gunnery Sergeant Bohr personally affected suppressive fires against the dismounted enemy with his M-16. He was personally responsible for no less than two confirmed killed enemy soldiers during the initial contact. The company was receiving heavy small arms and machine gun fire, as well as numerous RPG shots.

After the Company Commander's track was immobilized, and the company had to stop temporarily to effect repairs, Gunnery Sergeant Bohr moved his two HMMWV's behind the Company Commander's track and performed an administrative halt to confirm the vehicle's status, and he also checked his Marines to ensure they were not wounded. He gave words of encouragement and assisted in keeping his Marines calm and determined to achieve mission accomplishment.

On the move again to the palace, the company passed through a killsack from an enemy reinforced company. Alpha Company passed through the position to a dead end (a bridge that led into 3ᵈ ID's zone), which left the logistical vehicles exposed in the killsack as the company finally began to turn around. At this time, Gunnery Sergeant Bohr had the company logistical vehicles continue moving into the middle of the column, bypassing AAV's for protection from the intense RPG and small arms fire.

Gunnery Sergeant Bohr killed an unknown number of enemy soldiers while providing suppression and protection to his driver and the company log train. The company vehicles were riddled with small arms fire during the action and no less than two RPGs passed through the canvas of the Company Gunnery Sergeant's high-back. At no time on the radio with the Company Commander was there anything other than "we are good to go".

Gunnery Sergeant Bohr continued to deliver accurate rifle fire at the numerous enemy soldiers in his vicinity. When he was informed by squad radio that one of the headquarters Marines was wounded with a bullet through the wrist, he quickly called in a CASEVAC while still providing cover fire from his vehicle. During this attempt to save another Marines life, Gunnery Sergeant Bohr was killed instantly by small arms fire.

The support provided by Gunnery Sergeant Bohr was critical, as the battalion's logistical trains did not arrive until two days later at the presidential palace the battalion defended.

Gunny Bohr's moral and physical courage were inspirational to his Marines, in the logistics train and the men of the company. His volunteering to go on the mission was above and beyond the call of duty. His sound decision making under the most extreme fighting kept his subordinates alive and achieved mission accomplishment of combat re-supply for the company."

As Alpha Company pushed out through the gates of the palace, Marines aggressively engaged any suspected enemy positions, in an attempt to suppress RPGs prior to being fired. Marines kept a vigilant eye as they moved through the narrow alleyways. Every Marine kept a careful watch on windows, doors, and rooftops as they passed; covering high, low, front and back to reduce vulnerability in the tight urban environment. The tank platoon began receiving sporadic small arms fire as soon as they exited the palace gates. It intensified with each block, as RPG teams fired from streets and alleyways. An RPG hit platoon commander First Lieutenant Ashley Patrick's tank in the rear compartment, rendering the M1A1 a mobility kill. Staff Sergeant Johnnie Hughes in the number two tank and Master Sergeant Stephen Eisel in the number four tank pulled into the intersection to provide covering fire as Sergeant Daniel Belgrader pulled in to swap tanks with Patrick. Hughes' tank immediately took two RPG direct hits but continued to fight and suppress along with Eisel's. Lieutenant Patrick directed suppression into the source of the fires coming from houses and down streets.

The tight, restrictive streets and alleyways allowed only a single tank at a time, acting as the lead element and navigator. As intense enemy fire poured in from two and three story structures and alleyways no more than five meters wide, Master Sergeant Eisel skillfully navigated and simultaneously fought the lead tank. Upon reaching the mosque, the tank platoon maneuvered into an intersection and began providing main gun and machinegun suppression. The mechanized infantry company moved into position for the assault.

Alpha Company's Marines heard, saw, and felt AK-47 fire as they moved into the open town square adjacent to the mosque. Tank main guns, AAV up-guns (.50 cal and Mk-19), and M-16's all answered back with a deafening clatter and boom. Several Marines dismounted their AAV to set into a cold position for a SMAW shot. As they did, a loud explosion rocked the engineer's track and it lurched to a stop. Marines exited the damaged AAV to jump in the back of adjacent AAV's. The remaining engineers climbed into the bucket of an Armored Combat Earthmover (ACE) that was following in trace of the now useless AAV. As they did, others fired at the buildings, suppressing the fires that had interrupted their movement.

With significant enemy contact in several locations along Alpha Company's axis of advance and in the objective area, Grizzly 6 committed the RCT Quick Reaction Force (QRF) in support of 1st Battalion. First Lieutenant Keith Montgomery was leading the QRF into 1/5's zone when an RPG struck his tank. Undeterred he continued the attack as the company's lead element. Two more RPGs slammed into his tank while crossing an intersection en route to the eastern Presidential Palace. One RPG penetrated the tank, causing minor injuries to him, his loader and his driver. This one-in-a-million shot struck the tank in just the right spot between the heavy turret armor and the hull. Despite the shock and surprise it caused, Montgomery concealed his injuries from his crew and continued to lead the attack to clear a route through the city. At that point, the only functional weapon system was the coaxial machinegun, which could only be aimed by pivot steering the tank. Lieutenant Montgomery maintained his tank in the fight, although he was reduced primarily to battle command of his platoon and moral support. The QRF moved into position to support Alpha Company.

Lieutenant Patrick's tanks engaged the top and bottom story of every structure in the square as well as every alleyway with 120mm main gunfire in an attempt to suppress the enemy fire. Captain Blair Sokol the Alpha Company Commander, moved the company into a perimeter around the mosque and then coolly directed dismounted infantry to move in and clear the compound. The QRF completed the cordon by establishing positions on the roads leading to the mosque from the south. Alpha Company was facing a platoon size force from inside the mosque and a company size force surrounding it. The company was receiving intense fire from three sides, and the company's FAC, Captain Ray Lawler provided close and accurate CAS to relieve the pressure. The CAS included Air Force A-10 'Warthogs' firing within 65 meters with their 30mm cannon, and F-14's delivering 1000 lb bombs at 'danger close' range. The tank platoon and infantry company continued to engage RPG teams, enemy machine gun positions and troops for the next 4 hours. Staff Sergeant Avon Paul attempted to kill a sniper firing from a building with machinegun fire from his tank. When that proved ineffective, he received permission to use a Multi-Purpose Anti-Tank main gun round. The round silenced the sniper but started a fire in the building (which proved to be a Fedeyeen ammunition storage facility.) Soon after, the area was rocked by secondary explosions as ammunition stored in the building cooked off.

The dismounted Marines moved in to clear the mosque and capture any Regime figures who might be inside:

"As the company began isolating the mosque, the amount of enemy fire reached its peak, with RPG and small arms fire coming from both a platoon size element inside the mosque and a company size element in the surrounding buildings. The company received at least 90 shots from RPGs, 33 of which were confirmed hits on the AAVs. The platoon commander's AAV took a direct hit from an RPG, wounding four Marines. Staff Sergeant Coleman Kinzer dismounted his vehicle in the middle of the worst part of the fighting and moved to the platoon commander's vehicle in order to ensure face-to-face coordination prior to conducting a breach into the mosque.

The platoon moved to the southern side of the mosque and began dismounting in the vicinity of a gate in the wall surrounding the mosque. Staff Sergeant Kinzer was the first Marine at the gate, and exposing himself to RPK machine gun fire, coordinated the breach on the western building of the complex to include a SMAW Novel Explosive shot that killed an RPK gunner. He then took a squad up to the building, threw a grenade through the breach site, and gained a foothold. Once the foothold was gained, Staff Sergeant Kinzer remained at the lead of the platoon while conducting high-intensity MOUT room clearing with more hand grenades. The squad thoroughly cleared the building and ensured that proper security was posted.

At this time, coordination was made for the movement across a courtyard and into the main temple. Second platoon had now entered the complex and while his platoon commander was coordinating with them, Staff Sergeant Kinzer continued to press the attack. He led the attack across the courtyard and gained a foothold in the temple. Upon making entry into the temple, he came under small arms fire from the north side of a long corridor inside the building. The enemy fired several rounds before they fled out of the north exit. Recognizing that he did not yet have enough forces in the building to pursue, Staff Sergeant Kinzer held the team in place until he had enough forces to clear the large cavernous spaces of the main temple. He continued to coordinate the attack until the platoon commander arrived.

During consolidation, it became apparent that there were several weapons caches hidden on the north side of the mosque. After several attempts to gather the weapons failed under heavy sniper fire, the platoon commander tasked Staff Sergeant Kinzer with gathering the weapons. Staff Sergeant Kinzer coordinated the use of smoke and suppressive fires and successfully gathered the weapons. Staff Sergeant Kinzer was the last Marine out of the mosque and the last Marine to load into the AAVs, ensuring accountability for all his Marines."

With the assault on the last enemy hiding in the mosque underway, First Lieutenant Douglas Finn, the Company Executive Officer for the QRF, began to direct recovery operations for the numerous immobilized vehicles. Second Lieutenant Dennis Doyle dismounted his platoon and maneuvered through the alleys to establish perimeter security around Lieutenant Patrick's abandoned tank. Clearing the buildings around the company battle position, Doyle's platoon encountered several terrified Iraqi civilians caught up in the battle. The platoon detained an SUV containing several thousand dollars, Iraqi military uniforms, and documents. Another family they met voluntarily cooked for the Marines during the fight. For Lieutenant Doyle's platoon, the battle was known as 'The Battle for Macaroni Alley'.

Alpha Company did not capture any Regime leadership figures at the mosque, but the vigorous delaying tactics used by the fighters here may have given them time to escape through the maze of winding Baghdad streets. Once the mosque was secure, Alpha Company began the movement out of the square through the route cleared by the QRF. The tank platoon provided rear security. The platoon continued to take fire and engage enemy positions until completely out of the square and into the small cleared alleyway. The narrow alley had built for pedestrians and donkey carts, and could barely accommodate the behemoth M1A1s. As the tanks squeezed through the route, the main turrets could not traverse, and several of the buildings' external air conditioner units were ripped out of second-story windows. The tank commanders and loaders were eye level with the second level of the homes and looking up to the third and fourth stories. As they moved down the alleyway they were given one last task, to destroy an immobilized Amtrak. As rear security, Master Sergeant Eisel dropped a thermite grenade into the AAV commander's weapon station and destroyed the vehicle.

In securing their assigned objectives, 1st Battalion suffered 1 killed and 60 wounded Marines. They defeated the suspected enemy battalion in zone and killed several hundred Special Republican Guard and Saddam Fedeyeen fighters. Following 1/5's swift and decisive attack, thousands of Iraqis spontaneously took to the streets of Baghdad to cheer and thank the Marines and Sailors. Once again, destroying the last vestiges of the Saddam Regime earned the Division Marines the grateful thanks of the Iraqi people.

RCT-5 experienced another enemy tactic on the 10 April. As the Marines of 3/5 were establishing a checkpoint in the city, a suicide bomber attempted to kill as many of them as he could. First Lieutenant Timms from Lima Company 3/5 remembered the incident:

On 10 April we arrived in Baghdad. Second Platoon was tasked with manning a control point that was the border between three different units. The CP was located at a busy intersection between a main street, two side streets, a bridge over a highway, and an on-ramp. Second Platoon was reinforced by Machine guns, SMAWS, and a section of CAAT. Once we arrived we established security, which was challenging, as the crowds numbered in the hundreds and would not disperse. We set up on the right side of the end of the main street, 100 meters in front of the bridge. In the middle of the road on a median was an old Iraqi post made of sand bags. We set up a working party to move the sand bags over to our side of the road to block off the on-ramp and set up a post of our own. While this was being done an individual walked up through the crowd to the working party. Civilians shouted warnings as they ran away. The individual had an explosive device, and he detonated it near the Marines. Fortunately, The bomber's body absorbed much of the blast, but four Marines were severely wounded.

The heroes of RCT-5 responded with discipline and efficiency, as they evacuated the casualties, and restored the checkpoint.

Transition to Security Operations

No one was immune to the looting in Baghdad.

A rare photograph of an Iraqi family. The smiles on their faces reflect their joy in new found

The 'Battle of the Mosque' marked the end of major combat operations in eastern Baghdad, although the violence continued. Resistance remained, but the enemy encountered from here on was poorly organized and less capable than that encountered before. The Division did not have full control of all city sectors in zone, but anticipated having full control of the remaining sectors by the end of the day on the 11 April, thus securing the eastern half of Baghdad in its entirety. The Division's logistics posture was also steadily improving thanks to the heroic efforts of the 3d MAW and CSSG-11.

Upon entering Baghdad, 11th Marines was given two additional tasks that are not typically assigned to an artillery regiment; to establish its own sector for security operations, and to man and operate the 1st Marine Division CMOC. The 11th Marines approached these new tasks with the same gusto and professionalism that they had when providing artillery support for the last 22 days. The regimental headquarters immediately established two command posts, one in the 11th Marine's sector (for security and stabilization operations) and a second at the Palestine Hotel in downtown Baghdad (where the CMOC would operate.) At the CMOC, 11th Marines began building on the excellent groundwork laid by RCT-7.

In their sector, the regiment's task was to restore order and help ease the suffering. Within 24 hours the sector around Rasheed was greatly improved. Both 1/11 and 3/11 worked throughout the community conducting patrols, removing weapons caches, detaining looters, and showing the Iraqi people that Americans were not conquerors, but liberators. Through it all, 11th Marines maintained a firing capability to support counterbattery missions and calls for fire anywhere in the city.

Colonel Marletto was assigned as the Civil-Military Operations Coordinator for the 1st Marine Division, and he set the tone for accomplishing the daunting challenges in Baghdad. Through daily meetings with the RCT commanders, the 3d Civil Affairs Group liaison detachment, civic leaders, non-governmental organizations, and former Iraqi government officials, he organized the CMOC into functional departments to focus the humanitarian efforts throughout the 1st Marine Division's area of responsibility. Each day the CMOC accomplished more, achieving small 'victories' to improve the situation in Baghdad.

The rest of the Blue Diamond Marines continued to work at improving the security situation. During support to other clandestine operations in the Division's zone, elements of the Division were routinely tasked with clearing a temporary area of operations for raids. One mission in 3/4's zone was planned to hit an objective very early in the morning, not far from the Palestine Hotel. Throughout the night, the RCT-7 watch officer tracked the movement of the forces that were conducting the raid. He noted traffic on the raid force's tactical radio net indicating that there were bright lights shining on them from a building off to their right. Just then, the Division watch officer called to say that Dan Rather was doing a live feed from the Palestine Hotel where he reported heavy armor driving by. The international media was filming every minute of the 'clandestine' operation as they snuck through the urban area. The raid force operations officer called on the radio to report their current position. The RCT-7 watch officer responded, "I know right where you are. Dan Rather has got you on the news right now!" This successful mission resulted in the capture of an internationally known terrorist.

Abandoned Iraqi missiles are found throughout Baghdad.

Small Unit Actions Continue on the Baghdad Streets

After the initial push into the city, the RCTs divided their areas of operations into battalion and company sectors. In these zones, Marines established permanent traffic control points, set up spot check points and ran countless foot and vehicle patrols. Like good beat cops, the Marines were learning their neighborhoods. Each battalion created a collection matrix to record

The narrow maze of Baghdad streets make security operations difficult.

One of hundreds of caches of landmines in Baghdad

information the companies were generating on the streets. The matrix included locations and details of significant events, terrorist incidents, power plants, gas stations, hospitals, and anything else that could help restore stability and get the Iraqis back on their feet.

Even though major combat actions in the city had ceased, Baghdad remained a deadly and dangerous place. The vast majority of the five million residents welcomed the Marines, but the city also harbored those who did not. In a 'gun' culture, almost every household was armed, and the city was awash in automatic weapons and RPGs. The Marines of the Division worked to restore control during the day, and hunted the remaining Fedeyeen at night. The intensity of the resistance slowly diminished as the security situation improved, leaving only isolated incidents by the time the Division turned the Baghdad battlespace over to the US Army.

Kilo Company was tasked with conducting platoon-size urban security patrols in its area of operations as assigned by the battalion. First platoon conducted its patrol in the northern part of the company sector during the day without incident. Second Platoon was launched early on its night patrol to investigate random shooting roughly 700 meters west of the battalion CP. The shooting was random and, upon not being able to locate any suspects, 2d Platoon continued patrolling towards an intersection 600 meters south of the battalion CP.

It was at that checkpoint that the platoon came under the intense enemy fire of a near counter-mechanized ambush. Fighting ensued for just a few minutes as the platoon took automatic and RPG fire from enemy positions located all around the intersection and from as close as a few meters away. Sgt Navarro's 3d Squad was unable to effectively fight through the near ambush due to the enemy's use of natural obstacles and was ordered to pull out of the kill zone. Sgt Velis' 1st Squad, Sgt Hoover's 2d Squad, and the scout-sniper team quickly gained fire superiority over the enemy positions. Second Squad was instructed to cover the platoon's flank while 1st Squad bounded by fire team out of the kill zone with 3d Squad in over-watch. Finally, 2d Squad linked up with the rest of the platoon in consolidation and in securing a nearby field as an LZ after all enemy combatants fled or were destroyed. The company QRF was launched (under the command of Lieutenant Edwards) with two CAAT vehicles and linked up with 2d Platoon in order to effectively secure the LZ for CH-46s, which eventually evacuated the dead and wounded. The next morning, despite puddles of blood in the enemy's positions, the company discovered that all enemy casualties, shell casings, and weapons had been removed during the night. The local civilians were questioned but would not speak of the incident out of fear of retribution by the guilty parties. During the fight, Lance Corporal David Owens and Staff Sergeant Riayan Tejeda were both mortally wounded. SSgt Tejeda was ultimately recommended for the Silver Star (posthumously) for his heroic actions.

An example of the intensity of the lingering resistance was an ambush that occurred on 11 April on the northern side of the city. At night, the streets across Baghdad echoed with automatic weapons fire. It was often hard to tell the direction and target of this fire, but it was clear that the streets were still very dangerous, especially at night. To contain this threat, 3/5 (like other battalions in the city) conducted an aggressive patrolling posture. Only by hunting down the remaining violent elements that threatened the civilian population, would security be improved. First Lieutenant Michael Prato, from Kilo Company, 3/5, recounts:

The loss of Staff Sergeant Tejeda and Lance Corporal Owens was difficult for the Marines of Kilo Company. These Marines had fought heroically, and had died freeing another Iraqi neighborhood from the last of the Regime oppressors. Staff Sergeant Tejeda had died while selflessly exposing himself to enemy fire in order to move all of his Marines to safer ground. He lived long enough to make sure his fellow Marines promised to tell his children how much he loved them. The commitment of these Marines to the future of all the children of Iraq was heroic. The streets of Baghdad remained a deadly and violent place, and the loss of these Marines served as a warning to the rest of the Marines in the Division.

Newly freed Iraqis pour back into the city.

The elimination of enemy fighters in the urban area took several days longer. Marines from each of the Division's regiments and separate battalions were engaged in restoring order to the streets of Baghdad. Manning checkpoints, supporting raids, or conducting security patrols were common tasks to all of the Division's Marines. The Division continued to suffer casualties in the execution of these dangerous missions. Corporal Jesus Gonzalez was killed on 12 April while manning a security checkpoint. A tank crewman from 1st Tank Battalion supporting India Company of 3/4, Corporal Gonzalez had displayed specific heroism earlier during the mission to secure the southern bridge over the Diyala River. During that earlier mission, his company had come under heavy small-arms and RPG fire. Corporal Gonzalez had immediately identified the location of the enemy and marked the target by providing fires with his M-16 from atop the turret until his tank commander could call it in on the radio. During that fighting, Corporal Gonzalez had displayed a situational awareness and composure far beyond that expected from a Marine of his rank and experience. On 12 April, his platoon was providing security for a hospital compound in East Baghdad and Corporal Gonzalez managed a security checkpoint, controlling a frequently unruly crowd. While he was standing this post, a Fedeyeen gunman opened fire on the Marines at the checkpoint. Corporal Gonzalez immediately moved to the turret to assist in returning fire. As he moved to the turret he was struck by enemy fire and fatally wounded. His immediate response under fire revealed his dedication to the Marines around him and the mission at hand. Corporal Gonzalez was admired and respected by everyone who observed him.

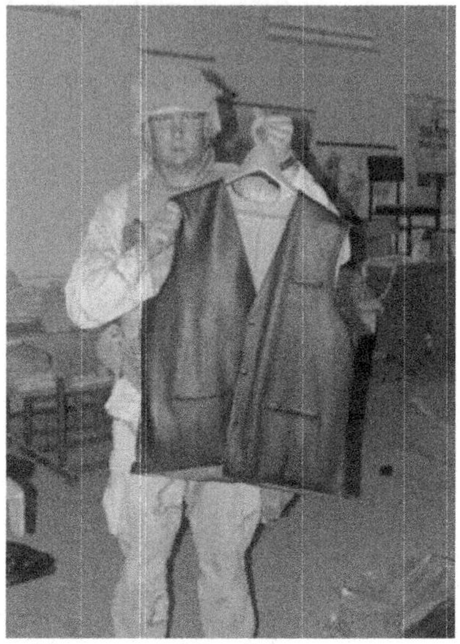

Lieutenant Matt Danner holding a suicide bomber vest.

In addition to combat patrols, the Marines of the Division worked ceaselessly to clear the city of remaining weapons caches and unexploded ordnance. As the former Regime had used their schools and mosques to hide weapons, the people were eager to cooperate with the Marines to have the threat removed. Acting on one tip, Marines from RCT-7 entered a school. Inside, they found a terrorist bomb production facility full of black leather 'suicide vests.' After taking control of the facility, 3/7 counted 280 of the suicide vests, two blind-timer briefcase bombs, and dozens of SA-14 SAMs. The suicide vests looked brand new, carefully hung from a dry-cleaning rack, each individually wrapped in a plastic bag. Across the city, in each RCT zone, the engineers and EOD teams worked around the clock to disarm, disable, and destroy the tremendous numbers of weapons, ammunition, and explosives that were uncovered.

Marines found that being 'No better friend' to the Iraqi people was rewarded with smiles and trust.

The logistics focus also changed when the Division crossed the Diyala and began operations in Baghdad. In addition to their primary focus on ammunition, food, and fuel; logisticians added quality of life items. Mail arrived for the first time in weeks. Health and comfort packs were provided for basic hygiene. Even though the Marines remained busy, the static nature of operations in Baghdad allowed the Division to get logistically healthy.

11 April
So what Happened to the Enemy?

As operations continued, the Division tried to reconcile pre-battle intelligence reports with the lack of a conventional military defense of Baghdad. The defense had been rigorous, but had lacked the coherence and firepower that had been expected. The Division had been told to expect a 'Chechnya', but had gotten something quite different. There were a number of likely causes of this apparent disparity. Probably the most significant, but least understood, was the internal political situation of the Saddam Hussein Regime. As always, it had been easier to count enemy equipment than it was to judge the enemy's will. The Regime leadership, never coherent to begin with, and motivated by fear, had begun to fracture even before the war began, and senior members of the Iraqi Regime had reportedly tried to escape the country to Syria or other destinations. In a quickly accelerating death-spiral, senior military leadership began to cut their own deals as they had opportunity. Some chose to deal with the Americans, but most chose to wring the last wealth they could from their position of power and attempt to go 'underground', out of the notice of military forces. Some of these individuals would not be captured for some months after the fighting had ceased. This spiral eventually took its toll on even the tactical leadership of the military units around Baghdad. With no senior military leadership remaining, and no Regime figures left in Baghdad to protect, the Republican Guard and Special Republican Guard troops had no reason to stay and fight. There was nobody left with enough authority to actually surrender. Centralized control had broken down, and all the pieces remaining were acting in self-interest in a bid for their own survival.

Murals of Saddam were defaced by Iraqis across the city.

The Division also expected to be targeted by chemical munitions at some point in the campaign, but the Iraqis had likely destroyed their last stockpiles to hide them from weapons inspectors. There was a good chance that those decision makers who could have executed the use of chemical munitions had either been killed, had gone 'underground', or had fled by this time. This would have aggravated the difficulty the Iraqis would have had in retrieving chemical weapons and making them ready for use. The rapidity of the American attack surely caught them by surprise, and did not allow them the time needed to reassemble the various agents and delivery systems into a tactical capability.

Iraqi conventional military forces received the greatest attention from the US shaping fires, and a great deal of their capability was decimated before they were able to use it. Knowing the survival rate of their equipment in a match-up with US Marines, few Iraqi soldiers were willing to go into battle at the controls of a tank or armored vehicle. The ease with which a 'soldier' could become a 'civilian' (by changing clothes and walking away) left very few with the discipline or commitment to face the alternative of certain death. The conventional military structure, gutted and misused by Saddam Hussein, was no longer a fighting force that could present a credible defense against a first order military.

The Fedeyeen and religiously motivated fighters were also greatly diminished by this time. To many of the *jihadis*, the concept of death at the hands of the Americans seemed less attractive when their human wave attacks against American firepower were mowed down mercilessly. Those that remained were of the same ilk as the anti-western terrorists that existed elsewhere. While cunning, neither their courage nor individual soldiering skills were sufficient to make them worthwhile enemies. Their lack of respect for laws of war and chivalry called into question their very manhood, as evidenced by the Fedeyeen's frequent use of women and children as human shields, and the murder of helpless non-combatants. For all the protestations of the 'thousands willing to die for Saddam', most chose a path of lesser resistance when actually given the choice.

At the base of the pyramid, the very Iraqi character seemed to work against a 'defense to the last' that some other cultures might adopt. At the crossroads of civilization, the Iraqis were dealmakers by heritage and culture, under the domination of a despotic regime. When the feared former Regime no longer offered the best deal, the Iraqis were happy to shift their position. Saddam had promised the Americans wouldn't come. Saddam had promised that if the Americans did come, they would be turned back with missiles. Saddam had promised that the Americans would not fight on the ground, and would be turned back by the strength of Arab manhood. Saddam had promised that the international community would somehow intervene. Promises had run out, hope was gone, and it was time to concede the end of the Saddam Hussein Regime.

Planning for the Kirkuk Operation...or Another

Planning for the Kirkuk operation continued apace. By the morning of the 11 April, less than 24 hours after receipt of the warning order, the Division had assembled a comprehensive plan to attack an additional 200 km into whatever remnants of another four enemy divisions that still existed. Brigadier General Kelly would command 'Task Force Tripoli' in the accomplishment of this mission. With the Division Main CP firmly established in eastern Baghdad, the Division Forward CP was reconfigured to the role of Task Force Tripoli headquarters, providing the command and control for General Kelly's mission. The force would be brigade-sized, largely LAR-based, with the addition of a company of truck mobile infantry and an artillery battalion. Logistics support would roll with the convoy to Kirkuk, but the seizure of temporary airstrips for both rotary wing aviation and KC-130 task force re-supply would be the linchpin of a successful effort. The G-2 had identified five airfields to choose from that offered intermediate re-supply points between Baghdad and Kirkuk. The routes had been identified, Kurdish linguist support had been arranged, and a helicopter had been dispatched all the way from Kuwait with the required maps. The critical infrastructure in the Kirkuk oilfields had been identified and objective folders had been built. The analysis of the enemy's disposition and willingness to fight was complete, and the recommendations for the scheme of maneuver developed. The LAR Battalions had conducted their physical preparation, and would be ready to roll in 24 hours. It appeared that all systems were go, so no one was surprised when the call came to change the mission.

As higher headquarters continued to adapt to the rapidly changing situation on the ground, the Division took mission changes in stride. In this case, the oilfields had fallen into friendly hands. The Kurds had been pounding on the door of the Kirkuk oilfields for years, and now the crumbling Iraqi Regime had weakened to the point that Kurdish forces, with US Special Forces advisors, were able to seize key nodes in the strategic oilfields. The Division had reacted quickly, but had been given the task about 48 hours too late to be needed. With Kirkuk in friendly hands, there was no longer a requirement for the Division to send a force there. But there was another mission.

The last Regime holdout and possible safe haven left in Iraq was the city of Tikrit, Saddam Hussein's hometown. On the afternoon of 11 April, the Division was ordered to attack to Tikrit instead of Kirkuk. Tikrit lies 200 km to the west of Kirkuk. Much of the mission planning for the Kirkuk mission remained unchanged, but the intelligence and evaluation of tactical objectives would have to be redone. The timeline for the mission, however, was to remain the same. The Division had less than 12 hours to assemble the plan, and would be expected to LD no later than 12 April. Having proven its ability to produce a mission plan for an attack of this magnitude in 24 hours, the Division proved its ability to do it again in only 12 hours. The entire process that had been completed in support of the Kirkuk mission was repeated for Tikrit, only in less time. The changed battlespace and enemy situation demanded a renewed analysis. The G-2 went to work immediately, and was still assembling target lists when the task force rolled up to the gates of Tikrit less than 48 hours later. Task Force Tripoli was born in the saddle, and would soon be riding to its first objectives.

The sun sets on another challenging day in Iraq.

Chapter 7
Task Force Tripoli

With Baghdad increasingly secure from conventional threats, the Coalition had gained all of its military objectives in southern Iraq. On 10 April, CFLCC tasked the MEF to send a task force to secure the Kirkuk oilfields, 200 km to the north of Baghdad. The mission was passed to 1st Marine Division, who formed an LAR-based task force under the Assistant Division Commander, Brigadier General John Kelly. Task Force Tripoli was formed and completed planning for an attack to the northern Kirkuk Oilfields. The following day, however, Kurdish forces, aided by US Special Forces and the US 173d Airborne Brigade, broke through the Iraqi defenses in the north and captured Kirkuk. With the Kirkuk oilfields in allied hands, the Coalition's attention turned to the last remaining holdout for Saddam Hussein's Regime, the northern city of Tikrit. CFLCC again turned to the 1st Marine Division to accomplish the task of taking this important objective. For the Division, the mission and timing for an attack north remained the same, but the objective became Tikrit instead of Kirkuk. With a requirement to move out later that same day, Task Force Tripoli completed its preparations without delay.

Task Force Tripoli took its name from the Marines' 1805 exploits against the pirates of the Mediterranean Barbary coast, which also included a long approach march through the desert against its objectives. Now, a set of modern-day pirates in northern Iraq required similar rough handling. In fact, Task Force Tripoli would be operating much closer to the Mediterranean Sea than the Arabian Gulf for most of its mission. As they set out on this long distance mission, the Marines of Task Force Tripoli intended to create the makings of another verse to the revered Marine Corps hymn. Now set in modern times, this operation would further demonstrate the reach of America's Marines operating from the sea.

Tikrit was Saddam's hometown, under the control of his Tikrit tribesmen and allies. During Saddam's rise to power, it had been a wellspring of support. With less than 12 hours notice, Blue Diamond redirected its Task Force Tripoli to seizing this last Regime stronghold, removing it as a possible place of refuge for Regime leadership, and eliminating it as a source of instability. The Task Force commander gave his planning guidance to the Task Force Tripoli Operations Officer, Lieutenant Colonel Robert Sinclair, the Intelligence Officer, Lieutenant Colonel Mike Groen, and the Fire Support Coordinator, Lieutenant Colonel Gary Smythe (USA). In less than 12 hours, the staff had developed a mission, routes, scheme of maneuver, and tactical objectives in the Tikrit area. By the evening of 11 April, Task Force Tripoli was prepared to launch its attack. The lead elements of the Task Force, including the Division Forward CP, began the march to the Tactical Assembly Area north of Baghdad. As the sun descended on the evening of 11 April, the Marines of Task Force Tripoli experienced an auspicious beginning to their mission. Passing north through the crowded streets of mostly Shia eastern Baghdad, the column was heralded with waves, shouts, smiles, and kisses from the newly liberated Iraqi people. The people lined the streets and continued to wave until it was too dark to see. The Task Force hoped to encounter a similar sentiment in the mostly Sunni citizens of the Tikrit region.

The Iraqis Prepare the Tikrit Redoubt

Unlike the Shia in the south, the people of Tikrit had benefited greatly during the days of the Regime. The Dictator's hometown had received more than its share of benefits, and the arrival of US forces implied the 'gravy train' was over. There was a great deal of uncertainty about the reception the Marines would get from the civilians who had benefited most from Saddam's Regime.

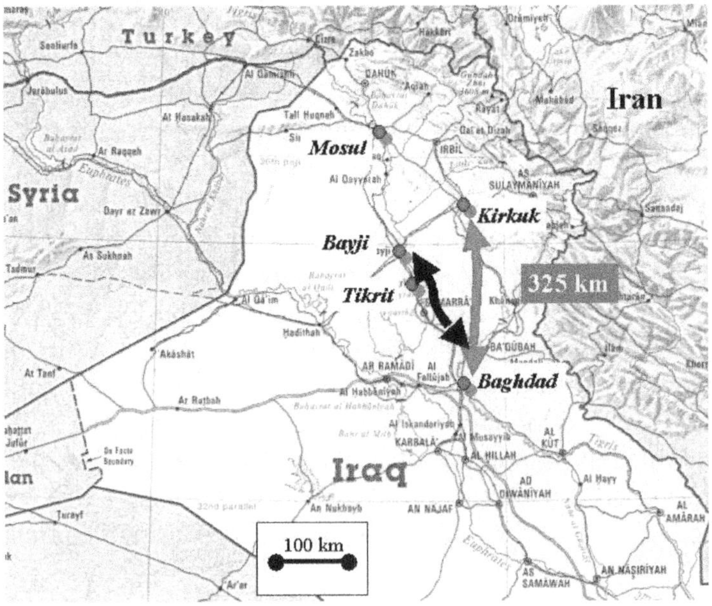

Tikrit was the final attack 1st Marine Division conducted during Phase III offensive operations. It was also the hometown of Saddam Hussein.

The Tikrit area had also been assessed as a probable location for the last stand of Regime forces. Even though the central governance of Iraq was located in Baghdad, a large portion of the citizens of the Baghdad area were Shia Muslims, and there was a constant threat of insurrection from the masses. If the Regime's control of Iraq began to slip, it was expected that Saddam Hussein, other key leaders, and the Special Republican Guard would seek to escape the threat from the mob. They would likely establish a final redoubt in Tikrit, surrounded by members of Saddam's own tribe and its allies. Here, it was thought, Regime figures could 'go to ground', and would be protected by thc pcople.

One brigade of the Special Republican Guard (SRG) was routinely garrisoned in the city of Tikrit. By 12 April, this Brigade had been reinforced with a number of remnants from the Republican Guard, Fedeyeen, and foreign fighters. Now that Baghdad had fallen, it was assessed that the last Regime die-hards were gathering in Tikrit. The 3[d] SRG Brigade formed the core of the conventional Tikrit defenses, but it was not a significant combat force. This was, literally, the 'palace guard' of the Regime in the Tikrit area, more accustomed to standing post and intimidating civilians than fighting. The 3[d] Brigade had reportedly been reinforced, however, with a mechanized battalion from the 4th SRG Brigade, which would give it a bit more punch. In addition, there were reported elements of the regular Republican Guard reinforcing the Tikrit defenses in approximately battalion strength. These units were supported by up to three battalions of artillery, as well as engineer and support assets from the Regular Army I Corps, who maintained its headquarters in the area. In addition to the conventional defense, Task Force Tripoli prepared to face up to eight hundred paramilitary fighters, to include foreign terrorists and the last of the Fedeyeen. Making the challenge even more interesting, Tikrit and the surrounding areas were allegedly home to much of Saddam's hidden stockpiles of missiles and chemical weapons. If the Regime had not used these by now, it likely never would; but Tripoli was prepared to encounter these weapons too.

Open source news reports from international media inside Tikrit reported the presence of a significant defensive force. The US intelligence community reported large Fedeyeen rallies being held in the downtown stadium area. Regime roadblocks and checkpoints were established along the major highways into and out of the city. This preparation was not entirely oriented on an attack from the south. As Tikrit lies near the Iraqi northern frontier, the Iraqi defenders had to be concerned with threats coming from multiple directions. There was a significant fear of the northern Kurds among the population of Tikrit. Because of the brutal treatment of the Kurdish people by Saddam's Regime, an invasion of vengeful Kurds was to be feared even more than an invasion by the Americans. When the Kurds seized Kirkuk on 11 April and proceeded to advance toward Tikrit from the east, the Tikrit defenders attempted to drop two spans of the Tigris River bridge that connected Tikrit to its 'suburbs' on the east side of the river. The Iraqis demonstrated their usual bridge-blowing prowess, however, and only managed to damage two of the spans beyond repair, while still leaving it passable by vehicles. The damaged bridge was to have a significant impact on Task Force Tripoli operations in the Tikrit area.

Despite the intelligence reports indicating a conventional-style defense of Tikrit, the TF-Tripoli intelligence section assessed the Iraqis would follow the pattern that the Division had seen previously. When attacked on the ground, the uniformed Iraqis would likely abandon their

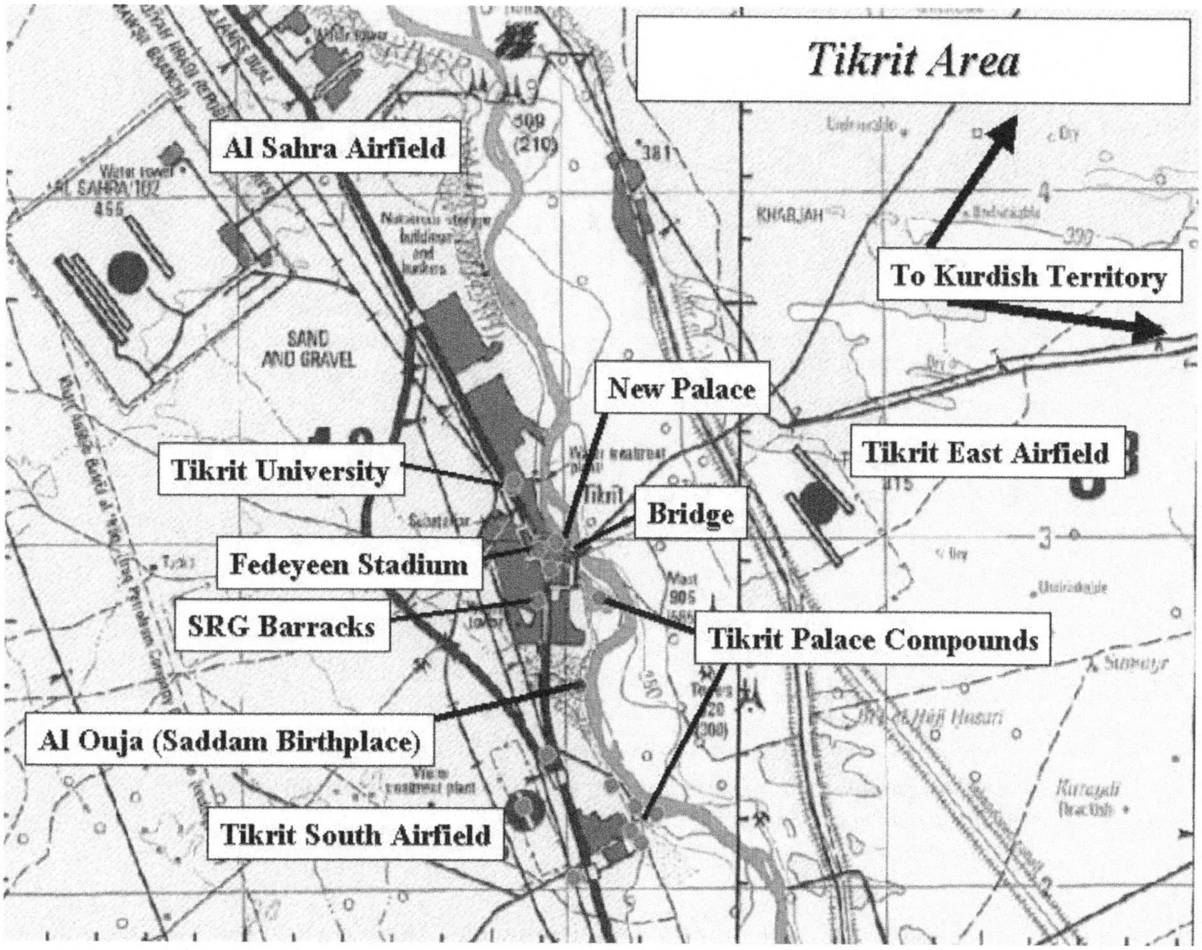

Marines occupied areas within the city of Tikrit and outlying areas to the north, south, and east.

posts while the Fedeyeen would put forward an ambitious but ineffective defense. The local populace, even though formerly among the Regime's most loyal subjects, would probably not take up arms or willingly support the paramilitaries. The Task Force was reminded of the 'deal maker' mentality of the population. The Regime no longer had much to offer in return for loyalty. Cooperation with the Americans would likely be the norm, but only after the last of the Fedeyeen and foreign fighters had been eliminated.

The Operational Design

The Highway 1 avenue of approach would be the one most expected by the Iraqis, but also offered the quickest route into the city. The Task Force judged that a rapid advance up Highway 1 would put the Marines at the outskirts of Tikrit before the Iraqis would have a chance to further prepare. The Light Armored Vehicles (LAVs) and other wheeled assets of the Task Force would support this rapid attack. The Task Force commander decided to bypass the numerous Regime targets along the way in Balad and Samarra, and sought to cordon off Tikrit as rapidly as possible. Because Tripoli's assembly area was on the northeastern side of Baghdad, the first task would be to get across the Tigris river. To gain operational speed right from the beginning, Tripoli would use one of two small bridges across the Tigris just to the north of Baghdad. These bridges were very small, and required a difficult approach over poor roads and through narrow village streets. The Tigris crossing options were limited, as the only other crossings were inside

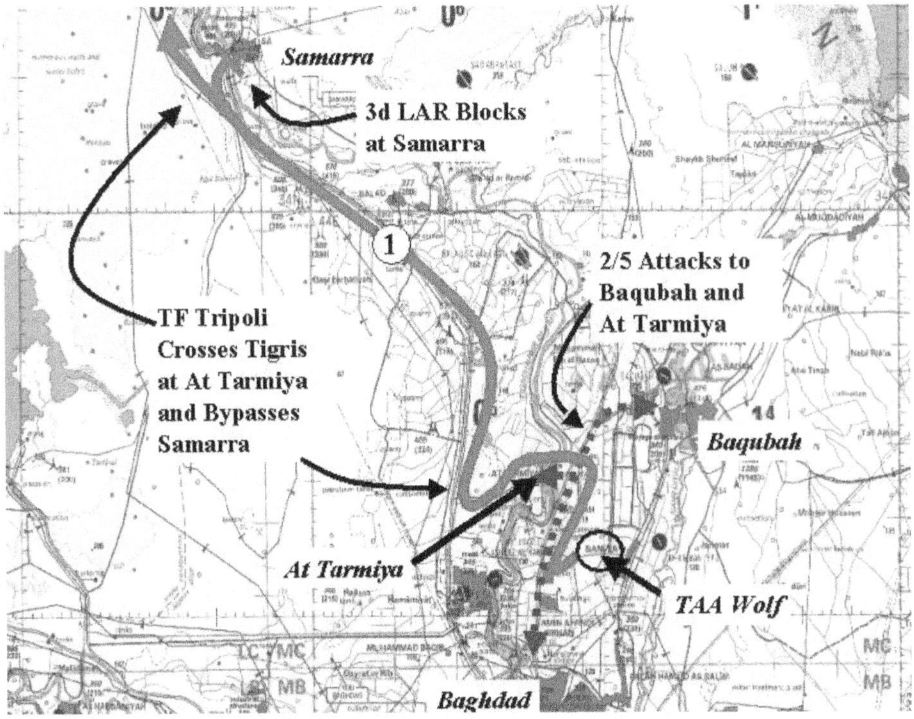

TF Tripoli's guidance was to make a swift and violent assault on the objective, bypassing urban resistance and closing the doors on the enemy before he could withdraw or defend.

Baghdad itself, or over 150 km to the north along Highway 5. Neither of these were acceptable. Crossing in the Baghdad urban area would require coordination for passage through the Army's zone in Baghdad. This area was reportedly still the scene of significant combat action, and passage would prove difficult. After some analysis, the Tripoli staff decided the risks posed by close ambush and restricted approaches were offset by the speed the Task Force would achieve by avoiding a cumbersome decision-making process required to pass through the Baghdad urban area. The US 3d Infantry Division on the ground was eager to pass the Tripoli Marines through, but the approval process would require coordination at higher levels. A long roundabout road march to the north (along Highway 5) would be even worse.

Once across the Tigris, Tripoli would conduct a rapid night march up Highway 1 to just outside Samarra. Samarra lies on the east side of the Tigris, with a bridge connecting it to the Highway on the western bank of the river. There had been some threats noted from the Samarra area, but the enemy disposition there was largely unknown. At Samarra, the 3d LAR 'Wolfpack', under Lieutenant Colonel Stacy Clardy, would be left to establish a blocking position near the bridge, while the rest of the Task Force bypassed the city on the west side. Thirty kilometers up the road, 2d LAR's 'Barbarians', under Lieutenant Colonel Ed Ray, would use a Tikrit highway bypass to establish a blocking position to the north of the city. Once the northern exit had been blocked, the 1st LAR 'Highlanders', under Lieutenant Colonel Duffy White, with a motorized infantry company (G/2/23) attached, would attack into the city. As the Highlanders pushed up, they would clear the southern Tikrit Airfield for the CSS element that was following in trace. CSSC Tripoli, under Major Mike Callanan, would establish the airfield as a logistics base for ground operations and a FARP in support of continued air operations. A

Despite an invasion and a ground war, Iraqi civilians attempted to carry on with their normal daily activities. The Iraqi people were reassured that there was "No Better Friend" than a United States Marine.

detachment of Naval Special Warfare SEALs accompanied the Task Force, and would be used in a reconnaissance role. The NSW liaison officer to the Division, Lieutenant Commander Brad Treadway, had been a key advisor to the staff throughout the campaign, and would serve in a similar capacity for Task Force Tripoli. An attached artillery battalion, 5/11 under Lieutenant Colonel Jerry Smith, would provide fire support. Delta Company of the Combat Engineer Battalion, two HET teams, a tactical PsyOps team, a civil affairs detachment and a TOW platoon rounded out the task organization.

Task Force Tripoli's organization for combat was unique in a number of ways. First, the Task Force headquarters was formed using the Division Forward CP as a nucleus. Over the many months of rehearsals, operations and planning, only minor changes had been required to the organization of the Forward CP. The Division Forward headquarters had been scaled to be closer to a brigade-sized headquarters rather than one built for division-sized operations. With a relatively small footprint, the Forward CP configuration was still a fully functional Division Headquarters, with the robust command and control required for this long-range operation. More importantly, however, was Task Force Tripoli's integration of three Light Armored Reconnaissance battalions working under a single commander. Under General Kelly (himself a former LAR battalion commander), Tripoli was, in effect, an LAR Regiment. The 'Task Force' was formed by reinforcing this core with its own CSS element, a company of truck-mobile infantry, combat engineers, and an artillery battalion. This proved the perfect combination of speed, sustainability, and lethality for this mission.

Besides north and south, the east offered an approach to Tikrit as well. Task Force approaches to Tikrit on the east side of the Tigris were carefully considered, then discounted, as the damaged bridge would prohibit maneuver of armored vehicles into Tikrit from that direction. Forces operating on separate sides of the wide Tigris River would be (tactically) non-mutually supporting. It was decided that even if the initial attack were to come only on the western bank of the Tigris, objectives on the east side of the River could later be attended to by HMMWV mounted patrols after the rest of Tikrit was isolated.

Logistics support to Task Force Tripoli also required flexibility and resourcefulness on the part of the Division's logisticians. Developing the sustainment plan for the Task Force and forming the CSSC that would support it occurred concurrently with combat operations in parts of Baghdad and stability operations in others. Supporting Task Force Tripoli's extended supply lines an additional 200 km to the north re-emphasized the criticality of MAW and CSSG support. At the height of TF Tripoli's attack, the Division's supply lines stretched over 800 km from the ports of Kuwait. To their tremendous credit, CSSG-11 was able to quickly task organize, and simultaneously support combat operations, stabilization operations, and the newly assembled attack force.

As part of his Commander's intent, General Kelly made it clear to the Tripoli leadership that speed of attack and operational tempo were the keys to victory at Tikrit. He specifically charged the three LAR battalion commanders with utilizing the speed of their movement to achieve tactical and operational surprise. Task Force Tripoli would rapidly block the routes of retreat from Tikrit, isolate the expected enemy forces inside, and conduct a swift and complete destruction of the remaining Regime loyalists.

11 April
Task Force Tripoli Assembles

During the night of 11 April, the various elements of Task Force Tripoli assembled at Tactical Assembly Area (TAA) Wolf. TAA Wolf was at the site of a former Iraqi maintenance, storage, and staging depot to the north of Baghdad. It was littered with abandoned equipment, ammunition, and missiles. T-72s, BMPs, SAMs, and ammunition were stored throughout the compound, most in deep revetments that had been untouched during the air shaping campaign. The local citizens guided members of the Task Force to a cache of unknown missiles located in a warehouse on the nearby Iraqi military training facility. With security support from ten Marines of the headquarters DASC led by Staff Sergeant Russ Normandin, the SEALs were able to locate ten FROG missiles and four live Abibil missiles. These weapons were destroyed in place by the Task Force EOD team led by Gunnery Sergeant Tracy Jones.

Over the course of 12 April, as the last elements of the Task Force arrived at TAA Wolf, scouting elements were dispatched along the routes and bridges the Task Force expected to use. The reports received regarding the bridges just north of Baghdad were not encouraging. The northern Tigris River bridge under consideration had a low overhead obstruction and undetermined weight class that would not allow passage of some of the Task Force's larger combat service support vehicles. The southern bridge was better, but had severely restricted approaches through a number of small villages. The 3^{d} LAR scouting elements continued to search and found a route to the southern bridge that would offer the best approach. The route wound through the small village of Al Swash, requiring the Task Force to navigate narrow village streets and clear low-hanging wires. The ambush threat in this tight terrain was unknown. On the west bank of the Tigris, opposite Al Swash, was a larger village called At Tarmiya. There the road improved. The movement would certainly be slow until the river was crossed, but it would still be faster than driving all the way back down through the Baghdad urban area.

12 April
Baqubah Re-visited by 2/5

Since moving into Baghdad, the Division received many reports of the whereabouts of U.S. POWs. Reports pointing to a prison in the vicinity of Baqubah were determined to be potentially reliable and the decision was made to dispatch a force to investigate. TF Tripoli planners also were interested in the pontoon bridge over the Tigris at the village of At Tarmiya as a potential crossing site. Lieutenant Colonel Dan O'Donohue's, 2^{d} Battalion, 5th Marines (2/5), with Echo Battery (2/11) in direct support, moved out early the morning of 12 April to conduct a reconnaissance in force to search for the POWs in Baqubah and to reconnoiter the bridge over the Tigris that connected Al Swash and At Tarmiya.

Second Battalion's movement up Highway 2 was without incident. Echo Company and the battalion's attached tanks (from Bravo Company, 2^{d} Tank Battalion) moved into Baqubah and determined there were no American POWs present at the prison. Captain Myle Hammond's Golf Company was sited along Highway 2, as the battalion's reserve. With the word that there were no POWs, Golf Company was ordered to move to the Tigris River to conduct a bridge reconnaissance at the bridge between Al Swash and At Tarmiya. The quiet morning would soon take on an entirely different face.

As Golf Company wound along the route to the Tigris, they passed through two small villages and were met with smiling faces, thumbs up signs and shouts of "Good, Good." The Marines were becoming accustomed to the vocal support of the Iraqi people ever since moving into Baghdad. As they approached the pontoon bridge at Al Swash, they found the single lane bridge clogged with civilian traffic, but a still receptive populace. Captain Hammond set in his overwatch and began to flow forces to the far side of the jam-packed bridge. The 2/5 after action report describes the sharp turn of events that soon followed:

"The traffic at the bridge site was fairly congested with a single lane available for transit across. Captain Hammond's instructions to Second Lieutenant Shawn Maurer, the First Platoon Commander, were to establish security on the far side, ensure that traffic flowed freely across until the battalion arrived, and then be prepared to block traffic on the far side in order to facilitate the battalion's movement. On the far side of the river there was a road that ran parallel to the river and perpendicular to the bridge, causing Second Lieutenant Maurer to position one AAV to cover down the southern avenue of approach and one AAV to cover to the north along the road axis. The Marines dismounted and began to set up positions to facilitate the movement of traffic across the bridge. Additionally, Marines were placed in over-watch positions to defend the bridge until it was time for the Battalion to cross.

Captain Hammond intended to secure the bridge until the remainder of the company and the battalion arrived to cross, which he expected to be in a couple of hours. Captain Hammond had not brought an interpreter with him, but his artillery forward observer, 1st Lieutenant Michael McDowell, was in the street attempting to communicate with the local people. Lieutenant McDowell had found an Iraqi civilian that spoke some English, and was attempting to find out as much information about the area as possible. During the course of Lieutenant McDowell's interaction with the people, the area on the east side of the bridge had become a "bottle neck," choked with about a hundred civilians and a dozen or so automobiles. At about 0850Z an Iraqi civilian drove across the bridge from the west side and motioned as if he were firing a gun and then pointed to the far bank of the river.

Captain Hammond called for Lieutenant McDowell to stop the driver and find out what the man had signaled. The English speaking Iraqi served as an interpreter, and informed Lieutenant McDowell that there were large weapons caches on the far side of the river as well as Republican Guard, Baath, and Fedeyeen forces. When Lt McDowell asked how many forces were on the far side the interpreter translated that there was a high-ranking general on the far side and that he had one thousand men. The company had encountered a number of weapons cache sites up to this date that had been abandoned by the Iraqi Army, so Captain Hammond asked the man how long it had been since they last saw the soldiers. The translator communicated that the man was saying that Iraqi forces were there now. At this point Captain Hammond had the translator go through the line of questioning again in order to confirm the information that the man had passed. After an exchange of dialogue the English speaking Iraqi confirmed the man's story, as the man sped away to the east. Captain Hammond then looked at

Lieutenant McDowell and observed that every single Iraqi person and automobile had immediately scattered back to the east. There was a surreal feeling that something was about to happen, and with this information Captain Hammond began walking to the FiST AAV. Captain Hammond ordered his Marines to get on their gear as he climbed onto the vehicle and got into the troop commander's hatch. He then reached for the communications helmet to alert the platoon that was on the far side, as well as the battalion that had intended to cross the bridge.

At 0857, just as the captain was about to get on the headset to alert First Platoon, a series of explosions detonated on the far side of the bridge. Two volleys of RPGs were fired at the AAV that sat to the left of the bridge on the far side of the river. Two RPGs scored direct hits to the front of the AAV, and two RPGs missed from the initial volleys. At that point all hell broke loose on the far side of the river. It was only a short time before the dismounted Marines on the far side were pinned down by a crossfire of small arms, machine gun fire, and waves of RPGs."

With his lead platoon split by the Tigris and the remainder of his company still on the road behind him, Captain Hammond found himself embroiled in a ferocious firefight with an enemy supported by mortars, RPGs and numerous machine guns. As Second Lieutenant Maurer led the Marines on the far side in their effort to get effective direct fires on the enemy in the village, Captain Hammond worked urgently to get supporting arms and to maneuver the First Platoon elements on the near side of the bridge into positions where they could suppress the enemy across the river. As the remainder of Golf Company moved forward, the Marines began to establish direct fire superiority over the enemy and the tide turned against the Iraqis. Approximately 20 minutes after the ambush was triggered, the guns of Echo Battery were fire capable and aircraft of 3^{d} MAW were in position to support. Corporal Chad Taylor of 3^{d} AA Battalion, wounded in the initial hail of RPGs was evacuated to the near side of the river under heavy fire and Golf Company began to take the attack to the enemy.

Captain David Yaggy, Golf Company's Forward Air Controller coordinated with an airborne FAC for air support. Within minutes, F/A-18s dropping laser guided bombs and a low level strafing run by the airborne FAC had silenced a mortar position that was delivering accurate fires in the vicinity of the bridge and had cut off Iraqi troops trying to reinforce the positions over looking the river. Golf Company began to flow across the bridge and press the attack, building by building through the village. Expertly employing mortars and artillery fire in support of their aggressive attack, Golf Company soon had the enemy on the run. Pushing to the far side of the village, a devastating fire mission from Battery E impacted a group of soldiers massing in a palm grove for a counter-attack. The Marines' ferocious attack was ready to become an exploitation.

Captain Hammond consulted with Lieutenant Colonel O'Donohue via radio and was given clearance to pursue the enemy, but was reminded that there was no intention of holding the ground at the crossing site. At this point in the battle, Lance Corporal Paul Gardner was severely wounded by a gunshot wound through the torso and required immediate evacuation. Captain Hammond decided to begin the retrograde across the bridge. Aviation and artillery continued to pound the enemy to cover Golf Company's withdrawal across the bridge. The battalion had suffered 4 WIAs. Having routed the enemy from their hideout in At Tarmiya, they moved back across the bridge, leaving approximately 60 enemy dead.

Re-assembling along Highway 2, 2/5 returned to Baghdad to continue stability and security operations. They were confident that the enemy had been roughly handled in At Tarmiya and would not be eager for a rematch.

Tripoli Launches a Night Attack

Tripoli launched its attack during the afternoon hours of 12 April. As the sun was going down, Task Force Tripoli approached the Al Swash bridge over the Tigris River, the same bridge that G/2/5 had fought its way across only hours previously. It was a very long but narrow bridge, with steep approaches on both banks. The approach to the bridge and passage through the small village of Al-Swash along the Tigris River was a surreal experience that none of the members of the Task Force would forget. The long columns of Task Force vehicles snaked through the tiny village, in many places squeezing between houses with barely room to spare on either side. The village had electricity, and in the dimly lit streets, houses, and cafes, the population turned out to welcome the Marines and wish them good luck. Many in the cheering crowd passed out candied dates and cold soda. Families with children lined every doorway and sidewalk, shouting "Hello", "George Bush good", and "What is your name?" The same phrases had been heard throughout the operation from An Nasiriyah to Baghdad, but in the surreal quiet of this recently liberated village along the Tigris, the words seemed to be shouted with a special enthusiasm. The local citizens had seen the reaction of the Marines to the presence of the enemy, and they rejoiced that the Regime fighters on the far shore had been forced to flee.

Iraqis in Al Swash welcome the Marines.

As the Task Force crossed through the village, the Marines began speaking with people along the roadside. Many of them reported that the infamous Ali Hassan al Majid (a.k.a. 'Chemical Ali') had been hiding out in the village across the river and had been forced to flee to Tikrit with an entourage of bodyguards. These were the forces that had been encountered by 2/5 on their mission to At Tarmiya earlier in the day. If true, Task Force Tripoli was hot on the heels of one of the most notorious of Iraqi war criminals. These reports were later echoed in Samarra and Tikrit, lending them credence. The elusive Chemical Ali seemed to remain just one step ahead

The street that took us through this small urban area was barely wide enough to support our LAVs. Most of the buildings were two to four stories tall. You could nearly reach out and touch the buildings on both sides of the road. At times, we literally had to duck to avoid the powerlines. It was a pitch-black night and the cultural lighting emanating from within the buildings would wash out the NVGs as we passed by each structure. Civilians lined up all along the street and down the alleyways. Most were waving and giving us the thumbs up, some were just curious, I imagine the sight of all these LAVs with its occupants staring down at them with these odd devices attached to their helmets, was something right out of their apocalypse. We crossed over a pontoon bridge and over to the urban area on the opposite side. This was a little weird as well but in a much different way. This town, At Tarmiyah, seemed totally abandoned. No lights, not even from inside the buildings, no people, not a single car. It looked like a ghost town right out of a western movie. We continued on to Highway 1 and headed north. It was a long road march. 'Li'l Sister', our driver, fell asleep twice so we replaced him with our comm tech, Corporal Pasqual. Corporal Walker, the mortar FO, stood in as the VC. The rest of the crew took the opportunity to get some sleep while we traveled north. It was a cold night, I remember that.

of the Marines. Task Force Tripoli passed through At Tarmiya on the far bank of the Tigris. The eerie silence of this Regime-loyal village was a stark contrast to the warm reception the Marines had experienced in the friendly village of Al Swash. (In fact, some of the village elders in Al Swash told of the bitter rivalry between the two towns, even before the war. The tribal factionalism of the Iraqi people lay just under the surface. Once the controlling hand of the Regime was removed, much of the violence that followed was a consequence of this factionalism.)

The restricted approaches were the cause of hours of delay, and Tripoli was well behind schedule when it finally cleared the Tigris. The decision was made to continue to press through the night all the way to Tikrit. The lead elements of the Task Force pushed out onto the hard surface road and immediately picked up speed. After the excitement of passing through the village, the drivers now had to stay up all night in a fast approach to Tikrit, denying the enemy a chance to react.

Abandoned vehicle on the Highway 1 approach to Tikrit.

13 April
Attack to Tikrit

As 3d LAR Battalion led the charge north, instead of encountering paramilitary ambushes or conventional defenses along Highway 1, they encountered only abandoned Iraqi artillery and armor. In addition to D-30 howitzers and S-60 AAA pieces, several abandoned T-72 tanks were also bypassed after an adrenaline-heightening encounter in the dark. The few civilians encountered during the night and early morning seemed shocked to see the Marines. They knew US forces were in Baghdad, but these were the first they had seen this far north.

Third LAR reached its blocking position in the vicinity of Samarra shortly after sunrise on 13 April. Samarra held special cultural and religious significance to the Iraqis, and the golden dome of its grand mosque shone brilliantly in the morning sun. The Wolfpack took up positions on the avenues of approach on the western side of the Tigris, and oriented its guns to the northeast. The remainder of the Task Force sped along the hard surface bypass road and continued on to Tikrit, their right flank secured by the Wolfpack.

Task Force Tripoli entered the outskirts of Tikrit under a distinctive mural showing Saddam Hussein on horseback, valiantly leading the Iraqi army in the attack to Jerusalem. There was no doubt that Task Force Tripoli had entered 'Saddam country', as every light pole and signpost contained a mural of the dictator. Unlike every other area of the country, these murals and paintings had not been defaced or touched by vandals. The streets were largely empty, and there was a conspicuous absence of cheering citizens. The citizens were clearly 'holding their breath' to see what the arrival of the Americans would mean, both for the Regime, and themselves.

Entering the gates of Tikrit along Highway 1

A smiling Saddam Hussein welcomes the Marines to Tikrit.

The Marines of Weapons Company, 1st LAR, made first contact with the enemy just to the south of Tikrit. The company established a screen +west of Highway 1 oriented north toward the city. Pathfinder Platoon established a roadblock on the western flank and immediately stopped a truck containing seven Iraqi soldiers carrying weapons, ammunition, documents, and caffeine pills to paramilitary fighters in Baghdad. The company came under sporadic small arms and RPG fire over a two-hour period. During one of these engagements, an LAV-25 took a direct hit from an RPG round. The RPG was a contact fused weapon, and striking the muffler caused the round to detonate before impacting the vehicle hull. The muffler was destroyed, but the crew was unharmed. The Company returned fire, resulting in five enemy KIA and two trucks destroyed. The Marines of Weapons Company, 1st LAR recounted:

"As we approached the southern outskirts of Tikrit, the sun came up revealing a pleasant landscape. I remember thinking to myself, ' this actually looks like a normal place!' We stopped for a quick break, grabbed some chow, and then continued north. South of town we identified a large area to the west of the highway with multiple revetments that contained enemy vehicles. Fixed wing air started checking in with us on the TAD nets. We were the only game in town today and 'Skychief,' the airborne DASC was sending everyone our way. 'Mud Duck' McCardle, our air officer, started directing the air onto the revetments and also had them take a look forward to the city. An F/A-18D from VMFA-225 reported numerous tanks, trucks, and other assorted vehicles in the revetted area. After passing a detailed brief to the Hornet we gave the clearance to engage the targets. Several other sections of fixed wing checked in and the FAC (A) F/A-18D began directing them onto the vehicles he had identified earlier. As they began dropping their ordnance we could feel the concussion of the 1000- pound laser guided bombs as they impacted their targets. We continued to a crossroads just south of the city and occupied a screen line at the 18 Northing, oriented north. Weapons Company was to the west of Highway 1 and Delta was tied in to our right flank. We weren't there very long when we began to take small arms fire from our direct front. The LAV-25s began returning fire. The volume began to increase and then there were several RPG shots. One of the grenades impacted an LAV less than 100 meters away from our position."

The Marines from Delta Company, 1st LAR had a similar view of the battlefield:

"The company arrived south of Tikrit and was directed to establish a blocking position across Highway 1 along the 18 Northing in order to prevent any enemy from escaping the town. Weapons Company tied in to our west flank while 2d Platoon controlled a section of Cobras to destroy AAA pieces and ammo caches identified just north of our position. We began taking small arms and RPG fire from the tree lines bordering Highway 1, and immediately answered with 25mm and 7.62. First Lieutenant William Parment began calling in organic 81mm mortars and rotary wing CAS to engage the enemy forces in that area. The company commander, Captain Seth Folsom, quickly task organized a composite platoon, which systematically engaged enemy fighters who were operating in small groups. An Iraqi cargo truck sped south and jumped the curb west of Highway 1 and immediately dismounted several enemy fighters. As 1st Platoon's Bravo section engaged the newly arrived enemy, they observed several of them fleeing into a culvert tunnel that ran perpendicular to and underneath Highway 1. As Sergeant Isaac McCorkle's team began clearing the tunnel from the west, the enemy personnel lobbed a grenade just outside the entrance. The grenade detonated and inflicted minor shrapnel wounds to Lance Corporal Chad Weise and Sergeant McCorkle. His team engaged with small arms and grenades, but the enemy retreated further inside the tunnel. Sergeant William DeWitt's team was tasked to clear the tunnel from the east, where Captain Folsom's LAV-25 was abreast of the tunnel. The team tossed grenades into the mouth of the tunnel and the entire team engaged with small arms. Captain Folsom dismounted his vehicle with a M240G to assist the team with additional firepower. As the team continued to engage with small arms, the enemy threw another hand grenade, which exploded at the tunnel's entrance inflicting minor wounds to Sergeant DeWitt. The team entered the tunnel to find three dead soldiers and one still alive wielding a grenade. The team had an animated discussion with the remaining Fedeyeen, who could speak a little broken English. Despite the team's orders to surrender, the soldier continuously refused, and attempted to throw his grenade. He was killed. The entire engagement that morning resulted in fifteen enemy KIA."

As Delta and Weapons Companies continued to take the fight to the paramilitary threat just south of Tikrit, 3d MAW was wreaking havoc on conventional weapons and vehicles in several areas south, west, and north of Tikrit. The Battalion was directing CAS onto no less than four different sites simultaneously while taking sporadic fire from the Fedeyeen to their direct front. South of the position, fixed wing aircraft were engaging enemy vehicles in the revetted complex. On the north side of town, at the Al Sahra airfield, A-10s were running an effort to destroy enemy targets there. Just west of the city, SA-2 and SA-3 missiles were discovered and attacked. To the direct front, AH-1s engaged the enemy Fedeyeen with rockets, 20mm, and Hellfire missiles. All the while, surface fires from artillery and 81mm mortars were integrated. In a matter of only a few hours, over 50 sections of fixed and rotary wing CAS were directed on the enemy in the vicinity of Tikrit. The orchestration of combined arms assembled by this air-ground team made a sweet sound to the Marines on the ground. The winged-FACs, familiar with the geometry, requirements, and dangers of coordinating this airspace were key conductors of the symphony of violence played on the enemy, destroying the last will to resist.

Rescue of the US POWs in Samarra

As the main effort encircled Tikrit, 3^d LAR remained in its blocking position on the western bridge into Samarra. HUMINT Exploitation Team 3, led by First Lieutenant Nathan Boaz, was with the Wolfpack. The HET entered the city of Samarra in HMMWVs and began to screen citizens to ascertain information about remaining Regime fighters or *jihadis* in the area.

By mid-morning, several of the Samarra tribal/city elders met with HET-3. Lieutenant Colonel Clardy met with the local leaders, and calmed their fears by explaining US intentions. The citizens in the Samarra and Tikrit areas were very fearful of US retaliation against them for their role in supporting the former Regime. Lieutenant Colonel Clardy was able to allay their fears, letting them know that as long as the Marines received no resistance from the town, the Marines would not attack it. He put the burden of peaceful coexistence on their shoulders, stressing their responsibility for controlling the actions of the populace.

Shortly thereafter, an Iraqi police officer approached HET-3 with information regarding the presence of US POWs in the town. Alerted to this information, the HET team chief, Staff Sergeant Randy Meyer, and HET translator, Raed al-Tarakmah, screened the source to ascertain the location of the POWs. Staff Sergeant Meyer gave the source a GPS receiver and instructed him on how to use it. The source was directed to proceed to the location of the POWs and record the precise coordinates.

About 45 minutes later, the source returned, and gave the grid location of LC9712985354. Staff Sergeant Meyer asked the source to provide a sketch of the location and a description of the enemy defenses around it. The source indicated that the POW guards were willing to turn them over to the Marines, and did not want to be killed. The guards would put down their weapons at the approach of the Americans.

Lieutenant Boaz contacted the Battalion S-2, Lieutenant Ben Keating, and the CO. Under pressure to continue the advance to Tikrit, Lieutenant Colonel Clardy made a quick decision, and ordered Company D, commanded by Captain Gordon Miller to conduct a raid to recover the POWs if they were actually there. Captain Miller sent a platoon under Second Lieutenant Brett Eubanks to the location indicated by the grid and the sketch, and they arrived at the house within 30 minutes. The LAR scouts and HET team members stormed the house. As promised, the guards had laid down their weapons, and the POWs were recovered without a shot being fired. The POWs were swiftly raced out to the waiting LAVs, and returned to the Battalion headquarters outside of town.

US Army POWs celebrate their rescue in Samarra by the Marines of the Wolfpack.

The seven POWs recovered were Chief Warrant Officer 2 R. D. Young, Jr. - 1st Cavalry (captured Karbala 23-24 Mar 03), Chief Warrant Officer 2 D. S. Williams - 1st Cavalry (captured Karbala 23-24 Mar 03), Sergeant J. J. Riley - 11th Brigade (captured An Nasiriyah 22 Mar 03), Specialist J. N. Hudson - 11th Brigade (captured An Nasiriyah 22 Mar 03), Private First Class P. W. Miller - 11th Brigade (captured An Nasiriyah 22 Mar 03), Specialist S. N. Johnson - 11th Brigade (captured An Nasiriyah 22 Mar 03), and Specialist E. A. Hernandez - 11th Brigade (captured An Nasiriyah 22 Mar 03).

The POWs were given medical attention and MEDEVAC'd to the rear, eventually back to the United States. The televised return home of the Army POWs to the United States was a tremendous tribute to the efforts of the US Marines, HET 3, and 3d LAR Battalion. After the liberation and evacuation of the former US Army POWs, 3d LAR rejoined the Task Force in action in Tikrit.

Securing Saddam's Palaces

Arriving at the gates of the city, Brigadier General Kelly looked to the G-2 to identify objectives in the city that would facilitate the discovery and defeat of remaining resistance. In addition, WMD sites and sites containing Regime intelligence information were to be secured. During the brief mission planning time for the operation, the G-2 had produced a list of over two dozen palaces, sensitive sites, underground facilities, military garrisons, weapons caches, and Regime headquarters facilities, all gleaned from master lists maintained by national intelligence agencies. The list had been refined overnight during the rapid march to the gates of Tikrit. This Intelligence would drive the operation for the next two days as the Task Force secured these sites and defeated the remaining Fedeyeen defenders along the way.

While 1st and 3d LAR concentrated on the southern side of Tikrit, 2d LAR established a blocking position on the north side of town, interdicting military targets that attempted to flee the city. The Battalion captured many Regime figures, surprised by the rapid approach of the Americans. These figures had waited too long to flee, and were caught up in Barbarian's net.

After a hastily convened commander's meeting at an abandoned market place alongside the highway, Brigadier General Kelly ordered 1st and 3d LAR Battalions to press the attack into the city. With Barbarian waiting at the back door, the remainder of the Task Force would kick down the front door. Keeping tempo, the CG wanted them to attack straight into the center of the city, destroy the remaining Fedeyeen, and secure a number of palace sites on the south side of town.

Tripoli Commanders meet at an abandoned market south of Tikrit to coordinate fires and maneuver prior to entering the city.

The Division was notified of a suspected gravesite for Saddam Hussein in Tripoli's area of operations, and was given a FragO to investigate. Brigadier General Kelly rounded up a personal bodyguard, and went to the site in an uncleared area of the city near Al Ouja (Saddam's birthplace). The report was a false alarm, and the courtyard to which the report led had no indications of recent activity. After a thorough search, the Marines returned to the task of rounding up the remaining Fedeyeen.

On the afternoon of 13 April, 1st and 3d LAR attacked through the southern section of town, including the southern airfield, the SRG barracks, and a military storage facility. As darkness fell, 1st LAR consolidated at a key road intersection near the abandoned SRG barracks. 1st LAR was arranged along the east-west road, when it uncovered and engaged a group of Fedeyeen soldiers hiding behind them near the southeast corner of the intersection. Coming up from the south toward the Fedeyeen positions, Golf Company, 2/23 also received fire from the Fedeyeen, and opened fire in the direction of the initial firefight, resulting in 1st LAR taking both friendly and enemy fire. A few sharp-eyed squad leaders from G/2/23 quickly identified the situation and narrowly averted a major engagement in the dark. The aggressiveness of both 1st LAR and G/2/23 to engage and destroy the enemy between them had nearly resulted in a major intramural firefight between the two units.

The next morning, Task Force Tripoli continued the attack into Tikrit. Second LAR remained in the northern blocking position, and secured the Al Sahra Airfield. With 1st LAR advancing on the east side of Highway 1, and 3d LAR on the west, the town was soon secured. Third LAR encountered scattered resistance in the urban area near the Fedeyeen stadium and the cemetery. The Fedeyeen had been camped out in the town cemetery, using the tombs and mausoleums as shelter. The resistance in this part of town was heavy, and the Fedeyeen continued to fight until daylight on 14 April. After the Fedeyeen had been cleared out, the cemetery looked like a junkyard of abandoned uniforms, half-cooked food, weapons, and equipment. Fighting positions and trenches had been dug throughout the cemetery. The irony of these allegedly 'holy' defenders desecrating a Muslim cemetery was not lost on the members of Task Force Tripoli or the local citizens.

Major Randy Nash from 1st LAR recounts:

That afternoon we kicked off the attack into Tikrit. Weapons Company attacked to seize an SSE objective just south of the main city. We continued north and consolidated at the SSE compound. There was some sporadic fire and a 2S3 was engaged just off to our right on the east/west road just south of the objective. The 2S3 exploded in a huge fireball and secondary explosions continued for nearly an hour as the artillery ammo continued to cook off. As the sun went down, we gathered the key leaders of the company and the battalion staff and discussed the day's operations and plans for the next morning. We had come a long way and fought hard most of the day on little to no sleep. We had also just learned of the rescue efforts of the seven POWs and everyone was feeling pretty good about Tripoli's efforts. Firm handshakes and knowing stares were the order of the evening. Few words were needed.

A little later that night an intense volley of small arms and machine gun fire were heard, as tracers were flying all around our position. Fortunately, I was inside the LAV-C2 at the time and listened on the radio to hear any reports coming in. There was only silence. Gunfire erupted again but nothing on the radio. After several minutes among the muffled shouting outside the LAV, I heard the always cool Battalion S-3, Major Byron Harper, come up on the battalion's tactical net. He contacted G/2/23 and asked them in a tactful manner if they would mind ceasing firing at the Highlanders to their front. When illumination was popped, you could see G/2/23 on line. One Marine was in a kneeling position ready to fire his AT-4 on Steelrain 12[an LAV-25]. Numerous attempts were made by voice and waving an American flag to let 2/23 know we were friendly. I learned later that Fedeyeen fighters had been uncovered between G/2/23 and the battalion. The geometry, as fate would have it, almost set us up for an intense friendly firefight. Two of our Marines, Sergeant Bryan Benson and Sergeant Nieves Avila were injured by Fedeyeen fighters during the engagement, but both were ambulatory. Sergeant Benson had narrowly escaped death. He received a couple of rounds in his lower leg, but one round had nailed him in the torso. One of his M-16 magazines on his vest had stopped the round from penetrating his chest. He was pretty beat up but in spite of the rounds in his leg and shrapnel from his magazine Sergeant Benson was a happy man.

Abandoned weapons litter the Tikrit cemetery. The Fedeyeen and foreign fighters had desecrated this Islamic cemetery in the center of the city.

This was the primary bridge people from the eastern countryside used to get into town. The TOW Platoon established a blocking position east of the bridge to control insurgent movement into the town.

All three LAR battalions conducted patrols through the city and surrounding areas.

First LAR secures the Tikrit New Palace. This palace sat on an ancient site on cliffs overlooking the Tigris River and the city.

First LAR occupies the square in the center of Tikrit.

The palace served as a meeting place for commanders and the setting for the Task Force Commander's meetings with local Sheikhs.

14 April
Attack on the Palace Compound and Tikrit Bridge

The next morning, the Task Force resumed the attack. 1st LAR Battalion attacked to seize the western approaches of the Tikrit bridge and the 'New Palace' complex. The New Palace was the most recent addition to Saddam's huge palace compound in Tikrit. The acres of the sprawling palace complex stretched down the west bank of the Tigris for several kilometers south of the Tikrit bridge. The compound contained dozens of palaces, large and small, around a series of artificial lakes. The entire compound was surrounded by high walls and guard towers, effectively isolating those inside from the reality of poverty stricken Iraqis just outside. On the north edge of the compound, overlooking the Tigris River, the New Palace had recently been completed as yet another monument to the megalomania and waste of the Saddam Regime. It was a large fortress-like complex standing on a dominating cliff that dropped off steeply to the Tigris. The New Palace was clearly built to impress, as it dominated the town from all directions. No one coming into the town from the east could avoid being intimidated (or perhaps sickened) by the immense structure on the hill. The statues and paintings were not enough for Saddam, he wanted his presence to imbue every facet of the Tikriti's existence. A Marine from 1st LAR recalls the attack to the palace compound:

Early the next morning we kicked off the attack to seize the palace compound and the Tigris bridge. The temperature was just right and the skies were clear, it was a perfect morning. That is, until our Company commander's vehicle took a well-aimed round that flattened one of the tires on his LAV. Without missing a step he jumped out of his vehicle and in to ours and off we went without delay. A section of AH-1 Cobras checked in with us and were directed to escort us through our attack. As we moved north then east around the man-made lake Saddam had built for pleasure boating, we encountered more of the same terrain that was perfect for the enemy ambush sites. It was comforting having the Cobra escort. They were masters at rooting around the enemy's potential hiding spots and building our situational awareness around every corner and further north to the bridge. We gained a deep respect for the Cobra pilots during the war. They had taken several hard hits throughout OIF, but they continued to support us in every way. Their devotion to the grunts was tremendous, and they often put themselves at risk flying low and well forward of our lines to find the enemy.

As we continued toward the bridge, we were all astonished at the sight of the palace grounds. The immaculate structures and serene landscape reminded us of a scene right out of 'Lifestyles of the Rich and Famous'. It was all the more striking because it stood in such stark contrast to the poverty and filth we had seen up to this point. The sweeping, grandiose compound just added to the disgust we already felt for Saddam and his cronies. They were living a life beyond belief while the majority of the Iraqi people lived as paupers, hopeless and desolate.

First LAR seized the Tikrit bridge and established a blocking position on the west side. The Battalion commander and the engineers made a quick assessment, and it was quickly established that the severely damaged bridge would not support the weight of a 13-ton 'Light' Armored Vehicle. A HMMWV could barely pass the damaged sections by riding with two wheels on the sidewalk and two on the four foot wide section of roadbed still intact. To ensure responsive fire support to forces guarding the east side of the bridge, the attached SEALs were assigned an overwatch position for snipers on top of one of the New Palace buildings. The attached Tow Platoon established a traffic control point on the east side of the Tigris bridge with their HMMWVs. Tripoli now controlled all three main entry and exit routes to the city of Tikrit, and began a rigorous inspection of all vehicles. All roadblock positions received orders to continue to search every vehicle and person going into or out of the city. This was fairly simple in the beginning, as the residents had gone to ground to escape the fighting. As combat operations began to taper off, the civilian crowds grew larger, becoming a significant problem over the next few days. Adding to the mayhem, dozens of vehicles fleeing from the fall of Kirkuk arrived at the bridge with reports of roving gangs of Kurds moving towards Tikrit, ravaging the countryside along the way. The Task Force pushed CAAT elements north along the Tikrit-Kirkuk road to the limit of its battle space, 20 km beyond the Tigris River. No hostile contact was made with the allegedly advancing Kurds. Reports indicated that the marauding bands of renegades withdrew upon hearing that the Marines were advancing north. Criminal gangs were to remain a problem throughout the operation.

The grounds of the new palace in Tikrit served as the Task Force Tripoli Headquarters during operations nearby.

The Task Force commander moved to join 1st LAR Battalion at the New Palace. The town had not yet been cleared of the Fedeyeen and SRG holdouts, and the CG's small escort of three vehicles was taken under fire at a barricade near the now abandoned SRG barracks. This happened to be just north of the site of the 'friendly fire' incident between 1st LAR and 2/23 a few hours before, and the ground remained littered with enemy equipment, uniforms, vehicles, and weapons. After a brief action, the CG redirected the Jump around the ambush point, and continued on to the New Palace. The rest of the Tripoli CP, led by Lieutenant Colonel Rory Talkington, displaced soon thereafter and established the Tripoli CP on the grounds of the New Palace.

Once established on the high ground just outside the main palace complex, the Task Force CP established communications and directed the organization of local security within the city. Through 14 April, elements of the Task Force consolidated their positions in and around Tikrit, and conducted aggressive security patrolling in zone. Sporadic light contact in the form of sniping attacks, mortar attacks, and small-unit hit and run ambushes were reported during the first two days of liberation, but tapered off completely by 16 April. The establishment of checkpoints in and around the city eventually allowed the LAR battalions to identify and detain former regime officials, confiscate illegal weapons, and maintain positive contact with local citizens on a regular basis.

The Tripoli CG reviews progress with the 1st and 3d LAR Battalions at his new command post.

First LAR, conducted reconnaissance in zone through all of eastern Tikrit, looking for enemy, weapons caches, and sensitive sites. As Charlie Company cleared a military hospital on the northeast side of the city, it uncovered a weapons cache consisting of more than 50,000 weapons. The 'hospital' had apparently served as a military compound, with rooms full of uniforms, NBC equipment, and other war-fighting gear. Several hospital buildings were stuffed floor-to-ceiling with crates of brand new rifles, machineguns, and RPGs.

The thousands of weapons, and hundreds of fully-functional (but recently abandoned) artillery pieces, tanks, and SAMs indicated how intense the fight for Tikrit could have been if the Iraqis had decided to stick with their Tikrit redoubt strategy. The absence of organized conventional resistance in and around Tikrit indicated that these units had collapsed as expected, and had melted away into the urban terrain or local countryside upon hearing of Task Force Tripoli's rapid advance. Information gained by the direct support HETs indicated that many of the Iraqi soldiers and officers had returned to their homes or fled north to an area close to the Syrian border known as Al-Jazira, or 'the island' in Arabic. Over the course of 14 April, organized military resistance ceased in the city of Tikrit.

An abandoned Iraqi missile at the Al Sahra airfield north of Tikrit.

One specific report received from an EPW gave significant insight into enemy procedures that made them vulnerable to the rapid attacks of Task Force Tripoli and the Division at large. The Iraqi forces were so concerned with Coalition radio direction-finding capability that they chose to utilize very low-power, short-burst radio transmissions. They established relay stations at regular intervals to pass messages from south to north. The Iraqi communications relay, however, was unable to keep up with the rapid movement of Blue Diamond. Conventional Iraqi military commanders were overwhelmed by the sudden appearance of American forces in unexpected places and departed their posts. The few soldiers that had not already deserted or fled followed closely behind.

15 April
Give Sheikh a Chance

On the morning of 15 April, representatives from the tribes in and around Tikrit approached the Marine checkpoint at the palace and requested a meeting with the Commanding General. During this initial contact, several of these tribal sheikhs sought to establish a tribal leadership council, and identified significant issues that affected their people. Major Estepp, the Task Force Civil Affairs officer, immediately began coordinating for the resumption of services in Tikrit, as the LAR battalions focused on establishing local security and weeding out remaining militia fighters. Gunnery Sergeant Camille Hamadani (a flawless Arabic linguist), and Captain Ben Connable (another accomplished linguist and Arabic FAO) played critical roles in this and subsequent negotiations between Brigadier General Kelly and the local political power structure.

The Tribal Sheikhs were an important part of the local power structure.

Brigadier General Kelly had some trepidation in dealing with the Sheikhs. Although nominally empowered with the authority over their tribes, they still represented a hereditary form of governance that was not representative. The sheikhs had many of the same repressive and inhibiting tendencies the dictatorship had. The sheikhs competed for influence with Brigadier General Kelly against another group, the 'technocrats' (educated Iraqis who had traveled internationally, and actually had the management skills to restore order to the daily life of the Iraqis.) On this micro-scale, the power struggles amongst competing would-be governors in Tikrit mirrored the struggles the entire Iraqi nation would have. Despite the fact that many of the tribal representatives had personal connections to the former regime, the sheikhs were allowed to form an interim council to support the restoration of services to Tikrit. Although not the ideal solution, the sheikhs offered an existing governmental mechanism that could restore order to the streets quickly.

Brigadier General Kelly meets with local leaders at the palace overlooking the Tigris.

General Kelly holds more talks with the Tikrit Sheikhs.

Brigadier General Kelly took the opportunity to talk the Iraqi leadership through a number of themes that explained the Marines' presence in the Tikrit area. First, the Marines were here to liberate, not to conquer. The Marines would protect the lives and property of the people by conducting a security mission. Second, the local leadership and the people had a responsibility to help hunt down any remaining Fedeyeen, criminals, or *jihadis* who violently opposed the American presence or threatened the citizens. Finally, the Iraqis would bear much of the responsibility for restoring services to themselves. The people were authorized to take whatever actions were necessary to restore power, water, food, and medical services. The Americans would help, but it was up to the sheikhs to identify people and equipment to affect the repairs to these systems.

This was not as easily done as said. For decades, the Regime had managed every aspect of Iraqi life through an entrenched Baath party bureaucracy. The people did not have a tradition of taking matters into their own hands or solving their own problems. Initiative was not a highly admired quality. The sheikhs received the message, but it was clear that the local leadership was much more concerned with establishing their own power position in a new Iraqi authority than they were about taking care of the needs of their people. The assembled leadership quickly fell to arguing over which of them would be in charge. They asked Brigadier General Kelly to establish by-laws for their ruling bureaucracy, and argued endlessly over the composition of their council. After years of fear and deprivation by the people, the tribal leaders' response was a clear indicator of the difficulties that lay ahead for the Iraqi people. Reversion to a hereditary tribal governance to replace the dictatorship might not represent forward progress for the Iraqis.

The first significant contact between the Task Force and the local populace took place on the Tigris River bridge immediately adjacent to the New Palace complex. The sheikhs had indicated that up to three-quarters of the Tikrit population had fled at the approach of Task Force Tripoli, and many had been stranded on the eastern side by the damaged bridge. As the residents of the city began to lose their fear of the Americans, they returned from hiding places in the countryside. A large crowd of pedestrians gathered on the eastern bridge abutment. By the end of the day on 14 April, a considerable crowd had gathered on the eastern side of the bridge. In keeping with the security mission, the Marines continued with thorough searches of every person and vehicle that attempted to cross the bridge. On the morning of 15 April, the CG told the sheikhs that he would relax some of the search requirements and open the bridge to pedestrian and light vehicle traffic. Searches of vehicles and packages would continue, but individuals would be allowed to cross with minimal intervention. For safety, the Marines would keep large civilian trucks off the bridge, and let civilian cars go only one at a time. Weapons Company, 1st LAR, had set up a blocking position on the west end of the bridge the day before under the command of Captain David Hudspeth. After word went out that the bridge was to be reopened, the crush of people trying to get back across the bridge to protect their property from looters was overwhelming. Despite the precarious nature of the bridge thousands of locals, by vehicle and foot, wanted to cross the bridge and enter Tripoli's security perimeter. The Marines of the TOW platoon were stationed on the eastern side of the bridge just before the damaged section, and were

LAR Marines guard the entrance to the Tikrit Palace Compound

trying to hold back the sea with their hands. The situation appeared critical, with the hard-pressed Marines getting closer to using warning shots to restore order to the crowd that threatened to surge past them.

As several thousand people massed along the road and against the easternmost checkpoint, the crowd began yelling and tussling with the Marines. Completely oblivious to the danger the collapsing bridge posed, the Iraqis grew impatient to cross it with their families, vehicles, and belongings. Tempers among the local populace began to flare. Observing the degrading situation from the cliff-top CP location, members of the Task Force headquarters moved on to the bridge to reinforce the security company. Members of the Task Force headquarters, HET 11, and others ran to the edge of the crowd and attempted to restore order. Using the linguists, PsyOps loudspeakers, and a small amount of physical persuasion, the Marines began imposing order on the people at the bridge. The HET linguists and some reinforcing Free Iraqi Force translators helped calm the crowd. Using humor and calm reassurance, the small group of Marines slowly regained control of the situation. The press of the crowd had surged halfway across the bridge, and now the Marines slowly worked them back. The Marines reestablished the checkpoint, keeping the crowd of people and vehicles off the near-collapse bridge. Captain Hudspeth set up a structured search and crossing system, and the crowd began to flow across the bridge in a (somewhat) orderly fashion. The surge of the crowd would take days to dissipate, as the Marines continued to search vehicles entering the city. Later that day, four Kurdish men, identified by locals as Peshmerga (the name given to Kurdish fighters), attempted to cross the bridge in a stolen vehicle. They were identified by former Tikriti police officers as common criminals, men who had dynamited homes and killed people in the town to the east of the river. As word passed through the crowd of their identity, several citizens attempted to beat the men, who attempted to escape by driving through the checkpoint. The individuals were stopped, and all four men were detained. An AK-47 rifle and explosive devices were discovered in the vehicle.

The Task Force Tripoli CP occupied key ground overlooking the Tikrit Bridge.

As the population started to filter back across the bridge, HET 11 reporting began to indicate a strong presence of former regime officials in the greater Tikrit area. Although it did not appear that one central figure was coordinating their efforts, these officials maintained a stranglehold of fear over the majority of the local residents through intimidation and coercion. The presence of these former Regime members would continue to influence the willingness of the population to support the Coalition, even after Task Force Tripoli had been relieved of the battlespace. The

'Shadow Regime' as it came to be called was a decisive influence on the people outside of the areas under direct US control.

With order being restored to the city streets, the Task Force continued aggressive patrolling into the countryside and in expanding areas within the city itself. Local Iraqis supporting the Marines identified numerous large weapons caches. The amount of weapons and ammunition found throughout the town was staggering. The challenge of destroying the weapons in the midst of the civilian populace was also an exercise in patience for all. The presence of these weapons had been a severe risk to the civilians, now their destruction would pose an additional inconvenience. The discovery of the military hospital by 1st LAR was one such. Another storage facility to the north of town was discovered to contain over 30 *warehouses* of weapons and ammunition. The EOD team from 5/11 worked ceaselessly to destroy this massive amount of munitions, and every thirty minutes another explosion was heard (the Division had established a policy of conducting EOD operations on the hour and half-hour only, so the Marines would be able to discern it from incoming fire.) It is always good to have Marines around, but they can sometimes be rather disruptive neighbors.

16 April
Restoring Civil Control

For the time being, direct combat action between Marine forces and enemy combatants in Tikrit had ceased. The cessation of hostilities, however, had not yet led to peace for the citizens living in the greater Tikrit area. Large numbers of automatic weapons and explosive devices looted from military installations were now in the hands of local troublemakers. In the areas under direct observation by Task Force Tripoli Marines, there was relative calm and an absence of violence. Beyond these zones of continuous physical presence of Marines, however, the situation deteriorated rapidly. The dissolution of the brutal and oppressive police force, although a short-term boon for Tikritis, left them without any form of law enforcement or security apart from the Marines. Because Tikrit had a reputation as being one of the wealthiest cities in Iraq, looters and criminals from outlying areas made regular trips into the greater urban area in order to steal property and conduct revenge attacks on the citizens. People within the city limits and in the immediate countryside continued to take justice into their own hands by murdering looters, seizing property at random, and shooting strangers that passed through their neighborhoods.

Hasty HLZ on Saddam's front lawn.

The Tripoli CG held another meeting with local tribal leaders in order to facilitate the restoration of local security and the resumption of services. These meetings were to continue on a daily basis to both highlight Tripoli's requests for assistance in establishing a local government, and to build rapport with the leaders. There were a large number of tribal leaders, and the Task Force assembled lists of the names, cross-checking them with the list of wanted Iraqi war criminals. Several rival tribal leaders accused each other of having intelligence contacts in the old Regime or of hiding Regime officials, and the HET teams aggressively worked to exploit this internal friction to gain information on the whereabouts of Regime figures who remained at large. The CG established a Tribal Council, assisted by the local professionals who had come forward to offer assistance in restoring electricity and water services to the city.

A Marine Human Exploitation Team disbands a vigilante checkpoint.

One of the first orders of business was the disestablishment of vigilante checkpoints. The Sheikhs had returned from the meeting the day before, and had instructed their people to set up independent checkpoints outside and out of sight of the Marine checkpoints. The stated intent of these checkpoints was to keep non-locals from looting the homes of local residents, but these unauthorized checkpoints took on a more sinister form. That morning, while traveling south along Highway 1, HET 11, under Warrant Officer Shawn Dunn, discovered four hasty checkpoints, each manned by approximately 40 individuals. A zigzag pattern of broken furniture and garbage was set at each point, and the men manning the point were armed with clubs, metal rods, and specially constructed batons with rubber tips designed for body beatings. At one checkpoint, a burned out car was located fifty feet from the zigzag and the men were in the process of smashing the windshield of a car stopped in the center. An elderly couple was cowering in the car. Without hesitation, the five members of the HET waded into the much larger crowd. The HET team stopped the attack, seized the weapons, and interrogated the attackers. At this checkpoint, and at each subsequent point, the ringleaders were former policemen claiming that they had received orders to establish the points by the tribal leaders at the behest of the Americans (this order was never given). Local vigilantes stated they had established these checkpoints for local security against looters and to prevent people coming from out of town to steal from their homes. Despite vigorous denials to the contrary, the thugs manning the checkpoints appeared to be stealing goods from the cars that passed through. The checkpoints were broken apart and the men were ordered to disperse.

These vigilante checkpoints were indicative of the lawlessness and fear that characterized much of Iraq in the interval between the end of the Baath regime and the restoration of law and order. In the period immediately following the liberation of Tikrit, the city police force had ceased to function, as officers changed into civilian clothes and fled into the countryside. The Task Force received numerous reports of reprisal killings and beatings from local citizens. Difficulties in re-starting the electrical plant left the area in darkness, facilitating looting and the activities of former regime paramilitary forces. Responding to the general lawlessness, Lieutenant Colonel White, CO of 1st LAR Battalion, began the creation of a civilian police organization to help restore order. Because many former police officers had committed human rights violations during the previous regime, 1st LAR set up a careful screening process for new police recruits. Within 24 hours of the initial meeting, the first traffic police recruits were on the streets, standing at checkpoints alongside the Marines. Expanding this small force into a larger, better-trained organization would fall to the relieving force once Tripoli departed.

The international media was everywhere in Tikrit.

While patrolling and escorting tribal leaders on the east side of the Tigris River, members of the Task Force command element were approached by agitated locals who stated that "Kurds" had murdered someone just up the road. When the patrol element reached the incident site, they discovered a small sedan taxi with numerous bullet holes in the windshield. The passenger was dead with wet blood on his face, and the driver was bleeding from glass fragment wounds. The distraught driver stated that four Kurds in a white pickup truck had attacked them. The Task Force Command element conducted a high-speed pursuit of the reported vehicle, but was unable to catch up to it. As the patrol moved past another intersection, another dead body was observed on the side of the road to the northeast. It was clear that to the east of the Tigris, the breakdown of law and order was having a chaotic effect. The Task Force continued to find examples of lawlessness throughout the countryside. Despite the local practice of blaming any disruptive activity on 'Kurds', most of the violence appeared to be perpetrated by the locals themselves. The breakdown of law in the countryside made it possible for locals to engage in the revenge killings, tribal disputes, and vendettas that were much a part of the local culture. Local civilians with pickup trucks were observed raiding a food warehouse and a grain storage facility. Although the local leaders continued to vehemently blame this behavior on outsiders and 'Kurds', this did not appear to be the case. A few of the more trustworthy tribal leaders accompanying the patrol confirmed this assessment. Restoring law and order to the Iraqis was going to be a challenge.

Second LAR continued with its operations on the northern side of town. One of the Battalion's primary objectives was the massive Al Sahra air base on the northwest side of Tikrit. When 'Barbarian' moved to clear this compound, they discovered vast revetment fields full of abandoned equipment. After scattering the ineffectual resistance in the area, they also discovered large warehouses containing everything from intact new (2002) French Roland missiles, to new Russian-made aerial cluster bombs. The Battalion reported the equipment, destroyed what they could, and documented the rest. Full clearing of this compound would require weeks of effort. Al Sahra was one of the facilities passed down to the relieving force when they arrived.

The excesses of Saddam's palaces shocked many Marines.

17 April
The Shadow Regime

There was never a dull moment on security in Tikrit. The locals responded well to the confident but non-aggressive posture of the Marines:

> *The Marines of 1st Platoon (Red Platoon), Company A (Apache Company) had the responsibility of establishing and operating an entry control point (ECP) on the northern main service road leading into Tikrit. Sergeant Timothy L. Williams, Mojave Valley, Arizona (Fort Mojave Indian Tribe) was the man on the scene. "The first few vehicles through the ECP were kind of hostile," recalled the sergeant, "but once they got used to us they were all pretty friendly." As the Marines conducted vehicle searches, Sergeant Williams noted that the occupants of the vehicles began to help them by teaching his Marines the Iraqi translations for trunk, hood, door, and other terms that would help his Marines in their mission. "The locals hospitality flowed the longer we were there." "They brought us pita bread, fresh eggs, and veggies." The Iraqis became so comfortable with Sergeant Williams and his Marines that when asked if they had weapons, would lift up their 'man dresses' and show the blushing Marines that they were not in fact armed.*

More disturbing than the overt violence and lawlessness were the unambiguous signs gathered by the Task Force indicating the presence of a 'Shadow Regime' in the greater Tikrit area. The old power elite remained active in the areas that Marines were not physically present. The Shadow Regime no longer had governance over the population, but clearly still retained the levers of power in the countryside. Thirty years of brutal dictatorship had left the Iraqi people incapable of independent actions in the absence of specific government direction and approval, and they retained a visceral fear of the former Regime. Organized anti-US/Coalition meetings were being held in outlying areas and secluded parts of the city. At these meetings, the former Regime members and Baathists organized a campaign to threaten the local populace with death if they cooperated with the efforts of the Marines. The 'Shadow Regime' was not a monolithic and centrally controlled opposition force, but instead was a loosely organized threat from former regime power brokers, coordinated primarily by a similar desire to return to their days in power. These men still wielded significant control over the local populace through fear and intimidation, and operated at the fringes of the areas under the Marine's control.

The Shadow Regime recognized its own inability to exercise military power against US forces, and had adjusted its tactics to work against the populace behind the scenes. By doing so, it sought to quietly undermine US efforts without provoking a military response. The Shadow

Regime posed a direct and present threat to the stability and the success of any future government. The actions and tactics observed by Task Force Tripoli would prove a reliable forecast of the actions of the last Regime holdouts, and a large number of other parties vying for power in the post-Saddam government.

The contribution of the Task Force's two HET teams continued to be a significant force multiplier. These Marines were seemingly everywhere, exploiting sources on the bridge, breaking up vigilante roadblocks, investigating local authority figures, and tracking down former Regime members who were part of the Shadow Regime that was working at cross purposes with the Task Force. The HET teams of Lieutenant Nathan Boaz and Warrant Officer Shawn Dunn were among the most important elements of the task force. These Marines worked tirelessly, often in dangerous situations, to serve as a go-between for the Task Force CG and the locals.

Locals help identify areas where the Shadow Regime remained in control.

Weapons caches included everything from small arms to Roland missiles and cluster bombs.

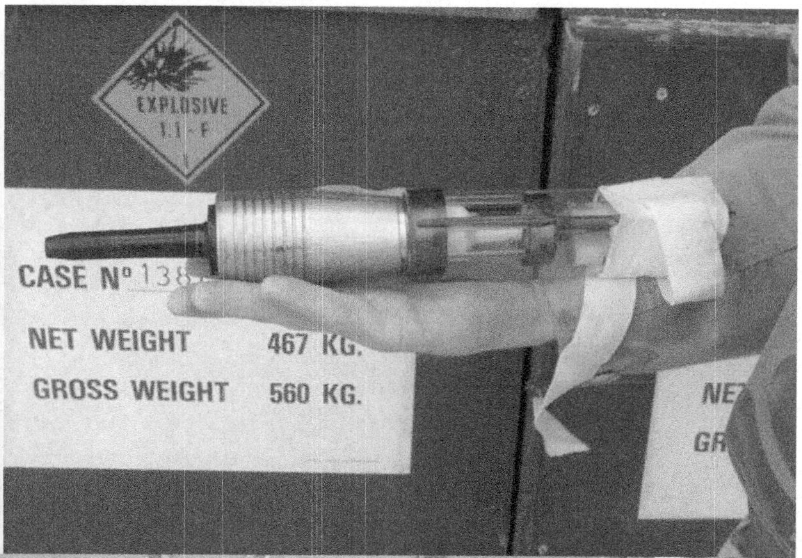

18 April
The 'Attack' to Bayji

General Kelly congratulates the newly appointed Police Chief in Bayji.

Shadow Regime members not withstanding, the situation in Tikrit increasingly became more stable. Capitalizing on the developing relationship with Fahran Al-Sudaid, a sheikh whose tribe included the areas north of Tikrit, Brigadier General Kelly coordinated a meeting with the tribal leaders of Bayji, a town approximately 40 km north of Tikrit. The arrangement, coordinated with Captain Connable's assistance, would help return the greater Tikrit area to a state of relative normalcy. There had been, as yet, no US forces entering the Bayji urban area. Demonstrating extreme trust in the relationship he had established with Fahran Al-Sudaid, the CG dictated that only a single LAR company accompany him, his staff, and Mr. Al-Sudaid to the center of Bayji.

The 'diplomatic' mission to Bayji was successful in all regards. No contact with hostile forces occurred (although a roadside arms bazaar was broken up by the Tripoli staff along the way) and a broad range of local leaders and citizenry attended the town meeting held in the center of the city. After delivering a liberation speech with the assistance of Al-Sudaid and Gunnery Sergeant Camille Hamadani from HET 11, Brigadier General Kelly approved the establishment of a locally elected council and police chief. The Task Force staff set communications and contact procedures with the council. Bayji was liberated with smiles and handshakes instead of with preparatory fires and tank main gun fire.

In addition to bringing Bayji under Task Force control, significant steps were taken to reduce former Regime presence in Tikrit and restore services to the populace. From the second day in Tikrit, the Marines of CSSC-Tripoli were purifying 20,000 gallons of sweet water from the Tigris River daily, and distributing it at five locations in the city. Major Estepp (CAG) continued to work closely with tribal leaders to coordinate the restoration of electrical power, moving two large cranes into the city from an outlying Kurdish area to assist in the repair of downed lines. After the local leadership had failed to do so, the Task Force CG ordered all images of Saddam Hussein in the town destroyed. Before initiating this action, however, he insisted that the local leadership request assistance in destroying the images. The local leadership had hesitated to take the initiative upon themselves out of fear that Saddam or his remaining local henchmen would return and deal harshly with them. They had come to trust the Marines, however, and now asked for their help. To keep this from becoming a mad exercise in vandalism, the CG assigned the task specifically to 1st LAR Battalion, who began what would become an ongoing and labor-intensive task with the assistance of EOD, CSSC, and HET Marines.

Marines conducted a turnover with the US Army, introducing the local leadership to the Army officers.

19-21 April
Relief in Tikrit

Task Force Tripoli was soon to be relieved by elements of the US Army's 4th Infantry Division. The lead elements of this Division began to arrive in Tikrit on the 19th, and were given a thorough orientation to the peaceful situation in town, and the continuing exercise in self-governance being worked with local leaders. This advance party gained an appreciation for the peaceful situation, and was pleased to take the word to their higher headquarters. The follow-on staff of the US 4th Infantry Division had a different perspective on the situation. The US 4th ID had missed the combat phase of OIF, and were determined to have a share in the 'fighting'. They characterized their recent road march to Tikrit (in trace of the Marines) as an 'attack', and remained convinced that the situation in Tikrit required a very stern military enforcement posture. The dichotomy between the two peacekeeping strategies was unsettling for the Marines, and many winced when Army Apache attack helicopters swooped in to the Division battlespace without coordination, and began to strafe seemingly indiscriminately on abandoned enemy armor directly between, and in close proximity to, 2d and 3d LAR's positions. Fortunately, they missed the LAVs. The Marine foot patrols, without flak vests and in soft covers, were replaced with menacing and physically destructive M1A2's and Bradley fighting vehicles on the streets. Stores that had reopened quickly closed back up as the people once again evacuated the streets, adjusting to the new security tactics. A budding cooperative environment between the citizens and American forces was quickly snuffed out. The new adversarial relationship would become a major source of trouble in the coming months.

The MEF Commander addressed the Marines in Tikrit and informs them of the end of full-scale combat operations.

On 19 April, Brigadier General Kelly and members of his staff conducted a final meeting with Fahran Al-Sudaid in order to further the ongoing working relationship and attempt a turnover with incoming 4th Infantry Division officers. At the invitation of Mr. Al-Sudaid, Kurdish and Arab tribal leaders from north of Bayji to the border areas with Syria and Turkey attended the combination meeting and traditional feast in order to initiate diplomatic contact with the Coalition. The design was to use this opportunity to pass down relationships based on trust and mutual respect for each other's agendas that had begun to blossom between tribal leaders and the Marines. Here was a tremendous opportunity to secure the peaceful cooperation of the northern Iraqi tribes. The meeting was successfully concluded, with plans for future contact with the northern tribes established. The meeting might have been even more productive had senior officers from 4th Infantry Division been willing to attend.

Third LAR had already been relieved of its battlespace in Tikrit, and sent back down to Samarra on 17 April. Wolfpack had relieved RCT 5 in Samarra, allowing the RCT to return to assume stabilization missions to the south in Ad Diwaniyah and As Samawah. Wolfpack resumed their former zone and watched as the peaceful Samarra countryside returned to normal. The peace was shattered on 19 April, as the US 4th Infantry Division attacked past their positions and deep into Tripoli's battle space. Wolfpack hunkered down at the sound of chain guns and .50 caliber machineguns, trying to avoid becoming friendly fire casualties. Apache attack helicopters were roving up and down the highway destroying military equipment, reopening military actions in an area that had been peacefully secured. Citizens who had been rejoicing in their new-found freedom were sent back, cowering in fear at this unexplained change in American tactics.

On this day, the Division CG's intentions message recorded the following lessons learned:

A critical lesson emerging in our handoff with V Corps is the mission essential requirement for dismounted infantry in civil-military ops. As we hand off our portion of Baghdad as well as the city of Tikrit, the lack of Army dismounts is creating a void in personal contact and public perception of our civil-military ops. Our forces need to project confidence in the security environment we have created. That is best exemplified in light, mobile force in contact with the local citizenry, sans helmets and flak jackets where appropriate, and without armored vehicles. If we cannot engender friendship and confidence in the local security environment, we cannot set the conditions for good order integral to a return to civil control. Future force structure analysis groups must recognize this imperative prior to cutting infantry from our Table of Organization. Mattis sends.

These words would prove prophetic, as later incidents in other parts of Iraq would demonstrate that a standoffish, overbearing occupation force would serve to alienate rather than endear the local populace. The individual discipline and overall numbers of dismounted Marine infantry proved well suited to the duties of both controlling and supporting a newly liberated people. Throughout Blue Diamond's zone, tanks and AAVs were rapidly withdrawn from use as firm and friendly Marines shifted to truck-borne and dismounted patrols.

Despite some misgivings about the danger posed to the peaceful civilians by the differing philosophies of the two American units, Task Force Tripoli carried out the relief in place with the US Army 4th Infantry Division from 19-21 April. The Task Force elements began to move south by echelon on 19 April as the Division Main CP displaced from Baghdad to Ad

Diwaniyah. Battlespace handover was executed smoothly, and on 21 April the last elements of the Task Force Tripoli repositioned from Tikrit to Ad Diwaniyah. By 22 April, all elements had returned to Ad Diwaniyah, and the Task Force had been stood down.

Marines investigate Saddam's elaborate palace complexes.

Closing the door on Saddam's last hideout.

2-5 May
Tripoli Rides Again

In late April, the Division continued its duties of stabilization and security for five of the southern Iraqi Governates. During the execution of this follow-on mission, the Division CG received intelligence that violent religious extremism and illegal drug shipments were finding their way across the Saudi Arabian border into the Division's southernmost area of responsibility. It was assessed that if these actions were allowed to continue unmolested they would help serve to undermine stabilization operations, directly threaten the restoration of order and inhibit the security conditions for a free Iraq. Additionally, the concern of local leaders about the influx of extremists of all types and the need to demonstrate the sovereignty of Iraq's borders dictated a response by the Marines. The CG determined that a show of force in the southern zone would help to assert coalition authority in an area previously untouched by combat or presence operations.

The CG and ADC determined that the pre-existing framework of Task Force Tripoli would best serve the mission requirements for this long-range reconnaissance and show of force. With minor changes, the Task Force (minus 2d LAR) was stood up again on 1 May. Colonel Dunford, the incoming Division Chief of Staff, took command of the Task Force. First and 3d LAR Battalions would serve as the combat elements of the Task Force. Combat Service Support Company Tripoli (CSSC-Tripoli) was reestablished, and began acquiring sufficient Class I, III, and IX-block supply items for five days in the desert without re-supply. As the Task Force would be operating over 150 miles from the nearest Division outpost, a Shock Trauma Platoon and a Forward Resuscitative Surgical Support (FRSS) element were attached in order to provide immediate on-site medical support. Facilitating MEDEVAC and local re-supply were two CH-53Es in direct support. Additionally, sufficient assets were attached to provide the full spectrum of data and voice communications to the commander and his subordinate elements.

On 2 May, a P-3 Patrol aircraft was assigned to support the mission, and immediately began to reconnoiter the battlespace. Based on imagery and map analysis, two approach routes were identified for the combat elements; designated Routes BLUE and WHITE. The G-3 tasked the combat elements to conduct a reconnaissance in zone to identify and detain smugglers, conduct route analysis, and establish a presence along the Iraqi-Saudi Arabian border. Task Force objectives included the airfield and town of Salman along Route WHITE, including a possible WMD site at the airfield.

On 4 May, the Task Force crossed the Line of Departure and began its reconnaissance down Routes BLUE and WHITE. First LAR, traveling down Route BLUE, did not encounter either enemy forces or any significant civilian population. The route was a difficult unpaved road, covered with sharp rocks and boulders. The poor quality of the road forced the battalion to keep the rate of movement at 15 kph or less for over 200 km. Highlander went through over 25 vehicle tires during this single movement to the border. Meanwhile, 3[d] LAR, the command

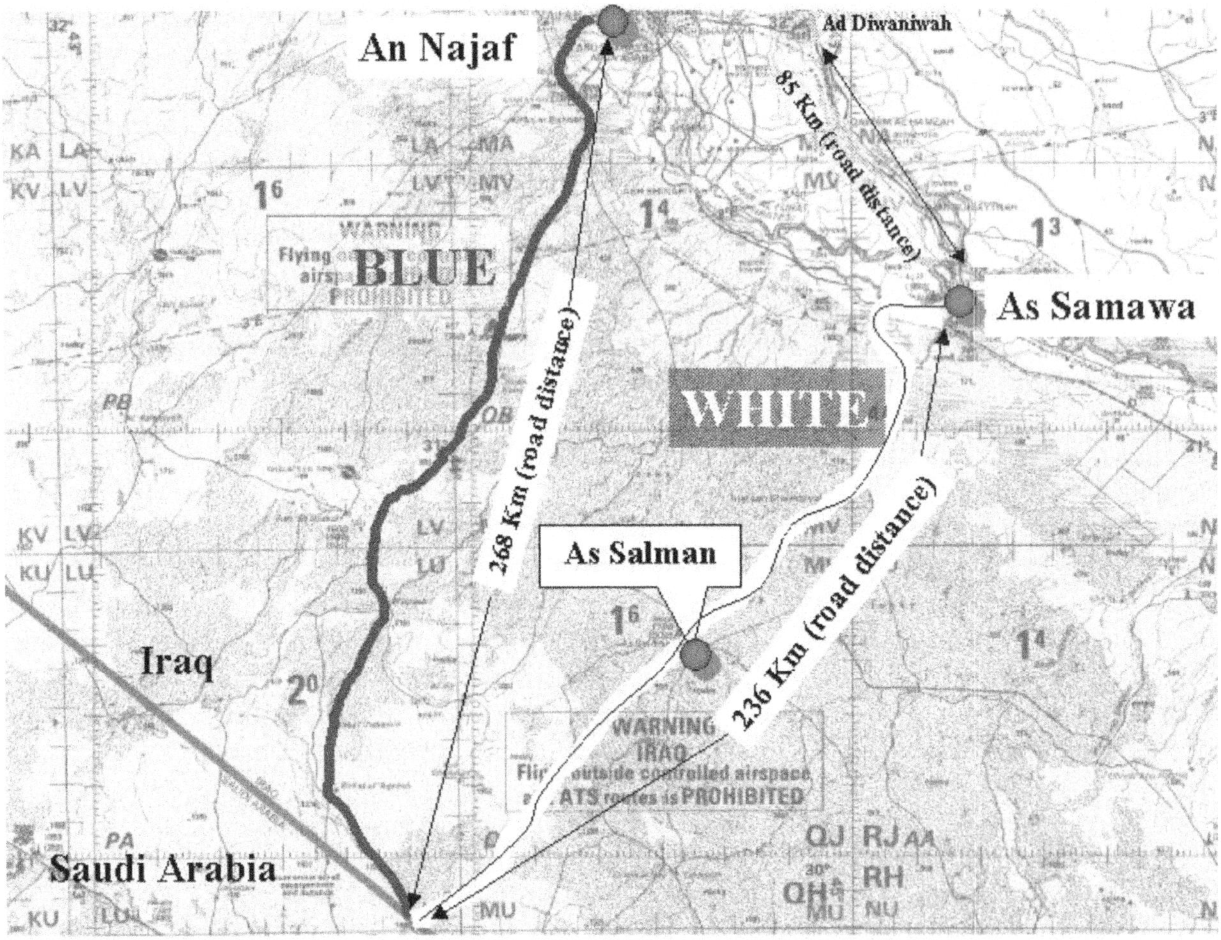

The routes depicted above (BLUE and WHITE) outline the patrol area for Task Force Tripoli's mission to investigate reported foreign insurgents moving into Iraq from the Saudi Arabian border.

element, and CSSC-Tripoli moved by serial down Route WHITE to secure the Task Force objectives, and eventually reached the border. The Marines were greeted intermittently by smiling and waving Bedouin animal herders on the sides of the road, but no enemy forces or smugglers were detected during the movement. Wolfpack moved on the airfield in force to discover that the facility had long-since been abandoned. Structures had been bulldozed to the ground, and Iraqi forces had deliberately cratered the airfield. Other elements of 3[d] LAR entered the town of As Salman, where more Bedouin greeted them. The town itself was assessed to be friendly, and no signs of enemy or illegal activity were present. The attached HET interviewed several people and helped to spread printed copies of (CENTCOM Commander) General Franks' message to the Iraqi people.

Upon reaching the border with Saudi Arabia, Wolfpack observed Saudi Border Patrol units aggressively patrolling on the other side of the berm. The Saudis spotted the Americans and waved. First LAR completed its movement along Route BLUE and established inspection and observation points along the main (illegal) crossing points on a cliff running along the border. A former Iraqi Border Patrol post was discovered in the 1st LAR zone. The two-story concrete structure was littered with abandoned boots, uniforms, and gas masks. No Iraqi forces were identified in the area. The general security situation was stable, and the Saudis had tight control of the border area.

After conducting a personal inspection of the border area on 5 May, Colonel Dunford determined that the mission of the Task Force had been accomplished. During the day of 5 May, both LAR battalions began retrograde operations. The Task Force began movement to Ad Diwaniyah along Route WHITE. Local Iraqi leaders were reassured that fundamentalists were not infiltrating the border, and the Marines had established a baseline for future security operations along the border. Task Force Tripoli was disestablished once again, this time permanently.

Tripoli Marines patrol the barren open expanses near the Saudi Arabian border

Chapter 8
Security and Stabilization Operations

By the middle of April the 1st Marine Division had fought spectacularly from the sands of Kuwait, across the dusty Fertile Crescent, and into the 'mean streets' of Baghdad. Task Force Tripoli had extended this success 150 miles further north to Bayji on the upper Tigris River, and removed the organized remnants of the Regime there. However successful these past operations were, however, the Division had paid a price. As the Marines contemplated their battlefield victory, they realized that that price, though heavy, might have been much worse. Now, the Division faced the possibly even greater challenges of providing security, stability, and humanitarian assistance to a country and a people devastated by decades of brutal repression and neglect. The Marines would experience this transition first in Baghdad, and then in several of the governates of southern Iraq. As many of the combat units of the Division began to return home, Marines who remained behind opened an entirely new chapter.

There was a doctrinal definition for the tasks that lay ahead of the Division Marines. 'Security and Stability Operations (SASO)' made the job seem very clear cut. Doctrine rarely captures the brutal realities of the situation on the ground, however, and the Marines of Blue Diamond were about to embark on a set of tasks that, although planned for in general terms, noone could have foreseen. With the removal of the Regime leadership, an enormous power vacuum opened up. Because of the ethnic, religious, tribal, and cultural fissures that run through the Iraqi people, it was quite possible that an unstable security situation could quickly degenerate into bloody civil conflict. Even before combat operations had ended, various factions in Iraq had begun jockeying for a piece of the post-Saddam power structure. The Marines had to fill the security void quickly, before this anarchy or civil conflict erupted. There was no time to rest on the laurels of a successful attack to Baghdad, as looters, factional violence, and even tribal vendettas had begun. The fissures that ran through the Iraqi people were already beginning to widen as the Marines stepped into this new mission. Fortunately, the Marines of Blue Diamond were prepared. The 1st Marine Division remained mindful of its motto, 'No better friend, no worse enemy.' The Division would help restore a sense of normalcy to the lives of ordinary Iraqis, and help set the conditions for restoring the public services, infrastructure, and food supplies. Yet, the attention paid to these new tasks did not detract the Marines from remaining combat tasks. There were still a great number of Former Regime Loyalists, foreign terrorists, and jihadis who actively worked to degrade the security situation. For those who sought a return to brutal repression or who sought to bring violence back to the streets of Iraq, Blue Diamond Marines would unhesitatingly unleash its offensive strength. The Marines and Sailors of the Division were mentally prepared for the new mission requirements, a mental agility that spoke highly of their professionalism, training, and humanity.

SASO Operations in Baghdad

The Marines began their mission transition by continuing to build on the successes they had already achieved in Baghdad. The 11th Marines CMOC had already been established, and was doing a booming business. In their zones, Marines continued their aggressive routine of patrols, searches, and roadblocks. The Marines remained a visible presence on the streets, deterring looters, maintaining order, and defusing potentially violent situations.

Our mission was to do security patrols, clear out weapon caches, and have security posts at a food bank. Three posts, two hours apiece, two hours relief and a quick reaction force (QRF) team ready. I was on the QRF, with my squad leader being Sergeant Capitol, and we got called early in the shift this day for a situation down at the food bank. The situation turned out to be three women and a man attempting to scale one of the walls to get some food from the food bank. We had just cleared this area earlier of looters and weren't too happy to be called back into action for this. Anyway, these four people became extremely shocked and frightened once they saw a squad of Marines running over to them with rifles at the ready. Instantly, it became a game of "charades," really the only way we had of communicating with them.... So after a few minutes of pointing and gesturing, the one man kept saying, "thank you, thank you!", to our squad leader, Sergeant Capitol. The man even went as far as attempting to kiss Sergeant Capitol on the cheek, which I guess is fairly common in their culture. Well, what started out as a strange day turned even stranger, as Sergeant Capitol, accepted and then returned the gesture. Our mouths all dropped, and nobody, not even Sergeant Capitol said a word until we got back to the gates at the CP. Finally, I just couldn't take is anymore, and I asked Sergeant Capitol, "So Sergeant, what was your boyfriend's name anyway?" Everyone died laughing, and I spent the following week pulling extra watch, but it was worth it. Since then we are quick to remind him of his "Arabian Night Love Affair" with the Iraqi man.

Marines have a hard time adjusting to some Arab customs.

Each RCT established checkpoints to regulate and screen movement within their sectors. They augmented these checkpoints with mobile and foot patrols, frequently relocating checkpoints to prevent the former regime holdouts or terrorists from targeting them. The patrols were intended to both promote good will between Marines and the residents of Baghdad, and provide a tactical function as well. According to Captain Sean Blodgett, Commander of Charlie Company, 1/5, "The Division needed to return a normal sense of life to people, as quickly as possible, but soon realized it could not import enough food and water for the entire city." The Division, consequently, worked at restoring Baghdad's public safety, food supplies, health services, transportation, and basic utilities, in order that the city might support itself. By restoring normalcy and rooting out remaining perpetrators of violence, the Division began to gain the trust of the people. As the citizens of Baghdad began to enjoy the fruits of the Marine's security operation, they began to help identify and eliminate those who still wished to promote violence against the Americans or within the civilian community. The Division systematically denied the remaining enemy sanctuary from which to operate against the Marines or the civilian population.

The task of coordinating the resumption of these critical services remained with the Civil-Military Operations Center (CMOC), established by the "cannon cockers" of 11th Marines. The 11th Marines found themselves solving problems ranging from neighborhood squabbles to reported terrorists, to alleged chemical weapons caches. The Marines really had their 'finger in the dike' as parts of one division of Marines tried to step in the place of a mature city government for several million people. Restoring a functioning city bureaucracy became one of the first orders of business. The Marines could not govern the city; that would have to be done by Iraqis. The Marines hoped to restore the conditions of normalcy that would allow the influx of Non-Governmental Organizations (NGOs), civil authority, and provisional civilian government. It was especially hoped that NGOs would coordinate their efforts with the CMOC to rapidly begin working on critical projects. Initial CMOC meetings were specifically designed to take place with Baghdad's secular and religious leaders, who were expected to be full partners in the effort to repair and rebuild the city. The first order of business was to contract English-speaking Iraqi citizens to serve as interpreters and advisers on local customs.

Meanwhile, during the course of their security patrols Marines uncovered huge caches of weapons, ordnance, and terrorist materiel. Most of the caches included a staggering number of small arms and RPGs, but many also included artillery, anti-aircraft guns, demolitions, missiles, and rockets. In fact, there were weapons stashed caches in almost every block and on every street in the city. Had the Iraqi military chosen to defend the city in any kind of organized and determined way, the cost would have been much higher as the regime certainly had the equipment to drive up the butcher's bill. A Marine from 3/4 recalled:

On 11 April the Main CP moved to the ministry compound, establishing what would become the central node of operations for the remainder of 3/4's time in Baghdad. Ultimately, CPs, Combat Trains, Bravo Company, HET, and a Force Reconnaissance platoon would be stationed there. Kilo moved to establish a company CP in a traffic circle southeast of the hotels. The mortar platoon consolidated around two hospitals in its zone, conducting patrols and liaison with local citizens for the remainder of the operation. The engineers soon undertook the demanding task of collecting and transporting those weapons caches they did not destroy themselves, and can very well attest to the staggering quantity of weapons the Iraqi regime had hidden throughout the urban area.

The RCTs shifted positions as the Division received tasks to assume additional missions. RCT-1 conducted a Relief in Place (RIP) with RCT-5 to allow Grizzly to move to the north of the city to provide support for Task Force Tripoli's operations in the Tikrit area. While this relief was the RIP was underway, RCT-7 secured several major sources of food. They coordinated the redistribution of these critical food supplies to the local population through the CMOC. During this time, units relentlessly hunted members of the former regime. Many of these figures began to turn themselves in, or were pointed out by local citizens. Included in the catch was Iraqi Lieutenant General Amir Hamudi Hasan Alssaadi, Presidential Advisor for Scientific and Technical Affairs. He was taken into custody at the Palestine Hotel in the center of eastern Baghdad. On 13 April, he was transferred into the hands of US governmental agency representatives and transported to Baghdad International Airport for further exploitation.

During all this the Iraqi people were clearly pleased the Marines were pursuing and eliminating the terrorist infrastructure, and providing the city much needed law and order. Lance Corporal Dane Jensen, India Company, 3/7 remembers:

> *In Baghdad, I witnessed many American flags being flown or displayed with pride and courage. In other ways of expressing gratitude, some Iraqis even went out of their way to greet Marine patrols with flowers and other souvenirs. I believe the media's assumption of the Iraqi citizens to be totally wrong. Yes, there were people some people who didn't want us in Iraq. However, I felt that this small fraction couldn't compare to the Iraqi's who are overwhelmed with joy for our presence.*

In this environment, the vast majority of Iraqis welcomed the Marines as liberators and friends, but a wary eye and lightening quick life-and-death decisions were required of every member of the force; regardless of rank, age, or occupational specialty. There was no front, no rear, no secured area; only reliance on your brother to cover your "six," and the professional maturity to know when and how to act.

Iraqi children are fascinated by the Marines, and the Marines return their attention with kindness and good-natured humor.

With literally no help or even guidance from an outside agency the CMOC established functional areas for police, fire, electricity, water, and medical care. It appointed an interim police chief and opened an Iraqi Police Academy in RCT-7's zone. The Red Cross assessed the water system at only 40% capacity due to the lack of power. Electrical engineers worked furiously to restore the power grid, as it was central to restoring the most basic of human services to the people of Baghdad. Agencies like CARE, Red Crescent, the World Food Program, and many others arrived and willingly joined the Division's efforts. These organizations were the first and most knowledgeable people on the ground to set the city and its people right. After RCT-7 secured the major Red Cross medical supply warehouse, the distribution of critical medical supplies to the major care centers began to flow to those who needed them most.

The Division also took on the pressing, and often-times very violent issue of crime, then running rampant throughout the city. After three decades of terror and repression, members from every segment of society sought revenge not only against symbols of the former government, but also against individuals associated with the previous regime by vandalizing their homes and businesses. As the looting spread from government buildings to private businesses, it was apparent that this wanton destruction would severely impede the nation's recovery. Crime spread rapidly to the commercial and banking centers of the city. Local thugs, using recently stolen military weapons, took advantage of the general breakdown in public order. Armed criminal gangs were formed, which began to rob banks, hijack vehicles, and steal whatever they could get their hands on. These gangs brazenly held proprietors at bay while they looted their establishments. Major combat operations with Iraqi military units were all but over, but the Marines of Blue Diamond began to combat this criminal element throughout the city.

The Division began to produce and disseminate Psychological Operations messages via broadcast and leaflets to assist in protecting the innocent civilians of eastern Baghdad. Lebanese Television, covering the daily CMOC-NGO meetings, provided reliable third person objective reporting regarding the level of effort and progress being made restoring services. Radio broadcasts also assisted local leaders and security forces by appealing to the citizens of Baghdad that for their own safety they remain in their homes between evening and morning prayers. Private First Class Devin Schmitt, Weapons Company, 1/7, recalled the situation in Baghdad at the time:

> *In Baghdad, people would always come to us for help. They would yell "Ali Babba" and point in a direction. We could not understand a word they said except that someone was stealing stuff.*

The CMOC's activities accelerated in the face of these and many more challenges. By identifying and meeting with key leaders, civil and military authorities began to plan the restoration of the city's services and repair of its infrastructure in a comprehensive way. Colonel Marletto also made liaison with the US Army's 3rd Infantry Division to ensure the synchronization of Civil-Military operations on both sides of the Tigris River.

Most important of these early contacts were with Iraqi electrical engineers who would eventually restore the power grid. The CMOC's 'electrical' section chaired a meeting with Dr. Karim Hasan, Directing General of Iraq for Electrical Affairs, along with 50 key plant managers and engineers. A major issue concerned jump-starting a 2-megawatt generator for the eastern Baghdad grid. Jump-starting the generator required more power than existed in I MEF's generation capacity. The engineers coordinated a plan to bring several substations on line to provide the massive jolt needed to re-establish Baghdad's service. The issue of electricity was significant, not only for the restoration of lights and power, but also as it was the required resource for generation of fresh water supplies. The Red Cross Water and Habitat managers reported that the water supply was adequate for the current Baghdad population, but would collapse under the strain as refugees returned to the city. With the onset of summer temperatures of 120 degrees, the demand for water and electricity for air conditioning would skyrocket.

The CMOC medical committee optimistically forecast that the medical system could provide basic emergency care and services if medical professionals of all categories returned to their places of work. Commander Kevin Moore, the Division Surgeon, coordinated with various NGOs and identified 14 hospitals that were necessary to be restored in order to provide even basic medical services to millions of people ravaged by war. The CMOC worked feverously to restore them to full operations. Requests for volunteers were made on local media broadcasts, and several hundred volunteers arrived at the Palestine Hotel to provide assistance. Medical support improved with the addition of four Qatari physicians, and the arrival of critically needed supplies.

By 15 April, the Division was well into preparations for the turnover of eastern Baghdad to the US Army's 3rd Infantry Division. The 358th Civil Affairs Command Deliberate Assessment Team arrived first and evaluated the needs of the city. As part of its preparations Colonel Jim Howcroft, the Division's senior Intelligence officer, prepared a detailed sector-by-sector analysis of the enemy threat for the commanders of the 'Rock of the Marne' Division. In each functional area, staff coordination surged forward. These two proud Divisions had fought side-by-side from Kuwait to Baghdad, and both had similar mindsets about both combat operations and SASO. Without written orders, instructions, or doctrine of any kind, the members of these proud Divisions who had fought side-by-side in the assault on Baghdad, quickly worked out the details of an urban reconstruction effort without precedent. After detailed staff work and commander interaction at every level was completed, the relief was initiated.

On 18 April, the relief in place with the 1st and 2nd BCTs of the 3rd Infantry Division began. Both 3/7 and 3/4 turned over their areas to the Army units, then and headed south to the RCT staging area just north of Al Iskandariyah. The following day, the remainder of RCT-7 completed its battle-hand over and moved to the assembly area. On 20 April, the remaining elements of Blue Diamond began movement to an intermediate TAA south of Baghdad. By 21 April, the Division planted its colors in the vicinity of Ad Diwaniyah, and continued its RIP with V Corps and movement into its Phase IV (SASO) positions. The Commanding General, 1st Marine Division personally coordinated the historical close out of the battle space with Major General "Buck" Blount, General Officer Commanding, 3rd Infantry Division. Blue Diamond continued to move to its new zone in southern Iraq.

Into the Shia Heartland

The men and women of Blue Diamond had already begun their journey south to their designated Phase IV (SASO) zone. In what amounted to be the entire southern half of the country (excluding only the British zone to the east around Basra), the area contained nearly half of Iraq's population. It was bounded by borders with Iran, Saudi Arabia, Kuwait, and the southern suburbs of Baghdad.

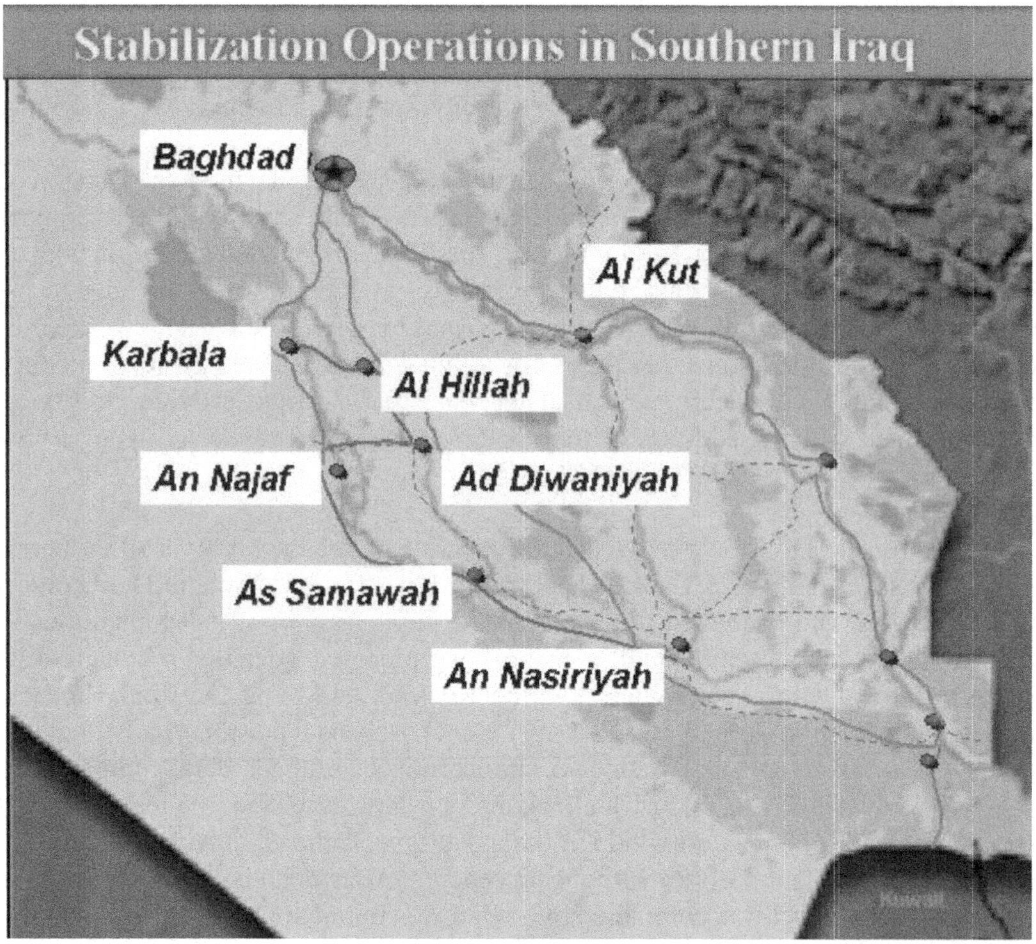

Blue Diamond conducts Security and Stabilization Operations in seven southern governates of Iraq. This zone included much of the Shia heartland, including the holy cities of Karbala and An Najaf. Most of the supply routes for US Army forces operating in Baghdad and to the north also ran through this zone.

The first unit to move into the southern zone was RCT-5, ordered to conduct a relief in place with the 2nd Brigade Combat Team (BCT) of the 82nd Airborne Division. The RCT, and subordinate battalion command groups, moved to an area southwest of Ad Diwaniyah and established an Assembly Area for the entire Division in an abandoned Iraqi military facility there. The Division's composite quartering party began to receive units in its area co-located with RCT-5. The next day, RCT-5(-) moved to an intermediate TAA south of Baghdad with its remaining infantry battalions, 2/5 and 3/5, moving from over 100 km north from the city of Samarra where it had supported TF Tripoli's actions, and joined on its parent unit the next day in Ad Diwaniyah and As Samawah. To avoid adding to the friction already caused by the thousands of Shia faithful participating in the Arba'een Pilgrimage, RCTs 1 and 7 delayed movement into their Phase IV positions until the Pilgrimage was completed. The fact that thousands of Shias were, for the first time in decades, able to participate freely in one of their most highly regarded traditions was a historic event. The smooth conduct of this pilgrimage, with US forces handing out water and providing medical services, greatly aided the relationship with the Shia populace in the south.

By 24 April, the Division successfully assumed all of its Phase IV battlespace and now

Iraqi civilians wave and cheer the Marines patrolling their streets.

controlled the Al Muthanna Province (2/5), Karbala Province (3/7), Babil Province (1/4), Al Qadisiyah Province (3/5), and An Najaf Province (1/7). Additionally, the provinces of Wasit and Dhi Qar were added when 3/23, LtCol Dave Couvillon commanding, and 2/25, LtCol Bob Murphy commanding, were attached from TF Tarawa. Both battalions were already in place in Al Kut and An Nasiriyah respectively, and we assumed the geographical space as well.

The battlespace given to the 1st Marine Division was a unique place. The population of the area was overwhelmingly Shia Muslim, and had suffered terribly under Saddam Hussein. The recent combat operations, combined with 12 years of UN sanctions and willful neglect on the part of the Saddam regime had left the infrastructure in the southern cities pathetically outdated and in a near state of collapse. The purposeful actions of the former regime had retaliated against the Shia's uncooperative attitude toward the regime by allowing only the most limited basic services. This created a major series of challenges for the newly designated military governors of the five (and eventually seven) provinces. There was no functioning water purification system, no intact power grid, and not even basic sanitation. The police had been part of the regime's system of repression. In addition, it was an overwhelmingly corrupt organization that provided little service to the community. The real power rested with the tribal sheiks, many of whom were also corrupt, and only interested in maintaining their own power and perquisites. In addition to tribal authority, the Shia community was made up of many different religious factions. Allegiance to one cleric or another defined the southern Shia almost as much as tribal affiliation. Religious spats between rival clerical groups often turned violent, and were to have a significant impact on stability in the zone.

The Blue Diamond battalion commanders who were thrown into this situation responded with heroic efforts. Supported by the Division, they set to work armed initially with only their initiative, imaginations, and the power of their wills to restore basic services throughout scattered cities, towns, and villages. They were hobbled in their efforts by the poor material condition of the infrastructure and a complete lack of competent technicians to operate it. Decades of neglect and fear inhibited even the most well intentioned efforts.

The Seven Governates

The Division Marines turned to organizing their zone. In what was jokingly referred to as the 'Blue Diamond Republic', the Division coordinated a strategy for each province/governate. This included an assessment of its infrastructure, identification of threats to peaceful good order, and identification of local authorities who could help bring order out of the current chaos. Each province was different, and each had its own unique challenges. To help focus the efforts in each province, understudies were assigned to help keep the battalion commanders on the ground informed of intelligence and information events effecting his area. The Division's approach was to decentralize the effort, with local commanders driving the effort as those closest to the point of contact. The overall goal was to quickly bring the maximum benefit to the greatest number of people. By starting their work in the largest population centers then moving outward into the countryside, the Marines sought to gain security and restore services through the entire countryside. Even this logical approach encountered problems from the beginning. This approach offended the rural population, as they often had competing interests with the urban populace. It seemed that every action taken to benefit one group would serve to alienate another. The battalions walked a tightrope, seeking to make an omelet by breaking as few eggs as possible. As the Marines grew more skilled in Iraqi diplomacy, wise compromises soon began to emerge. Improvements often benefited many groups. Electrical power in the city translated into pumps irrigating crops in rural areas. Functioning sanitation for urban dwellers meant raw-human waste was no longer trucked out into the country and dumped in the fields, and alongside roads and canals. As the days went on, the Marines continued to build trust with the local citizens across the battlespace.

The Al Qadasiyah governate was controlled by 3d Battalion, 5th Marines. Initially commanded by Lieutenant Colonel Sam Mundy, it was turned over to Lieutenant Colonel Pat Malay soon after SASO operations commenced. Ad Diwaniyah, the capitol city, had been a major military installation in southern Iraq during the Saddam years, and had seen a lot of combat action.

Babil province was the site of historic Babylon, and contained the MEF and Division headquarters. The governors of the province, however, were the Marines of 1st Battalion, 4th Marines, under the command of Lieutenant Colonel John Mayer. The capitol city of Al Hillah was a large city, and the outlying countryside harbored large numbers of residual regime loyalists.

Al Muthanna province was a largely frontier province on the edge of the southern desert. Although it had the character of a sleepy desert outpost, it had seen major Fedeyeen activity during the recent combat phase. In the capitol of As Samawah, Lieutenant Colonel Dan O'Donohue's 2d Battalion, 5th Marines, took charge.

Karbala governate was home to one of the holiest of Shia cities. The Marines stepped cautiously into this province with Lieutenant Colonel Mike Belcher's 3d Battalion, 7th Marines.

Karbala, with a population of nearly 300,000, was home to the shrine of Imam Abi Abdillah Al-Husain, the martyred grandson of Mohammed, and one of the founders of the Shia sect of Islam. The sensitivities of the religious community based in this holy city were of tremendous concern. Here, the predictions of cooperation were dire. The 'experts' advised the Division that as unbelievers, largely Christian, and (worst of all) Americans, this city would be a tinderbox of discontent and resistance.

An Najaf was another holy Shia city, and the capitol of the province of the same name. The experts gave an equally glum assessment of the Division's chances of success here. Najaf was also a desert outpost, and contained the tomb of Hussein, son-in-law of the prophet Mohammed. Najaf is situated on a high desert plateau, and is filled with religious sites, mosques, and revered tombs. As the spiritual center of Shiite Islam, it contained the offices of many of the important Shia clerics. An Najaf was controlled by 1/7 under Lieutenant Colonel Chris Conlin. Najaf was the holiest city in the Shia tradition, similar in every way to the Vatican or Canterbury for Catholics and Anglicans respectively. Throughout the Islamic world, the faith is very much a part of every aspect of life – including civil government to one degree or another. In Najaf there was no separation between church and civil authority. They were completely intertwined and inseparable. As with the other cities, however, Blue Diamond's approach was one that assumed the Marines already had the "hearts and minds" of the people. The Marines would conquer whatever reservations and animosities some might have towards them with respect for their culture, opinions, and religion. By treating the people with dignity, the Marines hoped to gain the trust and cooperation of even the most devout.

Later, the Division would inherit the province of Wasit from Task Force Tarawa. This was the province around Al Kut, and butted against the Iranian border. Third Battalion, 23d Marines, under Lieutenant Colonel Dave Couvillon, controlled this area. With elements of 4th LAR, this province would see creeping Iranian influence and activism during the period the Marines were in residence.

A second province added later was the Dhi Qar province around An Nasiriyah. As the scene of heavy fighting during combat operations, it was expected that there would be residual problems. In fact, 2d Battalion, 25th Marines were successful in maintaining order in this region under Lieutenant Colonel Bob Murphy. This governate lay astride the major supply routes between Kuwait and Baghdad.

Across the southern governates, Iraqi families return to their homes and try to return to a normal life. Even though relieved of Saddam's repression, the initial months would be difficult for many Iraqis.

Rebuilding the Police

Rebuilding police departments was one of the fist orders of business in each governate. Without Iraqi policemen walking the beat, the entire security effort would fall to the Marines. The restoration of police services was a challenge made difficult by the suspicion with which the population held every member of every department in the country. Corrupt and a full participant in the regime's application of terror, the police departments had squandered whatever trust they may have had before. Restoring public trust and confidence would be a daunting task. Formerly, the police were underpaid, with full expectation that these officers would make up for their low income by extortion and bribery. Even then, portions of police salaries were skimmed off the top and into the pockets of senior officials. In spite of the urgent requirement to put police on the streets right away, it was more important to send a signal to the officers themselves that the situation had changed. Improper conduct by police officers and officials would not be tolerated in the new force. In almost every case, the battalion commanders fired the entire force, then rehired those who they were personally convinced were salvageable. A few commanders invited vetted local officials and private citizens to participate in this process. The next immediate step was to get the police out of their old uniforms that were closely associated with the regime. The new police forces were given distinguishing new uniforms, even if it was only a ball cap initially. The new uniforms were often presented in a public ceremony for the benefit of the officers and the communities they served. In Karbala, the issuing of new uniforms was televised on 8 June at a City Hall parade. Most cities opened police academies, with a course of instruction that included ethics training, basic police techniques, weapons use and community relations. You can't shake down the locals anymore was lesson one. Joint Marine-police patrols began operating on the streets in this opening stage, building trust and confidence in departments that had for at least 35 years had not deserved either.

After the initial training programs came longer term initiatives that included expanded training opportunities, internal affairs functions, management, expanded ethics sessions, and instruction on the law. In As Samawah a ten day course was established that graduated over 700 officers. In many of the provinces commanders requested, and received, the services of Reserve Marines who were police officers in their civilian lives back in the United States. Army Military Policemen were instrumental throughout, adding their talents and expertise to the daily functioning of the departments. This once hated institution evolved quickly into relevant law-enforcement and service departments under the careful instruction and supervision of NCOs and junior officers. By July, the process was nearly complete with joint patrols all but eliminated. The streets were placed almost entirely into the hands of Iraqi policemen, with the police officers working in concert with the Marines for more complex operations such as raids and searches.

Supplying Energy Needs

Electricity (or more accurately the failure to provide sufficient and reliable quantities thereof) was the bane of existence for many commanders during the entire SASO period. The issue was larger than just electricity, as all sources of energy including diesel, benzene, and propane gas were the cause of significant troubles throughout the Division zone. The elements of power generation included the facility infrastructure itself, sources of fuel, and the distribution system.

Each element was burdened with corruption, inefficiency, and shortages. The infrastructure was decrepit, with no real maintenance conducted for years. For the Marines who inherited this system and were supposed to restore it to working order, it was unbelievable that a country so rich in these products had such a dysfunctional distribution system for fuel and natural gas. Commanders assaulted all three problems at once seeking out engineers and workers too afraid to come to work, and ensuring their protection. Evaluations were made of fuel distribution pipelines, most of which were not in use. Those pipelines that were functional were either temporarily out of service due to damage, or lost most of their contents to looters (who for a decade had tapped into the lines and stolen fuel for sale on the black market.) This was also the case with natural gas lines, although these lines suffered more from accidental explosions at the hands of unskilled thieves drilling taps into the system. Until pipelines were repaired, precious fuel was trucked from the production facilities around Al Basrah to the power plants generating electricity. This was not a very practical solution, as most plants required more than 100,000 gallons of fuel a day, and trucks and operators were at a premium.

The two major power generation plants in the Babil province were the Musayyib and Al Hillah facilities, both of which required extensive improvements and repairs to their supporting oil and gas lines. Overloading of the system caused an emergency shutdown of the Musayyib plant on 18 May. After significant work coordinated by the Marines of 1/4 and attached soldiers of the Government Support Teams (GST), the plant came back on line and slowly increased to nearly full capacity. On 16 June, the Al Hillah power station was placed on-line for the first time since the war after almost three months of inactivity. In Karbala, solutions included major repairs on the power lines throughout the province. One of the major problems throughout the zone were the looters who would actually tear apart high-tension power lines and strip them for the copper inside the lines. This 'normal' activity was an example of how far the Iraqi people had to go in order to regain a 'civil' society. The 3/7 Marines repaired a series of downed 200 foot high pylons, and worked out an equitable means to ration electricity between industrial and residential users. An Nasiriyah and 2/25 faced a unique power dilemma as a huge circuit involving a number of interconnected grids needed repair. With time, money, unrelenting perseverance, and installation of a huge transformer, the electric grid between Baghdad-Hillah-Nasiriyah was completed and power restored as far south as Basra and the Kuwaiti border.

Supplying the plants with fuel over the road was inefficient, unreliable, and was only a stop gap measure until the pipelines that feed the generators could be restored. Task Force Rio was organized by 1/7 in Najaf to provide inspection and security to all the lines within its zone. When they conducted an assessment along with the Director of Oil Distribution, only a partial inspection of the pipeline revealed 19 man-made breaks in the diesel fuel pipeline, one in the crude oil line, and another in the natural gas line. Brigadier General Terry "Guts" Robling, Assistant Wing Commander of the 3rd MAW, directed all his pilots to fly the lines whenever possible to report looters in order that ground reaction forces could respond. Thieves remained undeterred, however, by the patrols, arrests, and confiscations. A more permanent presence along the lines was required. The next step in facilitating the security of the lines was the establishment of a Facilities Protection Force. This group was assigned to the Ministry of Oil for use in pipeline and fuel point security. This repeated process of response and counter-response continued, as thieves grew more sophisticated and brazen.

Oil theft was profitable for those who played the game, but was very often deadly. On 1 July the Southern Oil Company reported a major pipeline explosion. First Battalion, 7th Marines dispatched a small force to the site. On scene was discovered a massive oil-line fire, destroyed tanker truck, Land Rover, and two portable Honda generators. Witnesses reported that several thieves tapping into the crude oil pipeline were killed when it exploded. The thieves were believed to be an organized, professional criminal syndicate that smuggled oil for re-sale outside Iraq. Forty-six civilian looters were arrested in Karbala in one day for this crime, and Iraqi police took the initiative with aggressive patrols. Captures of oil smugglers became a routine event. On 9 June, 3/7 surprised thieves as they were tapping into the diesel pipeline and arrested five individuals, confiscated four tanker trucks, pumping equipment, and AK-47s.

Reinventing Local Government

Much of the problem that faced the Division as it looked to the task of running half a country was the lack of a dependable public service bureaucracy. With over ten million inhabitants, and all the associated functions of public utilities, schools, medical care and food, this level of organization was critical. Restoring an organization of managers was not as easy as rehiring the Iraqis who had held these posts before. The previous regime's method for civil service was irrationally centralized and inefficient. This was a large component of how the former regime maintained power and exercised repression. The absence of the regime's corrupt henchmen, however, left the citizens and local leadership without any sense of how to exercise initiative and problem-solving in the interest of a smooth and functioning city. The town councils or city managers had no sense of how to do anything but take direction from Ba'ath Party directors. These had been corrupt and nepotistic, with no concern for the well being of their constituents.

In Al Hillah, Lieutenant Colonel Mayer and 1/4 held an initial meeting with the local government at which local technocrats and sheikhs pledged a shared commitment to restoring political power to thc pcople and the new civil government. He learned that government workers had not been paid for weeks, and received assurances from the acting-governor that salaries of all government employees would be disbursed by first paying each town council leader who would then distribute funds to the workers. This process had to be watched closely for corruption and skimming, 'business as usual under the former regime. Even as late as July, salaries continued to be the number one issue for local government workers. At a meeting with officials on 2 June, the commander convinced the workers to create a city council to provide a check and balance on the governor's actions. The battalion arranged meetings with religious, tribal, and civilian leaders to find independent delegates for the council. The goal was to establish a council that was not dominated by a single constituency.

In Karbala, the city council, led by Dr. Mohammed, met with representatives of 3/7 to seek the means to restore basic city services and critical infrastructure. Trouble developed, however, when hundreds of citizens demonstrated at the College of Arts against the appointment of Dr. Mohammed, whom they accused of being a former high ranking and corrupt Ba'ath Party official. Lieutenant Colonel Mike Belcher met with the demonstrators and reached a solution. On 14 May elections for a new interim City Council went as planned (with all positions being temporary and performance-based) until the establishment of more a permanent structure. The interim city council met for the first time and began organizing itself. Meanwhile, Belcher identified selected members of the council with the interests of the people at heart and real leadership ability. A more manageable group of city leaders emerged, and was soon making real progress.

As Samawah, on the other hand, was a complete success story in local governance. In this city, a unique set of circumstances made the environment more conducive to honest government. The most important factor was the high quality of local leadership, especially one senior tribal leader, Sheikh Sammi. Many members of the local leadership had suffered directly, and terribly, under the former regime. Under the firm direction of Lieutenant Colonel Dan O'Donohue and Sergeant Major Kim Davis, the government made a miraculous turnaround in its ethical service orientation to the populace. With the effort led by the battalion's legal officer, the first structure rebuilt was the local criminal and civil court system. The effort gained credibility by partnership with the twelve most prominent tribal leaders. On 8 May the new judiciary was stood up with four good and popularly screened judges sworn in the following day. City payroll procedures were next, with checks and balances in place to minimize extortion, corruption, and skimming by senior officials. The city council, once a tool of the regime, was reelected in a way that would make Americans proud. A viable, responsive, and all-inclusive, council went to work for the first time since 1968. As a check to the power of the sheiks, technocrats were included, as were a total of nine political and religious parties.

In An Najaf, Lieutenant Colonel Chris Conlin and Sergeant Major Henry Bergeron worked wonders facing the same challenges, but under significantly different conditions. In May there was absolutely no confidence in any of the sitting judges, so the commander removed twelve of them. The corrupt and murderous mayor in An Najaf also stood as an impediment to civil progress. The removal of the mayor became the first real test case of the Iraqi judicial system, newly reformed by American forces. Abu Haydar Abdul Mun'im was a former Iraqi Army Colonel placed in office by the 101st Airborne Division (after it had initially taken the city.) The process of removing him began almost immediately. The mayor was wildly unpopular before the appointment, and his popularity declined even more as he conducted himself in a way reminiscent of the old party hacks empowered for decades by Saddam. Less than three months after the mayor's appointment, an Iraqi investigative judge ordered his arrest on the crimes of holding hostages, pressuring government employees to commit financial crimes, and physically attacking a bank official. The 1/7 Marines strode into the mayor's compound (that was surrounded by his Praetorian Guard of special police and bodyguards) and placed him under arrest. A few days later, the governing council chose Haider Mahdi as an interim mayor. Warmly welcomed by the Marines, Mahdi was eager to institute reforms and provide stability in this sector of Iraq.

In virtually all of the cities and provinces it was the same. Corrupt governing councils and courts, the former regime's local means of control and repression, were replaced by decent and honest men (and, in a few cases, women.) These emerging leaders, vetted by the battalion commanders, assumed the role of popular representatives, a role denied to Iraqi citizens for longer than some of the new appointees had been alive. According to the people there had not been a single honest and fair election in Iraq in over 30 years, and the citizens looked to the commanders, and the tremendous Civil Affairs soldiers from the US Army attached to every battalion headquarters, for the way to govern themselves democratically. Just as importantly, the

Iraqis looked to those in American uniform to get their first glimpse of what selfless dedication, and ethical public service, were all about.

Securing Iraq's Territorial Integrity

A key responsibility the Division inherited after the collapse of the regime was to ensure the territorial integrity of Iraq. The Division sent a number of missions to the Iraq-Iran and Iraq-Saudi Arabia borders, to monitor the crossing stations. Task Force Tripoli had surveyed the Saudi Arabian border and found it under control.

The Iranian border was a more significant issue for the Marines. Tehran did indeed have an interest in a weak and destabilized Iraq, and sources indicated that it would attempt to exert its influence by controlling or coercing the Shia majority through influential clerics, particularly those in Karbala and Najaf. The most pressing issue for the Division on the border, however, was the violent, abusive, and criminal behavior of the Iranian border guards. These guards had taken to extorting and abusing Shia pilgrims who traveled back and forth between the two countries exercising their new-found religious freedom. For the first time since Saddam established his murderous regime, the holy cities of Najaf and Karbala were open to all Shias, without the controls and restrictions of the former regime. Pilgrims poured in from Iran by the thousands to visit religious sites previously denied to them. Additionally, the Division allowed refugees to return, as well as Iranians who desired to visit family members they had not seen in since well before the beginning of the Iran-Iraq war in the early 1980s. Screening every individual that attempted to pass in an effort to "keep the terrorists out," proved an impossible assignment without a system of passports, visas, and databases. The good will generated by allowing the free movement of Shias made closing the border ill advised.

The Division initially looked to the 4th LAR Battalion (already in place on the border) to provide security along the portion of the Iranian border in Wasit province. Later, 3d LAR Battalion joined with 4th LAR, and together they conducted a more detailed evaluation of the border region. The region was remote, sparsely populated, and very difficult terrain. The reestablishment of the Iraqi Border Guard force was a major step forward in the restoration of sovereignty. When the Commanding General made a personal inspection of the border on 29 May, he found:

> *Our sector of the Iranian border is locked tight. The now well-trained and motivated Iraqis are doing a good job manning the checkpoint with vigor. They appear to detest Iranians and are attentive to their border control point.*

As there were insufficient Marine forces to maintain a permanent presence on the border, the Division covered the requirement with regular aerial reconnaissance flights and intermittent short-tem missions by LAR units. What was learned from these initial assessments was that the primary traffic across the border in the Division's zone were pilgrims intent on visiting the Shia sites so long deprived to them by Saddam. As the Iranian border guards extorted passage money from pilgrims visiting the holy sites, and the border was porous and poorly patrolled, there was a brisk business of people smuggling by foot, cross-country around the official post. Once inside

Iraq, entrepreneurs in taxis or small busses transported the travelers to Al Kut, 100 miles to the south, where they then arranged transportation to their destination.

By July, sufficient forces were made available by the increasingly benign environment around Al Kut, which was ably controlled by Lieutenant Colonel Dave Corvillon, his Sergeant Major John Hugger, and the Marines and Sailors of 3/23. Colonel Couvilln found it possible to permanently position an infantry company to the north to observe the border. The task fell to Major John "J" Harris of India Company, 3/23, with personnel from Arkansas, Utah, and Pennsylvania. Setting up living spaces in one of the many abandoned Iraqi concrete forts that lined the border, the company began to work the very few villages in the region, mark the numerous minefields left unrecorded from the Iran-Iraq war, and generally provide a presence and 'stiffener' for the Iraqi border officials in the region. Cross-border activity increased exponentially during this period. Even though Iranian officials would only allow documented Iranians citizens to pass through their checkpoint, by mid-August the number of people passing daily through the border could be counted in the thousands. Travelers moved in both directions. Persons wishing to enter Iran took hired transportation to the crossing site, then walked into Iran as no vehicles of any kind were allowed in. This spurred a lively cart-transport business, as cargo was off-loaded on the Iraqi side, carried through the 100m buffer zone, then reloaded on trucks for further transport into Iran. The cargos of dozens of 18-wheelers were manhandled in this way every day by a small army of cart men.

The border remained benign with the presence force overseeing the Iraqi operation of the crossing, patrolling the few roads running parallel to the border, and monitoring the many smuggler routes through the hills and along the washes of the rough desert terrain. On 7 August, there was a sporadic incident of violence as Iranian soldiers began assaulting travelers waiting in the buffer zone seeking entry into Iran. The squad of Marines on the scene deployed and stood ready to protect the unarmed civilians, particularly when the Iranians began a large volume of AK-47 fire in the direction of the crowd, just above their heads. A number were injured by beatings with truncheons, but the Iranian guards fled back into Iran with the appearance of a squad of US Marines.

The Northern Babil Province Heats Up

During April and May, as US Army units continued to build up combat power at the end of the conventional stage of the conflict, Forward Logistics Base (FLB) Dogwood was established at a former military base in the Northern Babil Province astride the Euphrates River, 20 km southwest of Baghdad. This area of Iraq was a fault line between the predominately Shia south, and Sunni north. It was also the point at which the relative patronage of Saddam's regime ended, and his greatest abuse and repression began. Nearly all roads to Baghdad from the south passed through this area. The Main Supply Route (MSR) Tampa (Highway 1) from Kuwait, Alternate Supply Route (ASR) Jackson (Highway 8) from Al Hillah, and ASR Cleveland (Highway 9) from the west and Karbala, all ran through. Most notably, however, was ASR Sue, a narrow two-lane road that meandered through the lowland villages and agricultural fields for 35 road kilometers. ASR Sue was the connector between Dogwood and MSR Tampa. Sue was the logistical lifeline for the soldiers of the 1st Armored Division in Baghdad, and was used by up to

90 convoys a day. It was used heavily by the Army's theater command in its Herculean efforts to push all classes of supply north from the air and sea points of entry in Kuwait.

The Northern Babil Province, an area representing only a tiny fraction of the entire Division zone, had been all but ignored well into June. It was not an area that Lieutenant Colonel John Mayer and 1/4 had devoted much time or effort to, as they concentrated their attention on the provincial capital of Al Hillah. By early June, the Division began receiving frequent reports of convoy ambushes along ASR Sue, although the number of actual attacks was difficult to determine. At the time, there was no system in place to verify reports as they came through any of a number of Army command channels, turning incomplete and inaccurate initial reports, including double reporting, into 'fact'. For instance, in the first two weeks of June alone there were 51 reported attacks along this route, although the standard by which an attack was measured ranged from a single driver thinking he had been shot at, to the tragic wounding or death of a soldier and vehicles destroyed by combined automatic weapons and RPG fire. Thankfully there were few of these. Assuming the reports were accurate, there had been, on average, three attacks per day by small arms and two per day involving RPGs. During this period there was also another weapon that made an occasional appearance, the Improvised Explosive Device (IED) (this weapon was to gain momentum, and soon became the terrorist weapon of choice.) In response to the Army commander's immediate concerns, the Division organized a response it designated Task Force Scorpion.

TF Scorpion's mission and task organization evolved over time as the tactical situation changed, and the enemy reacted to the Division's successful initiatives. Initially the threat was against defenseless logistics convoys that were described as "manatees." The Army drivers typically wore CD headphones, assistant drivers were most often asleep, and few wore helmets or flak jackets as the convoys made their way along routes Tampa and Sue. There were few crew-served weapons mounts on the vehicles, and these were often unmanned as they were uncomfortably hot in the blazing Iraqi sun. Convoy personnel revealed they had never been briefed that the nature of the threat changed significantly once they put Ad Diwaniyah in their rear-view mirrors. Even convoys coming out of Dogwood, or down from Baghdad, were not taking the precautions prudent in such a high threat area. This made the manatees easy targets for a host of Iraqi predators. In the beginning, the ambush action was no more than one or two assailants standing next to the road in plain sight during the day emptying an AK magazine toward the approaching vehicles, and walking away. Occasionally an RPG might be fired, usually within 20 meters of the road itself. The convoy's doctrinal immediate action was to drive through the 'kill zone' and ignore the incoming fire. Seldom was fire ever returned.

On 13 June, Scorpion was organized under the initial command of Colonel John Toolan, Commanding Officer of the 1st Marine Regiment. This aggressive officer was perfect for the job, and conveniently still "in country" as his regiment was not yet fully redeployed. The Task Force's battle space was carved out of 1/4's zone and thereafter referred to as "Northern Babil." TF Scorpion's base unit was the 4th LAR Battalion(-), led by Lieutenant Colonel Andy Pappas and Sergeant Major Jose Sandoval. It was reinforced with a number of smaller units and capabilities. The first order of business was the obvious requirement to secure ASR Sue from the Iraqi thugs who preyed upon US convoys unmolested. This was accomplished in short order by saturating the immediate vicinity of the roadway with aggressive Marine patrols. Instead of manatees, the terrorists encountered aggressive warriors. Every would-be terrorist soon knew the capability of the LAV. Like the Cobra gunships flying overhead, the 'light horsemen' of the LAR community were respected – and mightily feared. That accomplished, the second task was to work with senior Army leaders in Dogwood and Baghdad to encourage them to improve the awareness and discipline of their convoy personnel when they entered Northern Babil. TF Scorpion had limited success with this initiative overall, but the 1st Armored Division responded with alacrity and their convoys 'hardened up' overnight. Unfortunately, only a fraction of the convoys were controlled by this Army combat organization, and the manatees from units without a warrior mentality continued to suffer.

After only a few days of operations the enemy adjusted tactics. They were no longer willing to stand next to the road and shoot at Marines, regardless of how slow they traveled and helpless they tried to appear. A new technique was attempted which was to place small intersperse LAVs in and among the vehicles in the convoys in the hopes that the ambushers might engage. The terrorists were ruthless, but not particularly suicidal, and this tactic failed to lure many from their hiding places. The next escalation in this 'cat and mouse' game was to run small 'Trojan Horse' convoys through the area. The 'bait' vehicles themselves had been hardened with a layer of MRE boxes filled with sand lining the outer edges of the vehicle beds, with Marines hidden inside ready to burst out and return fire. Terrorist fighters must have been amazed (just prior to being killed or wounded) as the first convoys of HMMWVs or MTVRs stopped, returned fire, and deployed infantrymen to hunt them down. After a few days of this, the ambushers moved further and further back from the road in order to survive, but this resulted in their fire becoming increasingly ineffective, as their marksmanship was so poor. By mid-June two changes had taken place in the north: Colonel Toolan had returned to CONUS; and the number of attacks had fallen off to a fraction of what they had been. By the very end of the month there were no enemy attacks at all.

As effective as the combat tactics were, the Marines found themselves treating 'symptoms', and not getting at the 'disease' of suspicion and hostility towards the Coalition in this contested region south of Baghdad. The TF Scorpion team implemented a parallel program of Civil Affairs activities in the population centers. As the Army combat units had fought through this area in April on the way to Baghdad, there was a massive amount of unmarked unexploded ordnance (UXO) lying about. A tremendously effective psychological operations and information campaign, developed by the Army soldiers attached to the Division from the 432d Civil Affairs Battalion (from Green Bay, Wisconsin) first warned local citizens of the dangers of UXO, and provided posters and handbills to help identify the objects. The second aspect of the campaign was to encourage the Iraqi citizens to contact the Marines for removal of dangerous materials from their homes, schools, and neighborhoods. The EOD Marines began a nonstop campaign that ultimately resulted in the removal of 250,000lbs of UXO across the zone. This common interest also began a process of extending the Marines' influence and trust into the population. The real intelligence sources would be found among the people, a process that had proven successful in the rest of the Division's battle space. Soon the people were pouring in with tips and warnings to the Marines with whom they had developed a new relationship of trust.

By very early July, the enemy once again changed tactics. They gave up on the ineffective, and highly dangerous, direct fire ambush. Instead, they switched to the almost exclusive use of IEDs and landmines. These were often mortar or artillery shells "daisy chained" together and command detonated. By now, the Marines of the Task Force had found their stride, and led by an exceptionally effective S-3, Major Dave Bellon, they very quickly figured out the enemy's routine and although IED attacks rose to a high average of three a day in the first two weeks of June, they fell to almost zero by 15 June. The Division's G-3, Lieutenant Colonel Norm Cooling, an officer of considerable intellect and extensive operational experience (and a former LAR company commander himself) worked this problem closely with Major Bellon. They both understood that the LAR centric Task Force had the mobile patrols covered, but clearly was in need of "grunts" to do the dirty work, to get out of the vehicles and hunt down and kill this elusive enemy.

On 29 June India Company, 3/7, commanded by Major Matt Grosz and First Sergeant Vic Martin, was attached to Scorpion. Major Grosz selected and occupied a company position off ASR Jackson just to the north of Al Mahmudiyah, and immediately south of the IED active 'mixing bowl' (intersection of Routes Jackson, Sue and Tampa). The company had not yet even unrolled their sleeping mats when they were out prowling the night on initial familiarization patrols of the area. Scorpion also welcomed Captain Ted Card and Golf Company, 2/5, redeployed from security duties at LSA-Matilda in Kuwait. Like Grosz, Card selected a position from which to operate, this time in the city of Al Yusufiyah. The Marines from Golf Company were glad to be back in the fight, and these warriors were into ambush sites the first night and enjoyed early contact. Additionally, Charlie, 1/4, was relieved of duties at the only fixed site security requirement in the northern part of the province by Puerto Rican National Guardsmen, freeing them for the hunt as well. The combined efforts of the Task Force of nearly 1,000 personnel from three services presented the enemy with so many dilemmas that they were forced to alter their tactics once again. They began to initiate sporadic mortar attacks from hastily laid or vehicle-borne tubes. This wildly inaccurate fire was even launched at the Companies' base sites, causing superficial wounds to a few Marines, but seriously injuring noone.

The final stage of operations for Task Force Scorpion, before their relief by a multinational Polish Division, was a preemptive spoiling attack against a large number of Iraqi terrorists just prior to a planned mass terrorist attack. This was enabled by the tremendous amount of actionable intelligence that Lieutenant Colonel Pappas and his executive officer, Lieutenant Colonel Dan Devine, were collecting from the many sheiks, clerics, police, and private citizens of Northern Babil. There was a build up of indications and warnings that were developed by Colonel Jim Howcroft and his G-2 section indicating a four-day period (corresponding to the anniversary of the founding of the Baath party in Iraq) in which the combined terrorist and Baathist forces in Iraq would attack American Forces. Blue Diamond planned a simultaneous spoiling attack against a number of these suspected individuals to disrupt the expected attack. Building up a target set, the Division held off striking suspected individuals until 12 July in order to minimize the period of time the enemy had to recover, and maximize the disruption of his plans. So successful was this initiative that the Division continued this tactic in an effort to round up senior regime officials, Fedeyeen, and terrorists. By continuing the attack, and taking the fight to the enemy, the Division greatly reduced terrorist influence in southern Iraq. The conditions were set to turn the battlespace over to a multinational division led by the Poles.

The Division Heads Home

By early summer, forces from as many as 24 nations began to descend on Babylon to begin the process of turning southern Iraq over to the Coalition Division. Among the first to arrive was an advance party from the Italian Defense Ministry arriving in the Dhi Qar Province to prepare the way for the arrival of the Garibaldi Brigade and the relief of 2/25. The Italian liaison team, led by Brigadier General Lopes, met with the Commanding General, 1st Marine Division, and began a reconnaissance of their zone. The Garibaldi Brigade advanced party arrived, followed by the main body on 26 June. Accompanying the Italian troops were police personnel and Romanian soldiers. On 26 June, the Battalion hosted the UK, Dutch, Italians, MEF, and

Division staff representatives for an informative meeting to plan for the relief. The Carabinieri advanced party conducted a reconnaissance and familiarization of law enforcement activities within An Nasiriyah and the surrounding areas. On 9 July, the relief progressed on schedule, with the Italians establishing command and control nodes at Qalat Sukkar and Ash Shattrah. The Dhi Qar province was turned over by 2/25 on 20 July.

Six more reliefs were accomplished in rapid succession. The pattern was similar for each of them. Planners and liaison officers were exchanged, followed by advanced parties arriving shortly thereafter. Finally, the main body of replacement troops would arrive, some at least slightly intimidated by the confidence of the young Marines from America. It was the heat they were already enduring, the threat of instant violence all around them taken for granted by those they were replacing, and the Spartan conditions these Marines considered "living large," that made the allies understand early on that they were replacing professionals. In often emotional goodbye ceremonies, the people of the towns that had grown to trust and love 'their' Marines bid them farewell.

The Polish Integration Group, who would lead the multinational Coalition Division, conducted a preliminary recon of the Blue Diamond fixed bases and received a brief on current operations in early June. The Polish Integration Group visited positions throughout the zone, and individual reliefs were accomplished. General Conway handed the formal responsibility for the zone to the commander of the Polish Division in a ceremony conducted in Babylon on 3 September.

End of Operations

With the end of operations in Iraq, Blue Diamond Marines waited in Kuwait for follow-on transportation for its return to Camp Pendleton. As they did so, they joined the thousands of veterans over the decades who awaited the journey home from wars in the Pacific, Korea, Vietnam, the Persian Gulf, and Somalia. Less than a year ago the vast majority of the young men and women of this, the Marine Corps' most decorated Division, could not even find Iraq on a map. Now, they knew the difference between Shia and Sunni, had been shot at, had seen friends die, and some had taken lives as well. They had volunteered to serve the cause of freedom, and picked the Marines because they wanted to be the best. These same young men and women carried a pride that would remain with them for the rest of their lives. They came because they were ordered to by the President, and fought because they were told to by the officers, SNCOs and NCOs to whom they entrusted their lives.

The Division Marines had conducted themselves differently in the south than SASO had been done in the north. What had worked for the Marines may not have worked in Baghdad, Tikrit, or Mosul, but the results had been startling. A people that had known only repression and violence had been given a new start toward a functioning democracy. The Marines flew from March Air Reserve Base in southern California, or floated from 32nd Street pier in San Diego, with the attitude that the Iraqi people were victims of a terrible disaster. Blue Diamond had been there to help. The Marines had already gained the 'hearts and minds' of the Iraqis during combat operations. The wild demonstrations of joy and celebration along every route north through eastern Baghdad had told them that. They also knew that these hearts and minds were theirs to lose, and the possession was a fragile thing. The Marines treated the people with dignity and respect, and it was returned. The Marines' first instinct had been to "do no harm" in this last phase, and they left a trail of goodwill behind them. They had made life-long friends here among ordinary Iraqis, who truly appreciated the immense sacrifices made on their behalf. The Marines had given them back their lives, and a new hope for the future.

Epilogue

With the October sunshine and cool breezes of southern California wafting over the silent crowd, Baghdad might have seemed far away. To the dignitaries, Marines, and those that loved Marines assembled, thoughts of Iraq were close to their hearts. The 1st Marine Division memorial services brought this family together to pay tribute to those comrades and loved-ones who had sacrificed so dearly for their fellow Marines, their Corps, and their Nation. It was an occasion to mourn. It was an occasion to remember shared sacrifice. It was the end of a long road.

Their mission accomplished, the warriors of the 1st Marine Division had begun to look home. From the streets of Baghdad, An Najaf, Ad Diwaniyah, and many other places, the Marines were relieved of their posts and embarked on the final leg of their journey. Returning the Division's men and equipment to the United States was as large a task as getting it into theater. A great deal of work remained to be done, but the Marines turned to the task with happy hearts and willing hands. Like their forefathers, they had fought in defense of liberty and freedom, and had earned their place in the history of the Marine Corps. They were eager to return to their families, and to reconstitute for the next time the nation called. The Marines prepared for the long journey home to American soil, like veterans of the 1st Marine Division had since its first battles overseas.

As the last pairs of dusty boots took their final footsteps in Iraq, Marines and Sailors of Blue Diamond reflected on their contribution to the legacy of the Division. Like warriors have for generations, they chose to serve their nation and the cause of freedom around the world. These men and women chose the Marine Corps for the special challenge. They chose the Marine Corps because they wanted to be the best. Those choices had carried them to a place called Iraq. A place of danger. A place of tyranny. The accomplishments of the Marines there would take their place in Marine Corps history beside places like Guadalcanal, Korea, Vietnam, Kuwait, Somalia, and many others.

Seldom does a generation have an opportunity to take part in such a noble and honorable quest. The Marines had made a difference. They had freed a people, crushed a tyrant, planted the seeds of democracy in a turbulent region. They had assaulted through the Mesopotamian marshlands between historic Babylon and the Garden of Eden, from the Euphrates to the Tigris, from Basrah to Baghdad and beyond. The places they had gone were household names in America, because of what the Marines had done there. They had leapt across the Diyala River into Baghdad, and had pushed so far north that they were ultimately closer to the Mediterranean Sea than the Persian Gulf. They understood the nature of Saddam Hussein's tyranny, because they had seen it. They had looked evil in the eye, and had destroyed it.

The military achievements were significant, and the list of accomplishments was long. The lasting legacy, however, was with the children of Iraq. The crowds of cheering Iraqis lining every road spoke of what the Marines and Sailors of 1st Marine Division had gained for the Iraqi people. Where there had been nothing but despair, there was now hope. The Marines had come in strength, but had acted with compassion. They had demonstrated a new kind of authority to the Iraqi people. An authority of ideals, not of tyranny. Soldiers of courage, not brutality. The school bells that the Marines of the Division installed in every school they restored will ring the story of their sacrifice for generations of Iraqis to come. The Marines had set the conditions for freedom in a portion of the world that had not tasted it before. The Marines of Blue Diamond had truly been "No better friend, no worse enemy."

The Marines returned home to a grateful nation, and the hugs, kisses, and tears of their families. Welcome home Daddy. Yellow ribbons lining the streets. The war had been difficult for those left behind too. Daily images of the rigors and dangers of their effort had been transmitted into homes across the country. Some of these homes had a special piece of their heart in Iraq as well. For those with fathers, sisters, children or friends in the 1st Marine Division, the waiting had been painful. In the days before the attack, Marines often spoke about how they missed their loved ones. Even then, they had known that returning without liberating the Iraqi people, without victory, was not an option. Now, they were proud of what they had done, and found comfort in the brotherhood of fighting men and women who had shared the dangers. They were the sons and daughters of those who stormed the beaches of Iwo Jima, scaled the walls of Inchon, and slogged through the jungles of Vietnam. They stood on the shoulders of the generations of the Marines that came before, and they had not let them down.

That family is what brought these people together on this day. Some of the seats returning home were empty. Although there was pride in accomplishment, there were also the painful memories of those who had not come home. The cost of freedom is high, many Marines and Sailors of the Division had given their lives in the cause of freedom. Hundreds more carried wounds received in combat. The mothers and fathers, sisters and brothers, children and spouses of these Marines were a special part of this family too. They all came together now to mourn the loss, and cherish the memory, of those special individuals who gave their lives for the freedom of a people they had never met. Their noble sacrifice will be remembered by their families, their comrades, and their country.

This day, an Iraqi child attended school. Tonight, an Iraqi family will gather together, free from the fear of the secret police. Tomorrow, a free people will change the face of the Middle East.

All of our days are numbered, but their deeds will live forever.

Semper Fidelis.

Appendix A

Command and Staff List

1st Marine Division

1. UNIT DESIGNATION
    a. Reporting Unit Code: M11001
    b. Table of Organization Number: 1986J
    c. Period Covered: 1 January – 30 June 2003
    d. Locations: Camp Pendleton and Twentynine Palms, California; Kuwait; Iraq

2. PERSONNEL INFORMATION

    a. Commanding General

MajGen James N. Mattis	1 January - 30 June 03

    b. Assistant Division Commander

BGen John F. Kelly	1 January – 30 June 03

    c. Subordinate Commanders

1st Marine Regiment	Col Joe D. Dowdy	1 January – 4 April 03
	Col John A. Toolan	5 April 03 – 30 June 03
5th Marine Regiment	Col Joseph F. Dunford, Jr.	1 January - 23 May 03
	Col R. Stewart Navarre	24 May 03 – 30 June 03
7th Marine Regiment	Col Steven A. Hummer	1 January – 30 June 03
11th Marine Regiment	Col Michael P. Marletto	1 January - 30 June 03
Headquarters Battalion	Col Robert J. Knapp	1 January - 30 June 03
1st Combat Engineer Battalion	LtCol Paul J. Cyr	1 January - 18 Feb 03
	LtCol Michael A. Micucci	19 Feb 03 – 18 April 03
	LtCol Paul J. Cyr	19 April 03 – 12 May 03
	Maj Michael J. Jernigan	13 May 03 – 30 June 03
1st Reconnaissance Battalion	LtCol Stephen A. Ferrando	1 January – 10 June 03
	LtCol Rory E. Talkington	11 June 03 – 30 June 03
1st Tank Battalion	LtCol James B. Chartier	1 January – 30 June 03
1st LAR Battalion	LtCol Duffy W. White	1 January - 12 June 03
	LtCol William R. Constantini	13 June 03 – 30 June 03
3rd LAR Battalion	LtCol Herman S. Clardy	1 January – 30 June 03
3rd AA Battalion	LtCol Robert S. Abbott	1 January – 30 June 03

d. Principal Staff Members

Chief of Staff	Col Bennet W. Saylor	1 January – 30 June 03
Assistant Chief of Staff, G-1	LtCol Cathy M. Powalski	1 January – 30 June 03
Assistant Chief of Staff, G-2	Col James R. Howcroft	1 January – 30 June 03
Assistant Chief of Staff, G-3	Col John A. Toolan	1 January – 4 April 03
	LtCol Clarke R. Lethin	5 April 03 – 30 June 03
Assistant Chief of Staff, G-4	LtCol John J. Broadmeadow	1 January – 30 June 03
Assistant Chief of Staff, G-6	Col Nicholas C Petronzio	1 January – 30 June 03
Assistant Chief of Staff, G-7	Col James W. Lukeman	1 January – 1 June 03
	Col Randolph P. Sinnot	2 June 03 – 30 June 03

e. Sergeant Major and Senior Navy Enlisted

SgtMaj Juan D. Duff	1 January – 30 June 03
MCPO Gary H. Labbs, USN	1 January – 30 June 03

3. <u>AVERAGE MONTHLY STRENGTH</u>

	USMC		NAVY		OTHER	
	OFF	ENL	OFF	ENL	OFF	ENL
JAN	1,349	19,355	42	907	0	0
FEB[1]	920	11,945	61	549	0	0
MAR	1,410	19,250	92	1,012	0	0
APR	1,410	19,303	96	1,045	0	0
MAY	1,461	20,906	114	1,196	0	0
JUN	1,007	18,441	76	851	0	0

4. <u>EQUIPMENT</u>

TYPE OF EQUIPMENT	NUMBER
M1A1 Main Battle Tank	120
Amphibious Assault Vehicles	525
Artillery	75
Light Armored Vehicles	263

[1] Does not account for ATF-West Division Marines.

Appendix B

Glossary

AA	Assault Amphibian
AAA	Anti-Aircraft Artillery
AAOE	Arrival and Assembly Operations Echelon
AAV	Amphibious Assault Vehicle
ACE	Aviation Combat Element
ADC	Assistant Division Commander
ADOCS	Automated Deep Operations Coordination System
APOD	Air Port of Debarkation
APOE	Air Port of Embarkation
ASLT	Air Support Liaison Team
ASOC	Air Support Operations Center
ASP	Ammunition Supply Point
ATARS	Advanced Tactical Air Reconnaissance System
ATO	Air Tasking Order
BCL	Battlefield Coordination Line
BCT	Brigade Combat Team
BDA	Battle Damage Assessment
BFT	Blue Force Tracker
BSSG	Brigade Service Support Group
C2PC	Command Control Personnel Computer
CBR	Counter Battery Radar

CE	Command Element
CEB	Combat Engineering Battallion
CENTCOM	U.S. Central Command
CFACC	Combined Forces Air Component Commander
CFLCC	Combined Forces Land Component Commander
CG	Commanding General
CGS	Common Ground Station
CIP	Combat Identification Panel
Class II	Batteries
Class VIII	Medical Supplies
Class IX	Repair Parts
CMOC	Civil-Military Operations Center
CPAO	Consolidated Public Affairs Office
CP	Command Post
CPX	Command Post Exercise
CRAF	Civil Reserve Air Fleet
CSS	Combat Service Support
CSSB	Combat Service Support Battalion
CSSC	Combat Service Support Company
CONPLAN	Contingency Plan
CONUS	Continental United States
COP	Common Operational Picture
DA	Dispersal Area

DAC	Division Admin Center
DASC	Direct Air Support Center
DIA	Defense Intelligence Agency
DOC	Deployment Operations Center
DSA	Division Support Area
EMCON	Emissions Control
EOD	Explosive Ordnance Disposal
EPW	Enemy Prisoner of War
FAC	Forward Air Controller
FAD	Field Artillery Detachment
FARP	Forward Arming and Refueling Point
FOB	Forward Operating Base
FOE	Follow on Echelon
FPOL	Forward Passage of Lines
FRAGO	Fragmented Order
FRSS	Forward Resuscitative Surgery System
FSCC	Fire Support Coordination Center
FSS	Fast Sealift Ships
FSSG	Force Service Support Group
GBS	Global Broadcasting System
GCE	Ground Combat Element
GOSP	Gas-Oil Separation Plant
HDR	Humanitarian Daily Ration

HET	Human Explotation Team
HF	High Frequency
HHA	Handheld Assay
HUMINT	Human Intelligence
IC	National Intelligence Community
IMINT	Image Intelligence
IMO	Information Management Officer
IO	Information Officer
IPSA	Intermediate Pumping Stations
JDAM	Joint Direct Attack Munition
JMEM	Joint Munitions Effectiveness Manual
JSTARS	Joint Surveillance and Target Attack Radar System
KAF	Kuwaiti Armed Forces
KI	Killbox Interdiction
KLF	Kuwaiti Land Forces
KMOD	Kuwaiti Ministry of Defense
LAR	Light Armored Reconnaissance
LASER	Light Amplification through Stimulated Emission of Radiation
LAV	Light Armored Vehicle
LD	Line of Departure
LOC	Line of Communication
LSA	Life Support Area Logistical Support Area

LTO	Logistics Tasking Order
LZ	Landing Zone
MACCS	Marine Air Command and Control Squadron
MAG	Marine Air Group
MAGTF	Marine Air-Ground Task Force
MANPAD	Man-Portable Air Defense
MARCORSYSCOM	Marine Corps Systems Command
MAW	Marine Air Wing
MCIA	Marine Corps Intelligence Activity
MCRE	Marine Corps Readiness Evaluation
MCWL	Marine Corps Warfighting Lab
MDACT	Mobile Data Automated Communications Terminal
MEB	Marine Expeditionary Brigade
MEF	Marine Expeditionary Force
MEFEX	Marine Expeditionary Force Exercise
MEG	MEF Engineering Group
MEWSS	Mobile Electronic Warfare Support System
MLC	Marine Logistics Command
MOD	Ministry of Defense (Kuwait)
MOI	Ministry of the Interior (Kuwait)
MOPP	Mission Oriented Protective Posture
MOS	Military Occupational Specialty
MOUT	Military Operations on Urban Terrain

MP	Military Police
MPF	Maritime Prepositional Force
MPSRON	Maritime Prepositioned Ship Squadron
MRLS	Multiple Rocket Launcher System
MSC	Major Subordinate Command
MSTP	MAGTF Staff Training Program
MWSG	Marine Wing Support Group
MWSS	Marine Wing Support Squadron
NBC	Nuclear, Biological, and Chemical
NBCRS	Nuclear, Biological, and Chemical Reconnaissance System
OCD	Obstacle Clearing Detachment
OMC-K	Office of Military Cooperation-Kuwait
OPCON	Operation Control
OPLAN	Operations Plan
OPP	Offload Preparation Party
OPT	Operational Planning Team
ORCON	Originator Controlled
OSW	Operation Southern Watch
PA	Public Affairs
PALT	Public Affairs Liaison Team
PIR	Priority Itelligance Requirement
PLI	Position Location Information
POL	Passage of Lines

POW	Friendly Prisoner of War
PRR	Personal Role Radio
QRF	Quick Reaction Force
RA	Regular Army
RCT	Regimental Combat Team
RFF	Request for Forces
RG	Republican Guard
RGFC	Republican Guard Forces Command
RIP	Relief in Place
ROZ	Restrictal Operation Zone
RRP	Refueling and Replenishment Point
RSO&I	Reception, Staging, Onward Movement and Integration
RUC	Reporting Unit Code
SAPOE	Sea and Aerial Ports of Embarkation
SAM	Surface-to-Air Missile
SASO	Security and Stabilization Operations
SIGINT	Signals Intelligence
SIPRNET	Secret Internet Protocol Routed Network
SLTLP	Survey, Liaison, and Reconnaissance Party
SMART-T	Secure Mobile Anti-Jam Reliable Tactical Terminal
SOP	Standing Operating Procedure
SRG	Special Republican Guard
SPINS	Special Instructions

SPOD	Sea Port of Debarkation
SPOE	Sea Port of Embarkation
SSE	Sensitive Site Exploitation
SSM	Surface-to-Surface Missile
TAA	Tactical Assembly Areas
TACON	Tactical Control
T/E	Table of Equipment
TEWT	Tactical Exercise Without Troops
TIO	Target Information Officer
TIP	Thermal Identification Panel
TMC	Tactical Movement Control
T/O	Table of Organization
TPC	Target Processing Center
TPFDD	Time Phased Force Deployment Data
TRANSCOM	U.S. Transportation Command
TTPs	Tactics, Techniques, and Procedures
UAV	Unmanned Aerial Vehicle
ULN	Unit Line Number
USCENTCOM	US Central Command
USNS	U.S. Naval Service Ship
WMD	Weapons of Mass Destruction

Appendix C

1st Marine Division Honor Role
October 2002–July 2003

LCpl Andrew J. Aviles
GSgt Jeffery E. Bohr, Jr.
SSgt James W. Cawley
2dLt Therrel S. Childers
PFC Ryan R. Cox
Cpl Mark A. Evnin
PFC Juan G. Garza, Jr.
LCpl Cory R. Geurin
Cpl Jesus A. Gonzalez
Cpl Bernard G. Gooden
PFC Christian D. Gurtner
LCpl Jose Gutierrez
PO2 Michael V. Johnson, USN
LCpl Nicholas B. Kleiboeker
Sgt Johnathan W. Lambert
LCpl Gregory E. McDonald
Cpl Douglas J. Marenco-Reyes
LCpl Francisco A. Martinez-Flores
SSgt Donald C. May, Jr.
SN Joshua Mcintosh, USN
1stLt Brian M. Mcphillips
Cpl Jesus M. Medellin
GSgt Joseph Menusa
Cpl Jason D. Mileo
PO3 David J. Moreno, USN
Maj Kevin G. Nave
LCpl Patrick T. O'Day
LCpl Eric J. Orlowski
LCpl David E. Owens, Jr.
Sgt Duane R. Rios
Cpl Robert M. Rodriquez
Cpl Erik H. Silva
LCpl Antonio J. Sledd
1stSgt Edward Smith
Sgt Kirk A. Straseskie
LCpl Jesus A. Suarez-Del Solar
SSgt Riayan A. Tejeda
LCpl Jason A. Tetrault
LCpl William W. White

Made in the USA
Las Vegas, NV
24 September 2022

55843058R20234